Ring[illegible] and Groups

An Introduction to Abstract Algebra

R B J T Allenby

Senior Lecturer, School of Mathematics, University of Leeds

Butterworth Heinemann

An Imprint of Elsevier

Rings, Fields and Groups
Allenby

Butterworth Heinemann
An Imprint of Elsevier
Linacre House, Jordan Hill, Oxford OX2 8DP
200 Wheeler Road, Burlington, MA 01803

Original ISBN : 0 340 54440 6

First Printed in India 2006

Indian Reprint ISBN-13 : 978-81-312-0123-7
Indian Reprint ISBN-10 : 81-312-0123-6

Published by Elsevier, a division of Reed Elsevier India Private Limited, 17-A/1, Main Ring Road, Lajpat Nagar-IV, New Delhi - 110 024, INDIA.

Printed and bound at Saurabh Printers Pvt. Ltd.
A-16, Sector IV, NOIDA 201 301.

Preface to the first edition

The overall aim of this book is to present a fairly leisurely introduction to some of the results, methods and ideas which are increasingly to be found in first and second year abstract algebra courses in British universities and polytechnics and in equivalent courses elsewhere.

There are several ways in which an author might present such an introduction. One is the (to some) aesthetically most pleasing take-it-or-leave-it purely axiomatic approach in which the reader is given a list of the appropriate definitions and is then led through proofs of those theorems universally agreed by those already in the know to be basic to the subject. The guiding spirit behind such an approach would almost certainly force the author to begin with a long and rather dry axiomatic development of the set-theoretical language needed. Although the purity of this particular approach would probably be preferred by a majority of practising pure mathematicians who already have a degree of familiarity with the material, the present author has found that it is appreciated by only a handful in every class of fifty or so beginners. Many students become restless at such an approach as they find difficulty in connecting the discussion with ideas with which they are already familiar. The sudden change from the world of concrete examples usually found in school mathematics to the abstract setting seemingly remote from the real world is one which can lead some students away from abstract algebra in particular and pure mathematics in general, a state of affairs which naturally saddens the author who much enjoys sharing with his beginning students the pleasure (even excitement!) to be obtained from following through some of the clever ideas and neat arguments to be found there.

A second approach, possibly more attractive to the beginner, would be to present a detailed historical account of the develolpment of algebra from, say, 1500 to the present day. To the student who is aware of some of the upheavals in school mathematics courses in recent years, the inclusion of an account of some of those theories which were once vigorously developed and were expected to become important but have generally failed to find favour or application might prove especially interesting! However, such an approach would leave, in a volume of reasonable size, little room for a really detailed account of any of the theories reviewed.

An intermediate course, and the one taken here, is to try to get fairly quickly into the spirit of abstract algebra, whilst at the same time interjecting occasional remarks and comments, either because of their historical content or because, quite simply, the author thinks they are fascinating (or both!). In particular, several complete sections are included more as light reading than as essential material. Amongst these I include Sections 3.5, 3.9, 4.4, 5.2, 5.12 and 6.7. The present approach, therefore, is a mixture of the formal and the informal. The author has certainly found this mixture acceptable in courses he has given on both sides of the Atlantic. On the one hand we take an informal approach to set theory because our concern is with the algebra; on the other hand we do not want to throw all formal working to the wind, present day algebra being, as it is, an axiomatic discipline. Indeed, one of the chief aims of the book is to develop the reader's critical faculties and we believe it preferable to begin this in Chapter 1 where, because of the student's (intuitive) familiarity with much of its content, the critical approach seems all the more prominent. (See also the Problems posed at the end of the Prologue.) Here formal definitions are given when some readers might feel that informal ones would suffice. (See, for instance, the development of the idea of polynomial from Sections 1.6 to 1.8). Such readers should not need encouraging to read, carefully, the reasons put forward for preferring the more formal approach. We also use this chapter to try and answer, by example, a question often asked by beginners, namely 'How much proof should I give?' The answer clearly depends upon the knowledge and maturity of the people trying to correspond. To help the reader through his first encounter with several of the proofs the author has been more expansive than he would normally be in communicating with a colleague. Those extra portions, which can be omitted as confidence grows and which to some extent are the answers to *the questions the reader should be asking himself* as he works through the text, have been put in square brackets.

We begin with a prologue in which we attempt to answer some of the questions students seem afraid to ask: What is abstract algebra? How did it develop? What use is it? The historical account of the development of algebra will include many words not familiar to the beginner, but we feel that in a new land it is preferable to possess a map, even one in a foreign language, than no map at all. In placing this material before rather than after the main body of the text we hope to whet the reader's appetite and heighten his sense of excitement with a description of the discoveries and inventions made by some of the mathematical giants of the past and that this excitement will fire him sufficiently to read this book avidly even when (as it probably will) the going gets a bit difficult.

The numbering of Chapter 0 indicates that we view it as a preliminary to the text proper. Chapters 1, 2, 3 and 4 concentrate on algebraic systems known as rings and fields (though these names are not formally introduced until Chapter 3) the concept of group not being mentioned until Chapter 5. A majority of texts on abstract algebra offer a study of the theory of groups before a study of rings, the reason often given being that groups, having only one binary operation, are simpler to begin with than are rings and fields which have two. The author (a group theorist!) feels that there is a rather strong case for reversing this order; the fact is that natural concrete examples of rings and fields (the integers, polynomials, the rational, real and complex numbers) are much better known to the beginner than are the equivalent concrete examples of groups (mainly symmetries of 2- and 3-dimensional figures). (This author will just not accept the complaint 'But the integers form a group under addition': so they do but that is not the natural way to look at them. Indeed the author rebels strongly against the argument which, briefly, runs: 'The integers under addition form a group. Therefore we must study group theory.') What in the author's opinion really clinches the argument for studying rings and fields before groups are the several exciting applications that can quite quickly be made to easily stated yet non-trivial problems in the theory of numbers and of geometrical constructions. (See especially Sections 3.8 and 4.6.)

The placing of ring theory before group theory will, it is true, give rise to a little more repetition of corresponding elementary concepts than might have been the case with the more usual presentation. The author does not, however, feel any need to apologise for that! (On the other hand Chapters 5 and 6 make little essential use of Chapters 3 and 4, so they can be studied directly after Chapter 2.)

Throughout the text problems, numbered only for ease of reference, have been inserted as they have occurred to the author. Some are reasonably easy, some solved later in the text and some are quite hard. I leave you to find out which! The purpose of these problems is (i) to set you thinking and then discussing them with a colleague or teacher; (ii) to get you into the habit of posing questions of this kind to yourself. Active participation is always much more exciting (and instructional) than passive reading!

In this book the numbering of theorems, lemmas, etc. in any one chapter is consecutive, thus: Theorem 5.5.4, Example 5.5.5, Notation 5.5.6, Definition 5.5.7. When referring to a theorem, lemma, etc. given elsewhere in the text usually only its number is given (e.g. 3.8.2). Reference to an exercise is however given in full (e.g. exercise 3.2.14) except when the exercise referred to is at the end of the section concerned. Thus, within Section 3.2, exercise 3.2.14 would be referred to as 'exercise 14'.

In producing this text I have received help from several people, especially from the secretaries in the School of Mathematics in the University of Leeds. In particular I should like to thank Mrs M M Turner, Mrs P Jowett, Mrs A Landford and Mrs M R Williams. Several colleagues in Leeds and elsewhere have offered helpful comments and gentle criticism on parts of the manuscript. Here I especially with to thank Drs J C McConnell, E W Wallace and J R Ravetz. For supplying me with photographs I thank the keeper of the David Eugene Smith collection at Columbia University, New York, and especially Prof Dr Konrad Jacobs of the University of Erlangen–Nürnberg who kindly donated the pictures of Emmy Noether and Richard Dedekind.

Leeds
1982

RBJTA

Preface to the second edition

Some while ago I decided to inflict, on the mathematical community, yet another algebra book. One reason was my belief that few texts, if any, then available both *proved* how exciting abstract algebra is *and* showed the reader 'behind the scenes' of the subject. Indeed it is not as easy to do this on paper as in the lecture theatre where one can laugh with or shout at the audience to convince them what a good time they are having (and there is no literary equivalent of enlivening the lecture by losing the chalk or falling off the podium). Nevertheless, I found myself writing as if I were lecturing, occasionally wanting to offer tit-bits of information or what I hoped would be helpful asides. So it was most gratifying to learn that this style was acceptable to many readers—students and teachers alike.

In asking me to prepare a second edition, the publishers kindly offered me extra pages which, after some consultation, I decided to fill with (i) a brief account of Galois Theory and (ii) hints/outline solutions to many of the 800 or so exercises. (I hope, from the teacher's [students'] point of view, that I haven't overdone [underdone] the latter.) Of course I have also taken the opportunity to consider if my original text said what I had intended. Where it didn't I've incorporated changes. The necessity of amending what Churchill might have called 'terminological inexactitudes' was pointed out to me by many friends and colleagues. In particular I should like to thank: Seymour Bachmuch (U.C. Santa Barbara), Allen Bell and Ken Goodearl (University of Utah), Al Hales (U.C.L.A.), Bob Gregorac (Iowa State University), John Meldrum (University of Edinburgh) and John Silvester (King's College, London) as well as some of my colleagues at Leeds. (Special thanks are due to Bell, Goodearl and Hales for sending me, unrequested, full solutions to the exercises they set in class. If they were thereby hinting at something, I'm afraid I've failed to grasp the message!) Incidentally I am not averse to receiving comments directly from students and I thank all who have written, including one from the University of Sana'a.

I have also updated the bibliography, including some books and papers which have appeared since the first edition of RFG went to press. However I (continue to) invite the reader to experience the pleasure of seeking out others I haven't found room to mention.

Finally I must thank (most effusively!) my colleague John McConnell who willingly surrendered the notes of his own successful lectures on Galois Theory and then (wilfully?) misunderstood my invitation to 'look over Chapter 7', taking it as a challenge to make me keep my account close to the way Galois Theory *should* be presented! Without his help the new chapter might well have contained more 'terminological inexactitudes' than did the whole of the original text.

Leeds
1991

RBJTA

Contents

How to read this book

I hope it won't be considered too presumptuous if I offer some advice to the reader who finds himself, perhaps for the first time, left to work on his own.

To most of us, reading a mathematics book is a lot harder than reading a novel. A certain amount of application is required. Begin by *getting up out of that armchair and, with pencil and paper, find yourself a nice clear desk and a comfortable but upright seat.*

Perhaps the first thing to realise is that learning is not a linear business. I have known students who feel obliged to know pages 1–65 almost by heart before they dare turn to page 66. This is silly. If your conscience allows you to say that you really are stuck at some point don't be afraid to pass on to the next paragraph. Things can often become clearer with hindsight. If, however, you find yourself carrying three or four such problems, it seems clear that there is nothing for it but to go back and attack the first difficulties again. Incidentally, one way to motivate yourself if things do get sticky is to imagine that you have to explain the subject to the class the next day. (Think of all the unpleasant questions your classmates might ask you and make sure you can answer them!)

Definitions are sometimes hard to grasp on first reading. One method I've found helpful is to learn the definition by heart. Once you have control over the words you can concentrate all your efforts on their meaning.

Perhaps the best approach to reading a proof for the first time is positively to disbelieve each assertion made. This commits you to examining each word and can be quite tiring. On the other hand the exhilaration following the successful completion of such a task is well worth the effort. Moreover, having mastered the proof as it were 'locally', it should now not be too difficult to master it 'globally' by identifying the key points that make the proof work. Explain the method—not the details—to yourself.

Regarding exercises: before attempting to answer a question do make sure you know the meaning of all the words in it! It may seem obvious but some appear to forget that if you don't understand what a question asks you have little chance of answering it! If a problem (or a proof) seems too difficult as it stands one can try looking at a special or a simpler case. This may point the way in regard to the original problem. In any event, if you can't manage a simpler case you are surely not going to succeed with the more difficult one. The message here is: Don't be afraid to examine lots of concrete examples. Drawing pictures, as I do in this text, can also be helpful.

One more (obvious) point. Don't expect the various parts of solutions to problems to occur to you in the 'right' order. (I can assure you that this book was not written straight through from beginning to end!) Having got a rough draft of a solution, now write it out neatly and in a logically developed manner so that you will be able to read and understand your solution in 6 months time.

Finally may I say that I hope that (most days) you enjoy reading this book as much as I (most days) enjoyed writing it.

Prologue

In this prologue we discuss some of the questions which few beginners seem to have the courage to ask and yet to which they would surely like some kind of answer. The questions considered are (i) what *is* algebra? (ii) what is its history? and (iii) what is it good for? We urge the reader to dip frequently into the historical outline below. The discoveries mentioned there of some of the world's best mathematicians should whet the appetite for, and also place in some kind of perspective, the mathematics covered in Chapters 1 through 6.

What is algebra?

The word algebra derives from the word *al-jabr* which appears in the title of a book written in the 9th Century by the Persian mathematician Mohammed Al-Khowarizmi (from whose name comes the word algorithm). This book, in a Latin translation, had great influence in Europe. Its concern with problems equivalent to those of solving polynomial equations, especially those of degree 2, led to the word algebra eventually becoming synonymous with the science of equations. This state of affairs persisted into the 19th Century, Serret, in 1849, observing that 'Algebra is, properly speaking, the analysis of equations.'

One possible definition of algebra [132]* is that it is the study of operations, of rules of computation. This is slightly unfair since, as we observe in Chapter 2, not all operations are interesting. In any case the word algebra is nowadays often prefixed, to indicate different stages of development, by adjectives such as *classical, modern* and *abstract*. Whilst there does not appear to be any universal agreement on the precise meaning to be attached to these prefixes (compare the definitions of modern algebra in [80, p. 669] and [92, p. 702)] we can fairly safely say that classical algebra is the present synonym for the theory of equations, a theory in which are manipulated symbols which invariably represent numbers, be they complex, real, or rational. The term modern algebra can then be used to describe that subsequent algebra, some of it arising from more detailed investigations within the classical theory, in which the symbols manipulated are no longer restricted to representing numerical quantities. (Cauchy, in 1815, had defined multiplication of permutations, whilst Gauss, in 1801, had combined pairs of binary quadratic forms and also integers 'mod p'.) Finally, we give the name abstract algebra to that

* Brackets such as these refer to the bibliography.

generalisation of modern algebra in which the main object is the study of algebraic systems defined solely by postulates or axioms (usually chosen not arbitrarily but with several concrete instances in mind), no particular meaning being attributed to the symbols being manipulated.

Abstraction and axiomatisation

The introduction of the concepts of abstraction and axiomatisation is attributable to the Greek mathematician/philosophers of the period 600–300 BC. Thus, neither is a recent invention.

The method of abstraction has several interconnected uses. By stripping concrete objects of their less essential features, they become less involved and hence more amenable to mathematical treatment. Also one increases the chance of revealing similarities between superficially distinct objects, so that a theorem from one area of mathematics may suggest an analogue in another area. Further, several concrete theories can be studied simultaneously under the umbrella of one general theory of which the concrete theories are special cases. Finally, one penetrates to the *real reasons* for the success (or failure!) of a theory. In this way a deeper understanding should result.

As a simple illustration of these remarks we note that the Babylonians posed many problems [80, p. 34] of the form: 'Find the side of a square if the area less the side is 870.' The verbal solution proceeds: 'Take half of 1, that is $\frac{1}{2}$; multiply $\frac{1}{2}$ by $\frac{1}{2}$, that is $\frac{1}{4}$. Add this to 870 to get $870\frac{1}{4}$. This is the square of $29\frac{1}{2}$. Add $\frac{1}{2}$ to $29\frac{1}{2}$ to get 30, the side of the square.' Of course this same procedure is applicable to other problems of a like nature. But how much deeper in content and how much easier to understand is the observation that infinitely many such problems can be dealt with all at once, as it were, by replacing numerical coefficients by literal coefficients of no initial specific numerical value. In short: if $ax^2+bx+c=0$ then $x=(-b\pm\sqrt{b^2-4ac})/2a$. (In the above problem $a=1$, $b=-1$, $c=-870$.)

The axiomatic method which rejects proofs based on intuition involves deducing by logical argument alone, from initial statements assumed without question (the axioms), other statements (the theorems). The most widely quoted example of the use of this method is Euclid's *Elements* (c. 300 BC) in which the author supposedly sought to give a consistent foundation to geometry. (For a different point of view see [121].) It is argued in [125] that Euclid's fifth axiom (The whole is greater than the part) was included specifically to eliminate from mathematics the paradoxes (c. 450 BC) of the infinite introduced by Zeno ([85, p. 35–7]). Furthermore, the 'crises' brought about by the discovery that $\sqrt{2}$ is not a rational number clearly called for rigorous investigation: after all, the intuitively 'true' had been proved false! (See problem 3 below.) Thus the axiomatic method was called upon to help confirm the basic consistency of mathematics.

For the next 2000 years the abstract/axiomatic approach was with a few exceptions replaced by a more concrete intuitive approach. However, in the headlong rush to develop the newborn calculus of Leibniz and Newton, absurdities arising from the free use of intuitive geometrical arguments (see problem 1) led to the so-called second great crisis and a call for analysis to be made more rigorous; in other words based on arithmetic (whose foundations were obviously (!) secure).

The publication by Lobachevsky in 1829 of a consistent (non-Euclidean) geometry, in which Euclid's parallel postulate is denied, should perhaps have turned mathematicians' attention back to a study of axioms, especially as it had long been appreciated that Euclid's use of the axiomatic method was, to say the least, inconsistent. ([121] contains stronger views.) However, this work apparently attracted little attention for several years.

In due course the 'arithmetisation' of analysis got under way: Dedekind insisted that 'what is provable should be proved', observing that even the equality $\sqrt{2}\times\sqrt{3}=\sqrt{2\times 3}$ had not yet (1858) been satisfactorily established.

Eventually the desire continually to express concepts in terms of yet more 'fundamental' ones led to Peano setting down in 1889 his symbolic and axiomatic description of the set of integers (in terms of the undefined concepts: set, belongs to, zero, number, successor of). Furthermore the whole numbers themselves were shown to be definable entirely in terms of Cantor's new notion of set (Section 0.1). Unfortunately Cantor's definition was too wide-ranging: intuition had failed once again at the most basic level (see Russell's Paradox in Section 0.2) and the third great foundational crisis was at hand. One result was Zermelo's attempt (1908) to build a formal set theory on an axiomatic basis.

One of the facets of 19th Century algebra, the ever-increasing number of concrete structures of distinct outward form but with similar underlying properties, encouraged their cataloguing and comparison by abstracting common features. Indeed Weber's book [40] of 1893 talks not just of (groups of) permutations, matrices, etc., but of (groups of) 'things' (*Dingen*, in German). Basing his proposals on the more commonly used properties shared by permutations, matrices, etc., he postulated that his 'things' be subject to similar rules (i.e. axioms). It would then follow (as Boole had said in 1847) that 'the validity . . . does not depend on the interpretation of the symbols . . . . Every system of interpretations . . . is equally admissible.' Note that whilst the axiomatic method permitted, in theory, the assuming of arbitrarily chosen sets of axioms, those adopted were not chosen at random, the aim being to reflect properties of concrete systems already deemed important.

Thus the abstract/axiomatic method has a role to play in classification and reorganisation; in making special results more intelligible by identifying common features. The method thus supplies greater transparency and insight and leads to a unified approach offering progress along a wide front. It also helps us avoid making intuitively obvious but unfounded assertions and hence 'proving' false theorems.

A final remark. Since Euclid's time the concept of axiom has changed somewhat. To Euclid axioms were unshakeable truths (e.g. every two distinct points uniquely determine a line). Today we interpret the word axiom differently. We do not ask whether or not an axiom is 'true'—just as we don't ask* if the rules of chess are 'true'. (One can of course question whether or not the axioms are appropriate if one is trying to model a concrete example.) The Greek word *axioma* originally meant 'request'; the reader is requested to accept the axioms unquestioningly as the rules of the game. All he can (and should) question is whether or not the asserted conclusions follow logically from the axioms.

Historical development of algebra

As stated above, the algebra now called classical concerned itself with (polynomial) equations, in particular with attempts to supply formulae for the roots of equations of degrees 3, 4, 5, etc. The solutions $x = \dfrac{-b \pm \sqrt{b^2 - 4ac}}{2a}$ of the general quadratic equation $ax^2 + bx + c = 0$ were known to the Babylonians via the process of completing the square (although only positive solutions were found acceptable).

A long search ensued for similar formulae in terms of radicals (that is, formulae involving just $+$, $\cdot$, $-$, $\div$, $\sqrt[n]{}$ and the coefficients of the given equation) for the roots of equations of higher degree, but none appeared until the 16th Century when a formula for the cubic was found by the Italians Scipione del Ferro, Niccolo Fontana (more commonly known as Tartaglia, 'the stammerer', because of a speech impediment brought about by injury in childhood) and Jerome Cardano. There is no need here to go into the rivalries of the various mathematicians involved in the search for the cubic formula—however, it makes for fascinating reading ([80], [84], [85], [89]).

Not long afterwards the solution of the general quartic (or biquadratic) equation was found by the Italian Lodovico Ferrari and in 1545 both the cubic and quartic formulas were published in Cardano's *Ars Magna*. After this double success hope must have been high that a solution in the case of the general quintic would soon be forthcoming. New ways of solving the quartic equation were discovered but still the quintic equation resisted attack.

In 1770/71 Lagrange set about analysing the various methods then known for dealing with the general equations of degrees 2, 3, 4 and he found that they all depended on the same general principle (see Section 5.2). In particular Lagrange showed cause for finding the number of formally distinct 'values' taken by a function, for example $xy + zt$, on permuting the symbols x, y, z, t, in all possible ways. Although Lagrange seemed to cherish hopes that his work would show the way to the solution of the general quintic, the results obtained indicated a distinct possibility that there might be no corresponding formula in the case of equations of degree greater than 4, and a supposed proof of this was given by Ruffini in 1799. It was apparently difficult to see if Ruffini's proof was complete and, despite a further attempt by him at the

*[78, p. 23].

problem, in 1813, the credit for supplying the first generally accepted proof of impossibility goes to Abel (pronounced 'Arbel') in 1824. Thus it was proven that no universal radical formula for obtaining all the roots of every quintic was available. Yet it was undeniable that such formulae were available for certain special quintic equations (for instance the five roots of x^5-1 are each radically expressible; see exercise 4.6.6). In 1832 Evariste Galois described, by associating with each equation a finite group, exactly which equations were treatable. This result is a mere corollary to a much more general theory, still a subject of research, called Galois Theory. (See Chapter 7). Galois, it is usually said,* coined the word group at this time and introduced the concept of normal subgroup. He was also the first to investigate fields with finitely many, p^n, elements where p is a prime and $n \geqslant 2$. (See Section 4.5.)

At the turn of the century other ideas, later to be seen as part of algebra, were coming from the pen of Carl Friedrich Gauss, one of the greatest mathematicians who ever lived. Before he was 19, Gauss had constructed, by straightedge and compass, a regular 17-gon, the first 'new' constructible regular polygon for 2000 years, and in 1799 he gave the first satisfactory† proof (where Newton, Euler and Lagrange had failed) of the fundamental theorem of algebra (4.8.1). Gauss' most influential contribution is probably his *Disquisitiones Arithmeticae* (1801), a work in which appear his 17-gon (see Section 4.6), his introduction of the notation of congruence (see Section 2.2), his proof that all the roots of x^n-1 are expressible in radicals, and a proof of the quadratic reciprocity law. (For an integer a and a prime r not dividing a, define $\left(\frac{a}{r}\right)$ to be 1 or -1 according as the congruence $x^2 \equiv a \pmod r$ can be solved for x or not. The reciprocity law states that, for distinct odd primes, $\left(\frac{p}{q}\right)\left(\frac{q}{p}\right)=(-1)^{\frac{p-1}{2}\cdot\frac{q-1}{2}}$. See, for example [42], Chapter 8.)

Despite the fact that at the turn of the century mathematicians were happily (in most cases‡) using complex numbers, there was some residual disquiet relating to them. First, although Wessel (1797) and Argand (1806) had tried to make complex numbers a little more respectable by showing how to interpret them, their addition and multiplication geometrically, there remained, possibly because of doubts concerning the intuitive use of geometrical arguments, the desire to put them on a firmer basis. Furthermore there was still the problem 'What exactly is $\sqrt{-1}$?' (see Section 4.4). Suffice it here to say that this problem was finally dispensed with in 1833, by banishing $\sqrt{-1}$ (!): Hamilton replaced the objectionable $a+ib$ by the ordered pair (a, b) of real numbers (see Section 4.4), thus duplicating an earlier (unpublished) work of Gauss. Second, unease was caused by the continued manipulation of letter symbols as if they were positive integers in situations in which

* Kiernan [108] thinks it might have been Galois' teacher L Richard.
† Taking into account the somewhat less rigorous demands of the period.
‡ There were still some who thought negative whole numbers 'absurd'!

they clearly were not. For instance, the equality $(a-b)(c-d) = ac+bd-ad-bc$, acceptable to everyone whenever a, b, c, d were whole numbers with b less than a and d less than c, appeared to remain valid when particular irrational or complex numbers were substituted for the letters. In addition, due to the efforts of Woodhouse, Babbage, Herschel and Peacock from 1803 onwards, Leibniz' notation and methods in the calculus had gradually replaced Newton's in England and as a consequence much of continental analysis had become available to the English. In particular the calculus of operations became an English preserve [109]. In this calculus one combines functions as if one is dealing with algebraic quantities. Thus in the differential equation $\frac{d^2y}{dx^2}+3\frac{dy}{dx}+2y=0$ one separates the 'operator' $\frac{d^2}{dx^2}+3\frac{d}{dx}+2$ from the operand y and even factorises it as $\left(\frac{d}{dx}+1\right)\left(\frac{d}{dx}+2\right)$. (Indeed, later (1843), Boole applied the method of partial fractions writing $\frac{1}{\frac{d^2}{dx^2}+3\frac{d}{dx}+2}$ as $\frac{1}{\frac{d}{dx}+1}-\frac{1}{\frac{d}{dx}+2}$.) So there was obviously raised the question as to the validity of doing this sort of thing. It is in fact the case that whereas not all the early investigators even tried to validate their reasoning, several, including Cauchy, Servois and Boole, certainly did. And it was in this connection that Servois, in 1815, introduced the notions of functions which are 'distributive' and 'commutative', terms still used today (see Section 1.2).

It was in this atmosphere that Peacock, his friends Babbage and Herschel having worked in the calculus of operations, introduced (1830, 1833, 1842) his two concepts of algebra: arithmetic algebra and symbolic algebra. Arithmetic algebra, declares Peacock, concerns the use of symbols representing positive integers so that, for instance, the expression $a-b$ is meaningless if $a<b$. Symbolic algebra is the same as arithmetic algebra except that its operations are universally applicable: $a-b$ has meaning for all a, b. The rules for operating in symbolic algebra are to be just those of arithmetic algebra. A consequence of this is Peacock's Principle of Permanence of Equivalent Forms which essentially says 'Whatever the occasion, $a \cdot b = b \cdot a$'. It was against this background, namely that it seemed preposterous to suppose that $a \cdot b$ could ever be other than equal to $b \cdot a$ in any consistent algebraic setting, that Hamilton looked for more than 10 years for an extension of complex numbers suitable for application to the physics of 3-dimensional space. At last, in 1843, he broke through the psychological barrier by founding a consistent set of 4-dimensional 'hypernumbers' in which all the usual laws of arithmetic hold with the exception that $a \cdot b \neq b \cdot a$ in general. He had invented the quaternions. (The reader who fails to see how any system of entities in which the laws $a \cdot b = b \cdot a$ and $a \cdot (b \cdot c) = (a \cdot b) \cdot c$ both fail can have any practical value has only to recall the familiar 3-dimensional vector calculus,

denoting vector multiplication. Subtraction on the set of integers is another example!)

At this very time, in Germany, H Grassmann was generalising complex numbers still further by inventing an unlimited number of arithmetics of n-tuples. Unfortunately his exposition was not easy to read and it was Hamilton's quaternions which gained the more attention. Hamilton had great hopes for quaternions but they were not quite what the physicists wanted. It is true that the great mathematical physicist Maxwell referred to them when making use (separately) of the 'scalar' and 'vector' parts of a quaternion, but the vector calculus of Gibbs and Heaviside (in the 1880s) seemed to suit the physicists better. For a time a battle raged between the supporters in each camp. Gibbs–Heaviside eventually won as regards applicability but the quaternions have the honour of being the first to demonstrate the existence of consistent number systems not satisfying the commutative law of multiplication.

During the 1840s, the quaternions soon inspired the manufacture of other consistent number systems which violated the most obvious laws of arithmetic. In 1845 Cayley described his (8-dimensional) octonions, still called the Cayley numbers. Here not only is $a \cdot b = b \cdot a$ not universally valid, the associative law $a \cdot (b \cdot c) = (a \cdot b) \cdot c$ is also broken.* In 1853 Hamilton invented biquaternions (quaternions with complex coefficients which turn out to be 2×2 complex matrices in disguise). Here $a \cdot b = 0$ is possible even when $a \neq 0$ and $b \neq 0$. That is, the biquaternions possess divisors of zero. The 'smaller' systems of hypernumbers introduced were put into some kind of order by the American B Peirce in a paper published posthumously in 1881. Peirce had set himself the task of methodically classifying, and looking for applications of, the n-dimensional systems for all $n \leq 7$.

Did Hamilton have to go to 4 dimensions to find his algebra? It appears that he did search for 3-dimensional examples but could find none without divisors of zero.† Indeed, Hamilton believed that all 3-dimensional systems had to have divisors of zero and considered the fact that the quaternions had no divisors of zero to be one of its chief merits. In 1861 Weierstrass proved (essentially) that the only finite dimensional extensions of the real numbers in which all the usual laws of arithmetic hold are the real numbers themselves and the complex numbers. In 1878 Frobenius showed that relinquishing the commutative law of multiplication adds only the quarternions to the list and, using algebraic topology, Bott, Milnor and Kervaire showed, in 1957, that relinquishing in addition the associative law adds only the Cayley numbers. For more on this see the article by C W Curtis in [2].

Returning to 1847, Boole invented another sort of algebra, since called Boolean algebra, in order to put logic on a symbolic mathematical basis. After his death his wife Mary wrote that the idea of symbolising logic had occurred

* Here the remaining laws of arithmetic still hold.

† See [37, p. 106].

to him at the age of 17 (Leibniz had had similar but less developed ideas as early as 1666), but several subsequent writers (see [109, p. 235]) have indicated that Boole's work on the calculus of operations in the early 1840s must have at least influenced his approach if not actually initiated it. In this book we have insufficient space to be able to do justice to Boole's ideas by indicating applications to logic, probability and computer design. Fortunately there are several introductory books on the subject; [74] and [75] are just two of them.

Looking at the 19th Century development of number theory, we return to Gauss and his reciprocity law. He was able to extend his law to cubic and biquadratic residues but, to state his results elegantly, he found it helpful to introduce numbers of the form $a+\rho b$ and $a+ib$ respectively where a, b are integers, ρ is a complex cube root of unity and, of course, i is the usual square root of -1. In this work he needed to know that these numbers factorised uniquely into primes (3.7.13) just as do the ordinary integers. Kummer endeavoured to study higher residues and considered, for the purpose, numbers of the form $a_0+a_1\zeta+\cdots+a_{p-2}\zeta^{p-2}$ where the a_i are integers and $\zeta = e^{2\pi i/p}$, p a prime. These numbers are also relevant for attempts to solve Fermat's Conjecture (see Section 3.5) and indeed the FC would be solved if only the uniqueness of factorisation theorem valid for numbers of the form $a+\rho b$ and $a+ib$ extended to them. Unfortunately, as Kummer knew,* $p=23$ provided the first instance (of infinitely many) of the failure of unique factorisation. To try and get round the problem Kummer introduced extra 'ideal' numbers (Section 3.9) to help him regain uniqueness of factorisation in many cases. His analysis showed that the FC is indeed true for all prime exponents $p<100$, except for 'irregular' primes $p=37$, 59, 67, which cases he dealt with later.

Starting in 1871, Dedekind extended Kummer's ideas further. Any complex number which is a root of an equation $a_0+a_1x+\cdots+a_nx^n=0$ where the a_i are integers will, said Dedekind, be called an algebraic number. (See exercise 3.2.14.) Those for which, in addition, $a_n=1$, will be termed algebraic integers ($\frac{-7+\sqrt{-11}}{2}$ is one such (!) since it is a root of $x^2+7x+15=0$). Dedekind introduced the term number field (*Zahlkörper*, in German) to denote a collection of complex numbers satisfying the field axioms to be found in Definition 3.2.2(10). One can prove that the algebraic numbers form a field, but not so the algebraic integers: given (algebraic) integers α, β we find that α/β need not be an integer. The concept introduced here, then, needs a name. It was called number ring (*Zahlring*, by Hilbert, in 1897). Our present concept of field grew out of Dedekind's work and also that of Kronecker, the basic notions being present already in the work of Abel and Galois. Dedekind looked at collections of numbers gathered into a completed whole, this concept being essential in his construction (Section 4.4) of the real numbers on the basis of the rational numbers. Kronecker would have none of this. He insisted

* There is some well known folk-lore relating to this. See [46, p. 80] and the references mentioned there.

that every entity asserted to exist in mathematics must be shown to exist by a finite set of instructions whose applications yield the entity. Even the statement that 'obviously any polynomial in x with integer coefficients must be expressible (by using a 'degree' argument) as a product of polynomials which cannot be further decomposed' was of little value to him unless it be accompanied by a method which in each instance would supply the indecomposable factors. He supplied such a method (see exercise 1.11.11).

Since Dedekind's algebraic numbers extend Kummer's concepts, his algebraic integers also lack uniqueness of factorisation. He reinterpreted Kummer's concept of ideal number in terms of collections of already existing numbers, called these collections 'ideals' (see Sections 3.4 and 3.9), and showed how every ideal could be expressed uniquely as a product of prime ideals (see Section 3.9).

In 1893 H Weber gave an account of Galois' theory which is 'applicable to every case ... from function theory ... to number theory' using the field concept 'without reference to any numerical interpretation'. Weber can thus be regarded as the founder of abstract field theory.

One further item which should be mentioned here since it has strong connections with present day ring theory is the subject of algebraic invariants. Algebraic quantities remaining essentially unchanged under a change in coordinates are important in coordinatised geometry since they correspond to intrinsic geometric properties. In number theory too there was interest in representing whole numbers by, amongst other things, binary quadratic forms $f(x, y) = ax^2 + 2bxy + cy^2$ and in quantities such as $b^2 - ac$ which remain unchanged when, in $f(x, y)$, x, y are replaced by $x = \alpha x' + \beta y'$, $y = \gamma x' + \delta y'$ where α, β, γ, δ are integers and $\alpha\delta - \beta\gamma = 1$. Motivated by an 1841 paper of Boole, Cayley, who was interested at the time in algebraic aspects of projective geometry, began seeking invariants of homogeneous forms of degree n in two and more variables. Cayley attracted his friend Sylvester into studying invariants (the term 'invariant' is due to Sylvester) and these two did so much research on the subject that they were named the Invariant Twins. Cayley's work soon encouraged many mathematicians to invariant theory, described by Sylvester as 'the essence of modern algebra', and so many invariants were found that, to bring some order to the subject, minimal systems of invariants were sought in terms of which all others could be expressed. Gordan (1868) proved that to each binary form $f(x, y)$ there is such a finite system. The proof is hard. The whole subject was brought to a sudden climax when Hilbert, in 1888, showed that for a form of any degree in any number of variables a finite 'basis' always exists—and this without giving any indication of how to find such a set in any particular instance! What Kronecker said doesn't seem to have been recorded;* Gordan called it 'Theology, not mathematics'. Hilbert's theorem in its ring-theoretic form is stated in Section 3.4.

Though it is sometimes said that Hilbert's theorem killed invariant theory, this is not entirely correct. Invariant theory continued—albeit at a reduced

* Hilbert's result, proved in 1888, appeared in print in 1890. Kronecker died in 1891.

level of intensity—and in recent years activity has begun to pick up again. (For more on the history see [102].)

A subject of which invariant theory formed a considerable part was that of algebraic geometry. Algebraic geometry which, very loosely speaking, is concerned with curves and surfaces in n-dimensional space which are defined by algebraic equations, is a meeting ground for several mathematical disciplines including geometry, complex analysis, topology and number theory as well as algebra. Around 1900 many deep results were obtained especially by the Italian geometers although the validity of some of their methods was not always apparent. The subject was set upon firm foundations around 1930 by Emmy Noether and van der Waerden using an abstract algebraic approach. Along with algebraic number theory, algebraic geometry can claim to be one of the main motivating factors behind an autonomous theory of current interest, that of commutative rings.

What about the development of group theory from Galois' time up to 1900? First, one should note that it was not until about 1846, when Liouville published two of his papers, that Galois' work became better known. As suggested in [108, p. 94], it was possibly in connection with the announcement of publication, made as early as 1843, that in the period 1844–6 A L Cauchy was especially active in developing the theory of permutations. Adding to his work of 30 years before he proved Galois' assertion that if a finite group G contains pm elements, p being a prime, then G contains a subgroup of order p. One should note however that Cauchy did not use the word group but talked rather of 'a system of conjugate substitutions'. Although Galois had introduced the word group, he had used it inconsistently. In 1854 Cayley explicitly defined the term 'group'. He thought only in terms of finite groups but, since he specifically insisted on the associative law being satisfied his ideas were similar to those of today. In this paper Cayley's concern was with systems of elements satisfying the equation $x^n = 1$ and for both $n = 4$ and $n = 6$ he showed that there is essentially just one other system besides the set of complex nth roots of unity.

Up to about 1867 only finite groups were considered. Furthermore the term group was not generally employed and the main interest was in finding the number of formally distinct values a function of n variables takes when the variables are permuted. In 1867 Jordan, motivated by earlier studies of crystal structure by the physicist Bravais, considered groups with infinitely many elements—in particular groups of movements. In 1870 Jordan produced his *Traité des Substitution's et des équations algébriques*. This work organised the known theory of permutation groups and its relationship with Galois Theory. It also introduced many new results and the concept of homomorphism (5.10.1) as well as providing the atmosphere for the eventual finding by Fedorov and Schönflies around 1890 of the 230 crystallographic space groups (see Section 5.12).

In 1872 the Norwegian L Sylow extended the Galois/Cauchy result mentioned above by replacing p by p^α in hypothesis and conclusion (see 6.2.8 and 6.2.12).

In the same year Felix Klein, in his famous inaugural address at the University of Erlangen, stated his aim of using group theory to bring a unity to the various classical geometries that had been found since the announcement of the first non-Euclidean one by Lobachevsky in 1829. Thus geometries would be classified by groups of transformations which left certain geometrical aspects invariant. (The concept of invariance was definitely the 'in' subject at the time and it was later to provide a central idea in the theory of relativity.)

Since rotations and translations of the plane can be arbitrarily small the notion arises of an (infinite) continuous group. In 1874 and 1883 Sophus Lie (pronounced 'Lee', Norwegian) used the idea to attempt a classification and simplification of the solutions to certain differential equations. In studying his continuous transformation groups (groups whose elements depend upon a system of continuously varying parameters satisfying certain differentiability conditions) Lie was led naturally to study some non-commutative, non-associative algebras subsequently named after him: Lie algebras. (In a Lie algebra multiplication and addition satisfy $a \cdot b = -b \cdot a$ and $(a \cdot b) \cdot c + (b \cdot c) \cdot a + (c \cdot a) \cdot b = 0$.) Lie groups and Lie algebras form a major component of the present day theoretical physicists' armoury.

We note, in passing, that initially there was no universal agreement about what exactly constituted a group. For example, whereas Cayley, in 1854, specifically demanded that the associative law should be satisfied, Lie and Klein, in their earlier work, did not feel obliged explicitly to mention the requirement; in all cases of interest to them the condition was automatically satisfied! As the group concept became yet more prominent, it became increasingly desirable to standardise terminology. In 1882 H Weber gave a set of postulates for abstract groups of finite order. These postulates are essentially those in use today.

Two other directions taken in the 19th Century by the theory of groups should perhaps be mentioned. One is Dyck's concentration on systems of generators for a group and on relations satisfied by these generators. These concepts came to be of prime importance with the introduction of non-abelian groups into topology, specifically via the fundamental group of a topological space. The second is the introduction of group representation theory in which groups are represented (via homomorphisms) by groups of matrices with complex number entities. Matrices have the advantage that they can be added together and multiplied by scalar quantities; further the concepts of determinant and trace are available to aid computation. This theory, developed by Frobenius, Molien, Schur and Burnside (see [104]) is of vital importance today ([104], [36, Chapter 12], [62]) in the theory of finite groups and also in representing certain groups which arise naturally from symmetry considerations in chemistry and physics.

Possibly inspired by Hilbert's (1899) full axiomatisation of Euclidean geometry, the 20th Century began with many attempts to find independent sets of axioms for fields and for groups, the main worker being E V Huntington around 1902–5. In 1905 J H M Wedderburn proved that every finite division ring is a field, a result which provides the only known proof that in a finite projective plane Desargue's theorem implies that of Pappus [19]. In 1907 Wedderburn proved one of the fundamental theorems of modern non-commutative ring theory. This theorem, proved for algebras over an arbitrary field, was extended by Artin in 1927 to more general rings. It describes, just as does the Fundamental Theorem for Finite Abelian Groups (6.4.4), the exact structure, in easy terms, of a wide class of rings (the so-called semisimple ones).

In 1908 K Hensel introduced for number-theoretic purposes a new type of number—the p-adic numbers. Inspired by Hensel's construction, E Steinitz, in a 140 page paper in 1910, undertook a classification of all fields, several apparently unrelated sorts of which had now come into existence. He proved that all fields fall into two categories: those whose unique minimal (so-called prime) subfield is essentially the same as the rational numbers and the others where it is essentially the (finite) Galois field $\mathbb{Z}_p$ (see Section 3.10). Further, every field can be obtained from its prime subfield by successively 'adjoining' elements.

1914 saw the first axiomatic declaration of exactly what constitutes a ring. (It is less general than the current one.) This axiomatising process was carried further in the 1920s and 1930s by Emmy Noether, the greatest of all women algebraists. Following her doctoral thesis which, supervised by Gordan, closed with a complete list of 331 covariant forms for a given ternary quintic, her strong inclination to unify, organise and generalise via axiomatisation took over. In 1921 she investigated differential operators in quantum mechanics by abstracting their essential properties, taking these properties as axioms and building all consequences thereupon. This approach she adopted in all her subsequent work thereby introducing a revolutionary style of attack on problems of algebra. In the period 1920–6 she brought within her general theory of ideals several theories which until that time had been seen as independent. Her methods also greatly simplified many of the earlier proofs, a number of them being very important in algebraic geometry. The rings she and her school investigated, in one sense dual to those of Artin mentioned above, are now called, naturally enough, Noetherian rings. Penetrating theorems about these rings were first obtained as recently as 1960 by A W Goldie.

Amongst the major achievements of the century in group theory one must mention the modular representation theory of groups by matrices over finite fields as pioneered by Richard Brauer and the subsequent use of this theory in investigations into finite simple groups (Section 6.6). In the theory of infinite groups various conjectures of Burnside have been established by the Russian mathematicians Golod, Kostrikhin, Novikov and Adjan. We refer the reader to [29] for precise statements.

The group-theoretic result of Golod (1964) is a consequence of joint work with Shafarevich (also 1964) in which was solved affirmatively an old problem of field theory closely related to the work described here in Sections 3.6, 3.7 and 3.9 (see [52, p. 125] for brief details).

Some uses of algebra

It is impossible for the author to know exactly what the reader will accept as being a 'proper' application of abstract algebra. Applications within pure mathematics other than those found in the text are far too numerous and diverse to mention here. Even excluding these the following represents only a small sample.

First, algebra offers its basic notations and concepts as a most convenient means for expressing, mathematically, certain concrete ideas. For instance, professional mathematicians often prefer the phrase 'ring of integers' to 'collection of all integers', even when no use is to be made of the ring structure: the algebraic terminology gives a more complete, a more faithful, picture. (In this text we make a similar use of set theory. That is, we use its more elementary concepts and notations as a language in which to express, succinctly, our algebraic notions.) Of course there is no obligation to use the language of algebra in this way any more than there is to use the symbol x in solving problems involving quadratic equations: as mentioned earlier, the Babylonians readily solved such problems writing down everything in longhand. It's just that the new terminologies and notations offer extra insight and clarity of expression and enable more powerful methods to be developed.

Regarding the more specialised applications which seem to use the algebra (of the type presented here)* in an essential way, the theory of groups, being (in a rather wide sense) the mathematical formulation of symmetry, is naturally widely employed in physics and chemistry—for instance in applications to crystallography, spectroscopy, general relativity, molecular vibrations, molecular orbitals, solid state physics and especially in the modern theory of elementary particles. (In February 1964 the Omega minus particle, which group theory had previously predicted should exist, was first identified.)

Ring theory is not quite as old a subject as group theory and direct 'practical' applications seem to be somewhat limited in number. This ceases to be the case when one considers rings which have extra structure, as for instance, when the ring is an algebra or a field. In particular algebras of matrices and of polynomials occur frequently: and in quantum mechanics use is made of the algebra of all polynomials in two (non-commuting) letters x, y between which the relationship $xy = yx + 1$ is assumed to hold. (Researchers in ring theory are still investigating this ring!) In other applications rings (algebras) also arise from groups. Both finite and infinite multiplicative groups give rise to so-called group algebras via attempts to 'represent' the group elements by matrices, and in studying a Lie group an investigation of the associated Lie algebra is invaluable.

The theory of fields naturally underlies all appearances of the fields of rational, real and complex numbers—but also finite fields (as well as finite groups, polynomial rings and power series rings) are put to practical use in

* Unfortunately there is no space in this book for that branch of algebra, namely linear algebra, which has found the most 'practical' applications.

the construction of efficient (from the point of view of cost!) error detecting and correcting codes in the area of data communications. Finite fields are also of importance in statistics via their association with (sets of orthogonal) Latin squares.

Finally, the algebra introduced by Boole to model logic mathematically has found application to the design of computers and telephone switching circuits, again via the very real problem of reducing construction costs.

The following problems are not algebraic in content. They are placed here mainly for your enjoyment and for subsequent discussion with friends and teachers. However, it is intended that they should extract from the reader that kind of critical attitude with which he should read this book from Chapter 1 onwards.

Problems

1 Sketch the graph of $y = |x|$. Clearly this function is continuous everywhere and fails to be differentiable only at the origin (i.e. $x = 0$ is the only point at which there is no tangent). Invent a function which is continuous for all x and yet not differentiable whenever x is a whole number. Can there exist a continuous function which is not differentiable for any x? (Try sketching such a function and then use your intuition. Finally, ask your teacher.)

2 Euclid defined a point as 'That which has no part'. Criticise this definition.

3 Discuss: Given any two straight line segments, say ———— ——————, it is obvious that there must exist some (perhaps very small) unit of length in terms of which the lengths of the above lines are m and n units respectively, m and n being whole numbers.

4 Draw a circle C. Call the interior of C 'the plane', each point inside C 'a point' and each chord of C (except for its end points) 'a straight line'. Defining two 'straight lines' to be 'parallel' if they do not meet in a 'point' (i.e. inside C) show that: Given a 'straight line' L and a 'point' P, not on L, in the 'plane', it is possible to draw through P infinitely many 'straight lines' which are 'parallel' to L.

0

Elementary set theory and methods of proof

0.1 Introduction

In 1895, at the beginning of his work *Beiträge zur Begründung der transfiniten Mengenlehre*, Georg Cantor* made the following definition:

> By a *set* we understand any collection M of definite, distinct objects m of our perception or of our thought (which will be called the *elements* of M) into a whole.

Thus examples of sets are: the set $\mathbb{Z}$ of all whole numbers, here called the *integers*; the sets $\mathbb{Q}$, $\mathbb{R}$ and $\mathbb{C}$ of all *rational*, all *real* and all *complex numbers* respectively; the set M comprising all moons of Mars; and even the set H of all ten-legged octopodes which visited Archangel last 1 April.

Cantor's need for such a definition had arisen around 1872 from his investigations concerning the possible uniqueness of representation of functions by trigonometric series. In due course it became apparent that all of mathematics could be made to rest upon a set-theoretic base. In particular Cantor and Richard Dedekind, in his *Stetigkeit und irrationale Zahlen* (1872), showed how the somewhat intangible *irrational numbers* (that is, those elements of $\mathbb{R}$ which are not in $\mathbb{Q}$) could, using the set concept, be made respectable in terms of $\mathbb{Q}$ (see Section 4.4) and Gottlob Frege (1884) demonstrated how the natural numbers $0, 1, 2, 3, \ldots$ (on which $\mathbb{Z}$ and ultimately $\mathbb{Q}$ can be based – see exercise 4.4.17 and 3.10.5(iii)) could be defined in set-theoretic terms. In addition, the concept of function can also be defined set-theoretically (see Section 2.6).

It is therefore not surprising that the notations, terminology and simpler notions of set theory now form an essential part of the language in which contemporary mathematical discussions are conducted. The next two sections introduce the simple set-theoretic ideas useful in this book.

0.2 Sets

We shall consider the words *set, collection, aggregate* as synonymous. The elements of a set we shall sometimes call its *members*. If A is a set and if an object a is an element of A we write $a \in A$. One usually reads the symbolism $a \in A$ as '*a belongs to A*'. If a is not a member of A we write $a \notin A$ and say

* Georg Cantor (3 March 1845 – 6 January 1918).

'*a does not belong to A*'. Thus $3 \in \mathbb{Z}$, $\frac{2}{3} \notin \mathbb{Z}$, $\frac{2}{3} \in \mathbb{Q}$, $\pi \in \mathbb{R}$, $\pi \notin \mathbb{Q}$, $\pi + i \in \mathbb{C}$, Phobos $\in M$, Isaac Newton $\notin H$.

Sets can be described by listing their members between pairs of curly brackets (also called braces). For instance, M may be written alternatively as $M = \{\text{Phobos, Deimos}\}$. It is of course impossible to describe sets containing infinitely many elements this way. On such occasions we might be tempted to write $\{x : P(x)\}$ where P is a property characterising those and only those elements of the set in question, $P(a)$ indicating that the object a has property P and hence lies in the set.* Thus the set $\mathbb{Z}^+$ of all positive integers may be written $\mathbb{Z}^+ = \{x : x \in \mathbb{Z} \text{ and } x > 0\}$. (This symbolism is read as 'z plus is the set of all x such that x belongs to z and x is greater than 0'.) However, the notation $\mathbb{Z}^+ = \{1, 2, 3, 4, \ldots\}$, where the three dots indicate the vague expression 'and so on', is sometimes used. Small finite sets can be exhibited in several ways. We have, for instance, $M = \{x : x \text{ is Phobos or } x \text{ is Deimos}\}$ or again $M = \{x : x \text{ is a Martian moon}\}$.

If A and B are sets and if each element of A belongs to the set B we say that A is a *subset* of B or that A is contained in† B, and we write $A \subseteq B$ (or $B \supseteq A$, the latter being read also as 'B contains A'). In particular $A \subseteq A$ for each set A and if $A \subseteq B$ and $B \subseteq C$ then $A \subseteq C$. Given $A \subseteq B$, if we know (and if we care!) that B contains elements not in A then we will write $A \subset B$ (or $B \supset A$) or even $A \subsetneq B$ ($B \supsetneq A$) if extra emphasis is required. A is then called a *proper subset* of B. Thus we write $\mathbb{Z} \subseteq \mathbb{Q}$, $\mathbb{Z} \subset \mathbb{Q}$ or $\mathbb{Z} \subsetneq \mathbb{Q}$ according to the emphasis required. If A is not a subset of B we write $A \not\subseteq B$ ($B \not\supseteq A$). Note that $A \not\subseteq B$ when and only when A contains at least one element which is not in B. We say sets A and B are *equal*, and write $A = B$, when and only when they contain precisely the same members. Thus $A = B$ when and only when both $A \subseteq B$ and $B \subseteq A$ hold simultaneously. It follows that a useful way to establish the equality of two sets A and B is to prove both $A \subseteq B$ and $B \subseteq A$.

Warning: Try not to confuse $\in$ and $\subseteq$. Somewhat roughly stated: $\in$ is used in relating a set to its elements; $\subseteq$ is used in relating a set to its subsets. To illustrate this, suppose $A = \{1, \{1\}, 2, \{3, 4\}, \mathbb{Z}\}$. Thus A is a set with five elements, namely 1, $\{1\}$, 2, $\{3, 4\}$ and $\mathbb{Z}$. Hence $1 \in A$, $1 \not\subseteq A$, $\{1\} \in A$, $\{1\} \subseteq A$, $\{2\} \notin A$, $\{2\} \subseteq A$, $\{3\} \notin A$, $\{3\} \not\subseteq A$, $\{3, 4\} \in A$, $\{3, 4\} \not\subseteq A$, $\mathbb{Z} \in A$, $\mathbb{Z} \not\subseteq A$. (Incidentally, sets like $\{2\}$ which contain exactly one element are called *singletons*.)

Now consider the sets $F = \{x : x \in \mathbb{Z} \text{ and } x^2 < 0\}$, $G = \{\ \}$ and H as mentioned earlier. It appears that each contains no members at all. A set with no members

* It is perhaps surprising that so simple a definition can give rise to contradictions. In fact even Cantor realised this, although the best known example is due to the British mathematician/philosopher Bertrand Russell who took $P(x)$ to be the condition $x \notin x$. Clearly $1 \notin 1$ and $\mathbb{Z} \notin \mathbb{Z}$ so that $R = \{x : x \notin x\}$ is surely non-empty. One can ask: 'Does R belong to R?' If so, then R satisfies the condition ($x \notin x$) for being an element of R. Hence $R \notin R$. If not, then R fails the test $x \notin x$ and we deduce $R \in R$. Thus $R \in R$ when and only when $R \notin R$. Various axiomatisations of set theory (in particular Zermelo's; see Prologue, p. xv) have been proposed to exclude the appearance of such paradoxes. The sets we shall consider will all be acceptable in Zermelo's scheme.

† Some mathematicians prefer to say 'included in' to avoid possible confusion with subsequent use of 'A contains a' to describe $a \in A$ This dual use of 'contains' is bad but common. Cf. above 'Warning'.

is called an *empty set*. Despite the differences in definition of F, G and H we can show that there is only one empty set. To prove this, and more, let $\varnothing$ be an empty set and let A be any set. Then $\varnothing \subseteq A$, for otherwise $\varnothing \not\subseteq A$ and so $\varnothing$ would contain an element not to be found in A. But this is silly since $\varnothing$ has no elements. Now suppose $\varnothing_1$ is another empty set. Since $\varnothing$ is empty $\varnothing \subseteq \varnothing_1$. Since $\varnothing_1$ is empty $\varnothing_1 \subseteq \varnothing$. These two inequalities, taken together, imply $\varnothing = \varnothing_1$. Thus we can talk of *the* (*unique*) *empty set*.

Problem 1 What is wrong with the following argument showing $\varnothing \not\subseteq A$?

To prove $B \subseteq A$ we must show that every element of B is also contained in A. Since $\varnothing$ has no elements we cannot show $\varnothing \subseteq A$. Hence $\varnothing \not\subseteq A$.

Problem 2 In connection with the first footnote on p. 2, can you think of a set A for which $A \in A$?

0.3 New sets from old

New sets can be made from old in several ways, one of which we saw above in defining $\mathbb{Z}^+$ in terms of $\mathbb{Z}$.

Let A and B be sets. Then the sets

$$A \cap B = \{x : x \in A \text{ and } x \in B\}$$

and

$$A \cup B = \{x : x \in A \text{ or } x \in B \text{ (or both)}\}^*$$

are called, respectively, the *intersection* and the *union* of A and B. Thus if $A = \{3, 1, 4\}$ and $B = \{\pi, 4, \text{Oliver Cromwell}, 1\}$ then $A \cap B = \{1, 4\}$ whilst $A \cup B = \{3, 1, 4, \pi, \text{Oliver Cromwell}\}$.

It follows immediately that for sets A and B we have

$$A \cap A = A, \quad A \cap B = B \cap A, \quad A \cap \varnothing = \varnothing$$

$$A \cup A = A, \quad A \cup B = B \cup A, \quad A \cup \varnothing = A$$

Intersection and union of sets can be thought of in terms of the shaded regions in Fig. 0.1; such figures are called *Venn*† *diagrams*.

If we introduce a third set C and shade the region common to $A \cup B$ and C (Fig. 0.2) it appears that the sets $(A \cup B) \cap C$ and $(A \cap C) \cup (B \cap C)$ are equal. This can be checked (exercise 5(b)) by an argument not depending upon pictures. [Pictures can be deceptive. What value has this pictorial proof if, say, $C \cap B = \varnothing$ or if $A \subseteq B$? See [124].]

There is no difficulty in extending the definition of union and intersection to larger finite or even infinite collections of sets. For example, if for each $n \in \mathbb{Z}^+$ we define

$$S_n = \left\{x : x \in \mathbb{R} \text{ and } -\frac{1}{n} < x < \frac{1}{n}\right\}$$

* In mathematics the word 'or' is taken to include the possibility of both. This is not always the case in ordinary conversation, for example 'Would you like tea or coffee (but *not* both)?'

† John Venn (4 August 1834 – 4 April 1923).

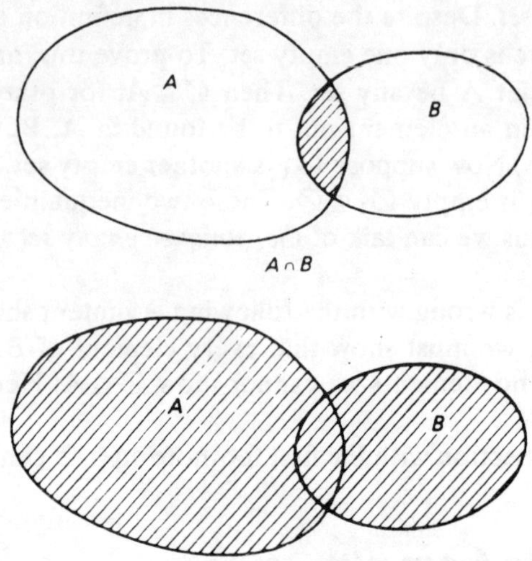

Fig. 0.1

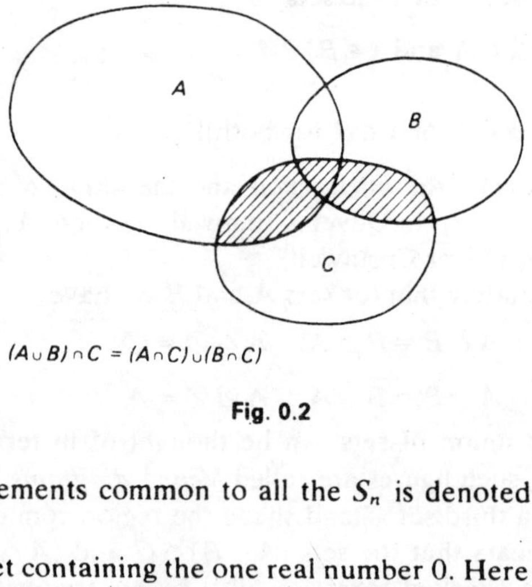

$(A \cup B) \cap C = (A \cap C) \cup (B \cap C)$

Fig. 0.2

then the set of elements common to all the S_n is denoted by $\bigcap_{n=1}^{\infty} S_n$. In fact $\bigcap_{n=1}^{\infty} S_n = \{0\}$, the set containing the one real number 0. Here we have 'indexed' the various S with the elements of the set $\mathbb{Z}^+$. We can also use the elements of $\mathbb{R}^+$ (see exercise 6) or indeed the elements of any set (see 3.4.5(F)) as an *indexing set*.

Another way of producing a new set from two old ones A and B is to define their *difference* $A \backslash B = \{x : x \in A \text{ and } x \notin B\}$. In particular the set of all non-zero real numbers is then denoted by* $\mathbb{R}\backslash\{0\}$.

The next definition is suggested by the way coordinates are introduced into the real plane. Points in the plane are made to correspond to pairs of

* Also denoted by $\mathbb{R}^{\times}$.

real numbers and *vice versa* in such a way that if points P and Q are given coordinates (a_1, a_2) and (b_1, b_2) respectively then P and Q coincide when and only when $a_1 = b_1$ and $a_2 = b_2$. In particular (1, 2) and (2, 1) correspond to distinct points. Since the only distinction between (1, 2) and (2, 1) is the order in which the numbers 1 and 2 are written down we refer to such pairs of numbers as *ordered pairs.*

Much more generally we make

Definition 0.3.1 Let A and B be sets. Then $A \times B$ denotes the set $\{(a, b): a \in A \text{ and } b \in B\}$ of *ordered pairs.* $A \times B$ is called the *Cartesian product* of A and B. The word 'ordered' implies that elements (a, b), (c, d) in $A \times B$ are defined to be equal when and only when *both* $a = c$ and $b = d$.

Notes 0.3.2
(i) The concept of ordered pair can be described in a purely set-theoretic manner. See exercise 8.
(ii) In a like manner one can define the set $A \times B \times C$ of *ordered triples* (a, b, c) where $a \in A$, $b \in B$ and $c \in C$. One can even form the Cartesian product of a collection of infinitely many sets (see exercise 6.3.20).

Example 0.3.3 If $A = \{1, \Pi\}$, $B = \{1, e, \hbar\}$ then $A \times B = \{(1, 1), (1, e), (1, \hbar), (\Pi, 1), (\Pi, e), (\Pi, \hbar)\}$. Note that $A \times B$ and $B \times A$ each have 6 (=2.3=3.2) elements but that $A \times B \neq B \times A$. [Why not?]

One final word on set production:

Definition 0.3.4 Let A be any set. By $\mathcal{P}(A)$ we denote the set of all subsets of A. $\mathcal{P}(A)$ is called the *power set* of A.

Example 0.3.5 If $A = \{a, Б, \gamma\}$ then

$$\mathcal{P}(A) = \{\varnothing, \{a\}, \{Б\}, \{\gamma\}, \{a, Б\}, \{Б, \gamma\}, \{\gamma, a\}, \{a, Б, \gamma\}\}$$

Note that A has 3 elements whilst $\mathcal{P}(A)$ has 2^3.

0.4 Some methods of proof

From the beginning of Chapter 1 onwards, the majority of this book is taken up with assertions described as theorems (or lemmas or corollaries) followed by explanations purporting to be 'proofs'. Just as we have adopted an intuitive approach to the concept of set so we shall allow our intuition to guide us in the matter of whether or not an explanation is logically acceptable: to formalise the notion of acceptability would take us too far afield, into symbolic logic. (The reader keen to see how this can be achieved is referred, for instance, to [57], [59].) On the other hand there are a few techniques, frequently used in constructing proofs, to which it is perhaps useful to draw attention as they have been known to give students the occasional difficulty. We do this via some concrete examples, simple enough to allow the logical principle involved to be easily seen.

Consider, for instance, how you would prove the following assertion:
(i) If x is an odd integer then x^2 is an odd integer.

You would probably say: 'Assuming x is odd, we may write $x = 2n + 1$ for some suitable $n \in \mathbb{Z}$. Then $x^2 = (2n+1)^2 = 4n^2 + 4n + 1$, which is clearly odd.'

Here we go directly from the hypothesis that x is odd to the conclusion that x^2 is odd. Such a proof is called a *direct proof*.

For a second example consider the following assertion about the integers x and y:
(ii) If $x^2 \neq y^2$ then $x \neq y$ (1).

A direct proof is possible: From $x^2 \neq y^2$ we deduce that $(x-y)(x+y) \neq 0$. It follows that $x - y \neq 0$ and hence that $x \neq y$.

However, the proof most of you would probably have given is:

If $x = y$ *then* $x^2 = y^2$ (2), a contradiction.

Hence the result.

This approach, using the method of *indirect proof*, conceals a number of logical points. Two we wish to draw attention to are most easily discussed in symbolic terms. Let A and B stand for '$x^2 \neq y^2$' and '$x \neq y$' respectively. Further let $\sim A$ and $\sim B$ denote their *negations* or *denials*. Thus $\sim A$ is '$x^2 = y^2$' and $\sim B$ is '$x = y$'. Then (1) is the assertion: If A then B. We write this briefly as $A \Rightarrow B$ and read it as 'A implies B'. On the other hand (2), apparently logically equivalent to (1), is the assertion $\sim B \Rightarrow \sim A$. In fact it is shown, in courses on elementary logic, that the assertions $A \Rightarrow B$ and $\sim B \Rightarrow \sim A$ are equivalent—as is the pair $B \Rightarrow A$ and $\sim A \Rightarrow \sim B$. What are clearly not (in general) equivalent are the assertions $A \Rightarrow B$ and $B \Rightarrow A$. [In terms of our concrete example this latter assertion reads $x \neq y \Rightarrow x^2 \neq y^2$, which is blatantly false, whereas (1) is (blatantly) true.] The assertion $B \Rightarrow A$ is called the *converse* of the assertion $A \Rightarrow B$.

To summarise: to prove that an asserted conclusion B follows from a hypothesis A one can proceed directly or prove, equivalently, that $\sim A$ is deducible from hypothesis $\sim B$. This kind of indirect proof is called *proof by contraposition*, $\sim B \Rightarrow \sim A$ being the *contrapositive* to $A \Rightarrow B$.

Another frequently adopted method of indirect proof is that of *reductio ad absurdum* or *proof by contradiction*. Here the negation of the hoped for conclusion is shown to lead to a contradiction (or absurdity). Possibly the first ever use of this method was in a proof of:

Theorem 0.4.1 If $x \in \mathbb{R}$ and $x^2 = 2$ then $x \notin \mathbb{Q}$. In other words, $\sqrt{2}$ is not a rational number.

Proof We suppose $\sqrt{2} \in \mathbb{Q}$ so that we may write $x = m/n$ where $m, n \in \mathbb{Z}$. Clearly we may assume m, n have no common divisor >1 in $\mathbb{Z}$; such a common divisor can be eliminated before proceeding. Then from $x = m/n$ we deduce that $m^2/n^2 = 2$, that is $m^2 = 2n^2$. It follows that m^2, and hence* m, is even. Writing $m = 2l$ we find $4l^2 = 2n^2$; that is $2l^2 = n^2$. But then n^2, and hence* n, is even. Thus our assumption, namely that $\sqrt{2} = x = m/n \in \mathbf{Q}$, has led to the contradiction (absurdity) that numbers m, n with no common divisor >1 are both even. Consequently $\sqrt{2} \notin \mathbf{Q}$.

Now consider the following two assertions.

(i) Every positive integer is the sum of four (non-negative) integer squares.
(ii) Every positive integer is the sum of eight non-negative integer cubes.

It is not very difficult to check that (i) is certainly true for each positive $n \leq 100$. For example $75 = 8^2 + 3^2 + 1^2 + 1^2$; $86 = 9^2 + 2^2 + 1^2 + 0^2$. You should note however that even if you check (i) for all positive integers up to, say, $10^{50\,000}$, (i) would still not be established *for every positive integer n*; indeed, for all you then know, (i) might fail for $n = 10^{50\,000} + 1$. No, to prove (i) you must devise a proof which covers *all* positive integers n.

In fact (i) is true; it was first proved by Lagrange in 1770. On the other hand (ii) is false: not all positive integers can be expressed as asserted. For instance 239 cannot so be expressed. 239 is thus a *counterexample* to assertion (ii). Actually there is only one other positive integer which is not expressible as a sum of eight cubes. Can you find it? It is less than 50. The point is that even though only two out of infinitely many positive integers fail to satisfy (ii) the existence of *even one* such 'nasty' integer is sufficient to destroy (ii)'s claim to be a valid assertion. In particular, finding the other 'nasty' number changes nothing—the challenge was set just for fun.

There are other principles of reasoning we could mention. For instance in showing that 2.5.1 follows from 2.5.1′ we split the proof into two cases (one when p divides n and one when it does not) and establish each case separately. This kind of proof is called *proof by cases*. Its use in the instance quoted is so straightforward that no special attention needed to be drawn to it. We refer the reader interested in a detailed account of different methods of proof to [59, pp. 30–46].

We close with one or two remarks on terminology and notation. We have already observed that the assertion 'If A then B' is written $A \Rightarrow B$ and read 'A implies B'. We also say 'A is a *sufficient condition* for B' (since A is enough, all by itself, to allow us to conclude B) or that 'B is a *necessary condition* for A' (since B necessarily follows from A—whether you like it or not). Alternatively, mathematicians say '*B if A*' or '*A only if B*'. If we know $A \Rightarrow B$ *and* $B \Rightarrow A$ we say '*A if and only if B*', an assertion we write briefly as $A \Leftrightarrow B$ or A *iff* B. Technically all definitions should be in 'iff' form. If, for instance, in 1.3.1 we had defined a to be a divisor of b *if* $b = ac$ this would have left open the question of whether or not we are to call a a divisor of b if no such c existed. Thus, use of 'iff' indicates that a will be called a divisor of b when *and only when* the required c exists. (Cf. Definition 1.3.3!)

* Exercise 11.

In symbolic logic much use is made of the signs $\exists$ (*there exists*) and $\forall$ (*for all*), although we don't often employ these here. As an example, note that one of the properties of the equality relationship on $\mathbf{Z}$ used in the proof of 1.2.1(i) may be stated succinctly as $(\forall x)(\forall y)(\forall z)(x = y \Rightarrow xz = yz)$, whilst axiom A3 in Section 1.2 includes the statement that $(\exists x)(\forall y)(x + y = y)$. Finally note that the negation of $(\exists x)(P(x))$, that is $\sim((\exists x)(P(x)))$, is $(\forall x)(\sim P(x))$ and that, similarly, $\sim((\forall x)(P(x))) \Leftrightarrow (\exists x)(\sim P(x))$.

Exercises

1 Which of the following assertions are true?
(a) $\frac{3}{5} \in \mathbf{Z}$; (b) $(\pi + \mathrm{i})^2 \in \mathbf{C}\backslash\mathbf{R}$; (c)* $\mathrm{e} - \pi \notin \mathbf{R}^+$; (d) $(\mathrm{e} - \pi)^2 \in \mathbf{R}\backslash\mathbf{R}^+$;
(e) $\pi^{\mathrm{e}} - \mathrm{e}^{\pi} \in \mathbf{R}^+$.

2 Let $A = \{\varnothing, \{\varnothing\}, 1, \{1, \varnothing\}, 7\}$. Which of the following are true?
(i) $\varnothing \in A$; (ii) $\{\varnothing\} \in A$; (iii) $\{1\} \in A$; (iv) $\{7, \varnothing\} \subsetneq A$; (v) $7 \subseteq A$;
(vi) $\{7\} \not\subseteq A$; (vii) $7 \subsetneq A$; (viii) $\{7, \{1\}\} \not\subseteq A$; (ix) $\{\varnothing, \{\varnothing\}, \{1, \varnothing\}\} \subseteq A$;
(x) $\{\{\varnothing\}\} \subseteq A$.

3 Let $A = \{x : x \in \mathbf{R}^+ \text{ and } x^2 > 7\}$, $B = \{1, 2, 3, 4\}$, $C = \{3, \text{Lewis Carroll}\}$. Find $A \cap B$, $B \cap C$, $A \cap B \cap C$, $A \cup B$, $(A \cup B) \cap C$, $(A \cap C) \cup (B \cap C)$. Draw a Venn diagram involving A, B and C.

4 Let $A = \{(x, y) : x \in \mathbf{R}, y \in \mathbf{R}, x^2 + y^2 = 1\}$, $B = \{(x, y) : x \in \mathbf{R}, y \in \mathbf{R}, y^2 = 4x\}$, $C = \{(x, y) : x \in \mathbf{R}, y \in \mathbf{R}, y^2 = x^3\}$. Find the intersections and unions asked for in exercise 3.

5 (a) For sets A and B show that $A \cup B = B$ iff $A \subseteq B$ and that $A \cap B = A$ iff $A \subseteq B$.
(b) For sets A, B and C prove that (i) $(A \cup B) \cap C \subseteq (A \cap C) \cup (B \cap C)$; (ii) $(A \cap C) \cup (B \cap C) \subseteq (A \cup B) \cap C$. Deduce that (iii) $(A \cup B) \cap C = (A \cap C) \cup (B \cap C)$. (Cf. exercise 3.)

6 For each $r \in \mathbf{R}^+$ and each $n \in \mathbf{Z}^+$ put $T_r = \left\{x : x \in \mathbf{R} \text{ and } -\frac{1}{r} \leqslant x \leqslant \frac{1}{r}\right\}$ and $R_n = \left\{x : x \in \mathbf{R} \text{ and } -\frac{1}{n} < x \leqslant n\right\}$. Find $\bigcap_{r \in \mathbf{R}^+} T_r$, $\bigcup_{n=1}^{\infty} R_n$, $\bigcap_{n=1}^{\infty} R_n$.

7 Let U be a set with subsets A and B. Define $A^c = U \backslash A$ etc. Show (i) $A^c \cap B^c \subseteq (A \cup B)^c$; (ii) $(A \cup B)^c \subseteq A^c \cap B^c$; (iii) $(A \cup B)^c = A^c \cap B^c$; (iv) $(A \cap B)^c = A^c \cup B^c$. ((iii) and (iv) are called *de Morgan's laws*.†)

8 Define $[a, b]$ to be the set $\{\{a\}, \{a, b\}\}$. Show that $[a, b] = [c, d]$ iff $a = c$ and $b = d$. (This explains note 0.3.2(i).)

9 (i) Show that if A and B are non-empty then $A \times B = B \times A$ iff $A = B$.
(ii) Let $S \subseteq A \times B$. Need S be of the form $C \times D$, where $C \subseteq A$ and $D \subseteq B$?

* $\mathbf{R}^+$ denotes the set of positive real numbers.
† Augustus de Morgan (? June 1806 – 18 March 1871).

10 How many elements has $\mathscr{P}(A)$ if: (i) A has 10 elements? (ii) $A = \varnothing$?

11 Let $m \in \mathbb{Z}$. Prove that: if m^2 is even then m is even.

12 Give another example of assertions A, B where $A \Rightarrow B$ is true but $B \not\Rightarrow A$ is false.

13 Prove that $e = 1 + \frac{1}{1!} + \frac{1}{2!} + \frac{1}{3!} + \cdots$ is irrational as follows. Suppose $e = \frac{m}{n}$. Then $t = n!\left(e - 1 - \frac{1}{1!} - \frac{1}{2!} - \cdots - \frac{1}{n!}\right) \in \mathbb{Z}$. But $t = \frac{1}{n+1} + \frac{1}{(n+1)(n+2)} + \cdots$ $\leqslant \frac{1}{n+1} + \frac{1}{(n+1)^2} + \cdots = \frac{1}{n}$, a contradiction. What kind of proof have we used here?

14 What kind of proof did we use in proving that $\varnothing \subseteq A$ for each set A?

15 Let x, y and $z \in \mathbb{Z}$. Show that $x^2 + y^2 + z^2$ cannot be of the form $8k + 7$ when: (i) exactly one of x, y, z is odd; and (ii) all of x, y, z are odd. Deduce that no integer of the form $8k + 7$ is expressible as a sum of three integer squares. What method of proof have we used here?

16 Write in words the following assertion, where x, y, $z \in \mathbb{Z}$: $(\forall x)(\forall y)(\exists z)(xy = z^2)$. Is the assertion true? Write symbolically: For all x and y in $\mathbb{Z}^+$ for which $x < y$, there exists z such that $x + z = y$.

17 Write in words the following assertions, in which x, $y \in \mathbb{Z}$: (i) $(\forall x)(\exists y)(y > x)$; (ii) $(\exists y)(\forall x)(y > x)$. Deduce that one may not necessarily be able to interchange $\forall$ and $\exists$ without the risk of changing the meaning.

1

Numbers and polynomials

1.1 Introduction

As implied in the Prologue, one of the central concepts of modern algebra is that of 'ring', two of the most fundamental examples being the ring $\mathbb{Z}$ of all integers (Sections 1.2 to 1.5) and the ring $\mathbb{Q}[x]$ of all polynomials in the 'indeterminate' x with coefficients in the field $\mathbb{Q}$ of rational numbers (Sections 1.6 to 1.11). The terms *ring* and *field* will be defined formally in Chapter 3. The main objectives of this chapter are as follows.

(1) To present some of the simpler properties of the sets $\mathbb{Z}$ and $\mathbb{Q}[x]$ in such a manner as to emphasise their similarities (and their differences!). Thus we shall have to hand important concrete examples and theorems which will help motivate, and can act as test-cases in, our later development.
(2) To introduce some terminology and notation in common use in the following chapters. One advantage of doing this at this stage is that the reader will probably feel able to devote a little extra effort to learning this terminology as the statements of the theorems themselves will take little remembering—many of them should be fairly familiar already.

Finally and by far the most important is:

(3) To introduce into this familiar setting a few notes of caution. Here we hope to develop the reader's critical faculties by showing, especially in relation to $\mathbb{Q}[x]$, that not everything is quite as straightforward as might be expected (see in particular Section 1.6). We hope the reader will examine proofs of theorems with one question continually in mind, namely 'Why can he (the author) say *that*'?

As promised earlier, many of the proofs in this chapter are written in expansive style with square brackets indicating those portions of proofs that could, without great loss, be omitted. The threefold purpose of these brackets is detailed in the preface.

1.2 The basic axioms. Mathematical induction

We start, then, by looking at the set $\mathbb{Z}$ of all integers. As noted in Chapter 0 a definition of integer can be given in set-theoretic terms (see, for example [58]). The algebraist is, however, little interested in what the integers *are*; he is mainly interested in the fact that these integers, whatever they may be, are added and multiplied together, two at a time, according to the following axioms (where we use the symbol '$\cdot$' to denote multiplication).

For every three integers a, b, c (distinct or not) we have:

A1 $a+b=b+a$

A2 $(a+b)+c=a+(b+c)$

A3 There exists in $\mathbb{Z}$ a unique integer, namely 0, such that $0+a=a+0=a$

A4 To each a in $\mathbb{Z}$ there exists a unique integer, namely $-a$, such that $a+(-a)=(-a)+a=0$

M1 $a \cdot b=b \cdot a$

M2 $(a \cdot b) \cdot c=a \cdot (b \cdot c)$

M3 There exists in $\mathbb{Z}$ a unique integer, namely 1, such that $1 \cdot a=a \cdot 1=a$

D $a \cdot (b+c)=a \cdot b+a \cdot c$; $(a+b) \cdot c=a \cdot c+b \cdot c$

P $\mathbb{Z}$ contains a non-empty subset N such that
(i) each element of $\mathbb{Z}$ belongs to exactly one of the sets N, $\{0\}$, $-N$ where $-N$ denotes the set $\{-x : x \in N\}$,
(ii) for all $a, b \in N$ we have $a+b \in N$ and $a \cdot b \in N$.

I If U is a subset of N such that $1 \in U$ and such that $a+1 \in U$ whenever $a \in U$ then $U=N$.

Remarks

(i) The axiom A1 is called the **commutative law of addition** in $\mathbb{Z}$: M1 is the **commutative law of multiplication**. A2 and M2 are the **associative laws** of addition and multiplication respectively. A3 and M3 announce the existence of additive and multiplicative **identity** (or **neutral**) **elements**. A4 asserts each integer has an **additive inverse**. D lists the **distributive laws**.

(ii) The reader will probably have recognised N as having properties usually ascribed to the set of positive integers and I as being the **principle of mathematical induction.**

(iii) Despite the fact that the algebraist is not interested in the nature of the integers themselves he certainly gets joy out of the fact, to be proved later (see Section 3.12), that there is essentially only one system of objects satisfying the above axioms A1 through to I. (The idea of two algebraic systems being 'essentially the same' will first be defined formally in 3.10.1. Before that, part (ii) of the Remark in Section 1.8 might prove helpful.)

(iv) It is notable that for the set $\mathbb{Z}$ there is no *multiplicative* analogue of A4. However, as is well known, the sets* $\mathbb{Q}$, $\mathbb{R}$, $\mathbb{C}$ *do* all satisfy a near analogue of A4, viz:

M4 To each *non-zero* a in $\mathbb{Q}$ (or $\mathbb{R}$ or $\mathbb{C}$) there exists in $\mathbb{Q}$ (or $\mathbb{R}$ or $\mathbb{C}$) a unique number, namely a^{-1}, $\left(\text{also written } \frac{1}{a}\right)$ such that $a \cdot a^{-1}=a^{-1} \cdot a=1$.

* We temporarily use $\mathbb{Q}$, $\mathbb{R}$ and $\mathbb{C}$ intuitively for illustrative purposes only. We shall construct them formally in Sections 3.10 and 4.4.

Problem 1 Assuming, for the moment, the truth of the assertion of uniqueness made in Remark (iii) above, one deduces that, amongst the axioms listed for $\mathbb{Z}$, there must be at least one which cannot be satisfied by $\mathbb{Q}$. Can you identify which axiom(s) from A1 through to I are not satisfied if one attempts to apply them to $\mathbb{Q}$ instead of $\mathbb{Z}$?

The reader can no doubt think of other properties usually ascribed to $\mathbb{Z}$ which we have so far failed to mention. For instance, we are all familiar with

Z: If $a, b \in \mathbb{Z}$ and if $a \cdot b = 0$ then $a = 0$ or $b = 0$ (or both; see the footnote on p. 3).
C: If $a, b, c \in \mathbb{Z}$, if $c \cdot a = c \cdot b$ and if $c \neq 0$ then $a = b$.
M: For all $a, b \in \mathbb{Z}$ it is always the case that $(-a) \cdot (-b) = a \cdot b$.

These properties might be called the **zero-divisor law**, the **cancellation law** and the **mysterious law** (see [5, p. 5]) respectively. Why do we not add them to the list of axioms given above? The answer is that we can, without using intuition, speculation or hearsay concerning $\mathbb{Z}$, *prove* that Z, C and M are logical consequences of the axioms A1 through to I. Now whilst algebraists do not regard it as their *prime* duty to reduce all such sets of axioms to a minimum size, it is part of an algebraist's function to investigate consequences of axioms such as those just referred to. In this manner the algebraist can hope to discover which features of a given system are essential and which only incidental.

We shall perform a number of consequence-seeking calculations of this type in a more abstract setting from Chapter 3 onwards. Let us content ourselves for the moment with finding out exactly which of the axioms given earlier are required to establish property M. First, then, we prove, giving all the details,

Lemma 1.2.1*

(i) For all $c \in \mathbb{Z}$ we have $0 \cdot c = 0$.
(ii) For all $a, b \in \mathbb{Z}$ we have $(-a) \cdot b = -(a \cdot b)$.
(iii) For all $c \in \mathbb{Z}$ we have $-(-c) = c$.

Comment If these results seem rather too trivial to bother about, let's see if we can make them more impressive (and less 'obvious'?) by stating their conclusions in words:

(i) The product of the additive identity with any integer always yields the additive identity.
(ii) The additive inverse of a product [that is $-(a \cdot b)$] is equal to the product of the additive inverse of the first [that is $-a$] with the second [namely b].

* A lemma is a result which helps in the proof of a future theorem (cf. the German word *Hilfsatz*) but is not deemed to be of sufficient importance by itself to warrant the title 'Theorem' (German *Satz*).

(iii) The additive inverse of the additive inverse of a given integer is equal to the given integer.

Proof

(i) [From the property of 0 as stated in A3 we have, on setting $a = 0$,]

$$0+0=0 \qquad \text{(A3)}$$

[Multiplying each side by c we have]
Hence

$$(0+0)\cdot c = 0\cdot c \qquad \text{(using the property } x=y \Rightarrow xz=yz \text{ of equality)}$$

[Now using the second axiom in D on the left-hand side of this equation]
Consequently

$$0\cdot c + 0\cdot c = 0\cdot c \qquad \text{(axiom D)}$$

[By A4 the element $-(0\cdot c)$ certainly exists: we add it to each side. Reflecting the fact that on the left-hand side it is being added to the element $0\cdot c+0\cdot c$ we write]
and so

$$(0\cdot c+0\cdot c)+(-(0\cdot c)) = 0\cdot c+(-(0\cdot c))$$
$$\text{(using the property } x=y \Rightarrow x+z=y+z \text{ of equality)}$$

[Now using the associative law on the left-hand side]
It follows that

$$0\cdot c+(0\cdot c+(-(0\cdot c))) = 0\cdot c+(-(0\cdot c)) \qquad \text{(A2)}$$

[And replacing $0\cdot c+(-(0\cdot c))$ by 0 on using A4 once on each side]

$$0\cdot c+0=0 \qquad \text{(A4 twice)}$$

[Finally using A3 on the left-hand side]

$$0\cdot c = 0 \qquad \text{(A3)}$$

as required.

(ii) Given a we obtain successively

	$a+(-a)$	$=0$	(A4)
$\therefore$	$(a+(-a))\cdot b$	$=0\cdot b$	(property of equality)
$\therefore$	$a\cdot b+(-a)\cdot b$	$=0$	(D on lhs; (i) on rhs)
But	$a\cdot b+(-(a\cdot b))$	$=0$	(A4)
$\therefore$	$(-a)\cdot b$	$=-(a\cdot b)$	[since both are additive inverses for $a\cdot b$ and] by A4 [there is a unique such inverse].

(iii) Given $c \in \mathbb{Z}$ we have

$c+(-c)=0$ and $(-c)+c=0$ (A4)

$\therefore$ $(-c)+c=0$ and $c+(-c)=0$ (Switching the equations over—or just using A1)

Hence c is an additive inverse for $-c$ (A4)

But this inverse is unique (A4)

[Now the unique additive inverse of any element x is denoted by $-x$. Hence the unique additive inverse of $-c$ is denoted by $-(-c)$. But c is this additive inverse.]
Hence $-(-c)=c$.

Theorem 1.2.2 For all $a, b \in \mathbb{Z}$ we have $(-a)\cdot(-b)=a\cdot b$.

Proof [Using 1.2.1(ii) on the elements a and $-b$ we get]

$(-a)\cdot(-b)=-(a\cdot(-b))$ (i) (1.2.1(ii))

But $a\cdot(-b)=-(a\cdot b)$ (ii) (Proof identical to 1.2.1(ii))

Combining (i) and (ii)

$(-a)\cdot(-b)=a\cdot b$ (using 1.2.1(iii))

Remarks
(i) If the reader feels we have gone to a lot of trouble to establish a result which is 'obvious' we ask him on what grounds he bases his belief in this result? Is it merely 'experience'? Or has he had it 'on good authority' that it is true? What *we* have done is to show that the semi-mystical assertion that 'minus times minus is plus' is deducible as a consequence of other ('more obvious') axioms of arithmetic.

In later chapters we shall prove several results of the above type where the symbols used will not necessarily stand for integers. Then, assertions of the above type will certainly be far from 'obvious', since we will lack the appropriate 'experience'.
(ii) We were trying to find out exactly what we needed to assume in order to prove 1.2.2. Including the (necessary) Lemma 1.2.1 it appears that our proof of 1.2.2 depends on various properties of the = symbol together with several applications of A4, A3 and D together with just one application of A2. Notice that A1, M1, M2 and M3 were not called upon.

Note 1.2.3 One other thing 1.2.2 does for us is to prove that $\mathbb{Z}$ can contain only one subset N with the properties listed in axiom P. (If M is a subset of

$\mathbb{Z}$ satisfying P then either 1 or -1 belongs to M. It follows from 1.2.2 that $1 = 1^2 = (-1)^2$ and then from axiom P(ii) that, in any case, $1 \in M$. Similarly $1 \in N$ and hence $1 \in M \cap N \subseteq N$. If, now, $a \in M \cap N$ then $a + 1 \in M \cap N$ since both M and N satisfy axiom P(ii). But then $M \cap N = N$ since $\mathbb{Z}$ satisfies axiom I. This means that $N \subseteq M$ in $\mathbb{Z}$. Clearly $N \subsetneq M$ is impossible [why?].)

The reader who is beginning to doubt the author's sanity should note that there *is* something needing proof here. The innocuous looking set $S = \{a + b\sqrt{2}: a, b \in \mathbb{Z}\}$ can support two distinct such Ns, namely the subset N_1 of all $a + b\sqrt{2}$ which are positive real numbers in the usual sense and the set $N_2 = \{c + d\sqrt{2}: c - d\sqrt{2} \in N_1\}$.

To deduce the familiar properties of ordering amongst the integers we make:

Definition 1.2.4 The (unique) subset N of $\mathbb{Z}$ described in axioms P and I will be called the **set of positive integers**; the subset $-N$ the **set of negative integers**. If $a, b \in \mathbb{Z}$ we say that a **is less than** b and write $a < b$ iff $b - a \in N$. If we only know (or care) that $b - a \in N \cup \{0\}$ we write $a \leqslant b$. ($b - a$ is the shorthand notation for the more accurate $b + (-a)$.) When convenient we shall also write $b > a$ ($b \geqslant a$) as an alternative to $a < b$ ($a \leqslant b$).

P(i) then says that each integer is either positive or zero or negative (and never has two of these properties simultaneously).

P(ii) says that the sum and the product of positive integers are both positive.

I says that any subset of positive integers which contains 1 and which contains $a+1$ whenever it contains a is precisely the set of *all* positive integers.

One can use axiom P and 1.2.4 to establish all the familiar properties of the $<$ sign. We treat just a couple, place some in the exercises and leave the rest for the reader to prove for himself or look up in, for example, [32].

Theorem 1.2.5 If $a, b, c \in \mathbb{Z}$ are such that $a < b$ and $b < c$ then $a < c$.

Proof [To say $a < b$ is,] according to 1.2.4, [to say that] $b - a \in N$. [To say $b < c$ is,] similarly, [to say that] $c - b \in N$. [But] from P(ii) we [can then] deduce [that $b - a + c - b \in N$, that is*,] that $c - a \in N$. Thus [using 1.2.4 again this simply says] $a < c$, as required.

Theorem 1.2.6 If $a, b, c \in \mathbb{Z}$ are such that $a < b$ and $0 < c$ then †$ac < bc$.

Proof (Briefly) We are given that $b - a$ and c are both in N. Hence, by axiom P(ii), $(b - a)c \in N$. It follows* that $bc - ac \in N$, that is $ac < bc$.

† From now on we shall often write ac (etc.) in place of $a.c.$ (etc.)

* These are a couple of points which need care. See exercises 1, 2 and 3 following.

Now seems a fairly appropriate time to make

Definition 1.2.7 Let $a \in \mathbb{Z}$. We define $|a|$ (called **modulus* of *a***) by

$$|a| = 0 \quad \text{iff} \quad a = 0$$

$$|a| = a \quad \text{iff} \quad 0 < a \quad (\text{i.e. iff } a \in N)$$

$$|a| = -a \quad \text{iff} \quad a < 0 \quad (\text{i.e. iff } a \in -N).$$

Thus

Example 1.2.8 $|-17| = 17$, $|31| = 31$.

We first use this concept in Theorem 1.4.5.

We concentrate for a while on I, the principle of induction.† Let us suppose that to each positive integer n there corresponds a statement which we denote by $S(n)$. For example, $S(n)$ might be '$1 + \frac{1}{2} + \frac{1}{3} + \cdots + \frac{1}{n} < 10 - \frac{1}{n^2}$'. Let us further suppose that: (i) We can show $S(1)$ to be a true statement; and that (ii) for each positive integer k we can prove, under the assumption that $S(k)$ is a true statement, that $S(k+1)$ is also a true statement. Then I tells us that for every positive integer n the statement $S(n)$ is true.

For: Let U denote the subset of N comprising all those positive integers n for which the statement $S(n)$ is true. That is, $U = \{n : n \in N \text{ and } S(n) \text{ is true}\}$. Then $1 \in U$ [since we are supposing that we can prove $S(1)$ to be true] and, whenever $k \in U$ [that is, whenever $S(k)$ is true] then $k+1 \in U$ [since we are supposing that we can then prove $S(k+1)$ true]. By principle I we see that $U = N$. Thus for each $n \in N$ we have $n \in U$; that is $S(n)$ is true, as required.

After talking about *proofs* by induction we should also mention the technique of *definition by induction*. Consider for example the famous *Fibonacci‡ sequence* $u_1, u_2, u_3, \ldots$ where (i) $u_1 = 1$, $u_2 = 1$ and (ii) for each integer $n \geqslant 3$, u_n is defined not directly but relative to previously defined terms of the sequence by the formula $u_n = u_{n-1} + u_{n-2}$. One is tempted to say that the Fibonacci sequence is defined by (i) and (ii) but there are some logical subtleties associated with this temptation. We shall not discuss them here as they can be overcome. The reader who is sufficiently intrigued by this warning is referred to a very readable article [106] by L Henkin.

We now look at two variants of I. Consider the following assertions:

I_1: Let V be a subset of N such that $1 \in V$ and such that $a+1 \in V$ whenever $x \in V$ for *all* x such that $1 \leqslant x < a+1$. Then $V = N$.

* Also called *absolute value*.

† The first acceptable statement of the principle is often credited to Blaise Pascal (19 June 1623 – 19 August 1662) after whom Pascal's triangle is named.

‡ Leonardo Fibonacci lived in the period c. 1170–1240.

W: Every non-empty set of positive integers contains a least member.
That is, if $T \subseteq N$ and $T \neq \varnothing$ then T contains an element t such that $t < z$ for any other $z \in T$.

Theorem 1.2.9 The statements I, I_1, W are equivalent. That is, each implies the other two.

Remarks
(i) On the face of it, it looks as if we have six results to prove, namely: $I \Rightarrow I_1$, $I_1 \Rightarrow I$; $I \Rightarrow W$, $W \Rightarrow I$; $I_1 \Rightarrow W$, $W \Rightarrow I_1$. In fact we only need to prove $I \Rightarrow I_1$, $I_1 \Rightarrow W$ and $W \Rightarrow I$ which we can write in abbreviated form as $I \Rightarrow I_1 \Rightarrow W \Rightarrow I$. (Compare this with the deduction of $a = b = c$ given that $a \leqslant b$, $b \leqslant c$ and $c \leqslant a$, $a, b, c \in \mathbb{Z}$.)
(ii) W is also expressed by saying '*The positive integers*, taken in their natural order, *are well-ordered*'. W is called the **well-ordering principle**—hence the letter W! W is manifestly untrue when applied to $\mathbb{Q}$ and $\mathbb{R}$. [Why? And what about $\mathbb{C}$?]

We shall use W a lot in this chapter in particular and call upon I_1 in 1.10.1.

Note 1.2.10 Despite the remarks above on economy of effort we shall prove here only the equivalence of I and W. We leave the reader to prove that $I \Leftrightarrow I_1$.

Proof of Theorem 1.2.9 $I \Rightarrow W$. [Here we wish to show that every non-empty set of positive integers has a least member so . . .] assume [to the contrary] that there is a non-empty set T, say, of positive integers such that T has no least member. Then $1 \notin T$ [since 1 is the smallest positive integer. Which axiom(s) prove this for us? See exercise 19]. Define the subset V of N by $V = \{x : x \in N \text{ and } x < t \text{ for all } t \in T\}$. Then $1 \in V$. Let $k \in V$. Then $k+1 \in V$: otherwise $k < s \leqslant k+1$ for some $s \in T$ so that [since $k \notin T$ and since no integer lies between k, and $k+1$ (why not?)] $s = k+1$ is the least integer in T [contradicting the assumption that T has no such least integer]. Thus, by principle I, $V = N$ whence $T = \varnothing$. This manifest contradiction ($T \neq \varnothing \Rightarrow T = \varnothing$) tells us that our one assumption is not tenable. Thus no non-empty set T exists and the required result is thereby proved.

$W \Rightarrow I$. Let U be a subset of N such that $1 \in U$ and $a+1 \in U$ whenever $a \in U$. Suppose $U \neq N$. Then the set $T = \{x : x \in N \text{ and } x \notin U\}$ is non-empty and so [since we are assuming W holds we can conclude that] T has a least element t, say. Thus $t-1 \notin T$ and so $t-1 \in U$ [is $t-1$ a positive integer?]. Then [by the given property of U] $t-1+1 \in U$. That is $t \in U$. But $t \in T$ [we chose it as the least element of T] and so $U \cap T \neq \varnothing$, a contradiction. Thus [once again] our one supposition [this time that $U \neq N$] must be invalid, so that $U = N$, as required.

Exercises

1 Let $a, b, c, d \in \mathbb{Z}$. Show that using axiom A2 alone one can prove $((a+b)+c)+d = (a+(b+c))+d = a+((b+c)+d) = a+(b+(c+d)) = (a+b)+(c+d)$.

[This shows that the 5 different ways one might set about working out the sum of the four integers a, b, c, d all yield the same answer which can therefore be unambiguously denoted by $a+b+c+d$. More generally one can show (2.7.7) that the sum of n integers $a_1, a_2, \ldots, a_n$ can be denoted unambiguously by $a_1+a_2+\cdots+a_n$. Similar remarks apply to products.]

2 Use the same sort of reasoning, together with careful use of axiom A1, to show that $b+(-a)+c+(-b)=c+(-a)$, as asserted in 1.2.5.

3 Prove, with the same degree of detail, that $-(a+(-b))=b+(-a)$, and that $(a-b)\cdot c = a\cdot c - b\cdot c$. [For this second part think what $a-b$ *means*.]

4 Show that the second axiom D is a consequence of the first axiom D and M1.

5 Prove that property C is a consequence of property Z (and which axioms?).

6 Using the style adopted in the comment following 1.2.1 state the result of 1.2.2 *in words*.

7 Let $C[0, 1]$ be the set of all real valued continuous functions* defined on the interval $[0, 1]$. For $f, g \in C[0, 1]$ define, for each $x \in [0, 1]$, $(f+g)(x) = f(x)+g(x)$ and $(f\cdot g)(x)=f(x)g(x)$. Show that with these definitions of $+$ and $\cdot$ $C[0, 1]$ satisfies A1, A2, A3, A4, M1, M2, M3, D but not Z. Thus Z is not a logical consequence of A1 through to D alone.

8 Prove, in full detail, that if $a, b, c \in \mathbb{Z}$ are such that $a<b$ then $a+c<b+c$.

9 Prove that if $a\in\mathbb{Z}$ and if $a\neq 0$ then $0<a\cdot a$. Deduce that $0<1$.

10 Prove directly from A3 and A4 that $-0=0$. Show also (Lemma 1.2.1(ii) might be helpful) that if $a\neq 0$ and $b\neq 0$ then $ab\neq 0$. (That is, prove property Z from axioms A1 through to I.)

11 Show that (i) if $a_1>b_1$ and if $a_2>b_2$ then $a_1a_2+b_1b_2>a_1b_2+b_1a_2$; and that (ii) if $a_1>b_1>0$ and if $a_2>b_2>0$ then $a_1a_2>b_1b_2>0$.

12 Prove that, for all $a, b \in \mathbb{Z}$, (i) $|ab|=|a|\cdot|b|$ and (ii) $|a+b|\leqslant|a|+|b|$.

[In problems 13 and 14 you are meant to get your hands dirty by experimenting. Just try any example that comes into your head—if it turns out to be no good, throw it away and start again! Why not try $\{3n+1 : n\in\mathbb{Z}\}$ for a start?]

13 Can you find a non-empty set of numbers which fails to satisfy I but satisfies all of the remaining axioms from A1 through to I?

* A formal definition of function is given in 2.6.1. You are asked to proceed informally here.

14 What about the same problem with A3, A4, M3, P(i) false (and the rest true)?

15 Let X denote the set $\{E, O\}$. On this set define addition (+) and multiplication (·) by

$$E+E=O+O=E; \qquad E+O=O+E=O$$

$$E\cdot E=E\cdot O=O\cdot E=E; \qquad O\cdot O=O$$

Which of the axioms A1, A2, A3, A4, M1, M2, M3, M4, D, Z are satisfied in this case? [Have you any ideas why we chose letters E, O rather than, say, a, b?]

16 Prove by induction (using principle I) that, for all positive integers n,

(i) $1+2+3+\cdots+n=\dfrac{n(n+1)}{2}$;

(ii) $1+3+5+\cdots+(2n-1)=n^2$;

(iii) $1^2+2^2+3^2+\cdots+n^2=\dfrac{n(n+1)(2n+1)}{6}$;

(iv) $1^3+2^3+3^3+\cdots+n^3=(1+2+\cdots+n)^2$.

17 Can you prove by induction that, for all positive integers n, $1+\frac{1}{2}+\frac{1}{3}+\cdots+\frac{1}{n}<10-\frac{1}{n^2}$?

18 Can you prove by induction that, for all positive integers n, $1+\frac{1}{2}+\frac{1}{4}+\frac{1}{8}+\cdots+\frac{1}{2^n}<2$?

[17 shouldn't cause you any trouble if you know something about the harmonic series, but a proof of 18, by induction on n, might cause you a few headaches.]

19 Prove, from principle I, that: for each positive integer n, $1\leqslant n$. Deduce that there is no integer c such that $0<c<1$ and that, for each positive integer k, there is no integer t such that $k<t<k+1$.

20 Suppose $a, b\in\mathbb{Z}$ with $a>0$ and $ab=1$. Show that $a=1$. [This assertion is called upon in Section 1.3. Hint: Since $0<a$ and $0<1$ we have $0<b$. Hence $1\leqslant a$, $1\leqslant b$ (by exercise 19) and hence $1\leqslant a\leqslant ab=1$.]

21 Prove $\mathrm{I}\Leftrightarrow\mathrm{I}_1$.

22 What is wrong with the following 'proof' that all triangles in the plane are pairwise congruent? Let $S(k)$ be the statement 'In every set of k triangles in the plane all the triangles are congruent to one another.' Clearly $S(1)$ is a true statement since every triangle is congruent to itself. Now suppose $S(n)$ is a true statement if $n=k$ and let $T_1, \ldots, T_{k+1}$ be any $k+1$ triangles in the plane. By the induction hypothesis $T_1, \ldots, T_k$ and also $T_2, \ldots, T_{k+1}$ are sets of mutually congruent triangles. Putting these two sets back together we see that *all* the T_i are congruent.

23 For the Fibonacci sequence prove: for each $k \in \mathbb{Z}^+$ (i) u_{5k} is a multiple of 5; (ii) $u_1^2+u_2^2+\cdots+u_k^2=u_k u_{k+1}$.

24 Evaluate $\dfrac{n(n-1)(n^2-5n+18)}{24}$ for $n = 1, 2, 3, 4, 5$. Make a conjecture involving 2^{n-1}. Try to prove this conjecture by induction.

1.3 Divisibility, irreducibles and primes in $\mathbb{Z}$

We move now to the central concept in this chapter by making

Definition 1.3.1 Let $a, b \in \mathbb{Z}$. We say that a **divides** b (or that a is a **divisor** of b) and we write $a|b$ iff there exists $c \in \mathbb{Z}$ such that $b = ac$. If a does not divide b we write $a \nmid b$.

Remark If we wish to emphasise that we are demanding $c \in \mathbb{Z}$ we might say *a divides b in* $\mathbb{Z}$. We shall have to be careful over this point in Section 3.8. (See also the paragraph following 1.9.15.) As trivial examples we offer

Examples 1.3.2 $3|-12$, $12 \nmid 18$, $0|0$ [is this one correct? I might be teasing to see if you are awake!], $5|0$, $0 \nmid 2$. Note that $12|18$ in $\mathbb{Q}$.

Definition 1.3.3 If $a, b, d \in \mathbb{Z}$ are such that $d|a$ and $d|b$ then d is called a **common divisor** of a and b.

The following result is easy but important.

Lemma 1.3.4 If $d|a$ and $d|b$ and if $s, t \in \mathbb{Z}$ then $d|sa+tb$.

Proof Since $d|a$ there exists $u \in \mathbb{Z}$ such that $du = a$. Since $d|b$ there exists $v \in \mathbb{Z}$ such that $dv = b$. Then $sa+tb = sdu+tdv = d(su+tv)$ so that $d|sa+tb$, as required.

The next definition, which we formulate from a desire to get at the fundamental building blocks as far as multiplication in $\mathbb{Z}$ is concerned, is intentionally unconventional, introducing, as it does, a familiar concept in an unfamiliar way. The reader will recall that one of the author's main aims in this chapter is to encourage a critical attitude on the part of the reader to statements made in this book. The main reason for adopting this definition will reveal itself in Sections 3.6 and 3.7.

Definition 1.3.5

(i) If $u \in \mathbb{Z}$ is such that $u|1$ (in $\mathbb{Z}$) then u is called a **unit**.

(ii) If $a \in \mathbb{Z}$ is neither 0 nor a unit we say that a is **irreducible** iff, whenever a is expressed as a product, $a = bc$ with $b, c \in \mathbb{Z}$, it follows that either b or c is a unit. [Both can't be units. Why not?]

(iii) If $a \in \mathbb{Z}$ is neither 0 nor a unit we say that a is **prime** iff, whenever a divides a product, that is, $a|bc$ where $b, c \in \mathbb{Z}$ it follows that $a|b$ or $a|c$ (or both).

(iv) If $a, b \in \mathbb{Z}$ are such that $a = bu$, where u is a unit, then a and b are **associates**.

Examples 1.3.6 1, −1 are the only units in $\mathbb{Z}$ (exercise 1.2.20); 3, −7 are irreducibles; 3, −7 are also primes; 8, −8 is a pair of associates.

Remarks The reader's first comments on the above definition might well include:

(1) What is the point of making 1.3.5(i) when the only two integers satisfying the property are 1 and −1?

(2) I thought the concept of primeness was *defined* by (ii) and that (iii) describes a well-known property of prime numbers.

(3) What is the point of introducing the two definitions, namely (ii) and (iii), when they express precisely the same concept?

(4) I can see why you exclude $a = 0$ from (ii) and (iii) but why debar a from being 1 and −1?

(5) I find from (iii) that −3 is a prime. Surely you don't allow negative numbers to be primes?

Points (1), (2) and (4) will be commented on later (see Remarks (i) after 1.9.3, Theorem 1.4.10 and the Remark following 1.5.1 respectively). For the moment we dismiss (5) with the answer 'Yes we do'! Regarding (3) we offer the reader the

Challenge: Are you *absolutely* sure these concepts *are* the same? If you *are* so sure, a proof shouldn't be hard to come by. So, go to it before reading on!

We now reveal that the concepts of irreducibility and primeness *do* coincide in $\mathbb{Z}$. It is, however, precisely because of the fact that these concepts do *not* coincide in every kind of number system we shall meet (see exercise 6) that the celebrated conjecture of Fermat remains unproven to this day. (See Section 3.5.)

Returning to $\mathbb{Z}$ we see that to prove the above assertion of coincidence we must show: (i) that every irreducible element of $\mathbb{Z}$ is necessarily a prime element; and (ii) that every prime element of $\mathbb{Z}$ is necessarily irreducible. The latter we can do immediately; the former will take a little longer (Theorem 1.4.10).

Theorem 1.3.7 Every prime element in $\mathbb{Z}$ is necessarily an irreducible one.

The strategy of this particular proof is to head directly from hypothesis to conclusion as follows: Let a be any prime element of $\mathbb{Z}$. We wish to prove that a is irreducible. This will (essentially) be achieved if we can show that, whenever we write a as a product, that is $a = bc$ with $b, c \in \mathbb{Z}$, then one of b, c is a unit.

Proof Let a be a prime in $\mathbb{Z}$. Then [by 1.3.5(iii)] $a \neq 0$ and a is not a unit, [and it only remains to establish the property described in 1.3.5(ii)]. [Thus] suppose $a = bc$ where $b, c \in \mathbb{Z}$. Then certainly $a|bc$. [But a is given to be prime and so, from 1.3.5(iii),] we deduce $a|b$ or $a|c$ (or both). WLOG we suppose $a|b$. Then [by definition] there exists in $\mathbb{Z}$ an element s, say, such that $as = b$. It follows that $asc = bc$. But $bc = a$ and so $asc = a = a \cdot 1$. Since $a \neq 0$ we may deduce from property C (Section 1.2) that $sc = 1$. Thus [since $cs = sc = 1$] c is a unit—as required.

Remarks
(i) Note that under the assumption that $a|b$ we have shown that c is a unit. If you think that we should now give another proof of the same length to show that the possibility that $a|c$ leads to the conclusion that b is a unit, you have cheated yourself in that you have accepted the statement 'WLOG* we can suppose $a|b$' above as a valid one when you don't even understand what it says! If you've been caught out here please do not be so careless again.
(ii) The word 'irreducible' is a good one for elements with the property listed in 1.3.5(ii) since according to that definition an irreducible element is one which cannot be represented as a product of two properly 'smaller' elements. Thus irreducible elements are seen to be the fundamental building blocks with respect to multiplication for the system $\mathbb{Z}$.

Let us suppose for the moment that we have shown that in $\mathbb{Z}$ irreducibles and primes are the same thing. A natural question, to which you probably know the answer (but not a proof?) is

Question 1.3.8 Are there infinitely many primes (i.e. irreducibles)?

To answer this we use:

Lemma 1.3.9 Let a be an integer such that $1 < a$. Then a can be expressed as a product of finitely many positive irreducibles (i.e. primes).

* Without Loss Of Generality. Roughly speaking this means: We need show you the proof only in one particular case. All other cases can be dealt with in an identical manner with only the most trivial modifications.

Remarks
(i) We extend the usual meaning of product to include the case of single numbers standing alone. Thus whilst $2 \cdot 3$ and $3 \cdot 2$ are the two ways of writing 6 as a product of (positive) irreducibles the expression 7 is regarded as the required product decomposition for the integer 7.
(ii) It follows immediately from 1.3.9 that if $a \in \mathbb{Z}$ and if $a < -1$ then a is expressible as a product of primes, one of which is negative.

Proof of Lemma 1.3.9 [We use the principle W.] Let S be the collection of all those integers (greater than 1), if any, which are *not* expressible in the desired form. [If $S = \varnothing$ the required result is immediate—there are no 'nasty' integers.] If $S \neq \varnothing$ then [S is a non-empty set of positive integers and] principle W asserts that S has a least member. Let this member be called m [m for minimum?]. Then m cannot itself be irreducible [since such an m would have a product decomposition of the required kind, namely m itself]. Thus m can [definition of irreducible] be expressed as a product, say $m = m_1 m_2$, where $1 < m_1 < m$ and $1 < m_2 < m$. But [since m is the least element of S we have] $m_1 \notin S$ and $m_2 \notin S$. Thus both m_1 and m_2 *can* be expressed as products of (positive) irreducibles, $m_1 = k_1 k_2 \ldots k_r$, $m_2 = l_1 \ldots l_s$ say. It follows that $m = m_1 m_2 = k_1 \ldots k_r l_1 \ldots l_s$ [which shows that m *is* expressible as a product of irreducibles after all]. Thus $m \notin S$, contradicting the assumption that $m \in S$. Hence [this assumption is wrong and] $S = \varnothing$.

Remark Whilst this proof uses the 'contradiction method' it gets to this contradiction by supposing the existence of a counterexample, hence, by principle W, a *smallest* counterexample and thence the contradiction. Accordingly this type of proof might be called **proof by minimum counterexample**.

We can now answer Question 1.3.8, giving a proof essentially due to Euclid.*

Theorem 1.3.10 There *are* infinitely many primes.

The proof is yet another 'by contradiction'. Recall that we are temporarily assuming that irreducibles and primes are one and the same thing in $\mathbb{Z}$.

Proof Suppose there is only a finite number, n, say, of positive primes. Let them be listed in increasing order as $p_1, p_2, \ldots, p_n$ [so that $p_1 = 2$, $p_2 = 3$, etc.]. Now form the integer $N_n = (p_1 p_2 \ldots p_n) + 1$. By 1.3.9, N_n can be written as a product of finitely many primes [i.e. irreducibles] $N_n = t_1 t_2 \ldots t_r$, say, where each t_i must be one of the primes from the complete list $p_1, p_2, \ldots, p_n$. Suppose $t_1 = p_m$. Then $p_m | p_1 p_2 \ldots p_n$ and $p_m = t_1 | N_n$. Hence [by 1.3.4, with $s = -1$ and $t = 1$, we see that] $p_m | N_n - (p_1 p_2 \ldots p_n)$. That is, $p_m | 1$. [But this is absurd

* Not much seems to be known about Euclid except that he taught mathematics in Alexandria c. 300 BC. He was supposedly a modest and kindly man who is alleged to have given money to a student who asked the use of studying geometry 'since he must make gain of what he learns'.

since by definition a prime cannot be a unit.] This absurdity completes the proof.

As an entertaining diversion and one on which you can again get your hands dirty we consider a small point which naturally arises here, namely: Is it in fact the case that each of the numbers N_n is itself a prime? After all with $p_1 = 2$, $p_2 = 3$, $p_3 = 5$, $p_4 = 7, \ldots$, etc. we get, successively, $N_1 = 3$, $N_2 = 7$, $N_3 = 31$, $N_4 = 211$, $N_5 = 2311$, all of which are primes. We make a

*Conjecture** If we set $N_n = (p_1 p_2 p_3 \ldots p_n) + 1$ then, for each positive integer n, N_n is itself a prime.

Can you amend the proof of 1.3.10 to obtain a proof of this conjecture—or can you supply a single counterexample which will kill off the conjecture?

Lemma 1.3.9 raises, and the remarks following it answer, an obvious question, namely: Is the way of expressing any integer (other than $-1, 0$ and 1) as a product of irreducibles unique? Clearly not, since we may write, for example, not only $6 = 2 \cdot 3 = 3 \cdot 2$ but also $6 = (-2) \cdot (-3) = (-3) \cdot (-2)$. Rather more to the point are questions like 'Are $13^2 \cdot 71 \cdot 103$ and $17 \cdot 19 \cdot 53 \cdot 73$ equal?' Disregarding variations of ordering and sign as in the decompositions of 6 as given above one can establish a uniqueness-type theorem (1.5.1 below).

We first prove 1.5.1 by employing the standard proof involving definite use of the concepts of irreducibility and primeness. Then, for comparison, entertainment and for later examination, we offer a more novel approach which apparently uses only irreducibility.

Exercises

1 Show that if $a|b$ and $b|a$ in $\mathbb{Z}$ then a, b are associates.

2 Show that if $a = bu$ where u is a unit in $\mathbb{Z}$ then $b = av$ where v is a unit in $\mathbb{Z}$.

3 Use the method of induction to prove that if a is a prime in $\mathbb{Z}$ and if $a|b_1 b_2 \ldots b_r$ with $b_1, b_2, \ldots, b_r \in \mathbb{Z}$ then $a|b_i$ for at least one value of i $(1 \leq i \leq r)$.

4 Show that if $x, y \in \mathbb{Z}$ and $3|x^2 + y^2$ then $3|x$ and $3|y$.
[Hint: If $3 \nmid x$ then x can be written in the form $3t + 1$ or $3t - 1$. Similarly for y.]

5 Show that if $x, y, z \in \mathbb{Z}$ and $5|x^2 + y^2 + z^2$ then $5|x$ or $5|y$ or $5|z$.

6 Let H be the set of all positive integers of the form $4k + 1$ where $k \in \mathbb{Z}$. Thus $H = \{1, 5, 9, 13, \ldots\}$. Call an H number h *H-irreducible* iff $h \neq 1$ and h

* Some work on this conjecture by an American undergraduate student can be found in volume 26, p. 567, of the *Mathematics of Computation* journal. See also volume 34, p. 303.

cannot be expressed as a product of two smaller H numbers. Call an H number p *H-prime* if $p \neq 1$ and whenever $p|ab$ with $a, b \in H$ we always have $p|a$ or $p|b$ in H. Write down the first ten H-irreducibles. Which, if any, of these are H-primes? Are all H-primes necessarily H-irreducible? Factorise 441 in two essentially distinct ways as products of H-irreducibles. [This example is usually credited to David Hilbert, below; hence the letter H. We shall meet other similar examples in Section 3.6.]

[In Questions 7 and 8 assume that primes and irreducibles are the same things.]

7 Show that there are infinitely many positive primes of the form $4k+3$ in $\mathbb{Z}$. [Hint: Suppose only finitely many, $p_1, p_2, \ldots, p_n$ say. Set $N_n = 4p_1 \ldots p_n - 1$. Show that N_n must be divisible by at least one prime of the form $4k+3$ and obtain a contradiction.]

8 Show that there are infinitely many positive primes of the form $6k+5$ in $\mathbb{Z}$. Where does this method break down when applied to numbers of the form $8k+7$? So there aren't infinitely many $8k+7$ primes, right?

David Hilbert *(23 January 1862 – 14 February 1943)*
Hilbert was born at Königsberg (now Kaliningrad), the son and the grandson of judges. In 1885 he obtained his doctorate with a thesis on the theory of invariants. After becoming a professor at Königsberg in 1892 he obtained the chair at Göttingen in 1895, a position he held until his retirement in 1930.

Hilbert's mathematical interests ranged widely, encompassing the theory of invariants, algebraic number theory, foundations of geometry, analysis and relativity theory. His outstanding longer works include the 370 page *Zahlbericht* (1895–7), in which he rewrote much of algebraic number theory, and his axiomatic approach to Euclid's geometry (1899). At the International Congress of Mathematicians in 1900 Hilbert presented his famous list of 23 problems to which he believed mathematicians should address themselves. Several of these still remain unsolved.

After his work in geometry perhaps Hilbert's greatest wish was to prove the consistency of arithmetic and thereby to resolve the 'foundations-crisis' that attracted philosophers such as Bertrand Russell. Some rejected his proposed method of procedure and in 1931 Kurt Gödel dashed all hopes by proving that in a consistent system formalising the natural numbers there is a theorem *A* such that neither *A* nor not-*A* can be proved within the system.

Around 1903 Hilbert, to help deal with a problem on integral equations, introduced that infinite dimensional extension of Euclidean space now called Hilbert Space.

1.4 GCDs

To get to the standard proof of 1.5.1 will take a while. We begin with

Definition 1.4.1 Let $a, b \in \mathbb{Z}$. An integer $c \in \mathbb{Z}$ is termed a **greatest common divisor (gcd)** or **highest common factor (hcf)** of a and b iff
(i) $c|a$ and $c|b$ and
(ii) *if* $d|a$ and $d|b$ *then* $d|c$.

Remark We talk of *a* gcd rather than *the* gcd of a and b since 1.4.1 says nothing about uniqueness (nor even about existence!). In anticipation of proving existence of a positive gcd we introduce

Notation 1.4.2 The positive gcd of a, b (not both of which are zero) is denoted by (a, b).

Thus

Examples 1.4.3
(i) $(60, 24) = 12$; $(17, -42) = 1$.
(ii) For $a, b \in \mathbb{Z}$ (not both zero) we have $(a, b) = (b, a) = (|a|, |b|)$. Thus we only ever need consider pairs of non-negative integers (see exercise 3 following).

Problem 2 It seems rather obvious that any pair a, b of integers (not both zero) must possess a unique positive gcd. Surely one simply takes from the set of all positive common divisors d of a and b the largest? It should be easy to prove that d is the required gcd. You may not even think there is a problem at all since the words 'greatest' and 'largest' are synonymous in the English language. Notice, however, that for us 'largest' is to be interpreted in the sense of ordering whereas 'greatest' (as used in 'greatest common divisor') is used in the sense of division. Bearing in mind these remarks you might try to prove *here and now* that the numerically largest common divisor d of a and b, as described above, is indeed their (positive) gcd. (Warning:

your proof had better not apply to the H-numbers introduced above (see exercise 15 below) nor to some of the number systems introduced in Section 3.6 (see exercise 3.6.12) where this assertion on gcds is false!)

Incidentally there is no problem in proving the uniqueness of the positive gcd—*given that a gcd actually exists at all.* The main difficulty seems to be with establishing existence.

A theorem which is very important for us and which proves not only that any two integers a, b (not both zero) do have a gcd, but much more besides is

Theorem 1.4.4 Any two integers a, b (not both zero) have a unique positive gcd. Further, if this gcd is denoted by c, we can find $s, t \in \mathbb{Z}$ such that $c = sa + tb$.

To help prove this theorem we shall need to call upon the following result, which is known as the Division Algorithm.

Theorem 1.4.5 (The Division Algorithm) Let $a, b \in \mathbb{Z}$ with $b \neq 0$. Then there exist unique $m, r \in \mathbb{Z}$ such that $a = mb + r$ where $0 \leq r < |b|$.

[The letters m, r are chosen as they are the initial letters of the words multiple and remainder respectively. A careful choice of notation can often be very helpful.]

Examples 1.4.6

(i) With $a = 17$ and $b = -5$ we find $m = -3$ and $r = 2$ so that $17 = (-3)(-5) + 2$. Note that $m \in \mathbb{Z}$, $r \in \mathbb{Z}$ and $0 \leq r < |b|$.

(ii) With $a = -19$ and $b = -12$ we find $m = 2$ and $r = 5$ so that $-19 = 2(-12) + 5$.

Our intuition and experience tells us that the division algorithm is clearly true. After all we have been working examples like this since age seven or so. However, let us note that the verification of 1.4.5 in as many as 20 million cases by no means establishes the truth of the theorem for *all* possible cases (cf. the remarks in Section 0.4 on sums of four squares). Furthermore, intuition is not always a good guide, as problem 3 in the Prologue demonstrates.

One difficulty the beginner might find on being asked to prove 1.4.5 is that of deciding exactly what to write down. It seems difficult to explain so obvious an assertion in simpler terms. A proof using $\mathbb{Q}$ may occur to the reader: Choose $m \in \mathbb{Z}$ such that m is the greatest integer not exceeding $\frac{a}{b}$. Then trivially $\frac{a}{b} - m < 1$ so that $a - mb < b$ (assuming for the moment that $0 < b$). Setting $a - mb = r$ we see that $a = mb + r$ where $0 \leq r < b$. This proof suffers from a slight defect in that one of our later aims is to show how to construct $\mathbb{Q}$ on the basis of $\mathbb{Z}$ so that officially $\mathbb{Q}$ does not yet exist. Another defect is raised by the question 'How do you know that the stated m exists?' (Exercise 3.12.10 gives an example not so unlike $\mathbf{Q}$ in which no such m can be found.) We base a proof on principle W.

Assuming for the moment that $0 \leq a$ and $0 < b$ we give

Proof of Theorem 1.4.5 Let S be the set of all non-negative integers belonging to $\{a - mb : m \in \mathbb{Z}\}$. Then S is not empty since $a \in S$. By the obvious extension of principle W from $\mathbb{Z}^+$ to $\mathbb{Z}^+ \cup \{0\}$, S has a least member, which we shall denote by r. Thus $r = a - m_1 b$ for some $m_1 \in \mathbb{Z}$. We claim: $r < b$. For otherwise $0 < b \leq r$. It then follows that $a - (m_1 + 1)b = r - b$ is an element of S smaller than the smallest element r of S. This absurdity leads us to conclude that $r < b$. Thus m_1 and r are such that $a = m_1 b + r$ where $0 \leq r < b$.

If $a < 0$ we can, by the above, find m, r such that $-a = mb + r$ where $0 \leq r < b$. Then $a = (-m)b - r$ where $-b < -r \leq 0$. But then $a = (-m-1)b + (b - r)$. Clearly $-m - 1 \in \mathbb{Z}$ and $0 < b - r < b$ if $0 < r$. (If $r = 0$ there is nothing to prove since then $a = (-m)b + r$ where $0 \leq r < b$, immediately.)

The proof of 1.4.5, in the case where $b < 0$, and of the assertions of uniqueness we leave to exercises 8 and 9.

We can at last give the

Proof of Theorem 1.4.4 Let S denote the set of all positive integers of the form $ma + nb$ where m, n are free to range over the whole of $\mathbb{Z}$. Then clearly at least one of a, $-a$, b, $-b$ belongs to S. [Why?] Thus S is not empty. Invoking principle W we see that S contains a least member—let us call it c. We claim: c is the required (unique) positive gcd of a and b.

In proving that c is a common divisor of a and b we in fact prove rather more, namely: if $w \in S$ then $c | w$.

Indeed, let $w \in S$ and use the division algorithm to write $w = kc + r$ where $k, r \in \mathbb{Z}$ and $0 \leq r < c$. Noting that if $r = 0$ we have nothing left to prove, we may assume that $0 < r$. Since w and c can be expressed in the forms $ua + vb$ and $m_0 a + n_0 b$ respectively, where $u, v, m_0, n_0 \in \mathbb{Z}$ we see that $r = (u - km_0)a + (v - kn_0)b$. Thus r [being greater than zero and of the correct form] belongs to S. But this contradicts the choice of c [as the least member of S; this contradiction shows that the assumption $0 < r$ just cannot hold]. Thus $r = 0$ and hence $c | w$, as required. It follows readily that $c | a$ and $c | b$.

Finally, c is a gcd for a and b. For, if $d | a$ and $d | b$ then $d | m_0 a + n_0 b = c$ [Lemma 1.3.4].

The uniqueness of c is left to the exercises.

Remarks

(i) The author would like to think that the reader got some enjoyment out of reading over that proof. It is a rather tidy kind of argument with no loose ends.

(ii) Whilst being a very agreeable proof of the universal *existence* of gcds the above proof is not very helpful in determining (a, b) for any particular pair of integers a and b. To do this we call upon the procedure known as the **Euclidean Algorithm***. In order to clarify the procedure we work the 'general case' first and only then give a couple of concrete examples.

Suppose, then, that $a, b \in \mathbb{Z}$ are given and that, to avoid trivialities, neither a nor b is zero. We use the division algorithm repeatedly as follows:

At step (1) we find $m_1, r_1 \in \mathbb{Z}$ such that $a = m_1 b + r_1$ where $0 \leq r_1 < |b|$
At step (2) we find $m_2, r_2 \in \mathbb{Z}$ such that $b = m_2 r_1 + r_2$ where $0 \leq r_2 < r_1$
At step (3) we find $m_3, r_3 \in \mathbb{Z}$ such that $r_1 = m_3 r_2 + r_3$ where $0 \leq r_3 < r_2$
$\vdots$

and in general
At step (t) we find m_t, $r_t \in \mathbb{Z}$ such that $r_{t-2} = m_t r_{t-1} + r_t$ where $0 \leq r_t < r_{t-1}$.

Since $|b| > r_1 > r_2 > \cdots$ and all the r_i are non-negative we must eventually reach a first integer l for which $r_l = 0$. Thus the lth step in the above then reads:

At step (l) we find m_l, $r_l \in \mathbb{Z}$ such that $r_{l-2} = m_l r_{l-1} + 0$

This last equality implies $r_{l-1} | r_{l-2}$ and as a consequence $r_{l-1} = (r_{l-2}, r_{l-1})$.
Now it is not difficult to see that $(a, b) = (b, r_1)$ (exercise 13) and that similarly $(b, r_1) = (r_1, r_2) = \cdots = (r_{l-2}, r_{l-1}) = r_{l-1}$. Consequently we see that $(a, b) = r_{l-1}$ which can (except in the case where $b|a$, that is where $r_1 = 0$) be described as 'the last non-zero remainder in the above process'.†
We offer two concrete examples where the gcds are perhaps not entirely obvious in advance!

Examples 1.4.7
(i) Find (10 113, 21 671).
We apply the above procedure to the pair 21 671, 10 113. Successively we get

$$\begin{aligned} 21\,671 &= 2 \cdot 10\,113 + 1445 \\ 10\,113 &= 6 \cdot 1445 + 1443 \\ 1445 &= 1 \cdot 1443 + 2 \\ 1443 &= 721 \cdot 2 + \textcircled{1} \quad \text{the last non-zero remainder.} \\ 2 &= 2 \cdot 1 + 0 \end{aligned}$$

Thus (10 113, 21 671) = 1. (Note that in fact neither of the given numbers is itself prime, since 10 113 = 3 · 3371 and 21 671 = 13 · 1667.
(ii) Find (30 031, −16 579).

* Obtained by Euclid in Book 7 of the *Elements*. Euclid's work was by no means concerned solely with geometry.
† G Lamé, whom we meet again in Section 3.5, proved that the number of steps required to find the gcd is at most 5 times the number of digits in the smaller of the two given numbers. See [56, p. 43].

We apply the procedure to the pair 30 031, 16 579. Successively we get

$$30031 = 1 \cdot 16579 + 13452$$
$$16579 = 1 \cdot 13452 + 3127$$
$$13452 = 4 \cdot 3127 + 944$$
$$3127 = 3 \cdot 944 + 295$$
$$944 = 3 \cdot 295 + \textcircled{59}$$
$$295 = 5 \cdot 59 + 0$$

Thus $(30\,031, -16\,579) = 59$, a result achieved without factorising either number! (Incidentally if the number 30 031 has no special significance for you, I think I can safely claim that you have not properly settled the conjecture following 1.3.10.)

1.4.4 asserts that there exist $s, t \in \mathbb{Z}$ such that $1 = 10\,113\,s + 21\,671\,t$. To find these we simply read the steps of the Euclidean Algorithm in 1.4.7(i) backwards, as follows.

$$\begin{aligned}1 &= 1443 - 721 \cdot 2\\ &= 1443 - 721(1445 - 1443) = 722(1443) - 721(1445)\\ &= 722(10\,113 - 6 \cdot 1445) - 721(1445) = 722(10\,113) - 5053(1445)\\ &= 722(10\,113) - 5053(21\,671 - 2 \cdot 10\,113)\\ &= 10\,828(10\,113) - 5053(21\,671)\end{aligned}$$

as required.

Thus we may choose $s = 10\,828$ and $t = -5053$. Of course these coefficients are by no means uniquely determined since we can also write

$$1 = (10\,828 - 21\,671\,k)(10\,113) - (5053 - 10\,113\,k)(21\,671)$$

where k is any integer.

On occasions it might prove convenient to compute the gcd and s and t simultaneously by putting the two procedures described above into one table, as shown in Table 1.1. We leave the reader to work out exactly how the various rows in the table are arrived at.

Remark The above working of the Euclidean Algorithm in the general case is easily seen to provide another proof of the existence of and the formula for the positive gcd of any two integers (not both zero). However, I am sure everyone would agree that the proof of 1.4.4 is much, much sweeter, the proof based on the Euclidean Algorithm being somewhat ungainly. Thus we have the interesting situation of having two proofs concerning gcds. The first offers a beautiful proof of existence but is fairly useless for calculations. The

Table 1.1

$sa+tb$	s	t
21 671	0	1
10 113	1	0
1445	−2	1
1443	13	−6
2	−15	7
1	10 828	−5053

second is perfect for specific calculations and also gives a proof of existence. On the other hand this proof seems to have little aesthetic merit.

Finally in this section we prove (at last!) that every irreducible element in $\mathbb{Z}$ is necessarily a prime element so that the concepts of primeness and irreducibility coincide in $\mathbb{Z}$. We need a definition and a trivial consequence.

Definition 1.4.8 Two integers a, b are said to be **relatively prime** (or **coprime**) iff $(a, b) = 1$.

Combining this definition with 1.4.4 we obtain immediately

Theorem 1.4.9 Let $a, b \in \mathbb{Z}$. Then a and b are relatively prime iff there exist, in $\mathbb{Z}$, integers s and t such that $sa + tb = 1$.

At last we have

Theorem 1.4.10 Let $a \in \mathbb{Z}$ be an irreducible element. Then a is a prime element.

Proof Suppose $b, c \in \mathbb{Z}$ and $a|bc$. [We want to show that $a|b$ or $a|c$ (or both).] If $a|b$ we are finished so suppose $a \nmid b$. It follows that $(a, b) = 1$. [Since a is irreducible its only divisors are $a, -a, 1, -1$, and since $a \nmid b$ the only divisors a can have *in common with those of* b are 1 and -1.] Thus by 1.4.4 there exist $s, t \in \mathbb{Z}$ such that $sa + tb = 1$. But then $sac + tbc = c$. Now $a|sac$ [you can see that it does] and $a|tbc$ [because . . . why?]. Thus $a|c$, as required.

[In summary: we have proved that either $a|b$ or, failing that, $a|c$. Thus by 1.3.5 (iii), a is a prime element in $\mathbb{Z}$.]

Remark As already noted in the system H of exercise 1.3.6, irreducible elements are not necessarily prime ones. Of course H is intended only as an easy illustration of what *might* go wrong—and go wrong it does in some of the important generalisations of $\mathbb{Z}$ that arise in Section 3.6.

Exercises (Do not use uniqueness of factorisation in $\mathbb{Z}$ in these exercises.)

1 Show that the pair 0, 0 has infinitely many common divisors but only one greatest common divisor.

2 Show that if c and d are gcds of a, b then c and d are associates. Deduce that (a, b) is unique (as implied by the Remark and by 1.4.2 on p. 26).

3 Show that for all a, b (not both zero) we have

$$(a, b) = (-a, b) = (a, -b) = (-a, -b) = (b, a)$$

4 Show that if $(a, c) = (b, c) = 1$ then $(ab, c) = 1$. [Use 1.4.9.]

5 Show that if $c|a+b$ and if $(a, b) = 1$ then $(a, c) = (b, c) = 1$.

6 Prove or disprove: For $a, b, c \in \mathbb{Z}$ (with $b \neq 0$), $(a, (b, c)) = ((a, b), c)$. [If the equality holds we could unambiguously denote either gcd by (a, b, c).] State and prove the analogue of 1.4.4 for three integers a, b, c.

7 Show that for each $n \in \mathbb{Z}$ $(5n+2, 12n+5) = 1$. [Hint: $c = (a, b) \Rightarrow c|12a - 5b$.]

8 Show that once a, b are given, the values m, r in 1.4.5 are uniquely determined. [Hint: Suppose $a = m_1 b + r_1 = m_2 b + r_2$.]

9 Complete the proof of 1.4.5 by proving it in the case of $b < 0$.

10 Let S be a non-empty set of integers such that $a \in S$, $b \in S \Rightarrow a - b \in S$. Prove that either S comprises 0 alone or else S contains a smallest positive member c and S comprises *precisely* all the integral multiples of c.

11 Show that when we write $c = sa + tb$, where $c = (a, b)$, then $(s, t) = 1$.

12 Given that $sa + tb = 1$ prove or give a counterexample to each of the following assertions
(i) $(sa, tb) = 1$; (ii) $(sb, ta) = 1$; (iii) $(st, ab) = 1$.

13 Show that if $a = mb + r$, as in the division algorithm, then $(a, b) = (b, r)$.

14 Find $(527, 901)$ and write it in the form $527\,s + 901\,t$. Find that pair of integers s, t for which s is as small as possible positive. Prove there is no solution s, t for which $25 < s < 50$.

15 Let H be the set of numbers defined in exercise 1.3.6. Show that 21 is numerically the largest common H-divisor of 441 and 693 but show that there is no gcd in the sense of 1.4.1.

16 Show that if $2^k - 1$ is a prime (where k is a positive integer) then (i) k is a prime and (ii) $2^{k-1}(2^k - 1)$ is a perfect number. [An integer greater than 1 is called *perfect* if it is equal to the sum of all its positive divisors, including 1 but excluding itself.]
• Find four perfect numbers. [Euclid proved (ii). Euler (see p.69) proved every even perfect number is of that form. No odd perfect number is known. The largest known is $2^{216090}(2^{216091} - 1)$.]

17 Let p be a prime. For each i such that $0 < i < p$ show that the binomial coefficient $\binom{p}{i}$ is divisible by p. Use this to prove, by induction on n, that for each $n \in \mathbb{Z}^+$ we have $p|n^p - n$. (This is **Fermat's Little Theorem**.)

18 Show that for all $n \in \mathbb{Z}^+$, $(u_{n+1}, u_n) = 1$, where $u_1, u_2, u_3, \ldots$ is the Fibonacci sequence (Section 1.2).

1.5 The unique factorisation theorem (two proofs)

Even though we have just proved the equivalence of the concepts of primeness and of irreducibility in $\mathbb{Z}$ we ask the reader to note that in the following theorem the concept of irreducibility is associated with the *existence* of a decomposition of the asserted kind whereas the property of primeness is used to establish the *uniqueness* of this decomposition.

Theorem 1.5.1* (The Unique Factorisation Theorem for $\mathbb{Z}$; also called the Fundamental Theorem of Arithmetic) Let a be a non-zero element of $\mathbb{Z}$. Then either a is a unit or a can be expressed as a product of a unit and finitely many positive primes. Further, if $a = up_1p_2 \ldots p_r = vq_1q_2 \ldots q_s$ where u, v are units and $p_1, \ldots, p_r, q_1, \ldots, q_s$ are positive primes then $u = v$, $r = s$ and the p_i and the q_j can be paired off in such a manner that paired primes are equal.

Proof Half of the theorem has been proved already: Lemma 1.3.9 and Remark (ii) following it show that every integer greater than 1 (respectively, less than -1) can be expressed as a product of (respectively, -1 times a product of) finitely many positive irreducibles (which we now know to be primes).

Now suppose there exists an integer a with decompositions as above but in which the p_i and q_j do not pair off. If a is a negative integer then $|a|$ is a positive integer with the same nasty property, and the set S of all positive nasty as is non-empty. Thus S contains a smallest member. WLOG let it be $|a|$. Now $|a| = p_1p_2 \ldots p_r = q_1q_2 \ldots q_s$ and clearly $p_1 | q_1q_2 \ldots q_s$. By exercise 1.3.3 we deduce [because p_1 is *prime*] that $p_1 | q_i$ for some i. But q_i is *irreducible* and so $p_1 = 1$ or -1 or q_i or $-q_i$. Since p_1 is not a unit and since p_1 and q_i are both positive we are forced to conclude that $p_1 = q_i$. Thus the above equality reduces to $p_2p_3 \ldots p_r = q_1q_2 \ldots q_{i-1}q_{i+1} \ldots q_s$. [Why are we allowed to cancel?] But this integer is clearly smaller than $|a|$ and so it does not lie in S. That is, unique factorisation *does* apply to this smaller integer and we can deduce that the remaining $p_2, \ldots, p_r$ and $q_1, \ldots, q_{i-1}, q_{i+1}, \ldots, q_s$ pair off (in particular $r - 1 = s - 1$ so that $r = s$) in the manner described in the statement of the theorem. Since p_1 and q_i have already been paired off in the appropriate manner (i.e. they are equal!) we find that $|a|$ is *not* a nasty integer after all. That is, S is empty, and this proves the theorem.

Remark If we admitted 1 and -1 as primes we would lose the uniqueness we've just obtained since, for instance, $7 = 1 \cdot 7 = 1 \cdot 1 \cdot 7$. Hence the restriction on primes and irreducibles being non-units.

* Euclid: Book 9.

Before closing this section we offer a delightful second proof of this last theorem in which (the reader is invited to verify) no use is made of the concept of primeness. The reader is also invited to ponder which of the two proofs gives him the greater personal satisfaction. (The author's preference should be obvious!) This time we shall work throughout with decompositions of integers $\geqslant 2$ into products of positive irreducibles.

Second proof of Theorem 1.5.1 Once again we quote 1.3.9 (which makes no use of the primeness property) to establish the *existence* of a decomposition (into irreducibles) in each case. To establish the uniqueness of this decomposition we proceed as follows.

If the assertion relating to uniqueness is incorrect then there exists a smallest positive integer ($\geqslant 2$) c, say, such that c is expressible in two essentially distinct ways. Suppose indeed that $c = p_1p_2 \ldots p_r = q_1q_2 \ldots q_s$ where the p_i and q_j are (positive) irreducibles. Since [by the minimality of c] no p_i can be equal to any q_j, we may assume [WLOG] that $p_1 < q_1$. Set $d = p_1q_2 \ldots q_s$ so that $d < c$. Now $c - d = (q_1 - p_1)q_2 \ldots q_s = w_1w_2 \ldots w_tq_2 \ldots q_s$ where $w_1w_2 \ldots w_t$ is the decomposition of $q_1 - p_1$ into irreducibles. (If $q_1 - p_1 = 1$ we get $c - d = q_2q_3 \ldots q_s$.) Now $p_1|c$ and $p_1|d$. Hence $p_1|c - d$. This means that $c - d$ has a decomposition [into irreducibles] of the form $c - d = p_1v_1v_2 \ldots v_m$ where $v_1v_2 \ldots v_m$ is a decomposition of $(c - d)/p_1$ into irreducibles. Thus $c - d$ has two essentially distinct decompositions into products of irreducibles. For: p_1 occurs in one decomposition but not in the other since (i) p_1 is not equal to any of $q_2, q_3, \ldots, q_s$, and (ii) p_1 cannot be equal to any $w_1, w_2, \ldots, w_t$ (or else $p_1|w_1w_2 \ldots w_t = q_1 - p_1$ whence $p_1|q_1$, which is impossible—why?).

Since $0 < c - d < c$ it appears that our assumption that there exists a smallest counterexample is untenable and so the theorem is proved.

Remark In Section 3.6 we shall discuss number systems in which the concepts of prime and irreducible definitely do not coincide and in exercise 3.7.16 we shall ask the reader to criticise the application of the above proof (since it leads to a manifest contradiction!) to these systems. See also exercise 2 below and exercise 3.11.1.

Exercises

1 Is $13^2 \cdot 71 \cdot 103 = 17 \cdot 19 \cdot 53 \cdot 73$? [Unhelpful hint: Both integers end in a 7.]

2 Consider once again the set H of exercise 1.3.6. Clearly H satisfies principle W. Every element of H has at least one factorisation into H-irreducibles and 441, for example, has the two distinct decompositions $9 \cdot 49$ and $21 \cdot 21$ into H-irreducibles. Determine where the above second proof of 1.5.1 fails for the set H.

3 Prove that if $d = (a, b)$ then $(a/d, b/d) = 1$.

4 Prove that if $d_1|a$, $d_2|a$ and $(d_1, d_2) = 1$ then $d_1d_2|a$.

5 Prove that if $a|bc$ and if $(a, b) = 1$ then $a|c$.

6 Prove that if $c^2 = ab$ where $(a, b) = 1$ and where $a, b \in \mathbb{Z}^+$ then a and b are both perfect squares.

1.6 Polynomials—what are they?

We now try to duplicate for polynomials some of the basic definitions and theorems set down earlier when investigating $\mathbb{Z}$. We shall find out that, despite the obvious differences in appearance between the elements of $\mathbb{Z}$ and those of $\mathbb{Q}[x]$, the underlying structures of these two systems are quite similar. Before beginning this attempted duplication we ought perhaps to remind ourselves exactly what polynomials *are*. This is easy. A polynomial in x with, let us say, rational number coefficients is simply an expression of the form

$$a_0 + a_1x + \cdots + a_mx^m$$

for some $m \in \mathbb{Z}$ and some $a_i \in \mathbb{Q}$. Further we define addition and multiplication of polynomials in the usual way. That is we define addition by

$$(a_0 + a_1x + \cdots + a_mx^m) + (b_0 + b_1x + \cdots + b_nx^n)$$
$$= (a_0 + b_0) + (a_1 + b_1)x + \cdots$$

and multiplication by

$$(a_0 + a_1x + \cdots + a_mx^n) \cdot (b_0 + b_1x + \cdots + b_nx^n)$$
$$= a_0b_0 + (a_0b_1 + a_1b_0)x + \cdots.$$

Before continuing we might just stop to clarify a few apparently trivial points.

(i) Are we to consider $2 + x^2$ as being of 'the same form' as $2 + 0x + 1x^2 + 0x^3$ or not?
(ii) What actually *is* x?
(iii) In the above definition of addition it seems that the + sign which is being defined (that is, the one *between* the brackets) is being defined in terms of itself! (Witness the same + signs *inside* the brackets.) Am I right? Is this fair?

It would appear then that not everything in the garden is quite as rosy as might have been suspected. Certainly one would like the answer to (i) to be 'yes'; and in fact that's easily arranged. And (iii) causes no real problem. One just changes the sign between the brackets to $\oplus$ and proceeds a little more carefully. Point (ii) however is a bit more problematical. To say that x is an 'indeterminate' is no good. (What on earth is an indeterminate?) No more can x be 'something which stands for a number' since one could equally well imagine it standing for a matrix or even the differential operator

D (meaning d/dx), where D^n, the 'product' of n copies of D, is interpreted as the operator d^n/dx^n.

[Actually the same sort of problem arises in the definition of a complex number as 'a number of the form $a+bi$'. *Questioner*: *What is* 'i'*?* Responder: It is $\sqrt{-1}$. *But what is that? Is it a number?* Well . . . yes. *It can't be a real number since its square is negative.* Of course. *So on what grounds do you assert that, for instance,* $2 \cdot i = i \cdot 2$*?* . . . No reply!

We shall deal with these questions in Section 4.4.]

There's an easy way to avoid embarrassing questions about x (just as there is about i). Dispense with x! Notice that the symbols $x, x^2, x^3, \ldots$ are really nothing more than place markers. This is especially noticeable (now you know to look for it!) when multiplying two polynomials in the usual manner. Taking into account the problem raised in (i) above we make the formal

Definition 1.6.1

(i) A **polynomial** with coefficients in $\mathbb{Q}$ is an infinite* sequence $(a_0, a_1, a_2, \ldots)$ in which the a_i all belong to $\mathbb{Q}$ and in which all the a_i are equal to 0 from some point onwards. (More formally: all the a_i belong to $\mathbb{Q}$ and there exists $N \in \mathbb{Z}$ such that $0 \leqslant N$ and $a_n = 0$ for all n such that $N \leqslant n$. Of course N is allowed to vary from one sequence to another.)

The set of all polynomials with coefficients in $\mathbb{Q}$ will, for reasons which will emerge in Section 1.8, be denoted, as usual, by $\mathbb{Q}[x]$.

(ii) The polynomial $(b_0, b_1, b_2, \ldots)$ and the above polynomial $(a_0, a_1, a_2, \ldots)$ are said to be **equal** iff $a_i = b_i$ for all non-negative integers i.

(iii) Addition is defined by

$$(a_0, a_1, a_2, \ldots) \oplus (b_0, b_1, b_2, \ldots) = (c_0, c_1, c_2, \ldots)$$

where, for each non-negative integer i, $c_i = a_i + b_i$.

(iv) Multiplication is defined by

$$(a_0, a_1, a_2, \ldots) \odot (b_0, b_1, b_2, \ldots) = (d_0, d_1, d_2, \ldots)$$

where, for each non-negative integer i, $d_i = a_0 b_i + a_1 b_{i-1} + \cdots + a_i b_0$. That is, we put† $d_i = \sum_{k=0}^{i} a_k b_{i-k}$.

Remarks

(i) The symbols $\oplus$, $\odot$ (the first of which is often called‡ 'hot cross plus'!) are used in preference to + and · to distinguish them from the signs by which we

* Wouldn't finite sequences be better? Think about it!

† Σ means 'sum'. Thus $\sum_{k=1}^{4} w_k$ is shorthand for $w_1 + w_2 + w_3 + w_4$.

‡ I am grateful to Dr R J S Crossley of the University of York for informing me of the alternative names 'plus bun' and 'dot bun'—appropriate terms since, as he says, $\oplus$, $\odot$ are examples of 'bunnery' operations! (See Section 2.1.)

are already combining elements of **Q**. The new symbols are sufficiently unfamiliar to remind us of their defining role but sufficiently similar to + and · to help us remember their meaning with a minimum of effort.
(ii) It is probably clear to the reader that, although we are trying to work formally and to assume nothing other than that given, we do not choose our definitions of $\oplus$ and $\odot$ in a perfectly random fashion, but rather make them reflect what we want to happen. Remember the aim here is to get rid of the problems concerning x, not to get rid of the concept of polynomial itself! Exercises 2 and 3 will convince doubters that our definitions of $\oplus$ and $\odot$ are the right ones!
(iii) According to the above definitions, the sum and product of two given polynomials are again polynomials. Clearly each sum and product is a sequence so the crux of the matter is to show that, in each case, it is a sequence with only finitely many non-zero terms (exercise 6 below).

Exercises (Recall: a polynomial is an infinite sequence)

1 Determine the polynomial z, say, which is such that, for all polynomials u, we have $u \oplus z = z \oplus u = u$. ($z$ is clearly uniquely determined and is called the **zero polynomial.**)

2 Let $u = (1, 3, 0, -5, 2, 0, \ldots)$ and $v = (0, 2, 0, -1, 0, 3, 0, \ldots)$ where all the entries not shown are to be assumed equal to 0. Find (a) $u \oplus v$ (b) $u \odot v$ (c) $u \odot u$ (d) $v \odot u$.

3 Informally using the more familiar notation suppose $f = 2x - x^3 + 3x^5$ and $g = 1 + 3x - 5x^3 + 2x^4$. Find (a) $f + g$ (b) $f \cdot g$ (c) g^2 (d) $g \cdot f$.

4 Did your find 3(b) easier to do than 2(d)? If so, can you explain why?

5 Find u, v each having at least two non-zero entries such that $u \odot v = (1, 3, 1, -1, 0, \ldots)$. [Recall: all we want is the answer. You may use any method you wish to get to that answer.]

6 Prove that the sum ($\oplus$) and product ($\odot$) of two polynomials is again a polynomial.

1.7 The basic axioms

We would now like to investigate which of the axioms A1, ..., M4, etc. listed in Section 1.2 hold for the set $\mathbf{Q}[x]$. We shall check a few and leave the rest for the reader. For instance:

For A2 (in great detail): Here we are given $A = (a_0, a_1, \ldots, a_n, \ldots)$, $B = (b_0, b_1, \ldots, b_n, \ldots)$ and $C = (c_0, c_1, \ldots, c_n, \ldots)$ where, for safety, we include the 'general' entries a_n, b_n, c_n. Now $A \oplus B = (a_0 + b_0, a_1 + b_1, \ldots, a_n + b_n, \ldots)$. Hence $(A \oplus B) \oplus C = (\{a_0 + b_0\} + c_0, \{a_1 + b_1\} + c_1, \ldots, \{a_n + b_n\} + c_n, \ldots)$. On the other hand $B \oplus C = (b_0 + c_0, b_1 + c_1, \ldots, b_n + c_n, \ldots)$. Hence $A \oplus (B \oplus C) = (a_0 + \{b_0 + c_0\}, a_1 + \{b_1 + c_1\}, \ldots, a_n + \{b_n + c_n\}, \ldots)$. The question is: Is the polynomial $(A \oplus B) \oplus C$ equal to the polynomial $A \oplus (B \oplus C)$?

The answer is: Yes, provided that for each non-negative integer n we have $\{a_n + b_n\} + c_n = a_n + \{b_n + c_n\}$. And this indeed we do since, in $\mathbb{Q}$, addition is associative.

Note, then, that the fact that $\oplus$ is associative depends only upon the actual definition of $\oplus$ and the fact that $+$ is associative on $\mathbb{Q}$.

For M3 (more briefly): Let I denote the polynomial $(1, 0, 0, \ldots)$. Then [by definition] $A \odot I = (d_0, d_1, d_2, \ldots, d_i, \ldots)$ where, for each i, $d_i = a_0 \cdot 0 + a_1 \cdot 0 + \cdots + a_i \cdot 1 = a_i$. Thus $A \odot I = (a_0, a_1, \ldots, a_n, \ldots) = A$. [How did I think of trying I? Answer: I asked myself 'What polynomial $g(x)$ (in the old-fashioned sense) is such that, for all polynomials $f(x)$, we always have $f(x)g(x) = f(x)$?' The answer, $g(x) = 1$ was obvious, and I is the sequence version of the polynomial $1 + 0x + 0x^2 + \cdots$. The point being made here is that, whilst one is forbidden to use intuition to *prove* theorems, one can use any means, and intuition is a good means, to feel one's way to an answer or proof. Once an outline has emerged it is often not too difficult to fill in the formalities.]

Note that I exists only by kind permission of $0, 1 \in \mathbb{Q}$. That I has the properties asked of it is due entirely to the fact that in $\mathbb{Q}$ we know that . . . what?

We can 'order' the members of $\mathbb{Q}[x]$ so that P(i) and P(ii) are satisfied, although one could not expect $\mathbb{Q}[x]$ to satisfy all the axioms A1 through to I. [Why not?] We shall leave this to exercise 4.

For the moment we concentrate on showing that property Z holds in $\mathbb{Q}[x]$.

For Z: Let $A = (a_0, a_1, \ldots, a_s, 0, \ldots)$, $B = (b_0, b_1, \ldots, b_t, 0, \ldots)$ where a_s, b_t are the last non-zero terms in A, B respectively. Setting $A \odot B = (d_0, d_1, \ldots, d_n, \ldots)$ we see that

$$d_{s+t} = a_0 b_{s+t} + a_1 b_{s+t-1} + \cdots + a_{s-1} b_{t+1} + a_s b_t + a_{s+1} b_{t-1} + \cdots + a_{s+t} b_0$$

$$[= \quad 0 \quad + \quad 0 \quad + \cdots + \quad 0 \quad + a_s b_t + \quad 0 \quad + \cdots + \quad 0 \quad = a_s b_t]$$

Thus $d_{s+t} \neq 0$ [since a_s, b_t are given to be non-zero in $\mathbb{Q}$]. In particular $A \odot B$ is not the zero polynomial $(0, 0, 0, \ldots)$.

Note that $\mathbb{Q}[x]$ satisfies Z essentially by kind permission of the fact that $\mathbb{Q}$ itself does.

Note 1.7.1 Although we don't need it just now we note that if we look at the terms d_i, for $s + t < i$, in the above product, we see that each is equal to 0. Thus d_{s+t} is the last non-zero term in the above product.

Finally for M4: With A, B as above, we have seen that $d_{s+t} \neq 0$ if a_s and b_t are respectively the last non-zero terms of A and B. Thus if $a_i \neq 0$ for any i

such that $0<i$ it is impossible to choose B such that $A \odot B = I$. Indeed we note that

Theorem 1.7.2 Given the polynomial A, there exists a polynomial B such that $A \odot B = I$ iff $a_0 \neq 0$ in $\mathbf{Q}$ and $a_i = 0$ in $\mathbf{Q}$ for all i such that $0<i$.

We leave the remaining details to the reader.

Exercises

1 Prove that for all $u, v \in \mathbf{Q}[x]$ we have $u \oplus v = v \oplus u$. That is, prove that $\oplus$ satisfies axiom A1.

2 Prove that $\oplus$, $\odot$ together satisfy axiom D.

3 Prove that $\odot$ satisfies axiom M2.

4 Find in $\mathbf{Q}[x]$ a subset N so that axioms P(i) and P(ii) are both satisfied. [If you have difficulty, why not team up with a colleague and argue over it?] You can't find *two* such subsets can you? (Cf. exercise 3.12.2.)

5 We define the set of all **power series over Q** to be the set $\mathbf{Q}[[x]]$ of all infinite sequences as in $\mathbf{Q}[x]$ except that the restriction that only finitely many of the a_i be non-zero is dropped. Show that the element $(a_0, a_1, a_2, \ldots)$ has a multiplicative inverse in $\mathbf{Q}[[x]]$ iff $a_0 \neq 0$.

1.8 The 'new' notation

Definition 1.6.1 is all very well in that it gets rid of any nasty problems concerning x, by getting rid of x! Further, exercises 1.6.2 and 1.6.3 indicate that these sequence-type polynomials do appear to behave as we wanted them to. (This is scarcely surprising since we defined $\oplus$ and $\odot$ so that they would!) But exercises 1.6.2 and 1.6.3 and perhaps, more especially, exercise 1.6.5 emphasise that sequence-type polynomials are a little difficult to handle. Why is this? The answer is simple. Unfamiliarity. We can work using the x-form faster than we can the sequence form simply because we've been using the former for years. Let's see how we can recover the familiar x-form from the sequence-type definition.

To begin with, consider the subset $\bar{\mathbf{Q}}$ of $\mathbf{Q}[x]$ comprising those polynomials (i.e. sequences) whose every term, excepting possibly the first, is zero. Thus we are looking at polynomials such as $(\frac{1}{3}, 0, \ldots)$ and $(0, 0, \ldots)$. For two such polynomials $(a, 0, \ldots)$ and $(b, 0, \ldots)$ we note that, according to 1.6.1, their sum and product are respectively $(a+b, 0, \ldots)$ and $(a \cdot b, 0, \ldots)$.

We see, then, that the brackets, commas, zeros and dots are just so much dead wood in that, if we write $(a, 0, 0, \ldots)$ more briefly as $\bar{a}$, then the above addition and multiplication can be written much more briefly as $\bar{a} \oplus \bar{b} = \overline{a+b}$ and $\bar{a} \odot \bar{b} = \overline{a \cdot b}$.

We achieve even more economy by giving the name 'x' to the polynomial $(0, 1, 0, \ldots)$. For then we observe that

$$\begin{aligned}(0, a_1, 0, \ldots) &= (a_1, 0, \ldots) \odot (0, 1, 0, \ldots)\\ &= \bar{a}_1 \odot x \qquad \text{in our new notation}\\ (0, 0, a_2, \ldots) &= (a_2, 0, \ldots) \odot (0, 1, 0, \ldots) \odot (0, 1, 0, \ldots)^{*}\\ &= \bar{a}_2 \odot x \odot x \qquad \text{in our new notation}\end{aligned}$$

and in general†

$$(0, 0, \ldots, a_n, 0, \ldots) = \underbrace{\bar{a}_n \odot x \odot x \odot \cdots \odot x}_{n\ x\text{s here}} \qquad \text{in our new notation.}$$

Finally we observe that

$$\begin{aligned}(a_0, a_1, a_2, \ldots, a_n, 0, \ldots) &= (a_0, 0, \ldots) \oplus (0, a_1, 0, \ldots) \oplus \cdots \oplus (0, 0, \ldots, a_n, 0, \ldots)\\ &= \bar{a}_0 \quad \oplus \bar{a}_1 \odot x \quad \oplus \cdots \oplus \underbrace{\bar{a}_n \odot x \odot x \odot \cdots \odot x}_{n\ x\text{s here}}\end{aligned}$$

and so, if we replace the $\oplus$ sign by $+$, if we simply drop the $\odot$ sign and if we then agree to write $x \odot x \odot \cdots \odot x$ as x^n we establish a notational way of writing polynomials that we might just find a little easier to work with! That is, we are now proposing to write $(a_0, a_1, \ldots, a_n, 0, \ldots)$ more intelligibly as $\bar{a}_0 + \bar{a}_1 x + \bar{a}_2 x^2 + \cdots + \bar{a}_n x^n$. It appears that the bars over the *a*s are also superfluous since (on using the new notations $+$ rather than $\oplus$ etc.) $\bar{a} + \bar{b} = \overline{a+b}$ and $\bar{a}\bar{b} = \overline{ab}$. So we drop the bars too and find that with the conventions described above we have shown that every polynomial can be expressed in the form $a_0 + a_1 x + a_2 x^2 + \cdots + a_n x^n$. Note that the ways of adding and multiplying polynomials in this new notation are precisely those you have always used.

Remarks

(i) The reader who feels both exhausted and cheated at this outcome (if any such reader there be) will perhaps comment: 'What a waste of time. All that just to get back to where we started.' In fact we haven't quite got back to where we started. We have shown that polynomials *can* be thought of in the way we have always thought of them, secure in the knowledge that uncomfortable questions about x can be circumvented.

(ii) In recognising that the sets $\mathbf{Q}$ acted upon by $+$ and $\cdot$ and $\bar{\mathbf{Q}}$ acted upon by $\oplus$ and $\odot$ are essentially the same, we are meeting for the first time a concept of prime importance in algebra, namely that of *isomorphism*. This concept can perhaps be more readily appreciated by looking at Fig. 1.1 which also clearly indicates that whilst the elements of $\mathbf{Q}$ are totally distinct from

* The triple product is unambiguous since $\odot$ is associative (exercise 1.7.3). Similar remarks apply to the other multiple products. See exercise 1.2.1, Theorem 2.7.7 and exercise 1 below.

† Note the neat sidestepping of a proper proof by induction. See exercise 1 below.

those of $\mathbb{Q}[x]$, nonetheless $\mathbb{Q}[x]$ contains the subsystem $\bar{\mathbb{Q}}$ which to all intents and purposes is the 'same' as $\mathbb{Q}$ so that, if it proves convenient, $\mathbb{Q}$ may be identified with it. Such identification of the elements of $\mathbb{Q}$ with those of $\bar{\mathbb{Q}}$ is in exactly the same spirit as the identification of the elements of $\mathbb{Z}$ with certain elements of $\mathbb{Q}$ (see Fig. 1.2), an identification to which you've probably never previously given much thought. (For more on this see Sections 3.10 and 5.9.)

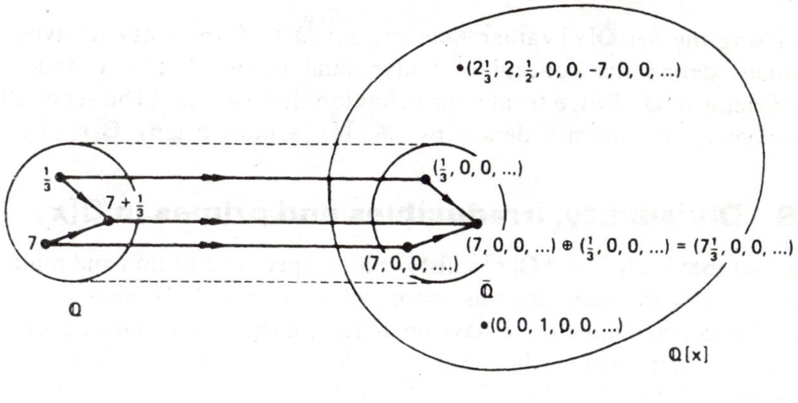

Fig. 1.1

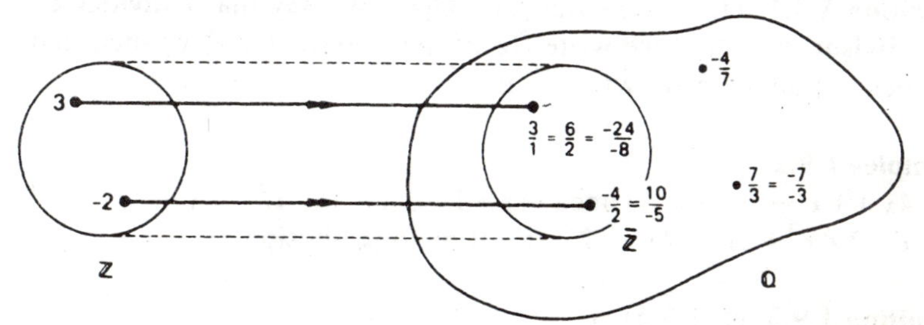

Fig. 1.2

(iii) At this point we should like to draw the reader's attention to a second aspect of polynomial algebra which is sometimes badly dealt with. The question arises: What are we to make of the equality $x^2-3x+2=0$? Since the left- and hence the right-hand sides of this equality are polynomials they are simply shorthand for $(2, -3, 1, 0, \ldots)$ and $(0, 0, \ldots)$ respectively. Hence 1.6.1(ii) tells us to deduce that $2=0$, $-3=0$ and $1=0$ in $\mathbb{Q}$! We leave the reader to think about this until we reach Section 1.11.

Note however that the assertion $x^2+2=(x-1)(x-2)+3x$ is certainly one which is meaningful—and also happens to be true! These remarks indicate that, if nothing else, we ought to proceed with circumspection.

Exercises

1 Since by exercise 1.7.3 polynomial multiplication is associative, we see (cf. exercise 1.2.1 and Theorem 2.7.9) that the definition of x^n as 'the product $x \odot x \odot \cdots \odot x$ with n occurrences of x' is unambiguous. In particular we can write x^{n+1} as $x^n \cdot x$. Using this equality prove, by induction, that, for each non-negative integer t, x^t is the polynomial $(0, 0, \ldots, 1, 0, \ldots)$ with all entries equal to 0 except for the lone 1 in the $(t+1)$st place.

2 Using the set $\mathbb{Q}[x]$ rather than the set $\mathbb{Q}$ to form sequence-type polynomials, define the concept of 'polynomial in two letters x and y with coefficients in $\mathbb{Q}$'. Prove from your definition that $xy = yx$. [The set of all such polynomials we naturally denote by $(\mathbb{Q}[x])[y]$ – more briefly, $\mathbb{Q}[x, y]$.]

1.9 Divisibility, irreducibles and primes in $\mathbb{Q}[x]$

The fact that both $\mathbb{Z}$ and $\mathbb{Q}[x]$ satisfy, with respect to addition and multiplication, most of the same axioms (both failing to pass M4) leads us to ask if other concepts defined for $\mathbb{Z}$ have analogues in $\mathbb{Q}[x]$. (Of course, even if they do, it will not necessarily mean that these concepts will have the same importance for $\mathbb{Q}[x]$ as for $\mathbb{Z}$.)

Following Section 1.3 we begin with

Definition 1.9.1 (cf. 1.3.1) Let $f, g \in \mathbb{Q}[x]$. We say that f **divides** g (or that f is a **divisor** of g) and we write $f|g$ iff there exists $h \in \mathbb{Q}[x]$ such that $g = fh$. If f does not divide g we write $f \nmid g$.

Examples 1.9.2

(i) $4x + 1 | x^2 - \frac{1}{12}x - \frac{1}{12}$ (since $(4x+1) \cdot \frac{1}{4}(x - \frac{1}{3}) = x^2 - \frac{1}{12}x - \frac{1}{12}$)

(ii) $x^2 + 2 \nmid x^3 + 2x^2 - 4x + 12$ (try it by long division!)

Definition 1.9.3 (cf. 1.3.5)

(i) If $u \in \mathbb{Q}[x]$ is such that $u|1$ then u is called a **unit**.

(ii) If $f \in \mathbb{Q}[x]$ is neither the zero polynomial 0 nor a unit we say that f is **irreducible** iff, whenever f is expressed as a product, $f = gh$ with $g, h \in \mathbb{Q}[x]$, it follows that either g or h is a unit. A non-zero non-unit polynomial f will be called **reducible** iff it is not irreducible.

(iii) If $f \in \mathbb{Q}[x]$ is neither 0 nor a unit we say that f is **prime** iff, whenever f divides a product, that is, $f|gh$ where $g, h \in \mathbb{Q}[x]$, it follows that $f|g$ or $f|h$ (or both).

(iv) If $f, g \in \mathbb{Q}[x]$ are such that $f = gu$, where u is a unit, then f and g are **associates.**

Remarks

(i) We emphasise that the requirement $u|1$ in 1.9.3(i) demands that we find an element v *in* $\mathbb{Q}[x]$ such that $uv = 1$. Thus according to 1.7.2 the units in

$\mathbb{Q}[x]$ are precisely the non-zero **constant polynomials**—and there are infinitely many of them. It follows that each non-zero element of $\mathbb{Q}[x]$ has infinitely many associates.
(ii) Instead of $\mathbb{Q}[x]$ we could equally well have considered $\mathbb{Z}[x]$, $\mathbb{R}[x]$, $\mathbb{C}[x]$, each defined in the obvious way. Note that in $\mathbb{Z}[x]$ the only units are the polynomials 1 and -1 and that whilst $2|x+2$ in $\mathbb{Q}[x]$, $2\nmid x+2$ in $\mathbb{Z}[x]$.
(iii) Whether or not a polynomial is irreducible depends upon which set of polynomials it is considered as belonging to. For example, the polynomial $2x^2+2$ is irreducible in $\mathbb{Q}[x]$ and $\mathbb{R}[x]$, but reducible in $\mathbb{Z}[x]$ and in $\mathbb{C}[x]$. Can you see why in each case?

Notes 1.9.4
(i) In Section 1.5 we proved that each non-zero non-unit element of $\mathbb{Z}$ factorises into a product of irreducibles in an essentially unique way. We naturally enquire whether or not the corresponding result holds in $\mathbb{Q}[x]$. Analysis of 1.5.1 shows that it does if $\mathbb{Q}[x]$ satisfies analogues of 1.3.9, 1.4.10, exercise 1.3.3 and property C of Section 1.2. Now 1.4.10 calls upon 1.4.4 which itself uses 1.4.5 which . . . ! The energetic reader is invited to supply the details. To give it maximum prominence the statement concerning factorisation in $\mathbb{Q}[x]$ is placed at the end of this section (1.9.18). However, instead of checking this chain of results for every concrete example where we seek to prove uniqueness of factorisation, it would be better if we could isolate the essentials behind these theorems leaving us in each specific case only these essentials to verify. This we shall do in Section 3.7.
(ii) Because (the analogue of) property C can be proved in $\mathbb{Q}[x]$, the analogue of 1.3.7 is easily seen to hold there. Together the analogues of 1.4.10 and 1.3.7 show that, in $\mathbb{Q}[x]$, *an element is prime if and only if it is irreducible.* Thus appears a very minor reason for our introducing the prime versus irreducible battle in 1.3.5 and 1.9.3: primes (i.e. irreducibles) in $\mathbb{Z}$ are usually called primes! Primes (i.e. irreducibles) in $\mathbb{Q}[x]$ are usually called irreducibles! Thus any difference implied by use of different words in $\mathbb{Z}$ and $\mathbb{Q}[x]$ is illusory.
(iii) One final point. The proofs of 1.3.9, 1.4.4 and in particular the Division Algorithm 1.4.5 make use of the fact that property W holds in $\mathbb{Z}$. Since, despite exercise 1.7.4, $\mathbb{Q}[x]$ does not seem to satisfy W (try it!), we are fortunate that the appropriate analogue for $\mathbb{Q}[x]$ of 1.4.5 depends upon $\mathbb{Z}$, rather than upon $\mathbb{Q}[x]$, satisfying W. We obtain this analogue in 1.10.1.

Somewhat more perplexing however is the following. In proving in $\mathbb{Z}$ and in $\mathbb{Q}[x]$ that all irreducibles are primes we make essential use of the appropriate division algorithm. However it is easily seen (exercise 1.10.2) that there can be no analogue of 1.10.1 in $\mathbb{Z}[x]$. This in itself does not immediately imply that there are irreducibles in $\mathbb{Z}[x]$ which are not primes. Of course if there are, the analogue for $\mathbb{Z}[x]$ of 1.5.1 could well fail since the proof as given certainly wouldn't then go over to $\mathbb{Z}[x]$. And yet *surely* factorisation in $\mathbb{Z}[x]$

is unique? What do you think? We pose

Problem 3 Are all irreducibles in $\mathbb{Z}[x]$ also primes?

Get your hands dirty by doing a bit of practical experimentation. You might even stumble upon a theorem or two in your researches.

We now ask the obvious question: Which polynomials in $\mathbb{Q}[x]$ are irreducible? The best reply we can give is: There is a test (see exercise 1.11.11) due to Kronecker* which will *always* tell in a finite number of steps (which may easily be large enough to require a computer) whether or not a given element of $\mathbb{Q}[x]$ is irreducible. Otherwise there is a criterion (1.9.16) which is easy to apply in practice and which describes infinitely many (but not all) polynomials irreducible over $\mathbb{Q}[x]$. Other tests are given by the remark following 1.11.7 and 4.2.10(ii).

To establish 1.9.16 we make some definitions and proofs relating to $\mathbb{Z}[x]$. For ease of recognition we denote polynomials in $\mathbb{Z}[x]$ by capital letters.

Definition 1.9.5 A polynomial $F = z_0 + z_1x + \cdots + z_nx^n \in \mathbb{Z}[x]$ is called **primitive** iff the positive gcd in $\mathbb{Z}$ of all the z_i is equal to 1.

Example 1.9.6 $10x^2 + 15x + 6$ is primitive, $72x^3 + 2x^2 - 42x + 8$ is not.

Notation 1.9.7 Extending the notation of exercise 1.4.6 we denote the positive gcd of $z_0, z_1, \ldots, z_n$ (not all zero) by $(z_0, z_1, \ldots, z_n)$.

Such a gcd always exists and is unique. A word for word copy of the proof of 1.4.4 shows this gcd is equal to the smallest positive integer in the set $\{k_0z_0 + k_1z_1 + \cdots + k_nz_n : k_0, k_1, \ldots, k_n \in \mathbb{Z}\}$.

A passage from $\mathbb{Q}[x]$ to $\mathbb{Z}[x]$ is provided by

Lemma 1.9.8 If $f \in \mathbb{Q}[x]$ then there exists in $\mathbb{Z}[x]$ an associate $F = \alpha f$ of f (where $\alpha \in \mathbb{Q}$ is chosen to be positive) such that F is primitive.

Proof I'll leave you to outline this. The following example shows you the way.

Example 1.9.9 If $f = \frac{3}{5} - 9x + \frac{18}{7}x^2 - \frac{12}{11}x^3 = \frac{3}{385}[77 - 1155x + 330x^2 - 140x^3]$ then $F = 77 - 1155x + 330x^2 - 140x^3$ and $\alpha = \frac{3}{385}$, a unit in $\mathbb{Q}[x]$.

The next result is due to Gauss (see p.68).

Theorem 1.9.10 If $F = y_0 + y_1x + \cdots + y_mx^m$ and $G = z_0 + z_1x + \cdots + z_nx^n \in \mathbb{Z}[x]$ are both primitive then so is FG.

Proof Let $FG = c_0 + c_1x + \cdots + c_{m+n}x^{m+n}$. If FG is *not* primitive then $c_0, c_1, \ldots, c_{m+n}$ have a common divisor (other than 1 and -1) in $\mathbb{Z}$ and hence

* Leopold Kronecker (7 December 1823 - 29 December 1891).

some common *prime* divisor p, say. Now p doesn't divide all the y_i nor all the z_j. [Why not?] Let s and t be the least suffices for which $p \nmid y_s$ and $p \nmid z_t$. Consider* $c_{s+t} = y_0 z_{s+t} + y_1 z_{s+t-1} + \cdots + y_s z_t + \cdots + y_{s+t} z_0$. Now $p \nmid y_s z_t$ [since $p \nmid y_s$, $p \nmid z_t$ and p is a prime]. On the other hand [by choice of s, t] p does divide every other term in the above equality including c_{s+t} itself. [This manifest contradiction shows that the assumption that FG is not primitive is untenable.] Thus FG *is* primitive, as required.

For the next theorem we need two definitions. First

Definition 1.9.11 Let

$$F = z_0 + z_1 x + \cdots + z_n x^n \in \mathbb{Z}[x].$$

The gcd $(z_0, z_1, \ldots, z_n)$ is called the **content** of F.

Example 1.9.12 The content of $-18x^2 + 45x - 30$ is $(-18, 45, -30) = 3$.

Second

Definition 1.9.13 Let $f = a_0 + a_1 x + \cdots + a_n x^n$ where $a_n \neq 0$. Then n is called the **degree** of f. We write $\deg f = n$. (We do not attribute a degree to the zero polynomial $f = 0$.)

We can now prove the important

Theorem 1.9.14 (Gauss) If $F \in \mathbb{Z}[x]$ and if we can write $F = gh$ where $g, h \in \mathbb{Q}[x]$ then we can write $F = GH$ where $G, H \in \mathbb{Z}[x]$, $\deg G = \deg g$ and $\deg H = \deg h$.

Remark The real meat in this theorem lies in the equalities of the degrees of G and g and of H and h. Without these equalities the theorem is true but absolutely trivial.

Proof

(i) We first suppose that F is primitive and write $g = \dfrac{a}{b} G$ and $h = \dfrac{c}{d} H$ where $a, b, c, d \in \mathbb{Z}^+$ are such that $(a, b) = 1 = (c, d)$, where $G, H \in \mathbb{Z}[x]$ and are primitive and where $\deg g = \deg G$ and $\deg h = \deg H$.

Thus $F = gh = \dfrac{ac}{bd} GH$. Hence $bdF = acGH$ in $\mathbb{Z}[x]$. Now F and GH are both primitive polynomials [by assumption and by 1.9.10 respectively] and so the content of bdF is bd [why?] whilst the content of $acGH$ is ac. [Since the polynomials bdF and $acGH$ are equal their contents must be.] Thus $ac = bd$ and $F = GH$ follows [from the equality above] as required.

* If $r > n$ we of course take z_r to be 0.

(ii) If now F is not primitive, write $F = zF_1$ where z is the content of F and where F_1 is primitive. Then, from $F = gh$ we get $F_1 = \frac{g}{z}h$. Hence, by (i), $F_1 = G_1H_1$ and consequently $F = (zG_1)H_1$ where G_1, zG_1 and $H_1 \in \mathbb{Z}[x]$.

As an immediate corollary we obtain a result you might have stumbled on in your researches into Problem 3.

Corollary 1.9.15 If $F \in \mathbb{Z}[x]$ and if one can't write F as a product of polynomials, each of smaller degree and with integer coefficients, then one still can't write F as a product of polynomials each of smaller degree even if one is prepared to use polynomials with *rational* coefficients. Equivalently: If $F \in \mathbb{Z}[x]$ is reducible in $\mathbb{Q}[x]$ then it is already reducible in $\mathbb{Z}[x]$.

We now give the solution to Problem 3. We take care in stating exactly where each division is taking place. (Recall that $2|3$ is false in $\mathbb{Z}$ but true in $\mathbb{Q}$.)

Solution to Problem 3 Let $F \in \mathbb{Z}[x]$ be irreducible (in $\mathbb{Z}[x]$). Then F is prime (in $\mathbb{Z}[x]$).

Proof Suppose $F|GH$ where $G, H \in \mathbb{Z}[x]$. [We must show that $F|G$ in $\mathbb{Z}[x]$ or $F|H$ in $\mathbb{Z}[x]$.] Now F is irreducible in $\mathbb{Z}[x]$. Thus, if $\deg F \geqslant 1$, F is primitive and is also irreducible in $\mathbb{Q}[x]$ [by 1.9.15]. Anticipating the proof (see 3.7.2 (ii) and 3.7.8) that in $\mathbb{Q}[x]$ irreducibles are necessarily primes we see that F is prime in $\mathbb{Q}[x]$. Hence [by the definition of primeness] either $F|G$ in $\mathbb{Q}[x]$ or $F|H$ in $\mathbb{Q}[x]$. Suppose WLOG the former. Then there exists $s \in \mathbb{Q}[x]$ such that $Fs = G$. Writing $s = \frac{a}{b}S$, where $a, b \in \mathbb{Z}^+$, $(a, b) = 1$ and $S \in \mathbb{Z}[x]$ is primitive, we obtain $aFS = bG$. Equating contents [noting that FS is primitive] we see that $a = b \cdot$ (content of G). Thus $\frac{a}{b} \in \mathbb{Z}^+$ whence $s \in \mathbb{Z}[x]$. This proves $F|G$ *in* $\mathbb{Z}[x]$.

If $\deg F = 0$ then F is irreducible, hence prime, in $\mathbb{Z}$ and hence prime as an element of $\mathbb{Z}[x]$. [Three assertions have just been made. Do you understand them? Do you believe them? Can you prove them? Cf. exercise 5.]

We now come to the first of the simpler tests for irreducibility over $\mathbb{Q}[x]$. This one is generally ascribed to Eisenstein.* As implied by 1.9.8 we can do all our working in $\mathbb{Z}[x]$.

Theorem 1.9.16 (Eisenstein's Test) Let $F = a_0 + a_1x + \cdots + a_nx^n \in \mathbb{Z}[x]$. If there exists a prime p in $\mathbb{Z}$ such that

(i) $p|a_0, p|a_1, \ldots, p|a_{n-1}, p \nmid a_n$,

* F G M Eisenstein (16 April 1823 – 11 October 1852), a pupil of Gauss. Gauss placed Eisenstein in the top three mathematicians of all time.

(ii) $p^2 \nmid a_0$,
then F is irreducible in $\mathbb{Q}[x]$.

Proof Supposing F is reducible in $\mathbb{Q}[x]$, we may assume [which theorem?] that $F = GH$ where $G = b_0 + b_1x + \cdots + b_rx^r \in \mathbb{Z}[x]$ and $H = c_0 + c_1x + \cdots + c_sx^s \in \mathbb{Z}[x]$ and where $r < n$ and $s < n$. We have

$$a_0 = b_0c_0$$

$$a_1 = b_0c_1 + b_1c_0 \quad \text{etc.}$$

Since $p|a_0$ we know that $p|b_0$ or $p|c_0$ [since p is prime], but this time not *both* since $p^2 \nmid a_0$. Suppose WLOG that $p|b_0$ and $p \nmid c_0$. Now not all the b_i are divisible by p or else all the a_i would be, contrary to $p \nmid a_n$. Let k be the smallest suffix for which $p \nmid b_k$. Then $0 < k \leq r < n$. Now $a_k = b_0c_k + \cdots + b_kc_0$ and $p|a_k$ since $k < n$. Also p divides all the terms b_ic_{k-i} except the last [since $p \nmid b_k$ and $p \nmid c_0$]. This contradiction establishes the theorem.

Example 1.9.17

(i) $f = \frac{1}{5} - 3x + \frac{6}{5}x^2 - x^3 + \frac{4}{5}x^4 + \frac{2}{15}x^5$ is irreducible in $\mathbb{Q}[x]$. For: We write $f = \frac{1}{15}\{3 - 45x + 18x^2 - 15x^3 + 12x^4 + 2x^5\}$. Noting that $3|3$, $3|45, \ldots, 3 \nmid 2$ and $3^2 \nmid 3$ we see that Eisenstein's theorem tells us that the polynomial in brackets is irreducible in $\mathbb{Z}[x]$. Consequently so is f, in $\mathbb{Q}[x]$.

The following type of polynomial plays an important role in Gauss' theory of constructible regular polygons (Section 4.6) and in the number theory described in Sections 3.5 and 3.9.

(ii) Let p be a prime in $\mathbb{Z}^+$. Then the pth **cyclotomic** (circle dividing) **polynomial** $C = 1 + x + x^2 + \cdots + x^{p-1}$ is irreducible in $\mathbb{Q}[x]$. For: If $C = fg$ in $\mathbb{Q}[x]$ then $C_{+1} = f_{+1}g_{+1}$ in $\mathbb{Q}[x]$ where, in each case, the polynomial h_a is defined as that obtained from h by replacing x by $a + x$. In particular $C_{+1} = 1 + (1+x) + (1+x)^2 + \cdots + (1+x)^{p-1}$. This is most easily simplified by noting that C can be written formally as $\dfrac{1-x^p}{1-x}$ and consequently

$$C_{+1} = \frac{1-(1+x)^p}{1-(1+x)} = \frac{1 - \left\{1 + \binom{p}{1}x + \binom{p}{2}x^2 + \cdots + \binom{p}{p}x^p\right\}}{-x}$$

$$= \binom{p}{1} + \binom{p}{2}x + \cdots + \binom{p}{p}x^{p-1}.$$

Here, each binomial coefficient displayed $\left(\text{except } \binom{p}{p} = 1\right)$ is divisible by p (exercise 1.4.17) and so Eisenstein's test says that C_{+1} is irreducible in $\mathbb{Q}[x]$. Thus there are no such $h, k \in \mathbb{Q}[x]$ (apart from trivial ones) whose product is

C_{+1}. Hence there are no $h_{-1}, k_{-1} \in \mathbb{Q}[x]$ (except trivial ones) whose product is C. Thus C is irreducible, as required. [All right?]

Remarks

(i) One might try to employ this sort of manoeuvre (exercise 10(c)) on any given polynomial one suspected of being irreducible. One trouble is that one wouldn't know for which integers a, if any, the replacement of x by $x+a$ would change the given polynomial into Eisenstein form.

(ii) Eisenstein's criterion shows that there exist in $\mathbb{Q}[x]$ infinitely many prime (i.e. irreducible) polynomials of each degree $\geqslant 1$.

(iii) As we did with the integers (1.3.9) we can use induction (but this time on the degree) to show that every non-zero non-unit polynomial in $\mathbb{Q}[x]$ can be expressed as a product of a finite number of irreducible ones. As implied in note 1.9.4(i) an analogue of 1.5.1 can be established in $\mathbb{Q}[x]$. This analogue is:

Theorem 1.9.18 (Unique Factorisation Theorem for Polynomials) Let f be a non-zero element of $\mathbb{Q}[x]$. Then either f is a unit or f can be expressed as a product of a unit and finitely many monic* irreducible polynomials. Further, if $f = ug_1g_2 \ldots g_r = vh_1h_2 \ldots h_s$ where u, v are units and $g_1, g_2, \ldots, g_r, h_1, h_2, \ldots, h_s$ are monic irreducible polynomials, then $u=v$, $r=s$ and the g_i and the h_j can be paired off in such a manner that paired polynomials are equal.

This result will finally be established in 3.7.13.

Exercises

1 Factorise x^4-4 into irreducibles in $\mathbb{Z}[x]$, $\mathbb{Q}[x]$, $\mathbb{R}[x]$ and $\mathbb{C}[x]$.

2 Write down all the divisors of $2x^2+6x+4$ in $\mathbb{Z}[x]$. [There are 16 of them.]

3 Is $2+2x+6x^2$ reducible or irreducible (i) in $\mathbb{Z}[x]$, (ii) in $\mathbb{Q}[x]$, (iii) in $\mathbb{R}[x]$, (iv) in $\mathbb{C}[x]$?

4 Find all the associates of $4x^2+2x+1$ in $\mathbb{Z}[x]$. Find five associates of $4x^2+2x+1$ in $\mathbb{Q}[x]$.

5 Does each prime in $\mathbb{Z}$ remain prime when considered as an element of (i) $\mathbb{Z}[x]$, (ii) $\mathbb{Q}[x]$?

6 Prove, by induction, that every polynomial of degree $\geqslant 1$ in $\mathbb{Q}[x]$ can be expressed as a product of irreducible polynomials.

7 Is $286x^2-187x+442$ primitive or not?

8 Represent as a product of a unit in $\mathbb{Q}[x]$ and a primitive polynomial (in $\mathbb{Z}[x]$) (a) $3x^2+6x+9$; (b) $\frac{x^2}{2}+\frac{x}{3}+7$.

* A polynomial in $\mathbb{Q}[x]$ is *monic* iff the coefficient of its greatest power of x is $+1$.

9 State (be careful!) and prove any relationship you can find involving deg f, deg g, deg $(f+g)$, deg $(f \cdot g)$ where $f, g \in \mathbb{Q}[x]$.

10 Determine whether or not the following polynomials are irreducible in $\mathbb{Q}[x]$.
(a) x^3+2x^2+4x+2; (b) x^3+2x^2+2x+4;
(c) $x^4+6x^3+12x^2+12x+7$. [Hint: look at f_{-1}.]

11 Use Eisenstein's method and the method of example 1.9.17(ii) to prove that x^4+1 is irreducible over $\mathbb{Q}$.

12 Show that $1+x^p+x^{2p}+\cdots+x^{(p-1)p}$ is irreducible in $\mathbb{Q}[x]$, p being a prime in $\mathbb{Z}$.

13 Prove the *back-to-front Eisenstein test* viz:

If $p|a_n$, $p|a_{n-1}, \ldots, p \nmid a_0$ and $p^2 \nmid a_n$ in $\mathbb{Z}$ then $a_0+a_1x+\cdots+a_nx^n$ is irreducible in $\mathbb{Q}[x]$. [Hint: you might care to look at $a_n+a_{n-1}y+\cdots+a_0y^n$ where $y=\frac{1}{x}$.] Hence prove that $7x^4+12x^3+12x^2+6x+1$ is irreducible in $\mathbb{Q}[x]$ (cf. exercise 10(c)).

14 Give a counterexample to show that the following converse to Eisenstein's test is false. (Students often seem to believe this 'converse' at examination time!)

Let $F=a_0+a_1x+\cdots+a_nx^n \in \mathbb{Z}[x]$. Suppose there is *no* prime p in $\mathbb{Z}$ satisfying conditions (i) and (ii) of 1.9.16. Then F is *not* irreducible in $\mathbb{Q}[x]$.

15 Give an example to show that the conclusion of Eisenstein's test cannot be replaced by: '... then F is irreducible in $\mathbb{Z}[x]$'. Identify where a proof of this, along the lines of 1.9.16, breaks down.

16 Let F ($\neq 1$) be a monic polynomial in $\mathbb{Z}[x]$. Show that the monic irreducible factors of F given by 1.9.18 all lie in $\mathbb{Z}[x]$.

17 Given $f, g \in \mathbb{Q}[x]$ not both zero, show that there exists in $S = \{sf+tg: s, t \in \mathbb{Q}[x]\}$ a unique monic polynomial m dividing every member of S. [Hint: look for the unique (?) monic polynomial of smallest degree in S—if such an object exists at all.]

1.10 The division algorithm

As implied in 1.9.4(i), we shall leave a proof of the uniqueness of factorisation in $\mathbb{Q}[x]$ until Section 3.7 where we shall get it as a special case of the theory developed there. However we do obtain here one of the ingredients needed for that proof, mainly as an aid to getting a second irreducibility test for polynomials in $\mathbb{Q}[x]$ in Section 1.11.

The analogue in $\mathbb{Q}[x]$ of 1.4.5 is:

Theorem 1.10.1 (The Division Algorithm) Let $f, g \in \mathbb{Q}[x]$ be such that $g \neq 0$.

Then there exist $m, r \in \mathbb{Q}[x]$ such that

$f = mg + r$ where either

(i) $r = 0$

or

(ii) $r \neq 0$ and $\deg r < \deg g$.

Proof

(1) If $f = 0$ we write $0 = 0g + 0$ which is clearly of the required form.

(2) Suppose $f = a_0 + a_1x + \cdots + a_mx^m$ and $g = b_0 + b_1x + \cdots + b_nx^n$ where $a_m \neq 0$ and $b_n \neq 0$. We set $d = m - n = \deg f - \deg g$.

(a) If $d < 0$: we write $f = 0g + f$ which is of the required form since $\deg f < \deg g$.

We deal with the remaining cases using induction (see 1.2.9) and starting with $d = 0$.

(b) If $d = 0$: Then $m = n$ and we write $f = \frac{a_m}{b_n} g + r$. Here r has its coefficient of x^m [recall $m = n$] and, trivially, of all higher powers of x, equal to 0. Thus even if $r \neq 0$ at least $\deg r < m = n = \deg g$.

(c) Assuming the result valid for $d = 0, 1, 2, \ldots, k - 1$ we take $d = k$: Set $f_1 = f - \frac{a_m}{b_n} x^{m-n} g$. [$f_1$ is constructed to ensure $\deg f_1 < \deg f$ (though it might happen by accident that $f_1 = 0$) so that we can apply the induction hypothesis to the pair f_1, g.] If $f_1 = 0$ we write $f = \frac{a_m}{b_n} x^{m-n} g + 0$ as required. If $f_1 \neq 0$ and if we set $t = \deg f_1 - \deg g$ then $-\deg g \leqslant t \leqslant k - 1$. Hence using case 2(a) [if $t < 0$] or case 2(b) [if $t = 0$] or the induction hypothesis of case 2(c) [if $0 < t \leqslant k - 1$] we see that, in each case, there exist m_1 and r_1 in $\mathbb{Q}[x]$ such that $f_1 = m_1 g + r_1$ where either (i) $r_1 = 0$ or (ii) $r_1 \neq 0$ and $\deg r_1 < \deg g$. It then follows from the definition of f_1 that $f = m_1 g + r_1 + \frac{a_m}{b_n} x^{m-n} g$, that is $f = \left(m_1 + \frac{a_m}{b_n} x^{m-n}\right) g + r_1$ exactly as required.

We give just one easy

Example 1.10.2 Given $f = 2x^5 - \frac{2}{3}x^4 - \frac{31}{9}x^3 - \frac{13}{12}x^2 + \frac{1}{9}x + \frac{1}{6}$ and $g = \frac{2}{3}x^2 + x + \frac{1}{6}$ we have, by long division,

$$
\begin{array}{r|l}
 & 3x^3 - \frac{11}{2}x^2 + \frac{7}{3}x - \frac{15}{4} \\
\hline
\frac{2}{3}x^2 + x + \frac{1}{6} & 2x^5 - \frac{2}{3}x^4 - \frac{31}{9}x^3 - \frac{13}{12}x^2 + \frac{1}{9}x + \frac{1}{6} \\
 & 2x^5 + 3x^4 + \frac{1}{2}x^3 \\
\hline
 & -\frac{11}{3}x^4 - \frac{71}{18}x^3 - \frac{13}{12}x^2 \\
 & -\frac{11}{3}x^4 - \frac{11}{2}x^3 - \frac{11}{12}x^2 \\
\hline
\end{array}
$$

$$\begin{array}{r} \frac{14}{9}x^3 - \frac{1}{6}x^2 + \frac{1}{9}x \\ \frac{14}{9}x^3 + \frac{7}{3}x^2 + \frac{7}{18}x \\ \hline -\frac{5}{2}x^2 - \frac{5}{18}x + \frac{1}{6} \\ -\frac{5}{2}x^2 - \frac{15}{4}x - \frac{15}{24} \\ \hline \frac{125}{36}x + \frac{19}{24} \end{array}$$

Thus $m = 3x^3 - \frac{11}{2}x^2 + \frac{7}{3}x - \frac{15}{4}$ and $r = \frac{125}{36}x + \frac{19}{24}$.

Note that $\deg r = 1 < \deg g = 2$.

Exercises

1 With f, g as given, find $m, r \in \mathbb{Q}[x]$ as in 1.10.1.
(a) $f = 4x^4 + \frac{3}{2}x^3 - \frac{3}{4}x^2 + x + \frac{2}{5}$, $g = x + \frac{1}{2}$.
(b) $f = 3x^6 - 2x^5 - \frac{13}{4}x^4 + \frac{7}{4}x^3 + \frac{9}{2}x^2 + \frac{9}{2}x + 3$, $g = 2x^3 - 3x^2 - x + 4$.

2 Give an example of $f, g \in \mathbb{Z}[x]$ ($g \neq 0$) such that there do not exist $m, r \in \mathbb{Z}[x]$ such that $f = mg + r$ with either $r = 0$ or $\deg r < \deg g$. [Thus there is in general no division algorithm in $\mathbb{Z}[x]$ except when g is monic.]

Show that there do exist $M, R \in \mathbb{Z}[x]$ and $z \in \mathbb{Z}$ such that $zf = Mg + R$ where either $R = 0$ or $R \neq 0$ and $\deg R < \deg g$. [Hint: work in $\mathbb{Q}[x]$ and get rid of fractions.]

3 Imitate the process of the Euclidean algorithm as defined for $\mathbb{Z}$ (p. 29) to find the unique monic gcd in $\mathbb{Q}[x]$ of the following pairs of polynomials:
(a) $x^5 + 2x^4 + x^3 + x^2 + 2x + 1$, $3x^4 + 9x^3 + 10x^2 + 5x + 1$.
(b) $x^6 - 6x^4 + 12x^2 - 8$, $x^3 - x + 2$.
[The proper way to define the gcd concept should be obvious. It's a word for word, copy of 1.4.1. The existence of the gcd is given by exercise 1.9.17.]

4 The Euclidean algorithm for $\mathbb{Q}[x]$ can be used, as in the case of $\mathbb{Z}$, to arrive at a proof of unique factorisation. Can unique factorisation in $\mathbb{Q}[x]$ also be proved by copying the second proof of 1.5.1 on p. 34? [Recall that the elements in $\mathbb{Q}[x]$ can be ordered; see exercise 1.7.4.] If not, identify the first place the proof breaks down. [Look, in due course, at exercise 3.11.1.]

1.11 Roots and the remainder theorem

Towards the end of Section 1.8 we noted that the only interpretation one could put upon the equality $x^2 - 3x + 2 = 0$ between the two polynomials mentioned is that $1 = 0$, $-3 = 0$ and $2 = 0$ in $\mathbb{Q}$. Thus the above equality is meaningful but blatantly false. It is because such equalities are therefore not worth investigating (they are all false!) that a totally different interpretation is usually attached to the statement $x^2 - 3x + 2 = 0$. This interpretation is expected to elicit from the reader the response: 'That's easy. $x = 2$ and $x = 1$ are the only solutions to this equation.' Any feeling of insecurity in the reader's mind concerning this dual interpretation of $x^2 - 3x + 2 = 0$ is not so much due to the perversity of the author but is rather due to the dual role that the public at large expects x to play. In popular language these two roles are those of 'indeterminate' and 'variable' respectively. This is perhaps best illustrated by considering the assertion that 'one way to show that $x-2$ is a factor of x^2-3x+2 is to put $x=2$ in x^2-3x+2 and show that the result is 0'. We *can't* put $x=2$ or anything else for that matter: x *is* the infinite sequence $(0, 1, 0, \ldots)$. [Nor can we apparently 'substitute 2 for x' since the result, namely $2 \cdot 2 \oplus (-3, 0, 0, \ldots) \odot 2 + (2, 0, 0, \ldots)$, is meaningless. On the other hand we can get round these difficulties, so that we *can* talk (loosely) as above without fear of any difficulties arising, by doing something which looks very much like substitution. In what follows we shall let J stand for any one of $\mathbb{Z}$, $\mathbb{Q}$, $\mathbb{R}$ or $\mathbb{C}$.

Definition 1.11.1

(i) Let us, for brevity, denote the polynomial $a_0+a_1x+\cdots+a_nx^n$ in $J[x]$ by f. For each $c \in J$ we define the **value** of f at c to be the element $a_0+a_1c+\cdots+a_nc^n$. We denote it briefly by $f(c)$ and refer to the process of obtaining $f(c)$ from f in this way by the ill-chosen but universally familiar expression 'substitution of c for x in f' since on the face of it we have just replaced every occurrence of x in f by c. It is then easy to check that, for $f, g \in J[x]$ and for $c \in J$, we have $(f \cdot g)(c) = f(c)g(c)$ and $(f+g)(c) = f(c)+g(c)$.

(ii) If $t \in J$ is such that $f(t) = 0 \in J$ we call t a **root** or **zero** of f in J.

(iii) The expression 'Solve the equation $a_0+a_1x+\cdots+a_nx^n = 0$ in J' means: 'Find all elements $c \in J$ such that $f(c) = 0$ in J.'

Notes 1.11.2

(i) What we are actually doing in 1.11.1(i) is to associate with the sequence $(a_0, a_1, \ldots, a_n, 0, \ldots)$ and the element $c \in J$ the element $a_0+a_1c+\cdots+a_nc^n$ in J.

(ii) In the expression $a_0+a_1x+\cdots+a_nx^n$ the a_i are abbreviated notations for infinite sequences; in $a_0+a_1c+\cdots+a_nc^n$ the a_i are actual elements of J.

(iii) The assertions that one can solve $x^2-3x+2=0$ in $\mathbb{Z}$ and that one cannot solve $x^2+1=0$ in $\mathbb{Q}$ should now be completely unambiguous (and both are true).

We can now prove a result which is often stated and proved in a rather casual manner. In order not to obliterate its intuitive meaning, and yet satisfy ourselves that we are not falling into a subtle trap, we denote the constant polynomial $(a, 0, 0, \ldots)$ in $J[x]$ once again by $\bar{a}$. It might be quite instructive for the reader to select an algebra text at random and see how many, if any, holes he can pick in the proof of the following theorem given therein.

Theorem 1.11.3 (The Remainder Theorem) Let $f \in J[x]$ and let $a \in J$. Then there exists $m \in J[x]$ such that $f = (x-\bar{a})m + \overline{f(a)}$. Further $x - \bar{a} \mid f$ in $J[x]$ iff $f(a) = 0$.

Proof By the division algorithm* we can surely find polynomials $m, r \in J[x]$ such that $f = m(x - \bar{a}) + r$. Further, either $r = \bar{0}$ [the zero polynomial] or $r \neq \bar{0}$ and $\deg r < \deg(x - \bar{a}) = 1$. In this case $\deg r = 0$ and r is (again) a constant polynomial. Thus set $r = \bar{b}$, where $b \in J$. If we now, according to 1.11.1, substitute a for x in each side of the above equality we obtain, using 1.11.1(i), $f(a) = m(a) \cdot (x - \bar{a})(a) + \bar{b}(a)$, from which one easily sees [if one thinks carefully what the symbols mean] that $f(a) = b$. Hence $r = \bar{b} = \overline{f(a)}$ as required.

We leave the second part of the theorem to exercise 1.

Remark If in 1.11.3 we adopt the notation finally settled on in Section 1.8, that is we drop the bars, we obtain the remainder theorem in its usual guise. Notice that then the symbol a is being asked to play two roles simultaneously, namely as an element of J and as a polynomial in $J[x]$ – unless (see Fig. 1.1) one is prepared to regard J and $\bar{J}$ ($\subseteq J[x]$) not just as isomorphic but as being one and the same (that is, identical) so that J is being regarded as a subset of $J[x]$.

Now we've shown that the sloppy approach can be put on a firm footing we shall, because of its greater familiarity, lapse into it!

There now follows immediately

Theorem 1.11.4 Let J be any one of $\mathbb{Q}$, $\mathbb{R}$, $\mathbb{C}$ and let $f \in J[x]$ such that $\deg f = n > 0$. Then f has at most n roots in J.

Proof We proceed by induction on n. If f has degree 1 then f has exactly one root in J. Now suppose f is a polynomial of degree $k+1$. If f has no root in J then since $0 \leqslant k+1$ the result holds. Otherwise suppose b is a root of f in J. Then, by 1.11.3, $x - b$ is a factor of f in $J[x]$. Write $f = (x-b)^t g$, where $t (\in \mathbb{Z}^+)$ is chosen as large as possible and $g (\in J[x])$ has degree $n - t$. By the induction assumption g has at most $n - t$ roots in J. Hence f has at most $(n-t)+1$ ($\leqslant n$) roots in J.

Note 1.11.5 If, in 1.11.4, we call t the *multiplicity of b in f* we can go further. Indeed, if $f \in J[x]$ has degree n and distinct roots $b_1, \dots, b_s$ with multiplicities $t_1, \dots, t_s$, then $t_1 + \cdots + t_s \leqslant n$. For example,

$$x^5 + 3x^4 - 4x^3 - x^2 - 3x + 4 = (x-1)^2(x+4)(x^2+x+1)$$

has roots 1, -4 with multiplicities 2, 1 respectively.

A second practical irreducibility test referred to in Sections 1.9 and 1.10 follows from

Theorem 1.11.6 (The Rational Root Test) If r/s is a rational root of the polynomial $z_0 + z_1 x + \cdots + z_n x^n \in \mathbb{Z}[x]$ where $(r, s) = 1$ then $r | z_0$ and $s | z_n$.

* Why can we use it? Cf. exercise 1.10.2.

Proof If r/s is a root then $z_0 + z_1(r/s) + \cdots + z_n(r/s)^n = 0$. Multiplying through by s^n yields $z_0 s^n + z_1 r s^{n-1} + \cdots + z_n r^n = 0$. Clearly s then divides $z_n r^n$ and hence $s | z_n$ [since $(r, s) = 1$] by exercise 1.5.5. In a similar manner one proves $r | z_0$.

Example 1.11.7 If $x^4 + 2x^3 + 2x^2 + x - 2$ has a rational root r/s with $(r, s) = 1$ then $r | -2$ and $s | 1$. Hence r/s has four possible values, namely 1, −1, 2 and −2. Substituting each of these in turn in the given polynomial never gives the value 0. Hence $x^4 + 2x^3 + 2x^2 + x - 2$ has no rational root.

Remark With regard to irreducibility, 1.11.6 is chiefly of use in showing certain cubic polynomials in $\mathbb{Q}[x]$ irreducible, because if a cubic polynomial is reducible at least one of the factors must have degree 1. Hence such a cubic polynomial has a root in $\mathbb{Q}$. The theorem is not of immediate use in checking polynomials of degree 4 or more for irreducibility. Indeed the above quartic which has no root in $\mathbb{Q}$ nonetheless factorises as $(x^2 + x + 2)(x^2 + x - 1)$ in $\mathbb{Q}[x]$.

Whilst talking about reducibility of polynomials we mention a very important result which we shall, using just a tiny bit of analysis, prove in full later (Theorem 4.8.1).

Theorem 1.11.8 (The Fundamental Theorem of Algebra)* Let $f \in \mathbb{C}[x]$ be such that $\deg f \geq 1$. Then f has a root in $\mathbb{C}$. It follows that f factorises completely into linear factors in $\mathbb{C}[x]$ so that if $\deg f = n$ then f has, including repeats, n roots in $\mathbb{C}$.

Applying this to $\mathbb{R}[x]$ we get

Theorem 1.11.9 If $f \in \mathbb{R}[x]$ then f can be expressed as a product of polynomials of degrees at most 2 in $\mathbb{R}[x]$.

Proof By exercise 8 below we see that if $a + ib$ is any complex root of f, then $a - ib$ is also a root of f. Thus non-real roots of f (if there are any) occur in complex conjugate pairs. The observation that

$$(x - (a + ib))(x - (a - ib)) = (x - a)^2 - (ib)^2 = x^2 - 2ax + a^2 + b^2 \in \mathbb{R}[x]$$

then suffices.

And as a corollary,

Corollary 1.11.10 An irreducible polynomial in $\mathbb{R}[x]$ is either of degree 1 or of degree 2.

* Gauss was the first to prove this. In all he gave six different proofs.

Remark Given a polynomial f in $\mathbb{R}[x]$, 1.11.10 restricts possible irreducible factors to having degree 1 or 2. This knowledge is unfortunately of little help in actually factorising f and in general we have to be content with approximation methods. Investigations along these lines naturally belong to numerical analysis.

Exercises

1 Complete the proof of 1.11.3.

2 Let $f \in \mathbb{Q}[x]$ be the polynomial $a_0 + a_1x + a_2x^2 + \cdots + a_nx^n$. Define the **derivative** f' of f to be the polynomial $a_1 + 2a_2x + \cdots + na_nx^{n-1}$. Show that if f has a double root a (that is f has a factor of the form $(x-a)^2$) with $a \in \mathbb{Q}$ then a is a root of f'.

3 Use exercise 2 to show that neither $x^4 - 4x^3 + 4x^2 + 17$ nor $x^4 + 4x^2 - 4x - 3$ has a repeated root in $\mathbb{Q}$.

4 Using the fact (see exercise 1.9.17) that the gcd d of any two polynomials $f, g \in \mathbb{Q}[x]$ can be expressed in the form $d = sf + tg$ for suitable $s, t \in \mathbb{Q}[x]$, show that the polynomial $x^4 + 4x^2 - 4x - 3$ of exercise 3 can't even have a repeated root in $\mathbb{C}$.

5 Let f and g be polynomials of degree n in $\mathbb{Q}[x]$. If $q_1, q_2, \ldots, q_{n+1}$ are distinct rational numbers such that $f(q_i) = g(q_i)$ for all i $(1 \leq i \leq n+1)$ show that $f = g$ in $\mathbb{Q}[x]$.

6 Let $u_1, \ldots, u_{n+1}$ be distinct rationals and $v_1, \ldots, v_{n+1}$ be rationals (not necessarily distinct). Show that there is exactly one polynomial $f \in \mathbb{Q}[x]$ such that (i) $\deg f \leq n$ and (ii) $f(u_i) = v_i$ for each i $(1 \leq i \leq n+1)$.

[Try defining f to be

$$\sum_{i=1}^{n+1} v_i \frac{(x-u_1)\ldots(x-u_{i-1})(x-u_{i+1})\ldots(x-u_{n+1})}{(u_i-u_1)\ldots(u_i-u_{i-1})(u_i-u_{i+1})\ldots(u_i-u_{n+1})}.$$

This is called **Lagrange's interpolation formula.**]

7 Find the roots, if any, in $\mathbb{Q}$ of

(a) $4x^4 - 2x^3 - 2x^2 - x + 1$; (b) $x^4 - x^3 - 2x^2 - 2x + 4$;
(c) $2x^4 + 7x^3 + 3x^2 + 12$; (d) $8x^3 - 6x - 1$.

8 Let f be a polynomial in $\mathbb{R}[x]$. Show that if $a + ib$ $(a, b \in \mathbb{R})$ is a root of f then so is $a - ib$. [Hint: Write out $f(a+ib)$ in full. Since $f(a+ib) = 0$ we have $\overline{f(a+ib)} = \bar{0} = 0$, the bar denoting complex conjugation. Show $\overline{f(a+ib)}$ is identical to $f(a-ib)$.]

9 The equation $x^4 - 3x^3 + 2x^2 + x + 5$ has $2 - i$ as one of its roots. Find all its complex roots.

10 Using any method you find convenient determine whether or not $x^4 - x^2 + 1$ and $x^4 + x + 1$ are irreducible in $\mathbb{Q}[x]$.

11 Kronecker's method for factorising $f \in \mathbb{Z}[x]$ is as follows. If $f = g \cdot h$ where $g, h \in \mathbb{Z}[x]$, then for each $z \in \mathbb{Z}$ for which $f(z) \neq 0$ we have $f(z) = g(z)h(z)$ so that $g(z) | f(z)$ in $\mathbb{Z}$. If $\deg f = n$ then $\deg g$ may be assumed to be at most* $[\frac{n}{2}]$. [Why?] Now, finding $[\frac{n}{2}] + 1$ integers $z_1, z_2, \ldots$ for which $f(z_i) \neq 0$, we determine all possible values of the $g(z_i)$. We then use exercise 6 to find all possible polynomials g. Next one simply tries all these in turn to see if any of them divides f in $\mathbb{Z}[x]$. This process is then repeated as necessary.

For example, if $f = x^4 + x + 1$ then f has no linear factor in $\mathbb{Z}[x]$. [Why not?] Suppose $f = gh$ where $\deg g = 2$. Then $g(-1) | f(-1) = 1$, $g(0) | f(0) = 1$ and $g(1) | f(1) = 3$. Thus g is such that $g(-1)$ is 1 or -1, $g(0)$ is 1 or -1 and $g(1)$ is 1 or -1 or 3 or -3. The finding and testing of all possible g is left to you!

* $[x]$ here denotes the greatest integer not exceeding x.

2

Binary relations and binary operations

2.1 Introduction

If we let X stand for any one of the sets $\mathbb{Z}$, $\mathbb{Q}$, $\mathbb{R}$, $\mathbb{C}$, $\mathbb{Z}[x]$, etc. mentioned in Chapter 1, then the operations of addition and multiplication defined on X may be described as *binary operations* on X in that, to each *pair* of elements of X, both $+$ and $\cdot$ produce a unique entity which is *again an element of* X.

As remarked in the prologue, present-day algebra might be defined as the study of (n-ary) operations on sets ($n = 2$, but also 0, 1, 3, 4, . . . etc.).* This would, however, be a little unfortunate since only relatively few n-ary operations are either interesting or important. In later chapters we shall investigate in depth some of those binary operations which have, by their repeated occurrence and usage, shown their importance. (For the sake of illustration we shall also introduce some which, to say the least, are of little significance!)

The general concept of binary operation on a set is defined in terms of that of 'function'. No doubt the reader can give what he feels is an adequate definition of 'function'. (Indeed we invite each reader, *here and now*, to make such a definition and compare his proposal with ours, given later.) However, in keeping with the somewhat critical approach adopted in Chapter 1, and because we also wish to indicate how our studies might be put upon a set-theoretical base (the soundness of which is the concern of set-theorists) we propose definitions rather more formal than many a reader might expect. Any reader worried by the prospect of such formality will be pleased to see how, as with polynomials in Section 1.8, the more informal notation to which he is more accustomed is soon restored.

The formal approach requires that we start with a study of binary relations.

2.2 Congruence mod *n*. Binary relations

We begin with some simple number theory which will help us feel our way and also lead us to some new and fascinating algebraic structures.

Let n be a positive integer, any one you wish, but fixed once chosen. Given two integers a and b, it is the case that either (i) $n|a-b$ or (ii) $n \nmid a-b$. [Equivalently: a and b (i) do have or (ii) do not have the same remainder (in the range 0 to $n-1$) on division by n.] In the first case we shall write $a \equiv b \pmod{n}$, in the latter $a \not\equiv b \pmod{n}$. These we shall read as *a is* (*is not*) *congruent to b*

* '2-ary' and 'binary' are synonymous.

modulo [more briefly, *mod*]n. Thus $7 \equiv -3 \pmod 5$ whilst $111 \not\equiv 40 \pmod 5$. Working modulo 5, then, we see that certain pairs of integers are related (by being congruent mod 5) whilst other pairs are not so related.

There are other more familiar ways in which certain pairs of integers appear to be related whilst other pairs are not. For example, since $3 < 7$, the ordered pair* $(3, 7)$ is related by $<$ whereas the ordered pair $(7, 3)$ is not. The symbol '$=$' yields another relation on $\mathbb{Z}$, two integers being related this time if and only if they are equal. As an example of doubtful mathematical value consider, on the set P of all human beings alive at, say, 10.05 a.m. (GMT) on 7 April 1978, the relationship 'is greatgrandfather of'. Still thinking aloud it appears that the relationships $\equiv \pmod 5$, $<$, $=$, and 'is greatgrandfather of' can be interpreted as subsets of $\mathbb{Z} \times \mathbb{Z}$, $\mathbb{Z} \times \mathbb{Z}$, $\mathbb{Z} \times \mathbb{Z}$ and $P \times P$ respectively. For example, $<$ can be identified with the subset of $\mathbb{Z} \times \mathbb{Z}$ comprising all ordered pairs (a, b) for which $a < b$. Similarly $=$ can be identified with the subset $\{(a, a): a \in \mathbb{Z}\}$ of $\mathbb{Z} \times \mathbb{Z}$ and $\equiv \pmod 5$ with the subset of all those (a, b) for which $a \equiv b \pmod 5$. That these identifications are circular (equality in $\mathbb{Z}$ being identified with the set of all pairs (a, b) for which a and b are equal!) is unimportant at the moment since we are still informally trying to feel our way. Now, being formal, but motivated by the above, we make

Definition 2.2.1 **A binary relation** on a set A is a subset of $A \times A$.

One easily solved and one unsolved problem in number theory are included in

Examples 2.2.2

(i) The binary relation $R_1 = \{(x, y): x \in \mathbb{Z}^+, y \in \mathbb{Z}^+, x \neq y \text{ and } x^y = y^x\}$ is a finite set comprising just the two elements $(2, 4)$ and $(4, 2)$. [Challenge: Prove it!]

(ii) The relation $R_2 = \{(x, y): x \in \mathbb{Z}^+, y \in \mathbb{Z}^+, y = x + 2 \text{ and } x, y \text{ both primes}\}$. Thus R_2 contains the pairs† $(3, 5)$, $(5, 7)$, $(11, 13)$, $\ldots$, $(1706595 \times 2^{11235} + 1)$, $(1706595 \times 2^{11235} - 1)$, $\ldots$

[It is an unsolved problem of number theory as to whether or not R_2 is a finite set. Each pair $\{x, y\}$ is called a pair of *twin primes*.]

(iii) The relation $R_3 = \{(x, y): x \in P, y \in P \text{ and } x \text{ is greatgrandfather of } y\}$. It seems unlikely that many readers will find themselves as first member (that is, in the x-place) of any element of R_3. (The author would certainly be pleased to hear from any such greatgrandfather! May 1991: I'm still waiting!)

Having shown (in 2.2.1) how we can give an unambiguous definition of binary relation in set-theoretic terms we make the notation more readable.

* See 0.3.1.
† As of September 1990.

Thus

Notation 2.2.3 Let R be a binary relation on a set A. If $(a_1, a_2) \in R$ we write $a_1 R a_2$ and say that a_1 **is related to** a_2 **under** (or **by**) R. Otherwise we write $a_1 \not R a_2$ and speak accordingly.

On any given set few (if any!) binary relations will have mathematical significance. Indeed, those binary relations occurring most frequently in practice possess special properties including some or all of:*

Properties 2.2.4

(r)	$\forall x \in A$	xRx
(s)	$\forall x, y \in A$	$xRy \Rightarrow yRx$
(t)	$\forall x, y, z \in A$	xRy and $yRz \Rightarrow xRz$
(a)	$\forall x, y \in A$	xRy and $yRx \Rightarrow x = y$

Definition 2.2.5

If R satisfies (r) we say that R is a **reflexive** binary relation.
If R satisfies (s) we say that R is a **symmetric** binary relation.
If R satisfies (t) we say that R is a **transitive** binary relation.
If R satisfies (a) we say that R is an **antisymmetric** binary relation.

If R satisfies (r), (s) and (t) we say that R is an **equivalence relation.**
If R satisfies (r), (t) and (a) we say that R is an **order relation.**

Note 2.2.6 To remember these latter two definitions the author uses the mnemonics rEst and rOta. You will, of course, remember your own mnemonics all the better.

Examples 2.2.7

(i) Since for all $x, y, z \in \mathbb{Z}$ we have: (r) $x \equiv x \pmod 5$; (s) If $x \equiv y \pmod 5$ then $y \equiv x \pmod 5$; (t) If $x \equiv y \pmod 5$ and $y \equiv z \pmod 5$ then $x \equiv z \pmod 5$ we see that $\equiv \pmod 5$ is an equivalence relation on $\mathbb{Z}$. $\equiv \pmod 5$ does not satisfy (a).

(ii) The relation $\leqslant$ on $\mathbb{Z}$ satisfies (r), (t) and (a) and thus is an order relation. [This basic example motivates the use of the words 'order relation' in 2.2.5.]

(iii) The relationship 'is brother of' on the set of all male human beings satisfies (s) but, in normal parlance, not (r). What about (t) and (a)?

(iv) The relation $\{(a, a), (a, b), (b, a), (b, b), (b, c), (c, c)\}$ on the set $\{a, b, c\}$ satisfies only (r).

(v) On the set of all triangles in the plane the relationships of similarity and of congruence are equivalence relations.

(vi) On $\mathbb{Z} \times (\mathbb{Z} \setminus \{0\})$ the relation R defined by $(a, b)R(c, d)$ iff $ad = bc$ is an equivalence relation of a kind which we shall meet again in the proof of 3.10.3.

* $\forall$ means 'for all'. See Section 0.4.

Problems 1

(i) Equality (on $\mathbb{Z}$, say) satisfies (r), (s), (t) and (a). Is it the only such binary relation on $\mathbb{Z}$?

(ii) Does the relationship 'is greatgrandfather of' on P satisfy (a) or not? [If not there must surely exist $x, y \in P$ such that xRy and yRx but $x \neq y$?]

Exercises

1 Show that on a finite set with n elements it is possible to define 2^{n^2} distinct binary relations.

2 Let $a, b, c, d, m, n \in \mathbb{Z}$ with $n > 0$. Show that if $a \equiv c \pmod n$ and if $b \equiv d \pmod n$ then (i) $a + b \equiv c + d \pmod n$; (ii) $ab \equiv cd \pmod n$; (iii) $ma \equiv mc \pmod n$. Given $ma \equiv mc \pmod n$ show that $a \equiv c \pmod n$ if $(m, n) = 1$. Give a specific example showing this conclusion can be false if $(m, n) \neq 1$.

3 Decide what it means to say that $x^5 + 3x^4 + x^3 + 5x^2 + x + 2 \equiv x^4 + 2x + 1 \pmod{x^2 + x + 1}$ and then decide if you believe this assertion.

4 For each of the following find, if possible, at least two solutions.
(i) $3x \equiv 7 \pmod 8$; (ii) $4x \equiv 7 \pmod 8$; (iii) $4x \equiv 6 \pmod 8$;
(iv) $71x \equiv 34 \pmod{31}$.

5 Which of the properties in 2.2.5 are satisfied by the following relations R on $\mathbb{Z}$?
(i) aRb iff $a < b$; (ii) aRb iff $a \leqslant b + 1$; (iii) aRb iff $ab = 0$; (iv) aRb iff $|a - b| \leqslant 1$.

6 Find binary relations R_1, R_2 on $\mathbb{Z}$ such that R_1 satisfies (s) and (t) but not (r), whilst R_2 satisfies (r) and (t) but not (s).

7 Which of (r), (s), (t), (a) does $R = \{(1, 1)\} \subseteq \mathbb{Z} \times \mathbb{Z}$ satisfy?

8 An equivalence relation E on $A = \{1, 2, 3, 4\}$ contains the pairs $(1, 1)$, $(1, 2)$, $(2, 3)$. Find E given that E is not the whole of $A \times A$.

9 Find as small a non-empty subset R of $\mathbb{Z} \times \mathbb{Z}$ as possible so that R is not (r), not (s) and not (t).

10 What is wrong with the following 'proof' that each symmetric and transitive binary relation on a set A is an equivalence relation?

$aRb \Rightarrow bRa$ (by (s)), aRb and $bRa \Rightarrow aRa$ (by (t)), therefore aRa so that R is (r). QED.

[Hint: look at your solution to R_1 in exercise 6 above.]

11 Let X be a non-empty set and $\mathscr{P}(X)$ the set of all its subsets. Show that the relation $\subseteq$ of inclusion between the elements of $\mathscr{P}(X)$ is an order relation on $\mathscr{P}(X)$. Let $X = \{a, b, c, d\}$. How many elements (i.e. ordered pairs) does the relation $\subseteq$ on $\mathscr{P}(X)$ have? Find $u, v \in \mathscr{P}(X)$ such that neither $u \subseteq v$ nor $v \subseteq u$.

12 Let R_1, R_2 be binary relations on a set S. Define their product $R_1 \circ R_2$ by: For $a, b \in S$, $a(R_1 \circ R_2)b$ iff there exists $c \in S$ such that aR_1c and cR_2b.

If R_1, R_2 are equivalence relations is $R_1 \circ R_2$? If R_1, R_2 are order relations is $R_1 \circ R_2$?

2.3 Equivalence relations and partitions

Equivalence relations are often more readily comprehended and also constructed in terms of the concept of partition.

Definition 2.3.1 Let A be any non-empty set and let ζ be any collection of non-empty subsets of A such that (i) the union of the sets in ζ is A and (ii) each distinct pair of sets in ζ has empty intersection. Then ζ is called a **partition** of A.

Examples 2.3.2
(i) The subsets of even integers (include 0) and of odd integers define a partition of $\mathbb{Z}$.
(ii) The sets of all males, of all females and of all joggers do not form a partition of the human race.
(iii) The concepts of similarity and congruence define partitions on the set T of all triangles in the plane.

Partitions give rise in a natural way to equivalence relations because of

Theorem 2.3.3 Let A be a non-empty set and let ζ be a partition of A. Define a relation R on A by setting, for $a, b \in A$, aRb iff a lies in the same element of ζ as b. Then R is an equivalence relation on A.

Proof
(r) Each a in A lies in some subset of A belonging to ζ. Then aRa follows immediately [since a lies in the same element of ζ as does a!].
(s) is also immediate. [If a belongs to the same element of ζ as b then b belongs to the same element of ζ as a.] Thus $aRb \Rightarrow bRa$.
(t) If a belongs to the same member of ζ as b and if b belongs to the same member of ζ as c then . . . [can you finish it?]. That is, aRb and $bRc \Rightarrow aRc$.

On the other hand, each equivalence relation (e.r. for short) gives rise naturally to a partition.

Theorem 2.3.4 Let R be an e.r. on the non-empty set A. To each $a \in A$ define $\hat{a}$ to be the subset $\{x : x \in A \text{ and } xRa\}$ of A. [Thus $\hat{a}$ is the subset of A comprising all elements x of A which are related to a under R.] Then the set $\zeta = \{\hat{a} : a \in A\}$ is a partition of A.

Proof Let $a \in A$. We know that aRa [why?]. Hence [immediately from the definition] $a \in \hat{a}$. [Thus ζ satisfies condition (i) of 2.3.1.] Next suppose $\hat{a}$ and $\hat{b}$ are such that $\hat{a} \cap \hat{b} \neq \varnothing$. Then there exists $c \in \hat{a} \cap \hat{b}$. It follows that cRa and cRb and consequently that bRc [why?]. Now, for each $d \in \hat{b}$ we have dRb. But then dRb and bRc and $cRa \Rightarrow dRa$ [Why? See exercise 8] whence $d \in \hat{a}$. Since d was any element of $\hat{b}$ we have shown that $\hat{b} \subseteq \hat{a}$. An identical argument* shows that $\hat{a} \subseteq \hat{b}$ and the equality $\hat{a} = \hat{b}$ follows.

We have thus shown that if two elements of ζ have non-trivial intersection then they are actually equal. In other words, distinct members of ζ have pairwise empty intersection. In particular if $y \in \hat{a}$ then $\hat{a} = \hat{y}$, even though a and y might not be equal (cf. 2.3.7).

There is a small point you might care to ponder.

Problem 2 We've seen that every partition on a set A gives rise to an e.r. on A and *vice versa*, each in a natural way. Suppose I give you a partition, you construct the corresponding e.r. and then I construct the partition corresponding to your e.r. Do I obtain the partition I began with?

Terminology 2.3.5 Given an e.r. R on a set A the members of the corresponding partition of A are called the **equivalence classes** of R.

Examples 2.3.6
(i) The relation R defined on the plane by setting $(x_1, y_1)R(x_2, y_2)$ iff $x_1^2 + y_1^2 = x_2^2 + y_2^2$ is an e.r. The equivalence classes are the circles with centre the origin.
(ii) The relation R defined on $\mathbb{Q}[x]$ by setting fRg iff $f(0) = g(0)$ is an e.r. on $\mathbb{Q}[x]$. Each class comprises all those polynomials with a particular constant term.

We now return to our most important example, that of **congruence mod *n*** on $\mathbb{Z}$.

Example 2.3.7 For each $n \in \mathbb{Z}^+$, $\equiv (\text{mod } n)$ is an e.r. In particular, when $n = 5$ there are five equivalence classes, namely

$$\hat{0} = \{\ldots -10, -5, 0, 5, 10, \ldots\}$$

$$\hat{1} = \{\ldots \ -9, -4, 1, 6, 11, \ldots\}$$

$$\hat{2} = \{\ldots \ -8, -3, 2, 7, 12, \ldots\}$$

$$\hat{3} = \{\ldots \ -7, -2, 3, 8, 13, \ldots\}$$

$$\hat{4} = \{\ldots \ -6, -1, 4, 9, 14, \ldots\}$$

* As an old schoolteacher of mine (DTC) used to say: 'Let symmetry work *for* you'.

Note that
(i) two integers a, b lie in the same (equivalence) class iff $a \equiv b \pmod 5$, that is iff $5|a-b$ or, again, iff $a = b + 5k$ for some $k \in \mathbb{Z}$; and that
(ii) although the notation $\hat{0}$, $\hat{1}$, $\hat{2}$, $\hat{3}$, $\hat{4}$ is most natural, these equivalence classes can also be denoted, respectively, by, for instance, $\widehat{75}$, $\widehat{101}$, $\widehat{-8}$, $\widehat{413}$ and $\widehat{-6}$.

Exercises

1 Define on $\mathbb{Z}$ a binary relation (other than $\equiv \pmod 5$) which also has exactly 5 equivalence classes. [This exercise is supposed to help substantiate the claim in the opening sentence of Section 2.3!]

2 Define, by means of a partition, an e.r. on $\mathbb{Z}$ which has, for each positive integer n, exactly one equivalence class with n elements. Describe this equivalence relation in the form 'xRy iff...'.

3 Describe the equivalence classes of the e.r. E in exercise 2.2.8.

4 Define in the form '$(x_1, y_1)R(x_2, y_2)$ iff ...' the equivalence relations on $\mathbb{R} \times \mathbb{R}$ whose equivalence classes are:
(i) all circles with centre (1, 1).
(ii) all squares with vertices (r, r), $(-r, r)$, $(r, -r)$, $(-r, -r)$; $r \in \mathbb{R}$.

5 How many distinct equivalence relations can be defined on the set $\{a, b, c, d, e\}$?

6 On $\mathbb{C}\backslash\{0\}$ define the e.r. R_1 by $z_1R_1z_2$ iff* $z_1|z_2| = z_2|z_1|$. Describe the equivalence classes geometrically. Do the same for R_2 defined by $z_1R_2z_2$ iff $z_1|z_2|^2 = z_2|z_1|^2$.

7 Let R be the e.r. defined on $\mathbb{R} \times \mathbb{R}$ by setting $(x_1, y_1)R(x_2, y_2)$ iff $x_1 - x_2$ and $y_1 - y_2$ are both integers. Show that in each equivalence class there is exactly one point lying in the region $0 \leq x < 1$, $0 \leq y < 1$.

8 Suppose that R is a transitive relation on A. If $a, b, c, d \in A$ are such that aRb and bRc and cRd, show that aRd.

2.4 $\mathbb{Z}_n$

Let n be any positive integer. Form the set of equivalence classes determined by the e.r. $\equiv \pmod n$. We call this set $\mathbb{Z}_n$. We now show how to manufacture very interesting 'number' systems by introducing a kind of addition and multiplication into $\mathbb{Z}_n$. These number systems are not mere curiosities. Indeed certain of them have found application in coding theory and in statistics as well as being important in geometry and in algebra itself. ([7], [61], [26] and 4.2.10.)

Using the notation of 2.3.4 let $\hat{s}$ and $\hat{t}$ be two elements of $\mathbb{Z}_n$. Thus $\hat{s}$ and $\hat{t}$ are equivalence classes of integers mod n. We define their 'sum' $\hat{s} \oplus \hat{t}$ and

* For $z = x + iy$, $|z|$ denotes the modulus of z, that is $|z| = \sqrt{x^2 + y^2}$.

'product' $\hat{s} \odot \hat{t}$ by

Definitions 2.4.1

$$\hat{s} \oplus \hat{t} = \widehat{s+t}$$

and

$$\hat{s} \odot \hat{t} = \widehat{s \cdot t}$$

Note that, whatever else, $\widehat{s+t}$ and $\widehat{s \cdot t}$ are certainly elements of $\mathbb{Z}_n$.

Taking the specific example of $n = 7$ we see that $\hat{4} \oplus \hat{6} = \widehat{10} = \hat{3}$ whilst $\hat{6} \odot \hat{6} = \widehat{36} = \hat{1}$. Information such as this is conveniently stored in the form of addition and multiplication tables as follows.

$\oplus$	$\hat{0}$	$\hat{1}$	$\hat{2}$	$\hat{3}$	$\hat{4}$	$\hat{5}$	$\hat{6}$
$\hat{0}$	$\hat{0}$	$\hat{1}$	$\hat{2}$	$\hat{3}$	$\hat{4}$	$\hat{5}$	$\hat{6}$
$\hat{1}$	$\hat{1}$	$\hat{2}$	$\hat{3}$	$\hat{4}$	$\hat{5}$	$\hat{6}$	$\hat{0}$
$\hat{2}$	$\hat{2}$	$\hat{3}$	$\hat{4}$	$\hat{5}$	$\hat{6}$	$\hat{0}$	$\hat{1}$
$\hat{3}$	$\hat{3}$	$\hat{4}$	$\hat{5}$	$\hat{6}$	$\hat{0}$	$\hat{1}$	$\hat{2}$
$\hat{4}$	$\hat{4}$	$\hat{5}$	$\hat{6}$	$\hat{0}$	$\hat{1}$	$\hat{2}$	$\hat{3}$
$\hat{5}$	$\hat{5}$	$\hat{6}$	$\hat{0}$	$\hat{1}$	$\hat{2}$	$\hat{3}$	$\hat{4}$
$\hat{6}$	$\hat{6}$	$\hat{0}$	$\hat{1}$	$\hat{2}$	$\hat{3}$	$\hat{4}$	$\hat{5}$

$\odot$	$\hat{0}$	$\hat{1}$	$\hat{2}$	$\hat{3}$	$\hat{4}$	$\hat{5}$	$\hat{6}$
$\hat{0}$	$\hat{0}$	$\hat{0}$	$\hat{0}$	$\hat{0}$	$\hat{0}$	$\hat{0}$	$\hat{0}$
$\hat{1}$	$\hat{0}$	$\hat{1}$	$\hat{2}$	$\hat{3}$	$\hat{4}$	$\hat{5}$	$\hat{6}$
$\hat{2}$	$\hat{0}$	$\hat{2}$	$\hat{4}$	$\hat{6}$	$\hat{1}$	$\hat{3}$	$\hat{5}$
$\hat{3}$	$\hat{0}$	$\hat{3}$	$\hat{6}$	$\hat{2}$	$\hat{5}$	$\hat{1}$	$\hat{4}$
$\hat{4}$	$\hat{0}$	$\hat{4}$	$\hat{1}$	$\hat{5}$	$\hat{2}$	$\hat{6}$	$\hat{3}$
$\hat{5}$	$\hat{0}$	$\hat{5}$	$\hat{3}$	$\hat{1}$	$\hat{6}$	$\hat{4}$	$\hat{2}$
$\hat{6}$	$\hat{0}$	$\hat{6}$	$\hat{5}$	$\hat{4}$	$\hat{3}$	$\hat{2}$	$\hat{1}$

Here the values of $\hat{a} \oplus \hat{b}$ and $\hat{a} \odot \hat{b}$ are exhibited at the intersection of the $\hat{a}$th row and the $\hat{b}$th column in the appropriate table.

Notes 2.4.2

(i) Our notation $\hat{s}$ is bad (but customary). It's bad because it does not indicate which $\mathbb{Z}_n$ one is looking at. However the context usually makes a more explicit notation (such as $\hat{s}_n$) unnecessary.

(ii) Corresponding tables for any other (fairly small) positive moduli n are easily constructed. What has perhaps escaped your attention is that our definitions of $\oplus$ and $\odot$ *might not make sense*! The problem is this: In $\mathbb{Z}_7$, say, we have seen that $\hat{4} \oplus \hat{6} = \widehat{10} = \hat{3}$ and that $\hat{6} \odot \hat{6} = \widehat{36} = \hat{1}$. Now $\hat{4}$ and $\hat{6}$ also go under the names $\widehat{39}$ and $\widehat{-925}$, for instance. Thus, according to our definitions,

$$\hat{4} \oplus \hat{6} = \widehat{39} \oplus \widehat{-925} = \widehat{-886}$$

whilst

$$\hat{6} \odot \hat{6} = \widehat{-925} \odot \widehat{-925} = \widehat{855625}$$

and so, unless $\widehat{-886}$ and $\widehat{855625}$ are just other names for $\hat{3}$ and $\hat{1}$ we are in the intolerable situation that a sum or product of two classes depends not only on which the classes *are* but also on *what we choose to call them*!

Fortunately everything works out all right (2.4.3), a situation we describe by saying that $\oplus$ and $\odot$ are **well defined**. Thus

Theorem 2.4.3 The $\oplus$ and $\odot$ of 2.4.1 are well defined.

Proof Suppose $\hat{a} = \hat{c}$ and $\hat{b} = \hat{d}$ in $\mathbb{Z}_n$. Then $a = c + un$, $b = d + vn$ for suitable $u, v \in \mathbb{Z}$. It follows that

$$\widehat{a+b} = \widehat{c + un + d + vn} = \widehat{c + d + (u+v)n} = \widehat{c+d}$$

whilst

$$\widehat{ab} = \widehat{(c+un)(d+vn)} = \widehat{cd + (cv + ud + unv)n} = \widehat{cd}$$

as required.

Problem 3 We could have avoided the well-definedness problem as follows. Denoting the elements of $\mathbb{Z}_n$ by $\hat{0}, \hat{1}, \ldots, \widehat{n-1}$ we define sum and product unambiguously by $\hat{x} \oplus \hat{y} = \hat{r_1}$ and $\hat{x} \odot \hat{y} = \hat{r_2}$ where, using 1.4.5, $x + y = m_1 n + r_1$ and $xy = m_2 n + r_2$ with $m_1, m_2, r_1, r_2 \in \mathbb{Z}$ and where $0 \leqslant r_1 < n$ and $0 \leqslant r_2 < n$. However, one rarely gets something for nothing in mathematics and another small difficulty now replaces the well-definedness one. Can you see what it is? [Hint: Try proving $(\hat{a} \odot \hat{b}) \odot \hat{c} = \hat{a} \odot (\hat{b} \odot \hat{c})$.]

We now see which of the (analogues of the) axioms A1, A2, . . . , M3, M4, and property Z, listed in Section 1.2, hold for $\oplus$ and $\odot$ on $\mathbb{Z}_n$. (We leave axioms D, P and properties C, M of Section 1.2 to exercise 11.)

For A1 we take any two $\hat{a}$, $\hat{b}$ in $\mathbb{Z}_n$ and ask: Is $\hat{a} \oplus \hat{b} = \hat{b} \oplus \hat{a}$? Indeed it is. For $\hat{a} \oplus \hat{b} = \widehat{a+b}$ [by definition] whereas $\hat{b} \oplus \hat{a} = \widehat{b+a}$. Since $a + b = b + a$ in $\mathbb{Z}$, we have $\widehat{a+b} = \widehat{b+a}$ in $\mathbb{Z}_n$ as required. The proof of M1 is similar. Let us offer a streamlined proof that M2 holds. We leave to the reader the explanation as to why each of the asserted equalities holds.

For all $\hat{a}, \hat{b}, \hat{c} \in \mathbb{Z}_n$ we have

$$\begin{aligned}(\hat{a} \odot \hat{b}) \odot \hat{c} &= \widehat{ab} \odot \hat{c} = \widehat{(ab)c} \\ &= \widehat{a(bc)} = \hat{a} \odot \widehat{bc} = \hat{a} \odot (\hat{b} \odot \hat{c}).\end{aligned}$$

The proof of A2 is similar.

Note 2.4.4 The rather pedantic observation that $(ab)c = a(bc)$ is included here only to emphasise that the associativity of $\odot$ depends heavily on that of $\cdot$ in $\mathbb{Z}$. Otherwise such pedantry relating to $\mathbb{Z}$ was left behind as long ago as 1.2.5.

Since for all $\hat{a} \in \mathbb{Z}_n$, we have $\hat{0} \oplus \hat{a} = \hat{a} \oplus \hat{0} = \hat{a}$ and $\hat{1} \odot \hat{a} = \hat{a} \odot \hat{1} = \hat{a}$ we see that the analogues of A3 and M3 hold in $\mathbb{Z}_n$. So does A4. For, given $\hat{a} \in \mathbb{Z}_n$, it is clear that $\hat{a} \oplus \widehat{-a} = \widehat{-a} \oplus \hat{a} = \hat{0}$.

We move on to M4 and Z. A quick glance at the $\odot$ table for $\mathbb{Z}_7$ shows that both hold in $\mathbb{Z}_7$. Consider however the $\odot$ table for $\mathbb{Z}_6$. It is, omitting circumflexes just for simplicity,

$\oplus$	0	1	2	3	4	5
0	0	1	2	3	4	5
1	1	2	3	4	5	0
2	2	3	4	5	0	1
3	3	4	5	0	1	2
4	4	5	0	1	2	3
5	5	0	1	2	3	4

$\odot$	0	1	2	3	4	5
0	0	0	0	0	0	0
1	0	1	2	3	4	5
2	0	2	4	0	2	4
3	0	3	0	3	0	3
4	0	4	2	0	4	2
5	0	5	4	3	2	1

It is clear that $\hat{0}$ and $\hat{1}$ are again the (unique) elements required by axioms A3 and M3. Further: (i) $\hat{2} \neq \hat{0}$ and $\hat{3} \neq \hat{0}$ whilst $\hat{2} \odot \hat{3} = \hat{6} = \hat{0}$, and (ii) there is no element $\hat{x} \in \mathbb{Z}_6$ for which $\hat{2} \odot \hat{x} = \hat{1}$. Thus both M4 and Z fail in $\mathbb{Z}_6$.

What about M4 and Z for other $\mathbb{Z}_n$s? It is pretty clear that the reason Z fails in $\mathbb{Z}_6$ is that 6 is a composite integer.* It is equally clear that Z fails in $\mathbb{Z}_n$ whenever n is composite. Further, if $n = n_1 n_2$ with $n_1 > 1$ and $n_2 > 1$ it is immediate that there can be no $\hat{x} \in \mathbb{Z}_n$ such that $\hat{n}_1 \odot \hat{x} = \hat{1}$.

We have, therefore,

Theorem 2.4.5 Let n be a composite positive integer. Then M4 and Z both fail in $\mathbb{Z}_n$.

On the other hand we can prove

Theorem 2.4.6 Let p be a positive prime. Then $\mathbb{Z}_p$ satisfies both M4 and Z.

Proof

(a) First let $\hat{a} \in \mathbb{Z}_p$ where $\hat{a} \neq \hat{0}$. Then $p \nmid a$ in $\mathbb{Z}$. [Why not?] Hence $(p, a) = 1$ [why?]. Consequently, by 1.4.9 there exist $r, s \in \mathbb{Z}$ such that $rp + sa = 1$. There then follows $\hat{1} = \widehat{rp + sa} = \widehat{sa}$ [why?] $= \hat{s} \odot \hat{a}$. Thus $\hat{a} \odot \hat{s} = \hat{s} \odot \hat{a} = \hat{1}$ so that $\mathbb{Z}_p$ satisfies M4.

(b) Now suppose $\hat{b}, \hat{c} \in \mathbb{Z}_p$ are such that $\hat{b} \odot \hat{c} = \hat{0}$. Then either $\hat{b} = \hat{0}$ [in which case there is nothing left to prove] or $\hat{b} \neq \hat{0}$. In this case we can, by part (a), find $\hat{t} \in \mathbb{Z}_p$ such that $\hat{t} \odot \hat{b} = \hat{1}$. Then $\hat{0} = \hat{t} \odot \hat{0} = \hat{t} \odot (\hat{b} \odot \hat{c}) = (\hat{t} \odot \hat{b}) \odot \hat{c} = \hat{1} \odot \hat{c} = \hat{c}$. Thus the assumption $\hat{b} \odot \hat{c} = 0$ leads either to $\hat{b} = \hat{0}$ or to $\hat{c} = \hat{0}$ whence property Z is seen to hold for $\mathbb{Z}_p$.

Problem 4 You may, even at first reading, have an uneasy feeling that the proof of (b) was not quite what you were expecting and that a more direct proof should be available. Indeed there is such a proof. Can you find it? (The reason for giving the present proof will appear in exercise 3.3.5.)

* A composite integer? Any integer which is different from 0, 1 and -1 and is not a prime.

Remarks We have just seen that in some ways the number systems $\mathbb{Z}_p$, where p is a prime, are more like $\mathbb{Q}$, $\mathbb{R}$ and $\mathbb{C}$ than is $\mathbb{Z}$ itself in that all of A1, A2, ..., M3, M4, D, Z hold in each of them whereas M4 does not hold in $\mathbb{Z}$. Indeed $\mathbb{Z}_p$, $\mathbb{Q}$, $\mathbb{R}$, $\mathbb{C}$ are all examples of *fields* (see 3.2.2(10)) whereas $\mathbb{Z}$ is not.

Exercises

1 Compute the $\oplus$ and $\odot$ tables for $\mathbb{Z}_2$, $\mathbb{Z}_3$, $\mathbb{Z}_4$. In the tables for $\mathbb{Z}_2$ replace every occurrence of $\hat{0}$ by the letter E and every occurrence of $\hat{1}$ by the letter O. Compare the result with exercise 1.2.15. Are you surprised? Should you be?

2 Solve $x^2 \equiv 1 \pmod 5$, $x^2 \equiv 1 \pmod{17}$ and $x^2 \equiv 1 \pmod 8$.

3 Find the number of distinct solutions to the equation $x^2 \oplus \widehat{-1} = \hat{0}$ in $\mathbb{Z}_5$, in $\mathbb{Z}_{17}$ and in $\mathbb{Z}_8$. In the light of 1.11.4 do you find the answer in the case of $\mathbb{Z}_8$ surprising? Why or why not?

4 (a) Carry out the proof of 1.11.4 with $\mathbb{Z}_p$ replacing every occurrence of J. Does the argument go through? What conclusion can you draw about roots of elements in $\mathbb{Z}_p[x]$?
(b) Comment on the assertion that the element $x^4 \oplus \hat{3}x^3 \oplus \hat{2}x \oplus \hat{7}$ of $\mathbb{Z}_{71}[x]$ has 9 roots in $\mathbb{Z}_{71}$.

5 Find the number of solutions in $\mathbb{Z}_7$ to (i) $x^2 \oplus x \oplus \hat{4} = \hat{0}$; (ii) $x^2 \oplus \hat{2}x \oplus \hat{4} = \hat{0}$; (iii) $x^2 \oplus \hat{3}x \oplus \hat{4} = \hat{0}$.

6 Show that in $\mathbb{Z}_4[x]$ there is a polynomial P such that $(\hat{2}x^2 \oplus \hat{2}x \oplus \hat{3}) \odot P = \hat{1}$.

7 Find all x such that $x^7 \equiv x \pmod 7$ and all x such that $x^6 \equiv 3 \pmod 7$. [Hint: For the second part, use the first part and the last but one sentence in exercise 2.2.2.]

8 Find, if possible, an element $\hat{a}$ such that $\hat{0}, \hat{a}, \hat{a}^2, \hat{a}^3, \hat{a}^4, \hat{a}^5, \hat{a}^6$ exhaust all of $\mathbb{Z}_7$.

9 Find an n and elements $\hat{a}, \hat{b}, \hat{c}$ in $\mathbb{Z}_n$ such that none of $\hat{a} \odot \hat{b}$, $\hat{b} \odot \hat{c}$ and $\hat{c} \odot \hat{a}$ is equal to $\hat{0}$ and yet $\hat{a} \odot \hat{b} \odot \hat{c} = \hat{0}$.

10 Find, if possible, a multiplicative inverse for $\hat{8}$ in each of (i) $\mathbb{Z}_5$; (ii) $\mathbb{Z}_{13}$; (iii) $\mathbb{Z}_{21}$; (iv) $\mathbb{Z}_{34}$; (v) $\mathbb{Z}_{341}$.

11 True or false?
(i) $\mathbb{Z}_7$ has elements $\widehat{-3}, \widehat{-2}, \widehat{-1}, \hat{0}, \hat{1}, \hat{2}, \hat{3}$ and the elements $\hat{1}, \hat{2}, \hat{3}$ show that $\mathbb{Z}_7$ satisfies P of Section 1.2.
(ii) Each $\mathbb{Z}_n$ satisfies property M and axiom D.
(iii) Each $\mathbb{Z}_n$ satisfies property C. [Hint: try n composite.]

12 The binary relation R on $\mathbb{Q}$ is defined by setting $a/bRc/d$ iff $a/b - c/d \in \mathbb{Z}$. Show that R is an e.r. on $\mathbb{Q}$. For equivalence classes $\widehat{a/b}$ and $\widehat{c/d}$ define $\widehat{a/b} \odot \widehat{c/d} = \widehat{ac/bd}$. Show that $\odot$ is not well defined.

Carl Friedrich Gauss *(30 April 1777 – 23 February 1855)*
Gauss, called by his contemporaries the Prince of Mathematicians, was one of the greatest scientists ever. Not only did he do stupendous work in many areas of pure mathematics but he also devoted much time to probability, theory of errors, geodesy, mechanics, electromagnetism, optics and even actuarial science! He was born in Braunschweig (Brunswick) Germany, to a poor family. His mother was intelligent but not fully literate, his father at various times a gardener, general labourer and merchant's assistant. It is said that, aged three, Gauss corrected an error in a wages list whilst, aged eight (some say ten), he wrote down in moments the answer to the following problem set in class: add together all the integers from 1 to 100. (One must presume he saw the answer *had* to be one half of 100×101.) Various sources claim that before age 20 he had, amongst other things, rediscovered and proved the law of quadratic reciprocity, discovered the double periodicity of elliptic functions, proved that every positive integer is a sum of 3 triangular numbers (these are integers of the form $n\dfrac{(n+1)}{2}$ where $n \in \mathbb{Z}^+$), formulated the principle of least squares and conjectured the prime number theorem (itself not proved until 1896). But what made him choose to devote himself to mathematics was his finding in 1795 of the criterion for a regular n-gon to be constructible using only compass and straightedge. Then for his doctoral thesis (1799) he gave the first ever proof of the Fundamental Theorem of Algebra (Section 4.8).

In 1801 he became famous in the eyes of the general public by locating the exact position of the newly discovered (and subsequently lost!) planet Ceres, this from very little information and where the best astronomers had failed. In 1807 he was appointed professor of astronomy and director of the Göttingen Observatory, a post he held until his death. 1801 also saw the publication of his classic book *Disquisitiones Arithmeticae*.

As one might expect, Gauss didn't collaborate much with others. One exception was a collaboration with Wilhelm Weber which produced in 1833 the first operating electric telegraph. In his unpublished notes Gauss anticipated major work of several mathematicians. In particular he considered non-Euclidean geometry before Lobatchevsky and Bolyai, quaternions before Hamilton, elliptic functions before Abel and Jacobi as well as much of Cauchy's complex variable theory. His revelations of priority were painful and, in the case of Bolyai, nearly disastrous.

It is reported in [134] that Gauss's friends found him cold, uncommunicative, ambitious for security and fame, and very conservative. It is said that he tried to dissuade his sons from studying mathematics so that the name of Gauss would remain synonymous with excellence.

There is a story that at the 50th anniversary celebrations of the award of Gauss' doctorate, Gauss was about to light his pipe with a page from the original *Disquisitiones*. Dirichlet, appalled by this, grabbed the page from Gauss and treasured it for the rest of his life.

2.5 Some deeper number-theoretic results concerning congruences

In this section we present (essentially) two number-theoretic results of which much use is made in many areas of mathematics. The second (2.5.4) may even surprise and delight you if you've not seen it before. We use 2.5.4 later (see 3.8.1) and 2.5.3 to solve exercises 2 and 3 in such a way as to show the power generated by a mere change in notation from $n|a-b$ to $a \equiv b \pmod n$. It was Gauss who first disclosed this power in his *Disquisitiones Arithmeticae*. The notation $\equiv$ was suggested to him because of the similarity of the properties of the relations of congruence and equality. The reader might care to recast all congruences in terms of divisibility and see if the results we obtain are then so obvious and their proofs so transparent. In particular he may care to flex his muscles by proving, without the use of congruences, that

(i) $341|2^{340}-1$; (ii) $641|2^{2^5}+1$;

(iii) the equation $x^2+y^2-15z^2=7$ has no solutions in integers x, y, z.

We begin by stating explicitly the famous Little Theorem of Fermat announced (but not proved) in a letter dated 1640. Euler* proved it in 1736 by the method of exercise 1.4.17 and later generalised it. Apparently Leibniz† was also in possession of a proof around 1683.

Theorem 2.5.1 (Fermat's Little Theorem) Let p be a (positive) prime and n a positive integer. Then $p|n^p-n$; in Gauss' notation $n^p \equiv n \pmod p$.

* Leonhard Euler (15 April 1707 – 18 September 1783); probably the greatest mathematician of the 18th Century. Even total blindness during the last 17 years of his life did not stop his tremendous output of mathematics.

† Gottfried Wilhelm Leibniz (1 July 1646 – 14 November 1716). Great mathematician and philosopher. Regarded as being a cofounder with Newton, his contemporary, of the calculus.

Note that if $(n, p) = 1$ we may deduce, using exercise 2.2.2, that $n^{p-1} \equiv 1 \pmod{p}$.

Conversely: if $n^{p-1} \equiv 1 \pmod{p}$ whenever $(n, p) = 1$, it follows from exercise 2.2.2(ii) that $n^p \equiv n \pmod{p}$ whenever $(n, p) = 1$. Since, trivially, $n^p \equiv n \pmod{p}$ whenever $p|n$, we see that 2.5.1 is equivalent to

Theorem 2.5.1′ Let p be a positive prime and let n be a positive integer such that $(n, p) = 1$ then $n^{p-1} \equiv 1 \pmod{p}$.

To get Euler's generalisation we first need

Definition 2.5.2 Let m be a positive integer. We denote by $\phi(m)$ the number of integers t such that (i) $1 \leqslant t \leqslant m$ and (ii) $(t, m) = 1$.

Thus $\phi(1) = 1$, $\phi(2) = 1$, $\phi(3) = 2$, $\phi(4) = 2$, $\phi(2520) = 576$ [clearly?] and $\phi(p) = p - 1$ for each prime p.

ϕ is called Euler's **ϕ-function** or **totient.**

Euler's Theorem can then be stated as:

Theorem 2.5.3 Let a and m be positive integers such that $(a, m) = 1$. Then $a^{\phi(m)} \equiv 1 \pmod{m}$.

Proof Let b_i $(1 \leqslant i \leqslant \phi(m))$ be the integers from 1 to m inclusive which are coprime to m. Since a is coprime to m so are all the ab_i $(1 \leqslant i \leqslant \phi(m))$ [why?]. If now for each i $(1 \leqslant i \leqslant \phi(m))$ we choose c_i so that $c_i \equiv ab_i \pmod{m}$ and $1 \leqslant c_i \leqslant m$, we see that each of the c_i is coprime to m [why?]. Further, if for $1 \leqslant j \leqslant k \leqslant \phi(m)$ we have $c_j = c_k$, then $ab_j \equiv ab_k \pmod{m}$. But then $b_j \equiv b_k \pmod{m}$ [why?] and hence $b_j = b_k$ [why?]. It follows that if $b_j \neq b_k$ then $c_j \neq c_k$. Since there are $\phi(m)$ distinct b_is there are $\phi(m)$ distinct c_is and so the c_is are just the b_is in some order.

Finally from all the $ab_i \equiv c_i \pmod{m}$ $(1 \leqslant i \leqslant \phi(m))$ we get on multiplying these congruences together,

$$ab_1ab_2 \ldots ab_{\phi(m)} \equiv c_1c_2 \ldots c_{\phi(m)} \pmod{m}$$

Since the b_is are the c_is in some order and since the b_is are coprime to m, we may cancel each b_i with the unique c_j equal to it to yield

$$a \cdot a \ldots a \equiv 1 \pmod{m}$$

that is $\quad a^{\phi(m)} \equiv 1 \pmod{m}$

Yet another proof accredited to Euler is found in exercise 9.

In 1770 the English mathematician Edward Waring* published his *Meditationes Algebraicae*. In it he announced, without proof, several new

* Edward Waring (b. 1736(?) – d. 15 August 1798).

results. One, which asserts that to each $s \in \mathbb{Z}^+$ there corresponds $k(s) \in \mathbb{Z}^+$ such that each positive integer is expressible as a sum of at most $k(s)$ positive integral sth powers, was first proved by Hilbert (in 1909!). There is a nice article on this in the *American Mathematical Monthly*, Vol. 78, 1971, pp. 10–36. We mentioned in Section 0.4 that Lagrange proved $k(2) = 4$ in 1770. In 1771 he proved another of Waring's assertions suggested to Waring by one of his students, John (later, Sir John) Wilson.*

Theorem 2.5.4 (Wilson's Theorem) Let p be a positive prime. Then $p \mid (p-1)! + 1$. In other words $(p-1)! \equiv -1 \pmod p$.

To make the proof easier to follow we first prove it in the case $p = 11$. In $\mathbb{Z}_{11}$ we have

$$\hat{2} \cdot \hat{6} = \hat{3} \cdot \hat{4} = \hat{5} \cdot \hat{9} = \hat{7} \cdot \hat{8} = \hat{1}$$

and, of course,

$$\widehat{10} = \widehat{-1}$$

Hence

$$\widehat{10!} = \hat{1} \cdot \hat{2} \cdot \hat{6} \cdot \hat{3} \cdot \hat{4} \cdot \hat{5} \cdot \hat{9} \cdot \hat{7} \cdot \hat{8} \cdot \widehat{-1} = \hat{1}^5 \cdot \widehat{-1} = \widehat{-1}$$

It follows that $10! \equiv -1 \pmod{11}$ as required.

The method of proof in the general case just copies this.

Proof of 2.5.4 The cases $p = 2, 3$ are easily dealt with directly. So we may assume $p \geq 5$. We consider the $p-3$ elements of the set $S = \{\hat{2}, \hat{3}, \ldots, \widehat{p-2}\} \subseteq \mathbb{Z}_p$ and we prove

(i) if $\hat{a} \in S$ then there exists $\hat{b} \in S$ such that $\hat{b} \neq \hat{a}$ and† $\hat{a}\hat{b} = \hat{1}$, and

(ii) if $\hat{a}\hat{b} = \hat{a}\hat{c} = \hat{1}$ then $\hat{b} = \hat{c}$. That is, the element $\hat{b}$ of (i) is unique.

We may then deduce that the elements of S resolve themselves into $\dfrac{p-3}{2}$ pairs, whose product in each case is $\hat{1}$. And since $\widehat{p-1} = \widehat{-1}$ in $\mathbb{Z}_p$ we shall be finished.

So let us prove (i) and (ii).

(i) Given $\hat{a} \in S$ we know from 2.4.6(a) that there exists $\hat{b} \in \mathbb{Z}_p$ such that $\hat{a}\hat{b} = \hat{1}$. Clearly $\hat{b} \neq \hat{1}$ and $\hat{b} \neq \widehat{-1}$ [why not?] and so $\hat{b} \in S$. Further $\hat{b} \neq \hat{a}$. For suppose it were. We should then have $\hat{a}\hat{a} = \hat{1}$, that is $a^2 \equiv 1 \pmod p$ or, again, $p \mid a^2 - 1$. Thus $p \mid a+1$ or $p \mid a-1$ in $\mathbb{Z}$. Recalling that $2 \leq a \leq p-2$ we see that the alleged divisions are impossible. This contradiction ensures $\hat{b} \neq \hat{a}$.

* John Wilson (6 August 1741–18 October 1793).

† For brevity we omit the multiplication sign $\odot$.

(ii) From $\hat{a}\hat{b} = \hat{a}\hat{c}$ we get $ab \equiv ac \pmod p$. Since $(a, p) = 1$ exercise 2.2.2 tells us that $b \equiv c \pmod p$. In other words $\hat{b} = \hat{c}$ in $\mathbb{Z}_p$.

Once again it seems that Leibniz knew this result 100 years before.

For an enjoyable journey through elementary number theory and its history the reader may consult [43]—or, perhaps, even [42]!

Exercises

1 (i) Find x such that $3^{60} = 29m + x$ where $0 \leqslant x \leqslant 28$. [Hint: $3^{28} \equiv ?$ (mod 29).]
(ii) Find x such that $0 \leqslant x \leqslant 18$ and $x \equiv 2^{100}$ (mod 19). [Hint: $2^{100} = 2^{64} \cdot 2^{32} \cdot 2^4$.]

2 Show that, mod 641, $5 \cdot 2^7 \equiv -1$, that $625 \cdot 2^{28} = 5^4 2^{28} \equiv 1$ and hence that $0 \equiv 1 + 16 \cdot 2^{28} = 1 + 2^{32}$. Thus $641 | 2^{2^5} + 1$. (This example, found by Euler,* shows Fermat's belief that $F(n) = 2^{2^n} + 1$ is prime for each $n \in \mathbb{Z}^+ \cup \{0\}$ is false. Fermat knew $F(n)$ to be prime for $n = 0, 1, 2, 3, 4$. No other prime $F(n)$ has ever been found.)

3 Show that $2^{10} \equiv 1$ (mod 11) and that $2^{10} \equiv 1$ (mod 31). Deduce that $2^{340} \equiv 1$ (mod 11), that $2^{340} \equiv 1$ (mod 31) and hence that $2^{340} \equiv 1$ (mod 341). (This provides a counterexample to the supposed claim of ancient Chinese mathematicians that n is prime if and only if $n | 2^n - 2$. On the other hand no smaller n provides a counterexample.)

4 Show that if n is an odd integer then $n^2 \equiv 1$ (mod 8). Deduce that the sum of 3 (integer) squares cannot be congruent to 7 (mod 8). Show that if $x^2 + y^2 - 15z^2 = 7$ then $x^2 + y^2 + z^2 \equiv 7$ (mod 8). Deduce that the equation $x^2 + y^2 - 15z^2 = 7$ has no integer solution.

5 Find the multiplicative inverses of $\hat{2}$, $\hat{3}$, $\hat{4}$, $\hat{5}$ and $\hat{6}$ in $\mathbb{Z}_{13}$. Hence verify Wilson's Theorem for $p = 13$.

6 Show that if $n \geqslant 2$ is an integer and if $n | (n-1)! + 1$ then n is prime. (This converse of Wilson's Theorem is useless as a means of finding large prime numbers.)

7 We know from 2.5.1′ that, for all a such that $p \nmid a$, $a^{p-1} \equiv 1 \pmod p$. Show that if d is the smallest positive integer such that $a^d \equiv 1 \pmod p$ then $d | p - 1$. [Hint: write $p - 1 = md + r$ where $0 \leqslant r < d$ and note that $a^r \equiv (a^d)^m a^r = a^{p-1} \equiv 1 \pmod p$.]

8 Let p be a positive prime. Show that $\phi(p^r) = p^{r-1}(p-1)$. [Hint: count!] It can be proved that if $m, n \in \mathbb{Z}^+$ and if $(m, n) = 1$ then $\phi(mn) = \phi(m)\phi(n)$.

* How on earth did Euler find this divisor? In fact he knew a theorem that told him that each prime divisor of $2^{2^5} + 1$ must be of the form $64k + 1$ (see [42, p. 85]).

Use this to prove that if $t = p_1^{\alpha_1} p_2^{\alpha_2} \dots p_s^{\alpha_s}$ then $\phi(t) = t\left(1 - \dfrac{1}{p_1}\right)\left(1 - \dfrac{1}{p_2}\right)\dots$ $\left(1 - \dfrac{1}{p_s}\right)$. Hence show that $\phi(2520) = 576$.

9 Prove Euler's Theorem directly from Fermat's as follows.

Let $m = p_1^{\alpha_1} p_2^{\alpha_2} \dots p_r^{\alpha_r}$ and let $(a, m) = 1$. Let p^α be any of the factors $p_i^{\alpha_i}$ of m. Then there exist $h_1, h_2 \in \mathbb{Z}$, such that $a^{p-1} = 1 + h_1 p$ and hence $a^{p(p-1)} = 1 + h_2 p^2$. Prove by induction on α that $a^{p^{\alpha-1}(p-1)} \equiv 1 \pmod{p^\alpha}$. Deduce that $a^{\phi(m)} \equiv 1 \pmod{p_i^{\alpha_i}}$ for each i $(1 \leqslant i \leqslant r)$. Then deduce that $a^{\phi(m)} \equiv 1 \pmod{m}$.

2.6 Functions

We have so far met several examples of what the reader would probably regard as a 'function', but we have as yet given no definition, formal or otherwise, of the concept. The reader may feel he can get along quite nicely without any definition! If so, we should first remind him that we shall be interested here in functions other than those between two sets of numbers. Further, he might find it interesting (and salutory) to read (see, for example, [80], [131]) how such great mathematicians as Euler, d'Alembert* and Daniel Bernoulli† came, around 1750, to arguing about their respective solutions to the 'vibrating string' problem essentially because their ideas as to what constituted a function did not coincide. In 1755 Euler wrote: 'If some quantities depend upon others in such a way as to undergo change when the latter are changed then the former are called functions of the latter.' For Euler the quantities were numbers: today we have need to discuss functions between more general sets.

The study of functions was originally an offshoot of the study of properties of curves, geometrically defined, so it is interesting to see how a modern definition of function, expressed in terms of the set concept, is equivalent to what you would, in the case of a real valued function of one real variable, naturally think of as the *graph* of the function.

One definition of function often found in present-day texts is: 'A function is a rule which associates with each element of some set A a single element of a second set B.'

You might care to discuss with your friends:

(i) Is the word 'rule' any more self-explanatory than the word 'function'?
(ii) Can different rules lead to the same function?

We thus eschew the word 'rule' and opt for a more‡ precise definition of the function concept.

* Jean le Rond d'Alembert (17 November 1717 – 29 October 1783).
† Daniel Bernoulli (8 February 1700 – 17 March 1782). Son of John Bernoulli; nephew of James Bernoulli. The Bernoullis were probably the most illustrious family mathematics has produced.
‡ A somewhat stricter definiton of 'function' can be found on p. 5 of [21, vol. I].

Definition 2.6.1 A **function** (also called **map, mapping** or **transformation**) from a non-empty set A to a non-empty set B is a non-empty subset f of $A \times B$ such that for each $a \in A$ there exists exactly one $b \in B$ for which $(a, b) \in f$.

Remarks

(i) The words map, mapping and transformation reflect the geometric origin of the function concept.

(ii) The words map and mapping seem to have gained greater favour in algebra than the word function; from Chapter 3 on we shall almost invariably use one of these two words.

Examples 2.6.2

(i) The set $\{(x, x^2+1): x \in \mathbb{R}\}$ is a function from $\mathbb{R}$ to $\mathbb{R}$.

(ii) The set $\{(x, x^2+1): x \in \mathbb{R}\}$ is also a function from $\mathbb{R}$ to $\mathbb{R}^+$.

(iii) The set $\{(w, \text{first letter of } w): w \in W\}$, where W is the set of all words in the English language, is a function from W to the set L of all letters of the alphabet.

(iv) The set $\{(x, x^3): x \in \mathbb{R}\}$ is a function from $\mathbb{R}$ to $\mathbb{R}$.

(v) The set $\{(a, 4), (b, 2)\}$ is a function from $\{a, b\}$ to $\{1, 2, 3, 4, 5\}$.

(vi) If J is any one of $\mathbb{Z}$, $\mathbb{Q}$, $\mathbb{R}$, $\mathbb{C}$, $\mathbb{Z}_n$, $\mathbb{Z}[x]$ etc., addition and multiplication are functions from $J \times J$ to J.

(vii) $\{(a, 4), (b, 2)\}$ is not a function from $\{a, b, c\}$ to $\{1, 2, 3, 4, 5\}$. [Why not?]

(viii) $\{(a, 4), (b, 2), (a, 1)\}$ is not a function from $\{a, b\}$ to $\{1, 2, 3, 4, 5\}$. [Why not?]

Several technicalities arising from 2.6.1 will be needed throughout the text. We do not try to motivate them but simply gather them together for ease of reference.

Notations 2.6.3

(i) If $f \subseteq A \times B$ is a function from A to B we emphasise this by writing $f: A \to B$ or $A \xrightarrow{f} B$.

(ii) Given $(a, b) \in f$ the uniquely determined b is most frequently denoted by either $f(a)$ or af.

Notes 2.6.4

(i) Which of the notations introduced in 2.6.3(ii) is used is merely a matter of personal preference.* On the whole, algebraists seem to have a slight preference for the latter whilst analysts tend to prefer the former, perhaps because of tradition. The latter has an advantage when discussing compositions of functions (see 2.6.7) but as both are in common use we shall encourage the reader to use both by *using both ourselves*! We invite the reader, in his

* In 1.2.7 we have a function symbol (namely $|\ |$) which straddles its argument; in 2.7.3 we introduce function symbols which bisect their arguments!

verbal communications, to refer to the symbols $f(a)$ and af as *the value of f at a* or, simply, *f of a* and *the image of a under f* respectively.
(ii) It then becomes rather palatable to describe f as the function given by $a \xrightarrow{f} b$ or again as the function given by $af = b$, for all $a \in A$.
(iii) 2.6.3(ii) emphasises the difference between the symbol f denoting the function and the symbol $f(x)$ denoting the value of the function f at the element x of A.

Definitions 2.6.5 Let $f: A \to B$ be given. A is called the **domain** of f. If $S \subseteq A$ we write Sf for the set $\{sf : s \in S\}$ and call this set the **image** of S under f. The subset Af of B is often called the **range** of f. (Note that $Af \neq B$ is permitted.) If $Af = B$ we frequently say that f is **onto *B*.** If we don't know (or if we know but don't particularly care) that f is onto B, we often just say that f maps A **into *B*.** (Thus 'into' is just another word for 'to'. Please note that 'into' does *not* mean 'not onto'.)

The map $f: A \to B$ is **one to one** (briefly 1–1) iff from $(a_1, b) \in f$ and $(a_2, b) \in f$ it follows that $a_1 = a_2$. That is, f is 1–1 iff distinct elements of A map to distinct elements of B. In the contrary case where there exist $a_1, a_2 \in A$ such that $a_1 \neq a_2$ and yet $a_1 f = a_2 f$ we sometimes say f is **many–one.**

Two functions $f: A \to B$ and $g: C \to D$ are **equal** iff the sets f and g of ordered pairs coincide. That is, $f = g$ iff $A = C$ and $af = ag$ for each $a \in A$ $(= C)$. Note that $f = g$ is possible even if $B \neq D$.

Let $f: A \to B$ and let S be any non-empty subset of A. The subset $f_S = \{(s, sf): s \in S\}$ is a subset of f and is a function from S to B. (Informally f_S is just f except that the action of f on elements of A lying outside S is ignored.) It is called the **restriction** of f to S and is commonly denoted by $f|S$. We shall, incorrectly, write f_S as f. No confusion will result.

If $b_0 \in B$ is such that $af = b_0$ for all elements a in A we say that f is a **constant function**. If $A \subseteq B$ and if $f: A \to B$ is such that $af = a$ for all a in A then f is called the **inclusion function** of A in B. If $f: A \to A$ is such that $af = a$ for all $a \in A$ then f is called the **identity function** on A.

Again let $f: A \to B$. If $T \subseteq B$ we write Tf^{-1} for the set $\{s : sf \in T\}$. [Remember this definition by 'equating' $sf \in T$ with $s \in Tf^{-1}$ in your mind.] Tf^{-1} is called the **inverse image of *T* under *f*.** In case f is 1–1 and onto B then, for each $b \in B$, $\{b\}f^{-1}$ comprises a singleton $\{a\}$, say, where a is the unique element of A such that $af = b$. The subset $\{(af, a): a \in A\}$ of $B \times A$ is a 1–1 function from B onto A called the **inverse of *f*.** We denote it (there is only one such inverse; see exercise 12) by f^{-1}.

If $f: A \to B$ is 1–1 and onto B, f is sometimes called a **1–1 correspondence**. A 1–1 correspondence $f: A \to A$ is called a **permutation** on A.

Note 2.6.6 The terms **surjection** and **injection** are sometimes used to describe maps which are respectively onto and 1–1. If $f: A \to B$ is both 1–1 and onto, f is then called a **bijection.**

All examples relating to 2.6.5 are relegated to the exercises.

Now suppose $f: A \to B$ and $g: B \to C$ are functions. We define a new function from A to C which we denote by $f \circ g$ to indicate its dependence on f and g by: for all $a \in A$, $a(f \circ g) = (af)g$, the image of af under g. This may be represented pictorially by Fig. 2.1.

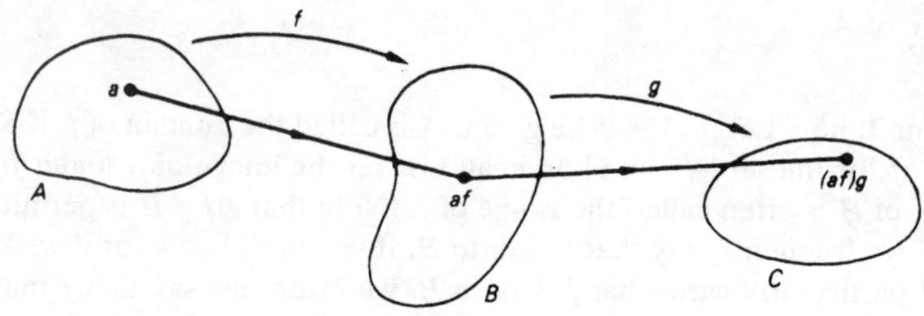

Fig. 2.1

We have

Definition 2.6.7 The function $f \circ g$ is called the **composite** of f and g. The operation $\circ$ combining f and g in this way is called **composition.**

Example 2.6.8 Let $f: \mathbb{R} \to \mathbb{R}$ and $g: \mathbb{R} \to \mathbb{R}$ be the functions given by $x \xrightarrow{f} x^3 + 1$, $x \xrightarrow{g} \cos x$ for all x. Then

$$x(f \circ g) = (xf)g = (x^3 + 1)g = \cos (x^3 + 1)$$

whilst

$$x(g \circ f) = (xg)f = (\cos x)f = (\cos x)^3 + 1$$

Thus $x \xrightarrow{f \circ g} \cos (x^3 + 1)$, $x \xrightarrow{g \circ f} \cos^3 x + 1$. It follows that $f \circ g \neq g \circ f$ since, in particular, $0(f \circ g) \neq 0(g \circ f)$.

Note 2.6.9 We now see one advantage of placing the function symbol to the right of the element it operates on, namely that the combined effect of f followed by g is naturally denoted by $f \circ g$. Putting the function symbol on the left leads to writing $g(f(a))$ as $(g \circ f)(a)$ so that $g \circ f$ denotes, once again, the action of f followed by g. This 'backwards' notation seems somewhat unnatural when dealing with permutations (see 5.3.6).

Exercises

1 Which of the following subsets of $\mathbb{R} \times \mathbb{R}$ are functions?
(i) $\{(x, y): x + y = 5\}$; (ii) $\{(x, y): x^2 + y = 5\}$;
(iii) $\{(x, y): x + y^2 = 5\}$; (iv) $\{(x, y): x^2 + y^2 = 5\}$.

2 (a) Let $A=\{1,2,3\}$, $B=\{a,b\}$. Which of the following subsets of $A\times B$ are functions?
(i) $\{(3,a),(1,b),(2,b)\}$; (ii) $\{(1,a),(2,a),(2,b),(3,a)\}$;
(iii) $\{(1,a),(2,a),(3,a)\}$.
(b) How many functions can be defined from A to B? From B to A?

3 Let $f=\{(x,x^2): x\in\mathbb{R}\}\subseteq\mathbb{R}\times\mathbb{R}$, $g=\{(x,x^2): x\in\mathbb{R}^+\}\subseteq\mathbb{R}^+\times\mathbb{R}^+$, $h=\{(x,x^2): x\in\mathbb{Z}^+\}\subseteq\mathbb{Z}^+\times\mathbb{Z}^+$. Determine which of the functions f, g, h are 1–1 and which are onto. What are the ranges of f, g and h?

4 Let $f:\mathbb{Z}\to\mathbb{Z}_p$ be given by $z\overset{f}{\to}\hat{z}$, $g:\mathbb{Q}\to\mathbb{Q}$ given by $xg=x^3-3x^2+3x+2$ and $h:\mathbb{R}\to\mathbb{R}$ given by $xh=x$ if x is a rational number, $xh=1-x$ if x is irrational. Which of the functions f, g, h are 1–1, and which are onto? [Hint for $g: xg=(x-1)^3+3$ so its graph (over $\mathbb{R}$) is not unlike that of $x\to x^3$.]

5 Given $f:A\to B$ some authors call B the **codomain** of f. Explain why, according to our definition, the use of the word 'the' is inappropriate.

6 Let $f:\mathbb{Z}_3\to\mathbb{Z}_3$ be given by $xf=x^4+x^2+\hat{1}$, $xg=\hat{2}x^2-\hat{2}$. Show that $f=g$. (Functions defined by polynomials are called **polynomial functions**.)

Suppose $f,g\in\mathbb{Q}[x]$ with $f\neq g$. Prove that the corresponding polynomial functions are distinct. [Hint: Use exercise 1.11.5.]

7 Let $f:\mathbb{R}\to\mathbb{R}$ be given by $x\overset{f}{\to}5-x^2$. Find $3f$, $\mathbb{R}^+f$, $\{0,7\}f$, $3f^{-1}$, $\mathbb{R}^+f^{-1}$, and $\{0,7\}f^{-1}$.

8 Let $f:A\to B$, let S, T be subsets of A, and U, V be subsets of B. Prove (let's change notation!): (i) $f(S\cup T)=f(S)\cup f(T)$; (ii) $f(S\cap T)\subseteq f(S)\cap f(T)$; (iii) $f^{-1}(U\cup V)=f^{-1}(U)\cup f^{-1}(V)$; (iv) $f^{-1}(U\cap V)=f^{-1}(U)\cap f^{-1}(V)$.

Show $f^{-1}(f(S))\supseteq S$. Is $f(f^{-1}(U))=U$? Give a specific example to show that $\subseteq$ may not be replaced by $=$ in (ii).

9 Let $f:A\to B$, $g:B\to C$ be functions.
(i) Prove that $f\circ g$ as defined above is indeed a function. Show that $f\circ g=\{(a,c)$: there exists $b\in B$ such that $(a,b)\in f$ and $(b,c)\in g\}$. (ii) Show that if f and g are 1–1 (respectively onto) then $f\circ g$ is 1–1 (respectively onto).

10 Let $f=\{(x,2x+1)\}\subseteq\mathbb{R}\times\mathbb{R}$ and $g=\{(x,x^2)\}\subseteq\mathbb{R}\times\mathbb{R}$. Show that $g\circ f=\{(x,2x^2+1)\}\subseteq\mathbb{R}\times\mathbb{R}$. Show $f\circ g\neq g\circ f$.

11 Let $f:A\to B$ be 1–1 and onto. Show that $f^{-1}=\{(af,a): a\in A\}$ is a function from B to A and that $f\circ f^{-1}=1_A$ and $f^{-1}\circ f=1_B$ where 1_A, 1_B are the identity functions on A and B respectively.

12 Let $f:A\to B$ and $g:B\to A$ be functions. Show that if $f\circ g=1_A$ then f is 1–1 and g is onto. Deduce that if, also, $g\circ f=1_B$ then f is 1–1 and onto B and $g=f^{-1}$ is the unique inverse of f.

13 Let A be a finite set. Show that $f:A\to A$ is 1–1 iff f is onto.

14 For $x \in \mathbb{R}$ let $[x]$ be the greatest integer less than or equal to x. Let $f = \{(z, 2z)\}$ and $g = \{(z, [z/2])\} \subseteq \mathbb{Z} \times \mathbb{Z}$. Show $f \circ g = 1_{\mathbb{Z}}$ but that $g \circ f \neq 1_{\mathbb{Z}}$. Can you find a similar example in which $\mathbb{Z}$ is replaced by a finite set?

15 It is clear that two finite sets A and B have the same number of elements iff there exists a bijection $f : A \to B$. Taking this over to infinite sets we say sets A and B are **equinumerous** iff there is a bijection $g : A \to B$. Show that the sets $\mathbb{Z}^+$ and $\mathbb{Z}^{(2)} = \{x^2 : x \in \mathbb{Z}^+\}$ are equinumerous. (This was observed by Galileo, 1564–1642.)

Any set equinumerous with $\mathbb{Z}^+$ is called **countable.** It can be shown that $\mathbb{Q}$ is and $\mathbb{R}$ is not countable (see [5, pp. 359–61]). Show that $\mathbb{Z}$ is countable.

Show that $\mathbb{R}$ and the interval $0 < x < 1$ are equinumerous.

Hint:*

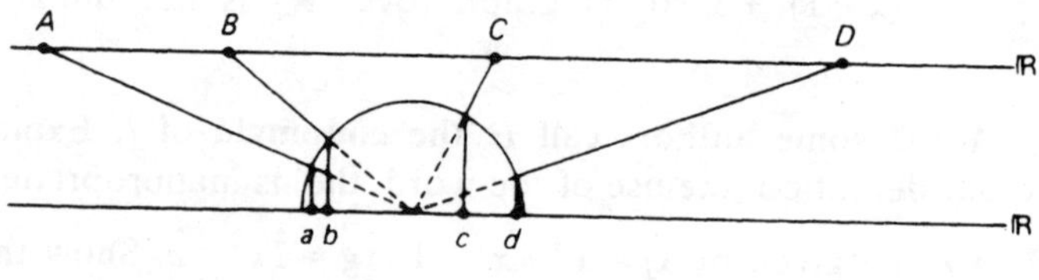

2.7 Binary operations

At last we arrive at

Definition 2.7.1 Let A be a non-empty set. A mapping $\rho : A \times A \to A$ is called **a binary operation** on A. If $S \subseteq A$ and if $(S \times S)\rho \subseteq S$ we say that S is **closed** under (or, with respect to) ρ.

Examples 2.7.2

(i) Addition, multiplication and subtraction on each of $\mathbb{Z}$, $\mathbb{Q}$, $\mathbb{R}$, $\mathbb{C}$, $\mathbb{Z}[x]$, $\mathbb{Q}[x]$, $\mathbb{Z}_n$, etc. are binary operations. Division is not a binary operation on any of these sets. [Why not?] Regarding $\mathbb{Z}^+$, $\mathbb{Q}^+$, $\mathbb{R}^+$, as subsets of $\mathbb{C}$, each is closed under the addition and multiplication on $\mathbb{C}$ but none is closed under subtraction.

Notation 2.7.3

(i) According to 2.7.1 and 2.6.3(ii) we should find ourselves writing $(3, 5)+ = 8$ instead of the more usual $3 + 5 = 8$. Since, due to familiarity, we are more at home with the second notation than the first we shall adopt it for all binary operations. That is, given ρ as in 2.7.1 we shall, from now on, write $a\rho b$ rather than $(a, b)\rho$.

(ii) If, as in 2.7.1, S is closed under ρ it follows that $\rho|S \times S$ is a binary operation on S. For psychological reasons and to stop pedantry getting out of hand, we replace $\rho|S \times S$ by the technically incorrect but intuitively more

* 'A diagram is worth ten thousand words' (A.W., Physics master).

palatable symbol ρ and speak accordingly of ρ as being a binary operation on S. Thus for s and t in S we write $s\rho t$ whether s and t are regarded as being in S or in A.

Examples 2.7.2
(ii) On the set V of all vectors in 3-dimensional space define ρ by: for all $v_1, v_2 \in V$, $v_1 \rho v_2 = v_1 \wedge v_2$, the vector product of v_1 and v_2.
(iii) On $\mathbb{Z}^+$ define ρ by $x\rho y = x^y$.
(iv) On $\mathcal{F}$, the set of all functions from $\mathbb{R}$ to $\mathbb{R}$, define $f_1 \rho f_2 = f_1 \oplus f_2$. (Here $f_1 \oplus f_2$ is that function from $\mathbb{R}$ to $\mathbb{R}$ defined by $x(f_1 \oplus f_2) = xf_1 + xf_2$ for all $x \in \mathbb{R}$.)
(v) On E, the set of all functions from the non-empty set X into itself, define ρ by $f_1 \rho f_2 = f_1 \circ f_2$ (as in 2.6.7).
(vi) On $\mathbb{Z}$ define ρ by $x\rho y = x + y - 7$.

The following chapters of this book deal with a variety of sets upon which are defined either one or two binary operations. In the most important cases these binary operations will satisfy several of the analogues of the axioms postulated for $\mathbb{Z}$ in Section 1.2. To ease the reader into the following chapters but, more important, to establish 2.7.6 and 2.7.7 we introduce, for general binary operations,

Definition 2.7.4 Let ρ be a binary operation on a set A. If for all $a, b \in A$ $a\rho b = b\rho a$ we say that ρ is **commutative**. If for all $a, b, c \in A$ we have $(a\rho b)\rho c = a\rho(b\rho c)$ we say that ρ is **associative**. If there exists in A an element e, say, such that $a\rho e = e\rho a = a$ for all $a \in A$ we say e is an **identity** (or **neutral**) **element** for ρ.

Examples 2.7.5 In examples 2.7.2, parts (ii) to (vi), ρ is commutative in parts (iv) and (vi), is associative in (iv), (v) and (vi) and possesses identity elements in (iv), (v) and (vi).
For example in (vi):

(a) $$x\rho y = x + y - 7 = y + x - 7 = y\rho x;$$

(b) $$(x\rho y)\rho z = (x + y - 7)\rho z = (x + y - 7) + z - 7$$

whereas $x\rho(y\rho z) = x\rho(y + z - 7) = x + (y + z - 7) - 7$;

Finally

(c) $$x\rho 7 = x + 7 - 7 = x = 7\rho x.$$

(Thus is 7 an identity element for ρ.)

Much more important is the verification that $\circ$ (in 2.7.2(v)) is associative.

Theorem 2.7.6 Composition of functions, when defined, is associative.

Proof Let $f: X \to Y$, $g: Y \to Z$, $h: Z \to T$ so that $f \circ g$, $g \circ h$ etc. all exist. Then $f \circ g$ is a function from X to Z and so $(f \circ g) \circ h$ is a function from X to T. Similarly $f \circ (g \circ h)$ is a function from X to T. Directly from the definition one finds, for all $x \in X$,

$$x\{(f \circ g) \circ h\} = (x(f \circ g))h = ((xf)g)h.$$

In a like manner $x\{f \circ (g \circ h)\} = (xf)(g \circ h) = ((xf)g)h$. Thus $(f \circ g) \circ h$ and $f \circ (g \circ h)$ act identically on all of X. Hence $(f \circ g) \circ h = f \circ (g \circ h)$ by the definition (see 2.6.5) of equality of functions.

That $\circ$ of 2.7.2(v) is associative follows on setting $X = Y = Z = T$.

Next a remark about associative binary operations. These comments have already been made in exercise 1.2.1 in connection with the binary operation $+$ on $\mathbb{Z}$.

Let ρ be a binary operation on a set A. For each ordered triple (a_1, a_2, a_3) of elements of A we can evaluate their product *in that order* either as $(a_1 \rho a_2) \rho a_3$ or $a_1 \rho (a_2 \rho a_3)$. If ρ is associative then, by definition, these two evaluations are equal and we can unambiguously denote the result by $a_1 \rho a_2 \rho a_3$. In the case of four elements there are five ways of evaluating their product, all of which yield the same result (copy the argument of exercise 1.2.1). We now prove generally

Theorem 2.7.7 (The Generalised Associative Law) If ρ is an associative binary operation on the set A then every way of evaluating the product of the ordered n-tuple of elements $a_1, a_2, \ldots, a_n$ of A taken in that order leads to the same result.

*Proof** We proceed by induction on n starting with the trivial case $n = 3$.

Now suppose the desired result proved for all n such that $3 \leq n \leq k - 1$ and let $a_1, a_2, \ldots, a_k$ be elements of A. The final step in any evaluation of a k-fold product is the evaluation of a product of the form $(a_1 \rho a_2 \rho \ldots \rho a_i) \rho (a_{i+1} \rho \ldots \rho a_k)$ for some i such that $1 \leq i \leq k - 1$ (see exercise 5). Because of the induction hypothesis, the products inside the brackets are unambiguous. Thus to prove the theorem we only need to show that for each i, j such that $1 \leq i \leq j \leq k - 1$ we have

$$(a_1 \rho \ldots \rho a_i) \rho (a_{i+1} \rho \ldots \rho a_k) = (a_1 \rho \ldots \rho a_j) \rho (a_{j+1} \rho \ldots \rho a_k) \tag{2.7.8}$$

If $i = j$ there is no problem. If $i < j$ we set $x = a_1 \rho \ldots \rho a_i$, $y = a_{i+1} \rho \ldots \rho a_j$ and $z = a_{j+1} \rho \ldots \rho a_k$. Now x, y, z are unambiguous by induction hypothesis. Thus 2.7.8 is valid provided that $x \rho (y \rho z) = (x \rho y) \rho z$. But this follows from the associativity of ρ, and the proof is complete.

* At the back of this proof is another logical subtlety. See the remarks concerning the definition of the Fibonacci sequence in Section 1.2 (p. 16).

Thus we have shown that for each n the product of n elements $a_1, a_2, \ldots, a_n$ can be defined unambiguously and can be denoted by $a_1\rho a_2\rho a_3 \ldots \rho a_n$ without brackets.

Finally we establish the so-called *laws of exponents* for positive integral powers. Writing a^2, a^3, etc. instead of $a\rho a$, $a\rho a\rho a$, etc., 2.7.8 leads to

Theorem 2.7.9 Let ρ be an associative binary operation on a set A and let $a \in A$. Setting a^n to be the (unambiguously defined) n-fold product $a\rho a\rho \ldots \rho a$ we have, for all $r, s \in \mathbb{Z}^+$,

$$a^r \rho a^s = a^{r+s}$$

and

$$(a^r)^s = a^{rs}.$$

Notes 2.7.10

(i) Of course, when $A = \mathbb{Z}$ and ρ is multiplication, the above merely repeats results we've known and used for a long time. The same is true when ρ is taken as addition except that, of course, a^n is then written na.

(ii) 2.7.9 will in general fail if the hypothesis of associativity is withheld. An example, already known to you from examples 2.7.2, is given in the following exercises.

Exercises

1 Show that the subset of odd integers of $\mathbb{Z}$ is closed under multiplication but not under addition. Can you find a subset of $\mathbb{Z}$ which is closed under addition but not under multiplication?

2 Confirm the assertions made in 2.7.5.

3 With ρ as in 2.7.2(iii), show that $(x\rho x)\rho x = x\rho(x\rho x)$ iff $x = 1$ or $x = 2$.

4 With ρ as in 2.7.2(v), show that ρ is commutative iff X has exactly one element.

5 Check the first assertion made in 2.7.7 about multiple products by removing all but the last two pairs of brackets from $((a_1\rho(a_2\rho a_3))\rho(a_4\rho a_5))\rho(a_6\rho(a_7\rho a_8))$.

6 Show by giving actual concrete examples that subtraction is neither commutative nor associative on $\mathbb{Z}$.

7 For each of the following binary operations on $\mathbb{Z}^+$ say whether or not it is associative, commutative, and whether or not it has an identity element. (i) $a * b = 2a + b^2$; (ii) $a * b =$ the larger of a and b; (iii) $a * b =$ the smaller of a and b; (iv) $a * b = 1$; (v) $a * b = a$; (vi) $a * b = |a - b| + 1$; (vii) $a * b = a^2 b$.

8 The following table defines part of the binary operation τ on the set $\{a, b, c, d\}$. Given that τ is commutative complete the table. [Hint: $a\tau c = a$ so $c\tau a = ?$]

τ	a	b	c	d
a	b	b	a	d
b	?	c	c	c
c	?	?	d	b
d	?	?	?	a

9 Show that there are, on a set containing n elements, precisely n^{n^2} binary operations. How many of them are commutative operations?

10 Let S be the set $\{a, N, x\}$ and let the operation $*$ on S be given by the table below. Determine whether or not $*$ is commutative, associative and whether or not it has a neutral element.

$*$	a	N	x
a	a	N	x
N	x	a	N
x	N	x	a

11 £ is a binary operation on the set W. Given that £ is commutative is it true that $((a_1£a_2)£a_3)£a_4 = ((a_4£a_3)£a_2)£a_1$, for all a_1, a_2, a_3, a_4 in W? A proof or a counterexample is required.

12 On $\mathbb{Z}\times\mathbb{Z}$ define & by $(a, b)\&(c, d) = (ac, bc + d)$. Check & for associativity, commutativity and identity element.

3

Introduction to rings

3.1 Introduction

In the Prologue we indicated some of the advantages of employing the abstract-axiomatic method in attempting to gain anything more than a superficial understanding of the many concrete examples (of numbers, of polynomials, etc.) which arise in an algebraic setting. A famous application of this method is found in the long and important paper *The Algebraic Theory of Fields* written by E Steinitz in 1910. In this paper Steinitz, beginning with an abstract definition of the concept of *field*,* attempted to bring some order to the multitude of concrete fields previously studied and set himself the task of finding all possible types of fields and the relationships between them. Aside from $\mathbb{Q}$, $\mathbb{R}$, $\mathbb{C}$ and all the $\mathbb{Z}_p$, the known fields included the algebraic number fields of Dedekind and the rational function fields of Kronecker (exercises 3.2.14 and 3.2.15), as well as the algebraic function fields of Dedekind and Weber and the fields of p-adic numbers of Hensel which we shall not consider here.

We observed in Section 2.4 that the set $\mathbb{Z}$ of all integers does not qualify as a field and there are many other concrete examples, each equipped with two binary operations of types akin to $+$ and $\cdot$ on $\mathbb{Z}$ which also fail for one reason or another to be fields. Amongst these are the sets $\mathbb{Z}[x]$, $\mathbb{Q}[x]$, etc., of polynomials, all the $\mathbb{Z}_m$ (m composite), Dirichlet's algebraic integers (exercise 3.2.14), the quaternions of Hamilton and the matrices of Cayley (Section 3.2), the algebras of Lie (exercise 3.2.3), Hensel's p-adic integers and the more general algebras of Grassmann and Peirce mentioned in the Prologue. Motivated by Hensel's and by Steinitz' work, A A Fraenkel in 1914 inaugurated a general investigation of the abstract structure underlying several of these examples. Such a system he called a *ring*,* following Hilbert's use of the term *Zahlring* (number ring) for sets of the form $\mathbb{Z}[\xi] = \{a + b\xi : a, b \in \mathbb{Z}\}$ where $\xi^2 + A\xi + B = 0$ with $A, B \in \mathbb{Z}$. (Since $\xi^2 = -A\xi - B \in \mathbb{Z}[\xi]$ we see that $\mathbb{Z}[\xi]$ 'closes up' (like a circle!) under multiplication.)

We shall begin our axiomatic study with a look at rings. After the remarks at the beginning of Section 2.6, it should not surprise you that little in mathematics is of totally permanent nature; even definitions are susceptible to change! Thus the notion of ring as currently defined is more general than that considered by Fraenkel. Since, however, technically speaking, the very

* Definition 3.2.2.

important Lie algebras do not qualify as rings (more accurately they are called non-associative rings) it may only be a matter of time before the present definition of ring is widened to incorporate them too.

To summarise briefly some features of this chapter: we begin with a formal definition of the concept of a ring and of several of its derivatives, including that of field. A long list of concrete examples indicates, in the spirit popular in the period 1900–1910, the independence of the axioms listed; although the real reason for giving this list is to show the reader that our theory does **cover a multitude of *essentially different* concrete examples. In Section 3.3 we** see first how a single abstract theorem can both summarise and verify a host of facts already intuitively known (3.3.1); and at a different level of sophistication supply an incisiveness hard to achieve by considering concrete examples alone (3.3.4). Later we shall see how the process of abstraction, by eliminating inessentials, suggests new concrete results which might otherwise have been overlooked, hidden by a mass of detail. We then apply these results to establish three assertions of the great number theorist P de Fermat. Finally, in Section 3.12, we shall see how the abstract method encourages us to ask the question: 'In what sense, if any, are the integers unique?' and how it helps us to supply a precise answer.

3.2 The abstract definition of a ring

As implied above, the set $\mathbb{Z}$ of integers is one of the 'models' for the abstract theory of rings. It is therefore no surprise that we shall make use of many of the analogues of the axioms listed in Section 1.2. It is apposite at this point essentially to repeat that list if for no other reason than to emphasise, by making a couple of notational changes, that we shall no longer necessarily be dealing with numbers or polynomials. Thus, in the remainder of this chapter, we shall be interested in sets of elements, often of an unspecified nature, upon which are defined two binary operations denoted (in deference to our chief example $\mathbb{Z}$ but also for want of anything better!) by $+$ and $\cdot$ and called (for the same reasons) *addition* and *multiplication*. For any particular set S, say, the operations $+$ and $\cdot$ on S may satisfy none, some or all of

3.2.1 The Axioms For any three elements $a, b, c \in S$, distinct or not,

A1 $a + b = b + a$

A2 $(a + b) + c = a + (b + c)$

A3 $\exists z \in S$ such that

$$z + a = a + z = a$$

A4 To each $a \in S$,
$\exists a^* \in S$ such that $a + a^* = a^* + a = z$

M1 $a \cdot b = b \cdot a$

M2 $(a \cdot b) \cdot c = a \cdot (b \cdot c)$

M3 $\exists e \in S$ such that

$$e \cdot a = a \cdot e = a$$

M4 To each $a \in S$ such that $a \neq z$
$\exists a' \in S$ such that $a \cdot a' = a' \cdot a = e$

D $a \cdot (b+c) = a \cdot b + a \cdot c$ and $(a+b) \cdot c = a \cdot c + b \cdot c$
Z *If* $a \cdot b = z$ *then* $a = z$ or $b = z$ (or both)

Remarks The axioms often go under the same names as in Section 1.2. For instance, M2 is the **associative law of multiplication.** For brevity, however, we use the words **zero** and **unity**† (rather than *additive* and *multiplicative identity*) to describe the elements z and e respectively. Any element s of S (including $s = z$) for which there exists a non-zero element t of S such that $s \cdot t = z$ or $t \cdot s = z$ is called a **zero divisor**. Some familiar systems satisfy all the above axioms, others only a few. Later, in Section 3.12, we shall consider analogues of axioms P and I.

For the present we make the following

Definitions 3.2.2 Let R be any non-empty set equipped with binary operations + and $\cdot$.

	If + and · satisfy										the triple $(R, +, \cdot)$ is called a
	A1	A2	A3	A4	M2	D	M1	M3	M4	Z	
1	√	√	√	√	√	√	—	—	—	—	**ring**
2	√	√	√	√	√	√	√	—	—	—	**commutative ring**
3	√	√	√	√	√	√	—	√	—	—	ring with unity
4	√	√	√	√	√	√	—	—	—	√	ring with no zero divisors‡
5	√	√	√	√	√	√	√	√	—	—	commutative ring with unity
6	√	√	√	√	√	√	√	—	—	√	commutative ring with no zero divisors
7	√	√	√	√	√	√	—	√	—	√	ring with unity and no zero divisors
8	√	√	√	√	√	√	√	√	—	√	**Integral Domain**§
9	√	√	√	√	√	√	—	√	√	√	**Division Ring**§ Skew field Non-commutative field. Sfield
10	√	√	√	√	√	√	√	√	√	√	**Field**§

Many concrete examples are given in 3.2.7 and in the exercises below.

Notes 3.2.3

(i) A ring, then, is a triple comprising a set R and two binary operations + and $\cdot$ satisfying at least the six axioms indicated. Frequently one 'forgets' the + and $\cdot$ and talks of *the ring R*. This is bad since R is only the underlying set. Further, R could well be the underlying set for two different rings (exercise 2). However this laziness is customary and rarely leads to confusion.

(ii) We have temporarily replaced the symbols 0, 1, $-a$, a^{-1}, of Section 1.2 by z (for zero), e (for *einheit*—a German word), a^* and a' initially to help us avoid unconsciously attributing unproven properties, based on our intuitive feelings about the numbers 0, 1, $-a$, and a^{-1}, to corresponding elements of a general ring (exercise 3.3.9(a)). On the other hand these changes should not be overdone (exercise 3.3.9(b)). After 3.3.2 we shall return to using 0,

† Do not confuse the words unity and unit (see 1.3.5(i) and 3.6.1(iii)).
‡ Meaning: no zero divisors *apart from z itself*.
§ See 3.2.3 (iv).

$1, -a, a^{-1}$ for all rings. Theorems 3.3.1 and 3.3.2 should encourage us to use our intuition, but *with care*.

(iii) Any ring R satisfying M4 (respectively A4) must of necessity satisfy M3 (A3). It is less obvious that if R satisfies M4 then it must of necessity satisfy Z (exercise 3.3.5). These two remarks explain why the above table contains only 10 rows and not 2^4.

(iv) In any ring R satisfying A3 and M3 we usually insist that $e \neq z$. Otherwise R is a ring with only one element (exercise 3.3.4). In particular, integral domains, division rings and fields are always assumed to have at least two elements.

(v) Despite the examples listed in Section 3.1 we do not insist that a ring must satisfy M3 (see 3.2.7(v)).

(vi) Why are the axioms A1, A2, A3, A4, M2, D taken as basic? Why not some other set? The answer is that a large enough number of important examples satisfy each of these axioms. The somewhat 'smaller' number of examples satisfying all ten axioms are of sufficient importance and richness for a similar independent study to be worthwhile. At the other extreme the theory of systems satisfying A1 and M3 (and possibly no more) would be very wide ranging but would include few more, if any, important examples than does the more restricted but structurally richer class of all rings.

As a matter of fact, at this very moment a theory of those systems satisfying A2, A3, A4, M2 and one of the axioms in D (systems which go by the name *near-rings*) is in the process of establishing itself as worthy of independent study. See [30] and [33], for example.

(vii) The reader will perhaps have noticed that in A3, M3, A4, M4 the word 'unique' has been dropped (cf. Section 1.2). In fact each uniqueness is given to us, gratis, with the compliments of the other axioms (Theorem 3.3.1).

Relationship 3.2.4 The relationship between the 10 types of rings listed in 3.2.2 can be summarised, without further explanation, by Fig. 3.1.

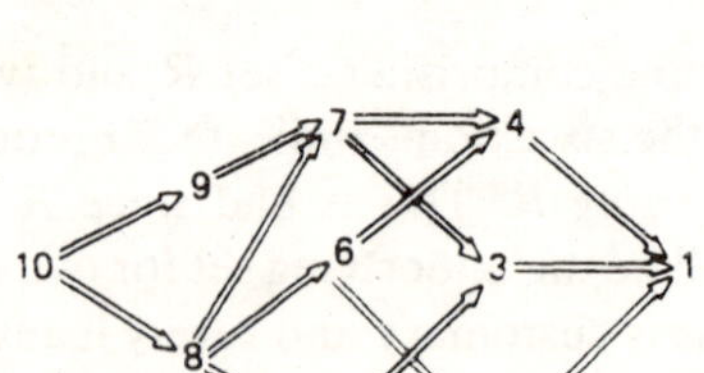

Fig. 3.1

Using 10 concrete examples we now show that none of these arrows is reversible. Perhaps the most interesting is the demonstration that there exist division rings which are not fields; that is $9 \not\Rightarrow 10$. The example, presented in a manner a little different from the original, is that of Hamilton's Quaternions.

William Rowan Hamilton (*4 August 1805 – 25 September 1865*)
Hamilton was born in Dublin, the fourth child of nine. When he was aged only one year his parents put him in care of his uncle, James. By the age of five he was reading Latin, Greek and Hebrew! In April 1827, whilst still an undergraduate at Trinity College, Dublin, he presented a paper which together with later supplements won him the 1835 Royal Medal of the Royal Society. In June 1827 he was appointed Astronomer Royal and Andrews professor of astronomy at Trinity. A few days after his 30th birthday he was knighted.

In 1833 he had presented a paper in which he eliminated the symbol 'i' from the complex numbers by defining them as 'algebraic couples' (see Section 4.4). For many years he tried unsuccessfully to find an extension to triples. Eventually the idea struck him—use 4-tuples (quaternions) and relinquish the commutative law of multiplication. He wrote several texts promoting the use of quaternions in physics. However, Gibbs' vector analysis was eventually preferred, perhaps not least because of the fact that much of Hamilton's work was difficult to read.

In undergraduate mathematics Hamilton's name appears in the Hamiltonian and Hamilton's equations in mechanics, and in the Cayley–Hamilton theorem in linear algebra.

First we make

Definition 3.2.5 A **2-by-2 matrix** with complex number entries is a square array $\begin{pmatrix} \gamma_{11} & \gamma_{12} \\ \gamma_{21} & \gamma_{22} \end{pmatrix}$, enclosed by brackets, of 4 complex numbers. Given two such matrices, $A = \begin{pmatrix} \alpha_{11} & \alpha_{12} \\ \alpha_{21} & \alpha_{22} \end{pmatrix}$ and $B = \begin{pmatrix} \beta_{11} & \beta_{12} \\ \beta_{21} & \beta_{22} \end{pmatrix}$, we define their *sum* $A \oplus B$

and their *product* $A \odot B$ by

$$A \oplus B = \begin{pmatrix} \alpha_{11}+\beta_{11} & \alpha_{12}+\beta_{12} \\ \alpha_{21}+\beta_{21} & \alpha_{22}+\beta_{22} \end{pmatrix} \text{ and}^{\dagger}$$

$$A \odot B = \begin{pmatrix} \alpha_{11}\beta_{11}+\alpha_{12}\beta_{21} & \alpha_{11}\beta_{12}+\alpha_{12}\beta_{22} \\ \alpha_{21}\beta_{11}+\alpha_{22}\beta_{21} & \alpha_{21}\beta_{12}+\alpha_{22}\beta_{22} \end{pmatrix}.$$

The set of all these matrices we denote, fairly naturally, by $M_2(\mathbb{C})$.

Theorem 3.2.6 $\langle M_2(\mathbb{C}), \oplus, \odot \rangle$ is a ring.

Proof The most arduous part of the proof is the checking of axiom M2. To do this, let A and B be as above and set $C = \begin{pmatrix} \gamma_{11} & \gamma_{12} \\ \gamma_{21} & \gamma_{22} \end{pmatrix}$. Then

$$(A \odot B) \odot C = \begin{pmatrix} \alpha_{11}\beta_{11}+\alpha_{12}\beta_{21} & \alpha_{11}\beta_{12}+\alpha_{12}\beta_{22} \\ \alpha_{21}\beta_{11}+\alpha_{22}\beta_{21} & \alpha_{21}\beta_{12}+\alpha_{22}\beta_{22} \end{pmatrix} \odot \begin{pmatrix} \gamma_{11} & \gamma_{12} \\ \gamma_{21} & \gamma_{22} \end{pmatrix} = \begin{pmatrix} X_1 & Y_1 \\ Z_1 & T_1 \end{pmatrix}$$

where
$$X_1 = (\alpha_{11}\beta_{11}+\alpha_{12}\beta_{21})\gamma_{11}+(\alpha_{11}\beta_{12}+\alpha_{12}\beta_{22})\gamma_{21},$$
$$Y_1 = (\alpha_{11}\beta_{11}+\alpha_{12}\beta_{21})\gamma_{12}+(\alpha_{11}\beta_{12}+\alpha_{12}\beta_{22})\gamma_{22},$$
$$Z_1 = (\alpha_{21}\beta_{11}+\alpha_{22}\beta_{21})\gamma_{11}+(\alpha_{21}\beta_{12}+\alpha_{22}\beta_{22})\gamma_{21},$$
$$T_1 = (\alpha_{21}\beta_{11}+\alpha_{22}\beta_{21})\gamma_{12}+(\alpha_{21}\beta_{12}+\alpha_{22}\beta_{22})\gamma_{22},$$

whereas

$$A \odot (B \odot C) = \begin{pmatrix} \alpha_{11} & \alpha_{12} \\ \alpha_{21} & \alpha_{22} \end{pmatrix} \odot \begin{pmatrix} \beta_{11}\gamma_{11}+\beta_{12}\gamma_{21} & \beta_{11}\gamma_{12}+\beta_{12}\gamma_{22} \\ \beta_{21}\gamma_{11}+\beta_{22}\gamma_{21} & \beta_{21}\gamma_{12}+\beta_{22}\gamma_{22} \end{pmatrix} = \begin{pmatrix} X_2 & Y_2 \\ Z_2 & T_2 \end{pmatrix}$$

where
$$X_2 = \alpha_{11}(\beta_{11}\gamma_{11}+\beta_{12}\gamma_{21})+\alpha_{12}(\beta_{21}\gamma_{11}+\beta_{22}\gamma_{21}) = X_1,$$
$$Y_2 = \alpha_{11}(\beta_{11}\gamma_{12}+\beta_{12}\gamma_{22})+\alpha_{12}(\beta_{21}\gamma_{12}+\beta_{22}\gamma_{22}) = Y_1,$$
$$Z_2 = \alpha_{21}(\beta_{11}\gamma_{11}+\beta_{12}\gamma_{21})+\alpha_{22}(\beta_{21}\gamma_{11}+\beta_{22}\gamma_{21}) = Z_1,$$
$$T_2 = \alpha_{21}(\beta_{11}\gamma_{12}+\beta_{12}\gamma_{22})+\alpha_{22}(\beta_{21}\gamma_{12}+\beta_{22}\gamma_{22}) = T_1.$$

Hence $(A \odot B) \odot C = A \odot (B \odot C)$, as required.

It is easily seen that axioms A1 and A2 are satisfied. Further, the matrix $O_2 = \begin{pmatrix} 0 & 0 \\ 0 & 0 \end{pmatrix}$ satisfies A3 whilst, given A above, the matrix $A^* = \begin{pmatrix} -\alpha_{11} & -\alpha_{12} \\ -\alpha_{21} & -\alpha_{22} \end{pmatrix}$ is clearly such that $A + A^* = A^* + A = O_2$. The checking of axiom D constitutes exercise 3.2.12.

We now consider the subset Q (for *quaternions*) of $M_2(\mathbb{C})$ comprising all matrices $\begin{pmatrix} s+it & u+iv \\ -u+iv & s-it \end{pmatrix}$ where $s, t, u, v \in \mathbb{R}$. It is not difficult to check [you should do it!] that Q is closed under $\oplus$ and $\odot$ as defined above so that $\oplus$

† The reason for the definition of $\odot$? Given $x_1 = ay_1 + by_2$, $x_2 = cy_1 + dy_2$ and $y_1 = Az_1 + Bz_2$, $y_2 = Cz_1 + Dz_2$ find x_1 and x_2 in terms of z_1 and z_2 and then you'll see. For an authoritative account of the introduction of matrices see [105].

and $\odot$ can be taken as binary operations on Q (see 2.7.3(ii)). Further $\langle Q, \oplus, \odot\rangle$ satisfies axioms A1, A2, M2 and D because these axioms already hold for *any three elements out of* $M_2(\mathbb{C})$. Clearly $O_2 = \begin{pmatrix} 0 & 0 \\ 0 & 0 \end{pmatrix}$ and $I_2 = \begin{pmatrix} 1 & 0 \\ 0 & 1 \end{pmatrix}$ belong to Q and satisfy A3 and M3 for Q. Further, A4 holds. We show that M4 (and hence Z) holds and M1 fails in Q.

For M4: given $A = \begin{pmatrix} a+ib & c+id \\ -c+id & a-ib \end{pmatrix}$ set $A' = \begin{pmatrix} \dfrac{a-ib}{s} & \dfrac{-c-id}{s} \\ \dfrac{c-id}{s} & \dfrac{a+ib}{s} \end{pmatrix}$

where $s = a^2+b^2+c^2+d^2$. Then $A \odot A' = A' \odot A = I_2$ (except when $s = 0$; that is, when $A = O_2$).

For M1: $\begin{pmatrix} i & 0 \\ 0 & -i \end{pmatrix} \odot \begin{pmatrix} 0 & i \\ i & 0 \end{pmatrix} = \begin{pmatrix} 0 & -1 \\ 1 & 0 \end{pmatrix} \neq \begin{pmatrix} 0 & 1 \\ -1 & 0 \end{pmatrix} = \begin{pmatrix} 0 & i \\ i & 0 \end{pmatrix} \odot \begin{pmatrix} i & 0 \\ 0 & -i \end{pmatrix}$

whence $\langle Q, \oplus, \odot\rangle$ is certainly not a field.

It is now an easy matter to give ten concrete examples which between them show that all the arrows of 3.2.4 are irreversible.

Examples 3.2.7

(i)	$\mathbb{Q}$ is a field.	It satisfies all the axioms.
(ii)	Q is a division ring and not a field, i.e.	$9 \not\Rightarrow 10$
(iii)	$\mathbb{Z}$	$8 \not\Rightarrow 10$
(iv)	$Q[x]$ (polynomials with quaternion coefficients)	$7 \not\Rightarrow 9$, $7 \not\Rightarrow 8$
(v)	$2\mathbb{Z}$ (the ring of even integers)	$6 \not\Rightarrow 8$
(vi)	$\mathbb{Z}_4$	$5 \not\Rightarrow 8$
(vii)	All elements of $Q[x]$ with zero constant term	$4 \not\Rightarrow 7$, $4 \not\Rightarrow 6$
(viii)*	$M_2(\mathbb{Z})$	$3 \not\Rightarrow 7$, $3 \not\Rightarrow 5$
(ix)	The set $\{\hat{0}, \hat{2}, \hat{4}, \hat{6}\}$ mod 8	$2 \not\Rightarrow 6$, $2 \not\Rightarrow 5$
(x)*	$M_2(2\mathbb{Z})$	$1 \not\Rightarrow 4$, $1 \not\Rightarrow 3$, $1 \not\Rightarrow 2$

Remark You should convince yourself that the above examples, which were the first to occur to the author, do what is claimed of them.

Problem 1 Can your replace each of (i) to (x) by an alternative concrete example?

Exercises

1 With the usual definitions of addition and multiplication, do the following sets form rings, integral domains, fields? Check the axioms in the order given in 3.2.2 and stop as soon as you come to the first (if any) which fails.

(a) The set $\mathbb{Z}^+$.

(b) The complex fourth roots, 1, i, −1, −i, of 1. [What is 1 + 1?]

(c) The set of all $a + ib$, where $a, b \in \mathbb{Q}$.

* See the end of the Notation Section.

(d) The set of all $a+ib$ where $a, b \in \mathbb{Z}$.
(e) The set of all a/b in $\mathbb{Q}$ where $(a, b) = 1$ and b is odd.

2 On the set $\mathbb{Z}$ define two new multiplications $\circ$ and $\Box$ by: for all $a, b \in \mathbb{Z}$, $a \circ b = 0$, $a \Box b = 1$. Show that $\langle \mathbb{Z}, +, \circ \rangle$ is a ring with zero divisors. Is $\langle \mathbb{Z}, +, \Box \rangle$ a ring?

3 Show that vector multiplication $\wedge$ on the set V of all vectors in 3-dimensional space is not associative. [Hint: Consider $\mathbf{i} \wedge (\mathbf{i} \wedge \mathbf{j})$ and $(\mathbf{i} \wedge \mathbf{i}) \wedge \mathbf{j}$.] Deduce that $\langle V, +, \wedge \rangle$ is not a ring, but that it satisfies axioms A1, A2, A3, A4 and D. [V is essentially a type of *non-associative ring* called a **Lie* algebra**.]

4 Let $S = \{s\}$, a set with one element. Define $s+s=s$ and $s \cdot s = s$. Prove that $\langle S, +, \cdot \rangle$ is a ring in which s satisfies both A3 and M3 (see exercise 3.3.4).

5 Let S be any non-empty set. On $\mathscr{P}(S)$ (see 0.3.4), the power set of S, define $\oplus$ and $\odot$ by $A \oplus B = (A \cup B) \backslash (A \cap B)$ and $A \odot B = A \cap B$. Show that the empty set $\varnothing$ satisfies axiom A3 and that for each $A \in \mathscr{P}(S)$, $A \oplus A = \varnothing$ and $A \odot A = A$. Go on to show that $\langle \mathscr{P}(S), \oplus, \odot \rangle$ is a ring. [That $\oplus$ is associative is easily 'verified' by using a Venn diagram.] Which elements of $\mathscr{P}(S)$ are zero divisors? [$\langle \mathscr{P}(S), \oplus, \odot \rangle$ is an example of a **Boolean ring**. Such rings are of interest in particular to logicians.]

If we define $\boxplus$ by $A \boxplus B = A \cup B$, is $\langle \mathscr{P}(S), \boxplus, \odot \rangle$ a ring?

6 Let F be the set of all functions $f: \mathbb{R} \to \mathbb{R}$. Defining $\oplus$ and $\odot$ by: for all $x \in \mathbb{R}$, $(f \oplus g)(x) = f(x) + g(x)$, $(f \odot g)(x) = f(x) \cdot g(x)$ show that $\langle F, \oplus, \odot \rangle$ is a commutative ring satisfying M3 but not Z. Which elements of F have multiplicative inverses? If $\odot$ is replaced by $\circ$ (i.e. 'composition') is $\langle F, +, \circ \rangle$ a ring?

7 Which elements of $\langle \mathbb{Z}_n, \oplus, \odot \rangle$ are zero divisors? Which have multiplicative inverses?

8 On the set $S = \{0, 1\}$ define $\oplus$ and $\odot$ by

$\oplus$	0	1
0	1	0
1	0	1

$\odot$	0	1
0	0	1
1	1	1

Show that $\langle S, \oplus, \odot \rangle$ is a ring. Is it a field?

9 Given rings $\langle R, +, \cdot \rangle$ and $\langle S, \boxplus, \boxdot \rangle$ define ρ and σ on $R \times S$ by $(r_1, s_1)\rho(r_2, s_2) = (r_1 + r_2, s_1 \boxplus s_2)$, $(r_1, s_1)\sigma(r_2, s_2) = (r_1 \cdot r_2, s_1 \boxdot s_2)$. Show that $\langle R \times S, \rho, \sigma \rangle$ is a ring. This ring is called the **direct sum** of R and S and is denoted briefly by $R \oplus S$.

10 Find a ring R and elements a, b, c all distinct from z such that $a \cdot b = a \cdot c$ and yet $b \neq c$.

* Marius Sophus Lie (17 December 1842 - 18 February 1899).

11 The equivalence classes $\hat{0}, \hat{2}, \ldots, \widehat{30}, \widehat{32}$ of even integers mod 34 form a ring with respect to the usual definitions of $\oplus$ and $\odot$. Show that $\widehat{18}$ is a unity element. Is it the only one?

12 Show that axiom D is satisfied in $\langle M_2(\mathbb{C}), \oplus, \odot \rangle$.

13 Is either of the rings $\langle M_2(\mathbb{Q}[x]), \oplus, \odot \rangle$, $\langle M_2(\mathbb{Z}_2), \oplus, \odot \rangle$ commutative?

14 A complex number α is called an **algebraic number** (respectively, an **algebraic integer**) iff α is a root of some monic* polynomial in $\mathbb{Q}[x]$ (respectively, $\mathbb{Z}[x]$). As particular examples show that, if $d \in \mathbb{Z}$, if $a, b \in \mathbb{Q}$ and if $u, v \in \mathbb{Z}$, then $a + b\sqrt{d}$ is an algebraic number and $u + v\sqrt{d}$ is an algebraic integer. [Hint: $a + b\sqrt{d}$ is a root of $\{x - (a + b\sqrt{d})\}\{x - (a - b\sqrt{d})\} \in \mathbb{Q}[x]$.]

Let $d\ (\neq 1)$ be *square-free*, that is: if $x \in \mathbb{Z}$ and $x^2 | d$ then $x^2 = 1$. Denote by $\mathbb{Q}(\sqrt{d})$ the set of all $a + b\sqrt{d}$ with $a, b \in \mathbb{Q}$ and by $\mathbb{Q}[\sqrt{d}]$ the set of all algebraic integers in $\mathbb{Q}(\sqrt{d})$, Let $\mathbb{Z}[\sqrt{d}] = \{u + v\sqrt{d};\ u, v \in \mathbb{Z}\}$. It can be shown (see [49, Vol. 2, p. 54]) that if $d \not\equiv 1 \pmod 4$ then $\mathbb{Q}[\sqrt{d}] = \mathbb{Z}[\sqrt{d}]$, whereas if $d \equiv 1 \pmod 4$ $\mathbb{Q}[\sqrt{d}] = \left\{\dfrac{u + v\sqrt{d}}{2};\ u, v \in \mathbb{Z}, u \equiv v \pmod 2\dagger\right\}$. Show that $\mathbb{Q}(\sqrt{d})$ is a field whereas $\mathbb{Q}[\sqrt{d}]$ is a ring but not a field. $\left[\text{Hint for } \mathbb{Q}(\sqrt{d}): \dfrac{1}{a + b\sqrt{d}} = \dfrac{a - b\sqrt{d}}{a^2 - db^2}.\right]$

15 Show that the set of all 'fractions' $\dfrac{f}{g}$ where $f, g \in \mathbb{Q}[x]$ forms a field under the obvious definitions of + and $\cdot$

$$\left(\text{i.e. } \frac{f}{g} + \frac{h}{k} = \frac{fk + gh}{gk}, \frac{f}{g} \cdot \frac{h}{k} = \frac{f \cdot h}{g \cdot k}\right).$$

[This example will be considered more formally in Section 3.10.]

3.3 Ring properties deducible from the axioms

Having seen that there is a limitless supply of concrete rings and even a large supply of different *types* of rings, it certainly appears worthwhile to try to develop a theory relating to abstract rings, the results of which will, by specialisation, be applicable to each and every concrete ring then encountered. Although we are allowed to draw on intuition in formulating statements of theorems and their proofs, we ensure that our results are logically sound by employing the axiomatic method in which hypotheses are clearly stated at the outset and a proof comprises a sequence of statements logically derived therefrom.

As a first example: It is clear that each concrete ring so far encountered contains exactly one zero element. Axiom A3 tells us that there is always at

* See 1.9.18.

† In other words u, v are both odd or both even.

least one. But is there always *exactly* one? In *every* ring? Surely! How could it be otherwise? But where is the *proof*? Here it is; and a little more besides.

Theorem 3.3.1 Let $\langle R, +, \cdot \rangle$ be a ring. Then R contains exactly one element satisfying axiom A3. Further, to each $a \in R$, there corresponds exactly one element a^* as given by A4.

Proof Suppose z and z_1 are elements of R each satisfying A3. [Thus $a + z = a$ and $b + z_1 = b$ for each a and b in R. In particular, taking a to be z_1 and b to be z we see that] then

$$z + z_1 = z_1 \text{ (since } z \text{ satisfies A3)}$$

and

$$z + z_1 = z \text{ (since } z_1 \text{ satisfies A3)}$$

It follows immediately that $z_1 = z$, as required.

For the second assertion, first note that R has at least one element of the required kind, by hypothesis. Next, suppose, given $a \in R$, that a^* and a_1^* are elements of R each satisfying A4. [We start from the equality $(a^* + a) + a_1^* = a^* + (a + a_1^*)$ which is true by A2.] Now

$$\begin{aligned} (a^* + a) + a_1^* &= z + a_1^* && \text{(since } a^* \text{ satisfies A4)} \\ &= a_1^* && \text{(since } z \text{ satisfies A3)} \end{aligned}$$

Similarly

$$\begin{aligned} a^* + (a + a_1^*) &= a^* + z && \text{(since } a_1^* \text{ satisfies A4)} \\ &= a^* && \text{(since } z \text{ satisfies A3)} \end{aligned}$$

Thus $a^* = a_1^*$, as required.

Similar results relating to axioms M3 and M4 can be proved (exercise 1 below). These assertions should not be regarded as obvious. Exercise 3.2.11 should emphasise this.

Without apology and further motivation we establish further consequences of the axioms. [Please read carefully and slowly, making sure you understand and believe each step in]

Theorem 3.3.2 Let $\langle R, +, \cdot \rangle$ be a ring. Then, for any $a, b \in R$:

(i) $z \cdot a = a \cdot z = z$ (ii) $(a^*)^* = a$

(iii) $a^* \cdot b = a \cdot b^* = (a \cdot b)^*$ (iv) $a^* \cdot b^* = a \cdot b$

If, further, R has a unity element e then

(v) $e^* \cdot b = b \cdot e^* = b^*$ (vi) $e^* \cdot e^* = e$.

Proof [Recall that we're only allowed to use axioms A1, A2, A3, A4, M2 and D.]

(i) $z + z = z$ [Why? Which axiom?]

$\therefore \quad a \cdot (z + z) = a \cdot z$ [Property of = sign.]

$\therefore \quad a \cdot z + a \cdot z = a \cdot z$ [Why?]

Now there exists in R an element $(a \cdot z)^*$ such that $a \cdot z + (a \cdot z)^* = z$. [Why?] Also

$\{a \cdot z + a \cdot z\} + (a \cdot z)^* = a \cdot z + (a \cdot z)^*$ [Why?]

$\therefore \quad a \cdot z + \{a \cdot z + (a \cdot z)^*\} = a \cdot z + (a \cdot z)^*$ [Why?]

$\therefore \quad a \cdot z + \quad z = z$ [Why?]

$a \cdot z = z$ [Why?]

The proof that $z \cdot a = z$ is similar.

[How are you going? Read the proof of (i) again if there was any step you didn't follow.]

(ii) By definition, $(a^*)^*$ denotes an additive inverse, unique by 3.3.1, of a^* in R.

But $a^* + a = a + a^* = z$. Hence a also satisfies axiom A3 for a^*. Thus $a = (a^*)^*$, by uniqueness.

(iii) By definition, $(a \cdot b)^*$ is the unique additive inverse for $a \cdot b$ in R. We show $a^* \cdot b$ is too! Now

$a \cdot b + a^* \cdot b = (a + a^*) \cdot b$ [Why? Which axiom?]

$= \quad z \quad \cdot b$ [Why?]

$= \quad z$ [Why?]

One can prove similarly (or use axiom A1) that $a^* \cdot b + a \cdot b = z$. Thus $a^* \cdot b$ is a (and hence *the*) unique additive inverse of $a \cdot b$. That is, $a^* \cdot b = (a \cdot b)^*$.

We leave proof that $(a \cdot b)^* = a \cdot b^*$ to the reader.

(iv) $a^* \cdot b^* = (a \cdot b^*)^* = ((a \cdot b)^*)^* = a \cdot b$. [Why? Why? Why?]

(v) Replacing a in (iii) by e we get $e^* \cdot b = (e \cdot b)^*$. But $e \cdot b = b$. [Why?] Hence $e^* \cdot b = (e \cdot b)^* = b^*$, as required.

(vi) Replacing b in (v) by e^* we get $e^* \cdot e^* = (e^*)^* = e$. [Why?]

Did the statements and proofs just given leave you asking 'What does all this mean? How did he think up the statements in the first place and how did he then find the proofs?' The answers are easily seen if, despite the fact that our elements may not be numbers—indeed their nature may not be specified—we replace the z, e and a^* notation by the symbols 0, 1, $-a$. In this notation statements (i) through (iii) above become (i) $0 \cdot a = a \cdot 0 = 0$; (ii) $-(-a) = a$; (iii) $(-a) \cdot b = a \cdot (-b) = -(a \cdot b)$; whilst (vi) becomes $(-1) \cdot (-1) = 1$. This illustrates how I allowed my knowledge of $\mathbb{Z}$ and the proofs in 1.2.1 to suggest the statements and proofs of 3.3.2 and shows the advantages (when treated with care) of a suggestive notation. One disadvantage (for the beginner) is highlighted in exercise 9(a).

We summarise our new notation and introduce more in:

Notation 3.3.3 From now on the symbols z, e, a^*, a' of 3.2.1 will be replaced by 0, 1, $-a$, a^{-1}. Further, $a+(-b)$ will be shortened to $a-b$. [It is then easy to show that $c\cdot(a-b)=c\cdot a-c\cdot b$ for all a, b, $c\in R$ (exercise 1(b)).] We shall also lapse into writing ab in place of $a\cdot b$.

We are now (already!) in a position to demonstrate the power of the abstract method. In 3.2.7(iii) we gave $\mathbb{Z}$ as an example of an integral domain which is not a field and in the subsequent problem you were asked to find a further such example. What did you offer? $\mathbb{Z}[x]$, $\mathbb{Z}_p[x]$, $\mathbb{Z}[\sqrt{2}]$, $\mathbb{Q}_2=\left\{\frac{a}{b}\in\mathbb{Q}: b \text{ is an odd integer}\right\}$? I can only guess. But one thing I *know*: each (valid) offering just had to be a ring with infinitely many elements. More explicitly,

Theorem 3.3.4 Each *finite* integral domain J is necessarily a field.

Proof [Since an integral domain satisfies all the field axioms except possibly M4 we only need show that J necessarily satisfies M4.]

Let the elements of J be labelled $a_0(=0)$, $a_1(=1)$, $a_2,\ldots,a_n$. Select any element a_i other than a_0. Consider the list a_0a_i, $a_1a_i,\ldots,a_na_i$ of elements, all in J. Suppose, for suffices j, k (where $0<j\leqslant k\leqslant n$) we have $a_ja_i=a_ka_i$. [Since J is a domain] it follows that $a_j=a_k$ [exercise 1(b)]. This means that the list of a_ja_is comprises $n+1$ distinct elements of J; that is, all of J. Since $1\in J$ there is an l such that $a_la_i=1$. But then $a_ia_l=a_la_i=1$ and a_i has a multiplicative inverse in J. [Thus axiom M4 holds, as required.]

Note 3.3.5 A much deeper result is: Every finite division ring is a field. This result, proved by Wedderburn† in 1905, supplies a proof of the geometrical assertion that every finite projective plane which is Desarguian is necessarily Pappian. (Prologue, p. xxiii.)

Exercises

1 (a) In the manner of 3.3.1 and 3.3.2 prove that in any ring (i) the equality $a+c=b+c$ implies the equality $a=b$; (ii) $a+x=b$ has the unique solution $x=a^*+b$; (iii) the elements e and a' of axioms M3 and M4 are unique—when they exist; (iv) if a' exists then $(a')'$ exists and $(a')'=a$.

† Joseph Henry Maclagan Wedderburn (26 February 1882 - 9 October 1948).

(b) Prove carefully, using the less suggestive notation, that $c \cdot (a-b) = c \cdot a - c \cdot b$. (That is, prove that $c \cdot (a+b^*) = c \cdot a + (c \cdot b)^*$.) Deduce that if† $c \neq z_J$ and if $c \cdot a = c \cdot b$ in an integral domain J then $a = b$ in J.

2 Let $\langle R, +, \cdot \rangle$ be a ring. For $a \in R$ and $n \in \mathbb{Z}^+$ we define (cf. 2.7.9) na and a^n to be respectively the sum and the product of n copies of a. We define $0a$ to be the element† 0_R (here $0 \in \mathbb{Z}$, $0_R \in R$) and, for $k \in \mathbb{Z}^+$, $(-k)a$ to be the element $k(-a)$.

Prove that, for all $a, b \in R$ and $m, n \in \mathbb{Z}$ we have (i) $(-m)a = -(ma)$; (ii) $(m+n)a = ma + na$; (iii) $(mn)a = m(na)$; (iv) $m(a+b) = ma + mb$; (v) $m(a \cdot b) = (ma) \cdot b = a \cdot (mb)$; (vi) $(ma) \cdot (nb) = (mn)(a \cdot b)$.

Assuming R to have a unity element e and a to have a multiplicative inverse a^{-1}, define $a^0 = e$ and, for $k \in \mathbb{Z}^+$, $a^{-k} = (a^{-1})^k$. State and prove the multiplicative analogues of (i), (ii) and (iii). Show by example that the multiplicative analogue of (iv), namely $(a \cdot b)^m = a^m \cdot b^m$, might not hold in R. Does (v) suggest any further identity? Still assuming that R has a unity element e show each ma (where $m \in \mathbb{Z}$, $a \in R$) can be written as $r \cdot a$ for suitable $r \in R$. [Hint: $r = me$ is suitable.]

3 Let $\langle R, +, \cdot \rangle$ be a ring. On $R \times \mathbb{Z}$ define $\oplus$ and $\odot$ by:

$$(r_1, z_1) \oplus (r_2, z_2) = (r_1 + r_2, z_1 + z_2)$$

$$(r_1, z_1) \odot (r_2, z_2) = (r_1 \cdot r_2 + z_1 r_2 + z_2 r_1, z_1 z_2).$$

Show that $\langle R \times \mathbb{Z}, \oplus, \odot \rangle$ is a ring with unity element $(0_R, 1)$.

4 Prove that if R satisfies (A3 and) M3 and if $e = z$ then R is a ring with one element. [Hint: For $a \in R$, $a = a \cdot e = a \cdot z = ?$] (See exercise 3.2.4.)

5 Prove that if R satisfies M4 then R satisfies Z. [Hint: generalise the proof of 2.4.6(b).]

6 Fraenkel,‡ in his definition of a ring, insisted that R should satisfy M3 but not necessarily A1. Expand $(e+e) \cdot (a+b)$ in two ways and using A3, A4, D show that $a + b = b + a$. That is, A1 is automatically satisfied in Fraenkel's rings.

7 Let $\langle R, +, \cdot \rangle$ be a ring in which $x^2 = x \cdot x = x$ for all $x \in R$. Prove that R is a commutative ring in which $2x = x + x = 0_R$. [Hint: expand $(a+a)^2$ and $(a+b)^2$.] Note: exercise 3.2.5 gives an example of such a ring.

8 Prove, or give a counterexample to, the assertion: for all x, y in any ring R, $(x+y)^2 = x^2 + 2xy + y^2$.

9 (a) Use the elements $a = \begin{pmatrix} i & 0 \\ 0 & -i \end{pmatrix}$ and $b = \begin{pmatrix} 0 & i \\ i & 0 \end{pmatrix}$ of Q to show that $(a \cdot b)^{-1} = b^{-1} \odot a^{-1} \neq a^{-1} \odot b^{-1}$. Hence pinpoint the error in the following

† $z_J(0_R)$ denotes the zero element of the ring $J(R)$.
‡ Adolf Abraham Fraenkel (17 February 1891 – 15 October 1965).

argument.* Writing $\frac{1}{a}$ in place of a^{-1} we get $(ab)^{-1} = \frac{1}{ab} = \frac{1}{a}\frac{1}{b} = a^{-1}b^{-1}$.

(b) To banish all possibility of making intuitive but unwarranted assumptions about + and · in an arbitrary ring let us replace + and · by ρ and τ.

Is $a\tau(b\rho(c\tau(a\rho b))) = ((a\tau c)\tau a)\rho((a\rho(a\tau c))\tau b)$ in $\langle R, \rho, \tau\rangle$?

Is $a \cdot (b + (c \cdot (a + b))) = ((a \cdot c) \cdot a) + ((a + (a \cdot c)) \cdot b)$ in $\langle R, +, \cdot\rangle$?

Which was easier to decide? Why? So is it easier to use ρ, τ or +, · ?

10 Let R be a finite ring with no (non-zero) zero divisors and at least two elements. Show that R is a division ring.

3.4 Subrings, subfields and ideals

Of the rings mentioned in Section 3.2, six fall into three pairs, the first in each pair being contained in a natural way inside the second. Thus the ring $2\mathbb{Z}$ is naturally thought of as being contained in the ring $\mathbb{Z}$, $M_2(2\mathbb{Z})$ as being part of the ring $M_2(\mathbb{Z})$ and, finally, Q as being in $M_2(\mathbb{C})$. Further, with a little bit of care (see Remark (ii) on p. 40) we can think of $\mathbb{Q}$ as being contained in $\mathbb{Q}[x]$ and of $\mathbb{Q}[x]$ as being inside $\mathbb{Q}[x, y]$ (exercise 1.8.2), etc. With the same care (Section 4.4) we can think of the succession $\mathbb{Q} \subset \mathbb{R} \subset \mathbb{C}$ of fields. Thus on the grounds of convenience alone it seems appropriate to introduce a terminology to describe the sitting of one ring nicely inside another. Note that in each case the elements of the 'smaller' ring form a subset of those of the 'bigger' ring and the results of adding and multiplying each pair of elements within the smaller ring coincide with the corresponding results when the same two elements are regarded as being in the bigger ring.

Definition 3.4.1 Let $\langle R, +, \cdot\rangle$ be a ring and suppose that S is a non-empty subset of R. We say that S is a **subring** of R if and only if (i) S is closed under + and · (so that† + and · are binary operations on S) and (ii) $\langle S, +, \cdot\rangle$ is a ring.

In addition to the motivating examples above we note that, in the ring $R \times \mathbb{Z}$ of exercise 3.3.3, the subset $R \times \{0\} = \{(r, 0): r \in R\}$ is a subring. Note that R itself is not a subring of $R \times \mathbb{Z}$ since R is not actually contained in $R \times \mathbb{Z}$. This state of affairs causes the algebraist no embarrassment at all (see Remarks in Section 3.10 and exercise 3.10.9).

How can one recognise when a non-empty subset S of a ring R is a subring of R? One can check 3.4.1(i) and (ii) of course; but easier is

Theorem 3.4.2 Let S be a non-empty subset of a ring R. Then S is a subring of R if and only if both (α) and (β) below hold.

* Due to a former student whose name I have conveniently forgotten!

† More accurately, the restrictions of + and · to S. See 2.7.3 (ii).

(α) *If* $a, b \in S$ *then* $a+b \in S$ and $a \cdot b \in S$
(β) *If* $a \in S$ *then* $-a \in S$ (Here $-a$ is the inverse of a *in* R.)

To prove the 'only if' part of this we shall need

Lemma 3.4.3 Let S be a subring of R. Letting 0_S and 0_R denote the zero elements of the rings $\langle S, +, \cdot\rangle$ and $\langle R, +, \cdot\rangle$ we have $0_S = 0_R$. Further, if $a \in S$ then $(-a)_S = (-a)_R$.

Remark This lemma is so obvious it doesn't need proof. Right? And just as obvious is the remark that if S and R have unity elements 1_S and 1_R then $1_S = 1_R$. Since this latter assertion is *false* (see 3.4.4 (iii) and exercise 7) we'd better take 3.4.3 seriously and offer proof.

Proof of 3.4.3 Let $s \in S$. Inside S we have $s + 0_S = s$. Inside R we have $s + 0_R = s$. Hence, in R, $s + 0_S = s + 0_R$. Thus $0_S = 0_R$ [by exercise 3.3.1(a)(i)]. Next, in S, $a + (-a)_S = 0_S$ whilst in R, $a + (-a)_R = 0_R$. Consequently, in R, $a + (-a)_S = 0_S = 0_R = a + (-a)_R$, whence $(-a)_S = (-a)_R$. [Why?]

We can now give the

Proof of 3.4.2 $\overset{\text{only if}}{\Rightarrow}$ Since S is a subring of R, $\langle S, +, \cdot\rangle$ is a ring. Hence for $a, b \in S$ we have $a+b \in S$, $a \cdot b \in S$ and $(-a)_S \in S$. But $(-a)_S = (-a)_R = -a$. This is enough.

$\overset{\text{if}}{\Leftarrow}$ [We assume conditions (α) and (β).] First, (α) implies that S is closed under $+$ and $\cdot$. Since $S \neq \varnothing$, there exists $s \in S$. By (β), $(-s)_R \in S$. By (α), $0_R = s + (-s)_R \in S$. Clearly 0_R is the element required, by A3, to be in S. Also for each $a \in S$, $(-a)_R \in S$ by (β). Clearly $(-a)_R$ is the element required by A4 to be in S. Finally, each of A1, A2, M2 and D holds for all elements of R; hence, in particular, all elements of S. [Thus we have shown that $+$ and $\cdot$ are binary operations on S and that $\langle S, +, \cdot\rangle$ is a ring.]

Examples 3.4.4
(i) Q is a subring of $\langle M_2(\mathbb{C}), \oplus, \odot\rangle$. For, given $x, y \in Q$, it is easily seen that $x \oplus y$, $x \odot y$ and $\ominus x$ are also in Q.
(ii) On $2\mathbb{Z}$ define $+$ as usual, but $\odot$ by: for all $a, b \in 2\mathbb{Z}$, $a \odot b = 0$. Then $\langle 2\mathbb{Z}, +, \odot\rangle$ is a ring but it is not a subring of $\langle \mathbb{Z}, +, \cdot\rangle$ since $a \odot b \neq a \cdot b$ for all $a, b \in 2\mathbb{Z}$.
(iii) $S = \{\hat{0}, \hat{2}, \hat{4}, \hat{6}, \hat{8}\}$ is a subring of $\mathbf{Z}_{10}$. Note that $1_{\mathbf{Z}_{10}} = \hat{1}$ whereas $1_S = \hat{6}$.
(iv) In the ring $\langle F, \oplus, \odot\rangle$ (exercise 3.2.6) of all functions from $\mathbb{R}$ to $\mathbb{R}$ the subset of all differentiable (respectively continuous) functions forms a subring, essentially because the sum and product of differentiable (continuous) functions is differentiable (continuous).

We now use 3.4.2 to establish the important

Theorem 3.4.5 Let S_1, S_2 be subrings of the ring R. Then the set-theoretic intersection $S_1 \cap S_2$ is also a subring of R.

Proof Since $0_{S_1} = 0_R = 0_{S_2}$ we see that $S_1 \cap S_2 \neq \varnothing$. Now suppose $x, y \in S_1 \cap S_2$. Then $x \in S_1$ and $y \in S_1$ and so $x + y$, $x \cdot y$ and $-x$ all lie in S_1 [by 3.4.2]. Similarly $x + y$, $x \cdot y$ and $-x$ all lie in S_2 and hence in $S_1 \cap S_2$. By 3.4.2, $S_1 \cap S_2$ is a subring of R.

This result clearly extends to intersections of arbitrary (possibly infinite) sets of subrings of a ring R (exercise 5).

There are important analogues of 3.4.1, 3.4.2 and 3.4.5 for fields. We have

Definition 3.4.1(*F*) A **subfield** of a field F is any non-empty subset T of F such that (i) T is closed with respect to the binary operations $+$ and $\cdot$ defined on F and such that (ii) $\langle T, +, \cdot\rangle$ is a field.

Theorem 3.4.2(*F*) Let T be a non-empty subset of a field F. Then T is a subfield of F iff T has at least two elements and both (α) and (β) below hold.

(α) *If* $a, b \in T$ *then* $a + b \in T$ and $a \cdot b \in T$

(β) *If* $a \in T$ *then* $-a \in T$ and *if* $a \neq 0$ *then* $a^{-1} \in T$

Theorem 3.4.5(*F*) Let $\{T_\alpha : \alpha \in A\}$ be a set of subfields of the field F. Then the set-theoretic intersection $\bigcap_{\alpha \in A} T_\alpha$ is also a subfield of R.

In particular the intersection P of the set of *all* subfields of F is a subfield (clearly the unique, smallest one in F) called the **prime subfield** of F. Clearly 0_F and 1_F lie in each subfield of F and hence in P. [Aren't you just a bit suspicious about 1_F? See Example 3.4.4(iii) above and exercise 3.4.9.] The exact nature of P is somewhat restricted (see 3.10.9).

In the theory of rings a special kind of subring arises in two disparate ways. Asking you to wait until Section 3.9 and 4.2.7 for motivation we make

Definition 3.4.6 Let I be a non-empty subset of a ring R. Then I is called an **ideal** of R if and only if (i), (ii) and (iii) below hold.

(i) *If* $a, b \in I$ *then* $a + b \in I$

(ii) *If* $a \in I$ *then* $-a \in I$

(iii) *If* $a \in I$ and $r \in R$ *then* ra and $ar \in I$

Clearly each ideal is a subring (in (iii) r can be any element of I). Note that $a \cdot r$ and $r \cdot a$ may well be unequal; we demand that *both* belong to I. Further, since each product ar and ra gets swallowed up by I we can think of I as a sort of ring-theoretic 'black hole'!

Examples 3.4.7

(i) Let $s \in \mathbb{Z}$. Then the set $\{sz : z \in \mathbb{Z}\}$ is an ideal of $\mathbb{Z}$. In fact every ideal—indeed every subring—of $\mathbb{Z}$ has this form, for suitable s. (See 3.7.16.)

(ii) Let $f \in \mathbb{Q}[x]$. Then the set $\{fm: m \in \mathbb{Q}[x]\}$ is an ideal of $\mathbb{Q}[x]$. In fact every ideal, though not every subring, of $\mathbb{Q}[x]$ has this form. (See 3.7.16 and exercises 1(b) and 13.)

(iii) In any ring R the subsets $\{0_R\}$ and R are ideals. If F is a field then $\{0\}$ and F are its only ideals (exercise 14).

(iv) Let R be any ring and let $a_1, a_2, \ldots, a_m \in R$. Then the set of all elements of the form $\sum_{i=1}^{m} z_i a_i + \sum_{i=1}^{m} s_i a_i + \sum_{i=1}^{m} a_i t_i + \sum_{i=1}^{m} \left(\sum_{k=1}^{n_i} u_{i,k} a_i v_{i,k} \right)$ where $m, z_i, n_i \in \mathbb{Z}$, $s_i, t_i, u_{i,k}, v_{i,k} \in R$ is an ideal. In fact it's the smallest ideal of R which contains $a_1, \ldots, a_m$. Hence it is called the **ideal generated by** $a_1, \ldots, a_m$.

If R is commutative and has a unity the above set reduces to the set $\{a_1 r_1 + \cdots + a_m r_m : r_i \in R\}$. [Why can $z_1 a_1 + \cdots + z_m a_m$ be written in this form?] This ideal we denote briefly by $[a_1, \ldots, a_m]$. If $m = 1$ the ideal $[a_1]$ is called the **principal ideal** generated by a_1. In particular $[1_R] = R$.

(v) The subset E of $\mathbb{Z}[x]$ comprising all polynomials with even constant term is an ideal of $\mathbb{Z}[x]$. In fact $E = [x, 2]$ [are you sure?] and is not principal. (See exercise 22(b).)

Remark A commutative ring in which every ideal can be generated by a finite number of elements is called a **Noetherian ring** (after Emmy Noether who undertook deep investigations of such rings). A theorem central to the subject of algebraic geometry is the so-called **Hilbert Basis Theorem**:* *Let R be a Noetherian ring. Then the polynomial ring $R[x_1, x_2, \ldots, x_n]$ in the n commuting letters $x_1, x_2, \ldots, x_n$ with coefficients in R is Noetherian.* In particular in the ring $R[x_1, x_2, \ldots, x_n]$ where $R = \mathbb{Z}, \mathbb{Q}, \mathbb{R}$, or $\mathbb{C}$, each ideal is expressible in the form $[f_1, \ldots, f_s]$, only finitely many polynomials f_i being required.

Amalie Emmy Noether (*23 March 1882 – 14 April 1935*)
Emmy Noether, the first child of Max and Ida Noether, was born in Erlangen, Bavaria, where her father, a famous algebraic geometer,

* The theorem Gordan called 'Theology, not mathematics'. For a proof see [9, Vol. 2, p. 401].

was a professor of mathematics. Although a discussion of the so-called Noetherian rings would be out of place in this book, we include this biography because she did so much in the 1920s and 1930s to point algebra in the direction it faces today. As the Russian mathematician Alexandroff said of her: 'Emmy Noether taught us to think in a simpler and more general way; in terms of homomorphisms, of ideals—not in terms of complicated algebraic calculations. She therefore opened a path to the discovery of algebraic regularities where previously they had been obscured by complicated specific conditions.'

Despite the prejudices in the early years of the century against women in universities, she obtained her Ph D in 1907. However, despite great efforts by Hilbert, she had, by 1922, still attained nothing more than an honorary professorship at Göttingen. And more difficulties were to come. Along with other Jewish colleagues she was dismissed from Göttingen in 1933. The USA gladly offered her a home and she spent the last two years of her life at Bryn Mawr and Princeton.

Emmy Noether was one of the greatest of algebraists and most probably the greatest of all women mathematicians.

Exercises

Subrings

1 Establish, using 3.4.2, whether or not the following subsets of the given rings are subrings:
(a) All positive integers in $\langle \mathbb{Z}, +, \cdot \rangle$.
(b) All polynomials with integer constant term in $\langle \mathbb{Q}[x], +, \cdot \rangle$.
(c) The sets $\left\{ \begin{pmatrix} x & y \\ 0 & 0 \end{pmatrix} : x, y \in \mathbb{R} \right\}$

and $\left\{ \begin{pmatrix} 0 & 0 \\ 0 & z \end{pmatrix} : z \in \mathbb{Z} \right\}$ in $\left\langle \left\{ \begin{pmatrix} a & b \\ 0 & c \end{pmatrix} : a, b, c \in \mathbb{R} \right\}, \oplus, \odot \right\rangle$.
(d) All integers not divisible by 3 in $\langle \mathbb{Z}, +, \cdot \rangle$.
(e) All polynomials of degree $\geqslant 6$ in $\langle \mathbb{Q}[x], +, \cdot \rangle$.
(f) All numbers $a + bi$ where $a, b \in \mathbb{Z}$ in the field $\langle \mathbb{C}, +, \cdot \rangle$.
(g) The set $\{75a + 30b : a, b \in \mathbb{Z}\}$ in $\langle \mathbb{Z}, +, \cdot \rangle$.
(h) All zero divisors in $\langle \mathbb{Z}_{14}, \oplus, \odot \rangle$; in $\langle \mathbb{Z}_{16}, \oplus, \odot \rangle$.

2 Find the smallest subring S of $\mathbb{Q}$ which (i) contains 1 [Hint: Clearly $1 \in \mathbb{Z} \subseteq S$. But $\mathbb{Z}$ is a subring.]; (ii) contains $\frac{1}{3}$ and $\frac{1}{2}$.

3 Let S be a non-empty set. If $A \in \mathcal{P}(S)$ show that $\{\varnothing, A\}$ is a subring of $\langle \mathcal{P}(S), \oplus, \odot \rangle$ (see exercise 3.2.5).

4 Show that a non-empty subset S of a ring R is a subring if and only if for all $a, b \in S$, $a - b \in S$ and $a \cdot b \in S$ (cf. 3.4.2).

5 Copying the proof of 3.4.5, extend it to arbitrary sets of subrings.

6 Show that the union $\{2m: m \in \mathbb{Z}\} \cup \{3n: n \in \mathbb{Z}\}$ of two ideals of $\mathbb{Z}$ is not even a subring of $\mathbb{Z}$.

7 Copy the proof of 3.4.3 to identify where a proof that $1_S = 1_R$ would break down.

Fields

8 Prove, using 3.4.2(F), that the subset $\{a + 0i: a \in \mathbb{R}\}$ is a subfield of $\langle \mathbb{C}, +, \cdot \rangle$.

9 Let T be a subfield of a field F. Show that $1_T = 1_F$ and that, for $a \in T$ $(a \neq 0_T)$, $(a^{-1})_T = (a^{-1})_F$ (cf. 3.4.3 and exercise 7 above.)

10 Prove 3.4.2(F) and 3.4.5(F).

11 How many different subfields can you find in (i) $\langle \mathbb{Q}, +, \cdot \rangle$; (ii) $\langle \mathbb{R}, +, \cdot \rangle$? [For (ii) recall all the $a + b\sqrt{p}$ where $a, b \in \mathbb{Q}$.]

Ideals

12 Show that, in $\mathbb{Z}$, $[15, 21] = [3]$. Is $[15, 35, -77] = \mathbb{Z}$?

13 Which subsets in exercise 1 above are ideals? [(b) tells you that, unlike in $\mathbb{Z}$, not all subrings of $\mathbb{Q}[x]$ are ideals.]

14 Show that if I is an ideal in the field F then either $I = \{0\}$ or $I = F$. [Hint: If $I \neq \{0\}$ then there exists $x \in I$ such that $x \neq 0$. Then x^{-1} exists in F and $1 = xx^{-1} \in I$.] Let R be a commutative ring with unity. Show that R is a field if and only if R has $\geqslant 2$ elements and $\{0\}$ and R are the only ideals. Exhibit infinitely many subrings of $\mathbb{Q}$ which contain $\mathbb{Z}$.

15 Let I be an ideal in the ring R. Show that $I[x] = \{a_0 + a_1x + \cdots + a_rx^r: a_i \in I, r \in \mathbb{Z}^+ \cup \{0\}\}$ is an ideal of $R[x]$.

16 Prove all the assertions in example 3.4.7(iv).

17 For ideals I_1, I_2 in a ring R define $I_1 + I_2$ to be the set $\{a + b: a \in I_1, b \in I_2\}$ and I_1I_2 to be the set $\{a_1b_1 + \cdots + a_nb_n;\ n \in \mathbb{Z}^+,\ a_i \in I_1,\ b_i \in I_2\}$. Show that $I_1 + I_2$, I_1I_2 and $I_1 \cap I_2$ are ideals of R. If R is commutative with unity, if $I_1 = [a_1, \ldots, a_m]$ and if $I_2 = [b_1, \ldots, b_n]$ show $I_1I_2 = [a_1b_1, a_1b_2, \ldots, a_1b_n, a_2b_1, \ldots, a_mb_n]$.

18 In $\mathbb{Z}[x]$ set $I_2 = [x, 2]$ and $I_3 = [x, 3]$. Show that $I = \{ab: a \in I_2, b \in I_3\}$ is not an ideal in $\mathbb{Z}[x]$. [Hint: $x^2 \in I$, $6 \in I$, $x^2 + 6 \notin I$.] (This explains the definition of I_1I_2 given in exercise 17.)

19 Let $I_1 \subseteq I_2 \subseteq I_3 \subseteq \cdots$ be a possibly infinite sequence of ideals in a ring R. Set $U = \{a: a \in R \text{ and } a \in I_n \text{ for some } n\}$. Prove that U is an ideal of R. [Hint: given $a, b \in U$ there exists t such that a and b are both in I_t.]

20 Show that $\begin{pmatrix} 0 & 1 \\ 0 & 0 \end{pmatrix} \odot \begin{pmatrix} a & b \\ c & d \end{pmatrix} \odot \begin{pmatrix} 0 & 0 \\ 1 & 0 \end{pmatrix} = \begin{pmatrix} d & 0 \\ 0 & 0 \end{pmatrix}$ in $M_2(\mathbb{R})$. After finding three similar equalities deduce that if I is an ideal in $M_2(\mathbb{R})$ then either

$I=\left\{\begin{pmatrix}0 & 0\\ 0 & 0\end{pmatrix}\right\}$ or $I=M_2(\mathbb{R})$. [A ring $R \neq \{0_R\}$ with no ideals other than $\{0_R\}$ and R is called a **simple ring**.]

21 Let S be a set $\{(a_i, b_i)\}$ of points in the plane $\mathbb{R}\times\mathbb{R}$. Show that the set $\{p(x, y)\}$ of all polynomials in $\mathbb{R}[x, y]$ for which $p(a_i, b_i)=0$ for all $(a_i, b_i)\in S$ is an ideal of $\mathbb{R}[x, y]$. (Such examples are at the heart of that part of algebraic geometry which studies the geometry of curves by algebraic means.)

22 (a) Show that in $\mathbb{Q}[x, y]$ the subset $S=\{xf+yg: f, g\in\mathbb{Q}[x, y]\}$ is an ideal. Show that S is not a principal ideal. [Hint: If $S=[h]$ then $x=hu$ and $y=hv$ for some $u, v\in\mathbb{Q}[x, y]$.]

(b) Show that in $\mathbb{Z}[x]$ the ideal $[x, 2]$ is not principal.

Pierre de Fermat (*20 August 1601 – 12 January 1665*)

Fermat's father was a prosperous leather merchant, and his mother came from a family of high social standing. As a young man he studied law and took a law degree in 1631, becoming a magistrate. He was fluent in Spanish, Italian, Latin and Greek. Serious interest in mathematics seems to date only from his late twenties. Fermat can, with Pascal, be regarded as a founder of probability theory. He also invented analytic geometry independently of Descartes, although he did not develop it so extensively, and contributed towards the invention of the calculus. But his main claim to fame is as the 'Father of modern number theory'. Most of his results were communicated by letter or written in his copy of Diophantus' *Arithmetica*, and almost invariably without proof. Formally he published almost nothing. It appears he must have had an amazing intuition because all of his definite assertions, except one, the famous conjecture (or Last Theorem), have been proved true. Even his belief that every integer of the form $2^{2^n}+1$ is a prime (this is false) was accompanied by his admitting that he couldn't actually prove it. (How much more intriguing does this make his unqualified assertion about the FC!)

Besides work mentioned above he is credited with finding the $f''(x)$ criterion for maximum and minimum values of functions, and in optics he enunciated the principle named after him.

3.5 Fermat's conjecture (FC)

In 1847 G Lamé* announced to the Paris Academy that he had finally proved the FC, otherwise known as Fermat's Last Theorem, dating from the 1630s. This is the assertion that to each integer $n > 2$ the equation $x^n + y^n = z^n$ has no solution in integers x, y, z for which $xyz \neq 0$. Fermat, given to writing statements, but not their proofs, in his copy of Bachet's 1621 translation of Diophantus' book *Arithmetica*, had asserted that 'I have found a truly remarkable proof but the margin is too small to contain it.' Fermat actually left a proof which covers the case $n = 4$ and an inadequate proof by Euler (1770) for the case $n = 3$ can be salvaged [46]. 1825 saw proofs by Legendre and Dirichlet for the case $n = 5$ and in 1839 Lamé gave a proof for the case $n = 7$.

Even today no proof of, nor counterexample to, FC is known and it is generally assumed that Fermat's proof contained a subtle error. Lamé's idea (and possibly Fermat's) in the case $n = p$, an odd prime (see exercise 3), is to factorise $z^n = x^n + y^n$ as $(x + y)(x + \zeta y) \ldots (x + \zeta^{p-1} y)$, $\zeta \neq 1$ being a complex pth root of 1. Since this product is a pth power, z^p, and since a greatest common divisor d of any pair $x + \zeta^i y$, $x + \zeta^j y$ can be shown to be a common divisor of all the $x + \zeta^k y$, it allegedly followed (as it does in $\mathbb{Z}^+$; see exercise 1) that each $x + \zeta^i y$ is itself a product of d and a pth power. From this a contradiction is to be obtained by Fermat's so-called *method of descent* (see 3.8.2 and exercise 3.8.6). The chief objection to Lamé's proof is in his implicit assumption that a theorem analogous to the Fundamental Theorem of Arithmetic (1.5.1) remains valid for the more inclusive sets of complex integers under consideration here. We amplify this objection in Section 3.6.

In fact, the unique factorisation theorem is not valid for numbers of the above kind, at least for $p = 23$.† This had apparently been known to Kummer since 1844 (see [46]). But this failure is a most fortunate one as far as progress in mathematics is concerned, for much high powered mathematics has resulted from attempts to circumvent it.

Kummer's efforts (c. 1847–57) extended the range of prime exponents p for which FC is known to be true from the set of all primes $p < 23$ to the set of all so-called *regular primes* (see [49, Vol. 2, p. 97], the first three irregular primes being 37, 59 and 67. Sadly, it is still not known if there are infinitely many regular primes though it has been known for almost 80 years that there are infinitely many irregular ones! Conditions sufficient to prove the FC in special cases have been discovered over the years. For example in 1909

* Gabriel Lamé (22 July 1795 – 1 May 1870) was considered, by Gauss, the foremost French mathematician of his generation.

† In fact, the unique factorisation theorem holds for all primes <23 and fails for all other primes. See *Crelle's Journal*, vol. 286–7, p. 248.

Wieferich showed that if $x^p+y^p=z^p$ and if $p \nmid xyz$ then $2^p \equiv 2 (\text{mod } p^2)$, the first prime satisfying this being 1093. Aided by results of this kind (and computers) it is now known that the FC is true for each exponent n which has a prime divisor less than 125 000 (!) (see [53]).

In addition, recent work for which Gerd Faltings won a 1986 Fields Medal (p. 266) shows that $x^n+y^n=z^n$ has, for each n, at most finitely many integer solutions.

Exercises

1 Show that if $z_1, z_2, \ldots, z_m, z, n \in \mathbb{Z}^+$ are such that $(z_i, z_j)=1$ when $i \neq j$ and if $z_1 z_2 \ldots z_m = z^n$ then each z_i is an nth power.

2 Look up a proof (you'll find one in, for example, [42] on pages 149–150) of the assertion that every solution of the equation $x^2+y^2=z^2$ is of the form $x=s(a^2-b^2)$, $y=2sab$, $z=s(a^2+b^2)$, or the same with x and y interchanged, where $s, a, b \in \mathbb{Z}$.

3 Show that to prove FC one only (!) needs to prove it for $n=4$ and for n an odd prime. [Hint: If (x_0, y_0, z_0) is a solution when $n=u \cdot v$ then (x_0^u, y_0^u, z_0^u) is a solution for $n=v$.]

3.6 Divisibility in rings*

In order to be more explicit about the remarks made in Section 3.5 concerning non-unique factorisation we must define the concept of division in a ring. We restrict ourselves to commutative rings and following Sections 1.3 and 1.9 we make, for any commutative ring R whatever,

Definition 3.6.1† Let $a, b \in R$.

(i) We say a **divides** b and write $a|b$ if and only if there exists $c \in R$ such that $ac=b$.

(ii) If $c, d \in R$, if $d|a$ and $d|b$ and if, from $c|a$ and $c|b$, it follows that $c|d$ then d is a **greatest common divisor** (gcd) of a and b. We denote any one of these gcds by (a, b).

Assuming, in addition, that R has a multiplicative identity:

(iii) $u \in R$ is a **unit** if and only if there exists $v \in R$ such that $uv=vu=1$.

(iv) a, b are **associates** if and only if there exists a unit u such that $a=ub$.

(v) $\pi \in R$ is **irreducible** if and only if $\pi \neq 0$, π is not a unit and from $\pi=\alpha\beta$ (with $\alpha, \beta \in R$) it follows that either α or β is a unit.

(vi) $\pi \in R$ is **prime** if and only if $\pi \neq 0$, π is not a unit and from $\pi \mid \alpha\beta$ (with $\alpha, \beta \in R$) it follows that either $\pi|\alpha$ or $\pi|\beta$ (or both).

(vii) $a, b \in R$ are **relatively prime** if and only if 1 is a gcd of a and b.

* Divisibility theory in fields is a rather dull topic. Why?

† 3.6.1 is easily adapted to rings which are not necessarily commutative but we shall have no need of such.

We have already given (Sections 1.3, 1.9) examples of these concepts in $\mathbb{Z}$ and in $\mathbb{Q}[x]$. Here we do the same for another type of ring of central importance in number theory, namely the ring $\langle \mathbb{Z}[\sqrt{d}], +, \cdot \rangle$, where d is square-free in $\mathbb{Z}$, as introduced in exercise 3.2.14. It is easy to check that $\langle \mathbb{Z}[\sqrt{d}], +, \cdot \rangle$ is a ring, indeed an integral domain.

To help determine the units in $\mathbb{Z}[\sqrt{d}]$ we formulate (in $\mathbb{Q}[\sqrt{d}]$: see exercise 3.2.14).

Definition 3.6.2 Let $\alpha = a + b\sqrt{d} \in \mathbb{Q}[\sqrt{d}]$. The **norm** of α, denoted by $N(\alpha)$, is defined by $N(\alpha) = |a^2 - db^2|$.

There follows easily

Lemma 3.6.3
(i) If $\alpha \in \mathbb{Z}[\sqrt{d}]$ then $N(\alpha)$ is a non-negative integer.
(ii) $N(\alpha) = 0$ if and only if $\alpha = 0$.
(iii) For $\alpha, \beta \in \mathbb{Q}[\sqrt{d}]$, $N(\alpha\beta) = N(\alpha)N(\beta)$.

Proof Exercise 11.

Now suppose $u = s + t\sqrt{d}$ is a unit in $\mathbb{Z}[\sqrt{d}]$. Then for some $v \in \mathbb{Z}[\sqrt{d}]$ we have $uv = 1 + 0\sqrt{d} = 1$. It follows that $N(u)N(v) = N(uv) = N(1) = 1$. By 3.6.3 (i), $N(u) = 1$.

If $d = -1$ then $N(u) = s^2 + t^2 = 1$. This implies that $s = 1$, $t = 0$ or $s = -1$, $t = 0$ or $s = 0$, $t = 1$ or $s = 0$, $t = -1$. Hence $u = 1$ or -1 or i or $-$i.

If $d < -1$ then $N(u) = s^2 - dt^2 = 1$. This implies $s = 1$, $t = 0$ or $s = -1$, $t = 0$. Hence $u = 1$ or -1.

If $d > 0$ the situation is quite different, for it is known (see [42]) that the equation $s^2 - dt^2 = 1$ has (for d not a perfect square) infinitely many solutions. For example, $u = 2 - \sqrt{3}$ is a unit in $\mathbb{Z}[\sqrt{3}]$ since $(2 - \sqrt{3})(2 + \sqrt{3}) = 1$. Further, the powers u^n are distinct and yet each is a unit (exercise 3).

Noting that for $d < 0$ the elements listed are indeed units in the appropriate $\mathbb{Z}[\sqrt{d}]$ we have, summarising the above,

Theorem 3.6.4 In the rings $\mathbb{Z}[\sqrt{d}]$, where d is square-free, the units are: (i) 1, -1, i, $-$i if $d = -1$; (ii) 1, -1 if $d < -1$; (iii) 1, -1 and infinitely many others if $d > 1$.

In Section 3.5 we asserted that the unique factorisation theorem fails in some rings of complex numbers. The example mentioned there, with $p = 23$, is too involved to reproduce here, so we describe the same phenomenon using an easier example, $\mathbb{Z}[\sqrt{-3}]$.

In $\mathbb{Z}[\sqrt{-3}]$ the equality $(1 + \sqrt{-3})(1 - \sqrt{-3}) = 2 \cdot 2$ holds. Suppose that $(1 + \sqrt{-3}) = (a + b\sqrt{-3})(c + d\sqrt{-3})$ in $\mathbb{Z}[\sqrt{-3}]$. Taking norms and using 3.6.3(iii) gives

$$4 = N(1 + \sqrt{-3}) = N(a + b\sqrt{-3})N(c + d\sqrt{-3}) = (a^2 + 3b^2)(c^2 + 3d^2)$$

Since $a^2+3b^2=2$ has no solution in $\mathbb{Z}$ we see that $a^2+3b^2=1$ or $a^2+3b^2=4$. In the former case $a=1$ or -1 and $b=0$, whence $a+b\sqrt{-3}$ is a unit. In the latter case $c^2+3d^2=1$ and $c+d\sqrt{-3}$ is likewise a unit. Thus, in expressing $1+\sqrt{-3}$ as a product, one of the factors is necessarily a unit. Since $1+\sqrt{-3}$ is clearly non-zero and is not a unit [why not?] it follows that $1+\sqrt{-3}$ is irreducible in $\mathbb{Z}[\sqrt{-3}]$. Similarly $1-\sqrt{-3}$ and 2 are proved irreducible in $\mathbb{Z}[\sqrt{-3}]$. We deduce that in $\mathbb{Z}[\sqrt{-3}]$, the number 4 has two distinct decompositions into products of irreducibles. In other words, *the analogue of the Fundamental Theorem of Arithmetic fails in* $\mathbb{Z}[\sqrt{-3}]$. (For other deductions see exercise 12. In view of Lamé's approach exercise 12(c) is especially interesting.)

Exercises

(Assume here that all rings mentioned, except matrix rings, are commutative.)

1 (a) Prove that if $a|b$ and $b|c$ in the ring R then $a|c$ in R.
(b) Prove that if $a|b$ and $a|c$ and if $x, y \in R$ then $a|bx+cy$ in R.

2 Let $c \in R \subset R[x]$, R being an integral domain. Prove that c is a unit (respectively prime, irreducible) in $R[x]$ if and only if c is a unit (respectively prime, irreducible) in R.

3 Let $u=2-\sqrt{3}$. Show that the set $\{u^n : n \in \mathbb{Z}\}$ is a set of infinitely many distinct units in $\mathbb{Z}[\sqrt{3}]$.

4 Find all the units in $M_2(\mathbb{R})$, in $M_2(\mathbb{Z})$ and in $\mathbb{Z}[x, y]$. Is $\hat{2}x^2+\hat{2}x+\hat{3}$ a unit in $\mathbb{Z}_4[x]$? Show that, for any field F, $u=a_0+a_1x+a_2x^2+\cdots$ is a unit in the power series ring $F[[x]]$ (exercise 1.7.5) if and only if $a_0 \neq 0$.

5 In $\mathbb{Z}[\sqrt{-5}]$ show that $3 \nmid 1+2\sqrt{-5}$. Deduce that 3 is irreducible but not prime in $\mathbb{Z}[\sqrt{-5}]$. [Hint: $3|(1+2\sqrt{-5})(1-2\sqrt{-5})$.] Show that if a^2+5b^2 is a prime in $\mathbb{Z}$ then $a+b\sqrt{-5}$ is irreducible in $\mathbb{Z}[\sqrt{-5}]$, but not conversely. [Hint: 3 is irreducible.]

6 Let $\mathbb{Q}[\sqrt{-3}]$ denote the ring of all complex numbers $\dfrac{a+b\sqrt{-3}}{2}$ where $a, b \in \mathbb{Z}$ and a, b are both even or both odd as introduced in exercise 3.2.14. Show that the six numbers 1, -1, and $\dfrac{\pm 1 \pm \sqrt{-3}}{2}$ are precisely the units of $\mathbb{Q}[\sqrt{-3}]$. Deduce that, in $\mathbb{Q}[\sqrt{-3}]$, 2 and $1+\sqrt{-3}$ are associates and that hence each divides the other.

7 If $a|b$ and $b|a$ in a ring R must a and b be associates? Prove, or give a counterexample. What if R has a unity element?

8 Identify the irreducible and the prime elements in (i) $\mathbb{Z}_{12}$; (ii) $\mathbb{Q}$; (iii) $\mathbb{Q}[[x]]$; (iv) the ring of exercise 3.2.6.

9 Show that if $\pi | abc$ in R and if π is a prime then $\pi | a$ or $\pi | b$ or $\pi | c$. Generalise this from 3-fold to n-fold products of elements of R.

10 Is 2 or any of its associates a square in $\mathbb{Z}[\sqrt{-1}]$? Show that associates of primes (irreducibles) in $\mathbb{Z}[\sqrt{-1}]$ are again primes (irreducibles).

11 Prove Lemma 3.6.3.

12 (a) Prove that in an integral domain each prime element is necessarily irreducible. [Hint: Copy the proof of 1.3.7.]
(b) Give an example of an irreducible in $\mathbb{Z}[\sqrt{-3}]$ which is not a prime.
(c) Show that, in $\mathbb{Z}[\sqrt{-3}]$, $\alpha = 1 + \sqrt{-3}$ and $\beta = 1 - \sqrt{-3}$ are relatively prime. Show that $\alpha\beta = 2^2$ and yet neither α nor β is a square in $\mathbb{Z}[\sqrt{-3}]$ (cf. exercise 3.5.1).
(d) Show that 2 and $1 + \sqrt{-3}$ are each (common) divisors of 4 and $2 + 2\sqrt{-3}$ in $\mathbb{Z}[\sqrt{-3}]$, but that 4 and $2 + 2\sqrt{-3}$ have no gcd in $\mathbb{Z}[\sqrt{-3}]$.

13 Show that in $\mathbb{Z}[\sqrt{-7}]$ the integer 8 can be expressed both as a product of two and as a product of three irreducibles.

14 Show that 1 is a gcd of x and y in $\mathbb{Q}[x, y]$. Do there exist $r(x, y)$ and $s(x, y) \in \mathbb{Q}[x, y]$ such that $r(x, y)x + s(x, y)y = 1$?

15 Let $R_2(\mathbb{R})$ denote the subring $\left\{\begin{pmatrix} 0 & a \\ 0 & b \end{pmatrix} : a, b \in \mathbb{R}\right\}$ of $M_2(\mathbb{R})$. Put $A = \begin{pmatrix} 0 & 1 \\ 0 & 0 \end{pmatrix}$, $B = \begin{pmatrix} 0 & 2 \\ 0 & 0 \end{pmatrix}$. Show that there exist infinitely many $C \in R_2(\mathbb{R})$ such that $AC = B$ but no $D \in R_2(\mathbb{R})$ such that $DA = B$.

3.7 Euclidean rings, unique factorisation domains and principal ideal domains

The fact that the analogue of the Fundamental Theorem of Arithmetic fails in $\mathbb{Z}[\sqrt{-3}]$ should make us all the keener to isolate the reasons why it holds in $\mathbb{Z}$. Uniqueness of factorisation in $\mathbb{Z}$ was proved as follows:

(A) Existence of factorisation: either $a \in \mathbb{Z}$ cannot be factorised or it can into smaller factors, and this process terminates after finitely many steps.

(B) Uniqueness depended upon: (α) every irreducible is a prime; (β) if p is a prime and if $p | ab$ then $p | a$ or $p | b$. Further, (α) was a consequence of the Division Algorithm.

This analysis indicates that we might be able to establish uniqueness of factorisation in *any* ring satisfying

Definition 3.7.1^0 Let R be a domain. Then R is called a **Euclidean Ring** (briefly ER) if and only if to each element a of R there corresponds a non-negative integer $\|a\|$ such that

(I⁰) For all $a, b \in R$, $\|a\| \cdot \|b\| = \|ab\|$

and

(II⁰) For all $a, b \in R$ with $b \neq 0$ there exist $m, r \in R$ such that $a = mb + r$ and $0 \leq \|r\| < \|b\|$

Of course $\mathbb{Z}$ is an ER if we define $\|z\|$ to be $|z|$. Unfortunately 3.7.1⁰ is not so obviously suitable for use with $\mathbb{Q}[x]$ since, for the most natural definition of $\| \ \|$, namely $\|f\| = \text{degree } f$, $\|0\|$ has no meaning. Further $\|fg\| \neq \|f\| \cdot \|g\|$ in general.

Remedies are available (see exercise 2) but it is somewhat more natural to modify 3.7.1⁰ without injuring $\mathbb{Z}$. Taking these points and 1.10.1 into account we arrive at

Definition 3.7.1 Let R be a domain. Then R is called a **Euclidean Ring** (ER) if and only if there exists a map δ, called a **norm** or **valuation**, from $R\backslash\{0\}$ into $\mathbb{Z}^+ \cup \{0\}$ such that

(I) For all non-zero $a, b \in R$, $\delta(a) \leq \delta(ab)$.

(II) Let $a, b \in R$ with $b \neq 0$. Then there exist $m, r \in R$ such that $a = mb + r$ where either (i) $r = 0$ or (ii) $r \neq 0$ and $\delta(r) < \delta(b)$.

There then follow easily

Examples 3.7.2

(i) $\mathbb{Z}$ is an ER if we set $\delta(z) = |z|$ for each non-zero $z \in \mathbb{Z}$.

(ii) $\mathbb{Q}[x]$ is an ER if we set $\delta(f) = \deg f$ for each non-zero $f \in \mathbb{Q}[x]$.

(iii) Each field F is an ER if we define $\delta(a) = 1$ for each non-zero $a \in F$.

We present less trivial examples in 3.7.4 below.

Notes 3.7.3

(i) A ring may well be Euclidean with respect to more than one norm (compare 3.7.2(i), (ii) with exercises 1(d), (e) below).

(ii) Another way of defining Euclidean rings, this time with $\mathbb{Q}[x]$ as its model, is given in exercise 2.

Further examples of ERs are given by

Theorem 3.7.4* The rings $\mathbb{Z}[\sqrt{-1}]$, $\mathbb{Z}[\sqrt{-2}]$ and $\mathbb{Z}[\rho] = \{u + v\rho : u, v \in \mathbb{Z}\}$ where $\rho = \dfrac{-1+\sqrt{-3}}{2}$ are all ERs.

* $\mathbb{Z}[\sqrt{-1}]$ is called the ring of **Gaussian Integers**. (See Prologue, p. xx.)

Proof For each α in $\mathbb{Z}[\sqrt{-1}]$ or in $\mathbb{Z}[\sqrt{-2}]$ we set $\delta(\alpha)=N(\alpha)$ as in 3.6.2. Condition (I) of 3.7.1 holds by 3.6.3 (iii) and the fact that if $\beta \neq 0$ then $N(\beta) \geqslant 1$.

For $\mathbb{Z}[\rho]$ we define $\delta(u+v\rho)$ to be $(u+v\rho)(u+v\rho^2)=u^2-uv+v^2$ (recall $\rho^3=1$ and $1+\rho+\rho^2=0$). Further, $\delta(u+v\rho)=\left(u-\frac{v}{2}\right)^2+\frac{3v^2}{4}>0$ unless $u=v=0$ whilst $\delta(\alpha\beta)=\delta(\alpha)\delta(\beta)$ since $\delta(\alpha)$, as in the other two cases, is in fact nothing more than the square of the modulus* of the complex number α.

So once again condition (I) holds.

Let us show that condition (II) holds in $\mathbb{Z}[\rho]$. Thus we are given $\alpha = u+v\rho$ and $\beta = s+t\rho \neq 0$. We form $\frac{\alpha}{\beta}=\frac{u+v\rho}{s+t\rho}=\frac{(u+v\rho)(s+t\rho^2)}{s^2-st+t^2}=l+m\rho$ for suitable $l, m \in \mathbb{Q}$. Now choose $L, M \in \mathbb{Z}$ such that* $|L-l| \leqslant \frac{1}{2}$ and $|M-m| \leqslant \frac{1}{2}$ and write $\frac{\alpha}{\beta}=(L+M\rho)+K$ where $K=(l-L)+(m-M)\rho$. Now $|K|^2=(l-L)^2-(l-L)(m-M)+(m-M)^2 \leqslant (\frac{1}{2})^2+\frac{1}{2}\cdot\frac{1}{2}+(\frac{1}{2})^2<1$ because both $|l-L|$ and $|m-M|$ are $\leqslant \frac{1}{2}$.

If we now write $\alpha=(L+M\rho)\beta+K\beta$ we see that both $L+M\rho$ and $K_1=K\beta$ are in $\mathbb{Z}[\rho]$. (Why is K_1?) Further $\delta(K_1)=|K|^2\delta(\beta)<\delta(\beta)$ and condition (II) of 3.7.1 has been shown to hold.

Similar working applies to $\mathbb{Z}[\sqrt{-1}]$ and $\mathbb{Z}[\sqrt{-2}]$ and is left to exercises 1(a) and 1(b).

We now set about obtaining those results which will indeed show that every ER has the uniqueness of factorisation property (3.7.13). Keeping one eye on the proof of the corresponding result (1.5.1) for $\mathbb{Z}$ we begin with

Theorem 3.7.5 (cf. 1.4.4) Let R be an ER with respect to the norm δ. Any two elements a, b not both 0 have a gcd c, say, expressible in the form $c=sa+tb$ for suitable $s, t \in R$. If, further d is a also a gcd of a and b then c and d are associates.

Proof Let S denote the set of all elements of R expressible in the form $ma+nb$ where $m, n \in R$. Since $1a+0b$ and $0a+1b$ belong to S we see that S does not comprise 0 alone. From all the non-zero elements of S choose one, $c=m_0a+n_0b$, say, with $\delta(c)$ as small as possible. Now suppose $w=ua+vb \in S$. By 3.7.1(II) there exist $k, r \in R$ such that $w=kc+r$ where either $r=0$ or $\delta(r)<\delta(c)$. Now $r=w-kc=(u-km_0)a+(v-kn_0)b \in S$. Thus the possibility $r \neq 0$ and $\delta(r)<\delta(c)$ is untenable [why?] and we are forced to deduce that $r=0$. But then $w=kc$. Consequently c divides every element in S; in particular $c|a$ and $c|b$. Thus c is a common divisor of a and b.

Next, if $c_1 \in R$ is such that $c_1|a$ and $c_1|b$ then $c_1|m_0a+n_0b=c$. This shows that c is a gcd of a and b.

Finally if c and d are gcds of a and b then $c=xd$ and $d=yc$ for suitable $x, y \in R$. It follows that $c=xd=xyc$, whence $1=xy$. Thus x is a unit, as required.

* See footnote p. 63.

Notes 3.7.6
(i) Applied to $\mathbb{Q}[x]$, $\mathbb{R}[x]$, etc., 3.7.5 lies at the heart of the proof of the well known results on partial fraction decomposition (see [32, pp. 165–70]).
(ii) For actually finding gcds in specific cases the analogue of the Euclidean Algorithm in Section 1.4 and exercise 1.10.3 is more useful than 3.7.5.

Example 3.7.7 Find, in $\mathbb{Z}[\sqrt{-1}]$, all the gcds of $10+11\mathrm{i}$ and $8+\mathrm{i}$.
To find $(10+11\mathrm{i}, 8+\mathrm{i})$ we look at

$$\frac{10+11\mathrm{i}}{8+\mathrm{i}}=\frac{10+11\mathrm{i}}{8+\mathrm{i}}\cdot\frac{8-\mathrm{i}}{8-\mathrm{i}}=\frac{91+78\mathrm{i}}{65}=(1+\mathrm{i})+\left(\frac{26}{65}+\frac{13}{65}\mathrm{i}\right)$$

$\left(\text{Here we write } \frac{91+78\mathrm{i}}{65} \text{ as } (1+\mathrm{i})+\cdots \text{ since 1 and 1 are the nearest integers to } \frac{91}{65} \text{ and } \frac{78}{65} \text{ respectively.}\right)$ Multiplying up by $8+\mathrm{i}$ we get

$$10+11\mathrm{i}=(1+\mathrm{i})(8+\mathrm{i})+(3+2\mathrm{i}) \qquad \text{(i)}$$

Next, to find $(8+\mathrm{i}, 3+2\mathrm{i})$ we look at

$$\frac{8+\mathrm{i}}{3+2\mathrm{i}}=\frac{8+\mathrm{i}}{3+2\mathrm{i}}\cdot\frac{3-2\mathrm{i}}{3-2\mathrm{i}}=\frac{26-13\mathrm{i}}{13}=2-\mathrm{i} \text{ exactly.}$$

Hence

$$8+\mathrm{i}=(2-\mathrm{i})(3+2\mathrm{i}) \qquad \text{(ii)}$$

Thus, as in the case of $\mathbb{Z}$, the last non-zero remainder, in this case $3+2\mathrm{i}$, in the sequence (i), (ii), ... of division algorithms is a gcd of $10+11\mathrm{i}$ and $8+\mathrm{i}$. The others are, of course, $-2+3\mathrm{i}$, $-3-2\mathrm{i}$ and $2-3\mathrm{i}$ [because ... why?].
From 3.7.5 we can deduce

Theorem 3.7.8 (cf. 1.4.10) Let R be an ER. If $\pi \in R$ is irreducible then π is prime.

Proof Suppose $b, c \in R$ and $\pi | bc$. If $\pi | b$ we are finished so suppose $\pi \nmid b$. Since π is irreducible its only divisors are of the form u and $u\pi$ where u is a unit. Since $\pi \nmid b$ the only divisors common to π and b are the units, one of which is 1_R. Thus $(\pi, b)=1_R$ and by 3.7.5 there exist $s, t \in R$ such that $s\pi + tb = 1_R$. But then $s\pi c + tbc = c$ whence, clearly, $\pi | c$ as required.

This is the theorem which will tell us that decompositions into products of irreducibles are unique. All we need now is to prove that such decompositions exist! For this we shall need

Lemma 3.7.9 Let a, b be non-zero elements in R, b being a non-unit. Then $\delta(a)<\delta(ab)$.

Proof By 3.7.1(I) we know $\delta(a) \leqslant \delta(ab)$. Suppose $\delta(a) = \delta(ab)$. By 3.7.1(II) there exist $m, r \in R$ such that $a = m(ab) + r$ where either $r = 0$ or else $r \neq 0$ and $\delta(r) < \delta(ab)$. If $r \neq 0$ then $\delta(r) < \delta(a)$. But $r = a(1 - mb)$ and so 3.7.1(I) implies $\delta(r) \geqslant \delta(a)$. These inequalities show that $r \neq 0$ is untenable. Hence $r = 0$, whence $mb = 1$ and b is a unit.

We can now have (cf. 1.3.9)

Theorem 3.7.10 Let R be an ER and let $a(\neq 0, \neq \text{unit})$ be an element of R. Then a is expressible as a product of irreducibles.

Proof Assume the desired result false. Then from amongst the non-empty set S of non-zero, non-unit elements a of R which are *not* expressible as products of irreducibles select one m, say, for which $\delta(m)$ is as small as possible. Now m is certainly not irreducible [why not?] and so $m = uv$ for suitable non-units $u, v \in R$. By 3.7.9 we find $\delta(u) < \delta(uv) = \delta(m)$. Similarly $\delta(v) < \delta(m)$. Hence $u \notin S$ and $v \notin S$. This means u, v *are* expressible as products of irreducibles. But then so is m [being equal to the product of these two products!], a contradiction.

We are now ready to prove the analogue for ERs of Theorems 1.5.1 and 1.9.18. Notice the change in format compared with the statements of 1.5.1 and 1.9.18. The reason for the change is given in the Remarks following the proof.

Theorem 3.7.11 (The Unique Factorisation Theorem for ERs) Let R be an ER and let a be a non-zero non-unit in R. If $a = \pi_1\pi_2 \ldots \pi_m = \pi'_1\pi'_2 \ldots \pi'_n$ where the π_i and π'_j are irreducibles then $m = n$ and there is a 1–1 correspondence between the π_i and the π'_j such that corresponding elements are associates.

Proof If the theorem is false there exists $a \in R$ with decompositions as given but in which the π_i and π'_j do not pair off. From the set of all such 'nasty' elements a, choose one for which m is as small as possible. Since $\pi_m | \pi'_1 \ldots \pi'_n$ and since π_m is prime (by 3.7.8) we see (exercise 3.6.9) that $\pi_m | \pi'_j$ for some j. Since π_m and π'_j are irreducible they are associates [proof?], $\pi'_j = u\pi_m$, say. But then, in R, $a = b\pi_m$ where

$$b = \pi_1\pi_2 \ldots \pi_{m-1} = \pi'_1 \ldots \pi'_{j-1}(u\pi'_{j+1}) \ldots \pi'_n$$

By choice of m, b is not a nasty element.* Hence the π_i $(1 \leqslant i \leqslant m-1)$ and the remaining π'_k pair off as associates. In particular $m - 1 = n - 1$, so that $m = n$. Since π_m and π'_j were also paired off we see that a is not nasty after all. This contradiction proves the theorem.

* What if $m = 1$? Certainly b won't be 'nasty' since it will be a unit and not a product of irreducibles. Does this affect the argument?

Remark The format of 1.5.1 and 1.9.18 could have been preserved in the statement of 3.7.11 if we had, in advance, selected within each set of associate elements in R a representative. In $\mathbb{Z}$ and in $\mathbb{Q}[x]$ obvious representatives stand out—namely positive integers and monic polynomials. In a general ER no such natural choice presents itself and the slight change of format seems preferable to having to choose a system of representatives at random.

We can now summarise 3.7.10 and 3.7.11 by introducing

Definition 3.7.12 Let D be an integral domain. D is a **Unique Factorisation Domain (UFD)** if and only if (i) every non-zero non-unit element a of D can be written $a = \pi_1\pi_2 \ldots \pi_m$, the π_i being irreducibles, and (ii) *if* $a = \pi_1 \ldots \pi_m = \pi'_1 \ldots \pi'_n$ (with irreducibles π_i, π'_j) *then* $m = n$ and the π_i and π'_j pair off as associates.

Then 3.7.10 and 3.7.11 combine into

Theorem 3.7.13 Every ER is a UFD. (In particular, $\mathbb{Z}$, $\mathbb{Z}[\sqrt{-1}]$, $\mathbb{Z}[\sqrt{-2}]$, $\mathbb{Z}[\rho]$ and* $F[x]$, F being any field, are all UFDs.)

It must be admitted straight away that 3.7.13 is not the best general theorem that can be proved. In example 3.4.7(iv) we introduced the notion of principal ideal. An integral domain in which every ideal is a principal ideal is called (naturally enough!) a **Principal Ideal Domain (PID)**. We give a very quick proof (see if you can follow it through) of

Theorem 3.7.14 Every PID is a UFD.

Proof
(i) Let $a, b \in R$, R being a PID and $b \neq 0$. Then $[a, b] = [d]$ for some $d = ar_1 + br_2 \neq 0$ in R. d is clearly a gcd of a and b. Following 3.7.8 we see that every irreducible in R is a prime.
(ii) Let $a \in R$ be a non-zero non-unit. If a is not a product of irreducibles then, certainly, a is not itself irreducible and so $a = a_1b_1$ where not both the non-units a_1, b_1 can be (products of) irreducibles. If a_1 (say) is not a product of irreducibles then $a_1 = a_2b_2$ where not both the non-units a_2, b_2 are (products of) irreducibles. Continuing in this way produces a sequence $[a] \subseteq [a_1] \subseteq [a_2] \subseteq \cdots$ of ideals in R. The set-theoretic union of this sequence is again an ideal (exercise 3.4.19) and therefore principal, $[z]$, say. But $z \in [z]$ and hence $z \in [a_j]$ for some j. Thus $[z] \subseteq [a_j] \subseteq [a_{j+1}] \subseteq [z]$ and $[a_j] = [a_{j+1}]$ follows. In particular $a_{j+1} = a_jr$ (for some $r \in R$). But $a_j = a_{j+1}b_{j+1}$, hence $a_{j+1} = a_{j+1}b_{j+1}r$. It follows that b_{j+1} is a unit—contradiction. Hence every non-zero non-unit $a \in R$ is a product of irreducibles.

Repeating the proof of 3.7.11 we obtain the desired result.

* $F[x]$ is pretty clearly an integral domain and we define $\delta(f) = \deg f$ just as for $\mathbb{Q}[x]$

Is there a relationship between ERs and PIDs? Indeed there is.

Theorem 3.7.15 Every ER is a PID.

Proof Let R be an ER and I an ideal of R. If $I=\{0\}$ then $I=[0]$ and is principal. If $I\neq\{0\}$ we choose in I an element a $(\neq 0)$ for which $\delta(a)$ is as small as possible. We claim $I=[a]$. If now $b\in I$ then by 3.7.1 we can find $m, r\in R$ such that $b=ma+r$ where $r=0$ or $r\neq 0$ and $\delta(r)<\delta(a)$. In this latter case we note $r=b-ma\in I$ [why?] and yet $\delta(r)<\delta(a)$, contrary to the choice of a. Thus $r=0$ is the only possibility. It follows that $b=ma$ and hence $b\in[a]$, as required. [Is there nothing else to prove?]

Confirming assertions made in examples 3.4.7(i), (ii) we have

Corollary 3.7.16 In each of the rings $\mathbb{Z}$, $F[x]$ where F is any field, $\mathbb{Z}[\sqrt{-1}]$, $\mathbb{Z}[\sqrt{-2}]$, and $\mathbb{Z}[\rho]$ each ideal is a principal ideal.

Proof 3.7.2 and 3.7.4 indicate that all these rings are Euclidean.

Remark If you re-read the proof of 3.7.15, keeping in mind exactly what demands are made on the ring R, you may be in for a surprise. For nowhere do we seem to have used the fact that condition 3.7.1(I) holds in R. That is, every integral domain for which 3.7.1(II) holds is necessarily a PID and hence (by 3.7.14) a UFD; the specially incorporated condition 3.7.1(I) is redundant! Observations of this kind have been at the basis of research papers written as recently as 1971 [119].

Problem 2 Let R be a commutative ring with no (non-zero) divisors of zero and suppose R satisfies condition 3.7.1(II). Confirm that R necessarily has a multiplicative identity, hence that R is an integral domain and consequently an ER.

In 3.7.15 and 3.7.14 we proved that every ER is a PID and every PID is a UFD. These assertions demand that we consider the converse questions as to whether or not every UFD is a PID and every PID is an ER. In fact the answer to each question is 'no'. Indeed we have already seen (exercise 3.4.22) that the ring $\mathbb{Z}[x]$ is not a PID whilst exercise 14 below invites you to prove that $\mathbb{Z}[x]$ is a UFD.

Problem 3 Here is a very short proof that $\mathbb{Z}[x]$ is a UFD. Is it a valid proof? Proof: $\mathbb{Z}[x]$ is a subring of the UFD $\mathbb{Q}[x]$. Hence $\mathbb{Z}[x]$ is a UFD immediately.

To see that there are PIDs which are not ERs (no matter how skilfully one tries to choose the mapping δ) one can consult an expanded account [130] of a result of Motzkin [113] which says that the set $\mathbb{Q}[\sqrt{-19}]$ of all algebraic integers in the field $\mathbf{Q}(\sqrt{-19})$ (see exercise 3.2.14) is a PID which is not an ER.

Concerning the number fields $\mathbf{Q}(\sqrt{d})$ where d is any (non-zero) square-free integer we have the following information. The algebraic integers lying in $\mathbf{Q}(\sqrt{d})$ form a subring denoted by $\mathbf{Q}[\sqrt{d}]$. If $d \equiv 2$ or 3 (mod 4), then $\mathbf{Q}[\sqrt{d}] = \{a + b\sqrt{d}: a, b \in \mathbb{Z}\}$, a set we have previously denoted by $\mathbb{Z}[\sqrt{d}]$. If $d \equiv 1$ (mod 4), then $\mathbf{Q}[\sqrt{d}] = \left\{\frac{a+b\sqrt{d}}{2}: a, b \in \mathbb{Z} \text{ and both are odd or both are even}\right\}$. It is known that $\mathbf{Q}[\sqrt{d}]$ is an ER iff $d = -11, -7, -3, -2, -1, 2, 3, 5, 6, 7, 11, 13, 17, 19, 21, 29, 33, 37, 41, 57, 73$. Further $\mathbf{Q}[\sqrt{d}]$ is a UFD for many positive values of d (it is conjectured to be a UFD for *infinitely* many positive d), whereas for $d < 0$, it is a UFD iff $d = -1, -2, -3, -7, -11, -19, -43, -67, -163$ (see [123, dated 1969]). Finally, if $d < 0$ and if $\mathbf{Q}[\sqrt{d}]$ is Euclidean then it is Euclidean with respect to the norm function as defined in 3.6.2. For $d > 0$ it seems to be unknown if $\mathbf{Q}[\sqrt{d}]$ can be Euclidean without being Euclidean with respect to the norm (see [118]).

The inquisitive reader will already be asking for values of d for which $\mathbf{Q}[\sqrt{d}]$ is a UFD but not a PID. This is readily answered: there is none (see Theorem 3.9.4).

After seeing the relative ease with which 3.7.15 and 3.7.14 were obtained the reader may ask why we didn't go straight to those theorems and omit the long journey to 3.7.13. The answer is partly one of learning to walk before one can run. The professional mathematician is usually quite happy to let an author pluck general definitions and theorems seemingly from thin air, so long as familiar examples do appear as special cases. But this text is not written for one with such maturity.

Exercises

1 Show that the following rings together with the given norms are ERs:

(a) $\langle \mathbb{Z}[\sqrt{-1}], +, \cdot\rangle$ where $\delta(a+ib) = a^2 + b^2$;

(b) $\langle \mathbb{Z}[\sqrt{-2}], +, \cdot\rangle$ where $\delta(a+b\sqrt{-2}) = a^2 + 2b^2$;

(c) $\langle \mathbb{Z}[\sqrt{2}], +, \cdot\rangle$ where $\delta(a+b\sqrt{2}) = |a^2 - 2b^2|$;

(d) $\langle \mathbb{Z}, +, \cdot\rangle$ where* $\delta(z)$ is the number of digits in a binary representation of $|z|$ (thus $\delta(-6) = 3$ since, in binary notation, 110 represents 6);

(e) $\langle \mathbf{Q}[x], +, \cdot\rangle$ where $\delta(f) = 2^{\deg f}$;

(f) $\langle \mathbf{Q}[[x]], +, \cdot\rangle$ where $\delta\left(\sum_{i=0}^{\infty} a_i x^i\right)$ is the smallest i for which $a_i \neq 0$;

(g) $\langle \mathbf{Q}[\sqrt{-7}], +, \cdot\rangle$ where $\mathbf{Q}[\sqrt{-7}] = \left\{a + b\left(\frac{1+\sqrt{-7}}{2}\right): a, b \in \mathbb{Z}\right\}$ and where $\delta\left(a + b\left(\frac{1+\sqrt{-7}}{2}\right)\right) = a^2 + ab + 2b^2$;

* Taken from [119, p. 289].

2 The following definition, clearly modelled upon $\mathbb{Q}[x]$, has been given ([32, p. 147]): Let R be an integral domain. Then R is called a Euclidean ring if and only if there exists a map η: $R\backslash\{0\}\to\mathbb{Z}^+\cup\{0\}$ such that, for $a\neq 0$ and $b\neq 0$ in R, (i) $\eta(ab)=\eta(a)+\eta(b)$; (ii) $\eta(a+b)\leqslant\max\{\eta(a),\eta(b)\}$ if also $a\neq -b$; (iii) there exist $m, r\in R$ such that $a=mb+r$ and either $r=0$ or ($r\neq 0$ and $\eta(r)<\eta(b)$). Is it true that a ring Euclidean in this sense is also Euclidean in the sense of 3.7.1? What about the converse question? That is, are the definitions equivalent?

3 Apply the technique of 3.7.4 and 3.7.7 to find m, r satisfying 3.7.1(II) when
(i) $a=29+20i$, $b=2+19i$ in $\mathbb{Z}[\sqrt{-1}]$;
(ii) $a=10+69\sqrt{-2}$, $b=10+18\sqrt{-2}$ in $\mathbb{Z}[\sqrt{-2}]$.
Find the gcds of each of these pairs of elements.

4 Show that $-4+i=(-1+i)[5+3i]+(4-i)=-1[5+3i]+(1+4i)$. Deduce that, in the division algorithm for $\mathbb{Z}[\sqrt{-1}]$, m and r are not unique. (They *are* in $\mathbb{Z}$: exercise 1.4.8.) Can you explain this?

5 Prove that if $f, g\in\mathbb{Q}[x]$, where $\deg f\geqslant 1$ and $\deg g\geqslant 1$, and if $(f,g)=1$ then there exist $r, s\in\mathbb{Q}[x]$ such that $rf+sg=1$ and $\deg r<\deg g$ and $\deg s<\deg f$.

6 Show that $\mathbb{Z}[\sqrt{-6}]$, $\mathbb{Z}[\sqrt{10}]$ cannot be made into ERs no matter how skilfully you may try to choose δ.

7 Verify that, in $\mathbb{Z}[\sqrt{2}]$, $(5+\sqrt{2})(2-\sqrt{2})=(11-7\sqrt{2})(2+\sqrt{2})$, each of these elements being irreducible since their norms are primes in $\mathbb{Z}$. Surely this contradicts the assertion of 1(c) above that $\mathbb{Z}[\sqrt{2}]$ is an ER and hence a UFD?

8 Corollary 3.7.16 says that $\mathbb{Z}$ and $\mathbb{Q}[x]$ are PIDs. Find two distinct values of z and a monic polynomial f such that (i) $[z]=\{77z_1+91z_2+143z_3: z_1, z_2, z_3\in\mathbb{Z}\}$, (ii) $[f]=\{(x^3+2x^2+3x+2)g+(x^4+x^3+3x^2+4x+1)h: g, h\in\mathbb{Q}[x]\}$.

9 Let R be an ER. Show that (i) $\delta(1)\leqslant\delta(a)$ in $\mathbb{Z}$ for all $a\neq 0$ in R; (ii) $\delta(a)=\delta(1)$ iff a is a unit in R; (iii) $\delta(a)=\delta(b)$ if a and b are associates in R.

10 An element m in a commutative ring R is a *least common multiple* (*lcm*) of elements $a, b\in R$ iff (i) $a|m$ and $b|m$ in R and (ii) $a|k$ and $b|k$ in R imply $m|k$ in R. Show that in a UFD each pair of elements, not both zero, has a gcd and an lcm.

11 It can be shown ([6, p. 96]) that in a domain each pair of elements (not both zero) has a gcd iff each pair (not both zero) has an lcm. Show that in $\mathbb{Z}[\sqrt{-3}]$ the elements 2 and $1+\sqrt{-3}$ have gcd 1 and yet have no lcm at all. [Hint: Use norms.] Doesn't this contradict the theorem just referred to?

12 Let D be a domain in which each non-zero non-unit element has at least one factorisation into irreducibles. Prove that D is a UFD iff each irreducible in D is a prime.

13 Let $\mathbf{Q}_{\mathbb{Z}}[x]=\{f: f\in\mathbf{Q}[x]$ and f has integer constant term$\}$. Given $f, g\in \mathbf{Q}_{\mathbb{Z}}[x]$ let d be a common divisor of maximum possible degree in $\mathbf{Q}_{\mathbb{Z}}[x]$. Show that $f_0=f/d$ and $g_0=g/d$ (both in $\mathbf{Q}_{\mathbb{Z}}[x]$) have gcd 1 in $\mathbf{Q}[x]$. Deduce that $z=mf_0+ng_0$ for suitable $z\in\mathbb{Z}$ and m and n in $\mathbf{Q}_{\mathbb{Z}}[x]$. Write $f_0=f_1+\alpha$, $g_0=g_1+\beta$ where α, β are the constant terms in f_0 and g_0. Show that $\alpha = f_0-\left(\frac{f_1}{z}\right)z$ lies in the ideal $[f_0, g_0]$ in $\mathbf{Q}_{\mathbb{Z}}[x]$ and similarly for β. Set $c=(\alpha,\beta)$ in $\mathbb{Z}$. Deduce $[f_0, g_0]=[c]$ and hence that $[f, g]=[cd]$, a principal ideal in $\mathbf{Q}_{\mathbb{Z}}[x]$. Deduce from $x=\frac{1}{2}x\cdot 2=\frac{1}{4}x\cdot 2^2=\cdots$ that x cannot be expressed as a product of irreducibles in $\mathbf{Q}_{\mathbb{Z}}[x]$ so that $\mathbf{Q}_{\mathbb{Z}}[x]$ is not a UFD. This surely contradicts 3.7.14 (every PID is a UFD)? [This example is due to P M Cohn.]

14 Prove that $\mathbb{Z}[x]$ is a UFD as follows: Use exercise 3.6.2 and the concept of 'content' to write every non-zero non-unit element f as a product $u_1u_2\ldots u_rf_1f_2\ldots f_s$ of irreducibles in $\mathbb{Z}[x]$ such that $\deg u_i=0$ and $\deg f_i\geqslant 1$. Now use the solution to Problem 3 of Section 1.9 on the f_i and then exercise 12 above.

15 Show that a subring of a UFD need not be a UFD. [Hint: $\mathbb{Z}[\sqrt{-3}]\subseteq\mathbb{Z}[\rho]$.] Compare this with Problem 3 in Section 3.7.

16 Try to copy the second proof of 1.5.1 (p. 34) to prove that $\mathbb{Z}[\sqrt{2}]$ and $\mathbb{Z}[\sqrt{-5}]$ are UFDs. (Careful!)

3.8 Three number-theoretic applications

We interrupt our discussion of uniqueness of factorisation to give three applications of the theory. To be sure these results can be obtained without leaving the ring of integers (see, for example, [42], [44]). Since, however, number theory has given rise to a great deal of present day algebra it is nice to repay the compliment and re-obtain these number-theoretic results by algebraic means. First we prove

Theorem 3.8.1* Let p be a (positive) odd prime in $\mathbb{Z}$. Then $p=a^2+b^2$, for suitable $a, b\in\mathbb{Z}$ if and only if p is of the form $4k+1$. Further if $p=a^2+b^2=c^2+d^2$ then either (i) $a=\pm c$ and $b=\pm d$ or (ii) $a=\pm d$ and $b=\pm c$. That is, the representation of p as a sum of squares is essentially unique.

Proof $\overset{\text{only if}}{\Rightarrow}$: If $u, v\in\mathbb{Z}$ and if $n=u^2+v^2$ is odd then n is of the form $4k+1$ (exercise 1).

* Stated but not proved by Fermat in a letter to Mersenne dated 25 December 1640. Euler gave a proof in 1754.

$\overset{if}{\Leftarrow}$: Suppose p is a (positive) $4k+1$ prime. Then, by exercise 2, $p|1+x^2$ in $\mathbb{Z}$ where $x=\left(\frac{p-1}{2}\right)!$ Thus $p|1+x^2=(1+ix)(1-ix)$ in $\mathbb{Z}[\sqrt{-1}]$. Since, clearly, $p\nmid 1+ix$ and $p\nmid 1-ix$ in $\mathbb{Z}[\sqrt{-1}]$, p is not prime and hence not irreducible in $\mathbb{Z}[\sqrt{-1}]$. Thus we can write $p=(l+im)(u+iv)$ as a product of non-units in $\mathbb{Z}[\sqrt{-1}]$. Since $\delta(l+im)\delta(u+iv)=\delta(p)=p^2$ we see that $l^2+m^2=\delta(l+im)=p$ in $\mathbb{Z}$, as required.

'Further': From $p=(a+ib)(a-ib)$ we find $\delta(a+ib)=\delta(a-ib)=p$, a prime in $\mathbb{Z}$. Hence (cf. exercise 3.6.5) $a+ib$, $a-ib$ are primes (i.e. irreducibles) in $\mathbb{Z}[\sqrt{-1}]$. Similarly for $c+id$, $c-id$. By uniqueness of factorisation in $\mathbb{Z}[\sqrt{-1}]$ (Theorem 3.7.13) the equality $(a+ib)(a-ib)=(c+id)(c-id)$ yields either (α) $a+ib=u(c+id)$ or (β) $a+ib=v(c-id)$ where $u,v\in\{1,i,-1,-i\}$ are units. Conditions (i) and (ii) above follow immediately.

Next we give, essentially, Gauss' proof* of the FC for $n=3$.

Theorem 3.8.2 The equation $x^3+y^3+z^3=0$ has no solution in non-zero elements of $\mathbb{Z}[\rho]=\mathbb{Z}\left[\frac{-1+\sqrt{-3}}{2}\right]$. In particular it has no solution in non-zero integers (see exercise 5).

Proof Suppose that $\alpha,\beta,\gamma\in\mathbb{Z}[\rho]$ do satisfy $x^3+y^3+z^3=0$. We may suppose α,β,γ have no common prime (i.e. irreducible) factor. [Otherwise cancel it and start afresh.] Then $(\alpha,\beta)=(\beta,\gamma)=(\gamma,\alpha)=1$ in $\mathbb{Z}[\rho]$. [Why?] Set $a=\beta+\gamma$, $b=\gamma+\alpha$, $c=\alpha+\beta$. Then $(a+b+c)^3-24abc=(b+c-a)^3+(c+a-b)^3+(a+b-c)^3=0$. Setting $\pi=1-\rho$, we get $\delta(\pi)=3$ (see proof of 3.7.4). Hence π is irreducible (i.e. prime) in $\mathbb{Z}[\rho]$. [Why?] Also $-\rho^2\pi^2=-\rho^2(1-\rho)^2=-\rho^2(1-2\rho+\rho^2)=3$. Hence $\pi|3$ in $\mathbb{Z}[\rho]$, and so $\pi|24abc=(a+b+c)^3$. Consequently $\pi^3|(a+b+c)^3=24abc$, whence $\pi|abc$. [For, if $\pi^3|24$ then $\pi^3|24-3^3=-3=\pi^2\rho^2$, whence $\pi|\rho^2$, a contradiction since ρ is a unit in $\mathbb{Z}[\rho]$.] If $\pi|a$ then $\pi|a^3=(\beta+\gamma)^3=\beta^3+\gamma^3+3\beta^2\gamma+3\beta\gamma^2$. Since $\pi|3$ we have $\pi|\beta^3+\gamma^3=-\alpha^3$. Hence $\pi|\alpha$, whence $\pi\nmid\beta$ and $\pi\nmid\gamma$.

Now it can be shown (exercise 8) that there are $c,d\in\mathbb{Z}$ such that† $\beta\rho^c\equiv 1$ or $\equiv -1 \pmod 3$ and $\gamma\rho^d\equiv 1$ or $\equiv -1 \pmod 3$ in $\mathbb{Z}[\rho]$. Since $(\beta\rho^c)^3=\beta^3$ and $(\gamma\rho^d)^3=\gamma^3$ we may assume that the β and γ given at the outset satisfy $\beta\equiv 1$ or $\equiv -1$ and $\gamma\equiv 1$ or $\equiv -1$. Now we cannot have $\beta\equiv 1$ *and* $\gamma\equiv 1$. For then $\beta^3+\gamma^3\equiv 1+1 \pmod 3$ in $\mathbb{Z}[\rho]$. Now $\pi|\alpha$, hence $\pi|\alpha^3$. Since $\pi|3$ we would have $\pi|\alpha^3+\beta^3+\gamma^3-2=0-2=-2$, and hence $\pi|-2+3=1$, a contradiction. The congruences $\beta\equiv -1\equiv\gamma$ lead to a similar contradiction. Thus WLOG we assume that $\beta\equiv 1$ and $\gamma\equiv -1$, that is $\beta=1+3\lambda$, $\gamma=-1+3\mu$ where

* Published posthumously.

† $x\equiv y \pmod z$ in the ring R means: there exists $w\in R$ such that $wz=x-y$. That is, $z|x-y$ *in* R.

$\lambda, \mu \in \mathbb{Z}[\rho]$. It follows that $\beta^3+\gamma^3=(1+3\lambda)^3+(-1+3\mu)^3$ is a multiple of 3^2 and hence of π^4 in $\mathbb{Z}[\rho]$. Set $A=\dfrac{\beta+\gamma\rho}{\pi}$, $B=\dfrac{\beta\rho+\gamma}{\pi}$, $C=\dfrac{(\beta+\gamma)\rho^2}{\pi}$. [Do A, B, C belong to $\mathbb{Z}[\rho]$?] Then $A+B+C=0$ whilst $ABC=\dfrac{\beta^3+\gamma^3}{\pi^3}=\left(\dfrac{-\alpha}{\pi}\right)^3$. Recalling that $\pi^4|\beta^3+\gamma^3$ we see that $\pi|ABC$ and hence $\pi^3|ABC$.

Next $\beta=-\rho A+\rho^2 B$ and $\gamma=\rho^2 A-\rho B$ so that $(A, B)=1$ in $\mathbb{Z}[\rho]$. [Otherwise we would have $(\beta, \gamma)\neq 1$ in $\mathbb{Z}[\rho]$.] Thus from $A+B+C=0$ we find $(B, C)=(C, A)=1$ in $\mathbb{Z}[\rho]$.

From $(A, B)=(B, C)=(C, A)=1$ and $\pi^3|ABC=\left(\dfrac{-\alpha}{\pi}\right)^3$ we deduce that (i) except for the possible presence of units, A, B, C are all cubes in $\mathbb{Z}[\rho]$ (cf. exercise 3.5.1) and (ii) π divides exactly one of A, B, C— let us suppose C. Set $A=u_1\phi^3$, $B=u_2\chi^3$, $C=u_3\psi^3$ where u_1, u_2, u_3 are units in $\mathbb{Z}[\rho]$. From $A+B+C=0$ we get, on multiplying through by u_1^{-1}, $\phi^3+u_4\chi^3+u_5\psi^3=0$ where u_4, u_5 are units. Since $\pi^3|C$ we have $\phi^3+u_4\chi^3\equiv 0 \pmod{\pi^3}$. Since $\pi\nmid\phi$ and $\pi\nmid\chi$ we may write $\phi^3\equiv\pm 1 \pmod{\pi^3}$ and $\chi^3\equiv\pm 1 \pmod{\pi^3}$. (For each element ν, say, of $\mathbb{Z}[\rho]$ is of the form $z_1+z_2\rho$ and hence of the form $z_3+z_4\pi$ where $z_1, z_2, z_3, z_4\in\mathbb{Z}$. From $\pi\nmid z_3+z_4\pi$ we deduce that $\pi\nmid z_3$. Since $\pi|3$ we see that z_3 is not a multiple of 3, in $\mathbb{Z}$. Hence $z_3\equiv 1$ or $z_3\equiv -1 \pmod 3$ in $\mathbb{Z}$. It follows that z_3, and hence ν, is congruent to 1 or $-1 \pmod{\pi}$ in $\mathbb{Z}[\rho]$. From $\nu=\pm 1+\sigma\pi$ $(\sigma\in\mathbb{Z}[\rho])$ we find $\nu^3=\pm 1+3\sigma\pi\pm 3\sigma^2\pi^2+\sigma^3\pi^3\equiv\pm 1 \pmod{\pi^3}$, as required.) We deduce that, mod π^3, $0\equiv\phi^3+u_4\chi^3\equiv\pm 1\pm u_4$. But, by exercise 3.6.6, $u_4=\pm 1$, $\pm\rho$ or $\pm\rho^2$. The congruence $\pm 1\pm u_4\equiv 0 \pmod{\pi^3}$ shows that $u_4=1$ or $u_4=-1$ are the only possibilities. Since $u_4=u_1^{-1}u_2$ we find $u_2=\pm u_1$. But $u_1u_2u_3$ is a unit and a cube. Hence $u_1u_2u_3=\pm 1$ whence $u_3=\pm\dfrac{1}{u_1u_2}=\pm\dfrac{1}{u_1^2}=\pm u_1$.

Thus the equalities $A+B+C=0$ and $ABC=\left(\dfrac{-\alpha}{\pi}\right)^3$ yield the equations $\phi^3+(\pm\chi)^3+(\pm\psi)^3=0$ and $(\phi\chi\psi)^3=\left(\dfrac{\pm\alpha}{\pi}\right)^3$.

Finally note that if the total number of πs (or associates of π) occurring in the decomposition of $\alpha\beta\gamma$ into a product of primes is n then, since they all occur in α, their total number in $\phi\chi\psi$ is $n-1$. Repeating the above process a further $n-1$ times we arrive at a solution to $x^3+y^3+z^3=0$ in which no πs are present. But this contradicts the remarks of the first paragraph of this proof.

Remark The method of 'reducing' a supposed solution to one which is in some sense 'lesser' was invented by Fermat and is called the **method of descent** (see exercise 6).

The third result, again due to Fermat and mentioned in a letter of 1657 is

Theorem 3.8.3 The only positive integer solution to the equation $x^2+2=y^3$ is $x=5$, $y=3$.

Proof We consider the equation inside the ER $\mathbb{Z}[\sqrt{-2}]$ where we may factorise x^2+2 as $(x+\sqrt{-2})(x-\sqrt{-2})$. Assume π is a prime (i.e. irreducible) in $\mathbb{Z}[\sqrt{-2}]$ which divides both $x-\sqrt{-2}$ and $x+\sqrt{-2}$. Then $\pi|(x+\sqrt{-2})-(x-\sqrt{-2})=2\sqrt{-2}=-(\sqrt{-2})^3$. Since $\sqrt{-2}$ is irreducible in $\mathbb{Z}[\sqrt{-2}]$ we see that $\pi=\sqrt{-2}$ (or $-\sqrt{-2}$). [Why no other possibilities?]

This means that $\sqrt{-2}=\pi|x-\sqrt{-2}$ whence $\sqrt{-2}|x$ and hence $2|x^2$, all in $\mathbb{Z}[\sqrt{-2}]$. Since $x\in\mathbb{Z}$, the assertion $2|x^2$ in $\mathbb{Z}[\sqrt{-2}]$ implies the assertion $2|x^2$ in $\mathbb{Z}$. [Why?] Thus $x\in 2\mathbb{Z}$ whence $x^2\in 4\mathbb{Z}$. This means y is even so that $y^3\in 8\mathbb{Z}$. Since $x^2+2\notin 8\mathbb{Z}$ [why not?], the assumption on π is false. That is, $(x+\sqrt{-2}, x-\sqrt{-2})=1$ in $\mathbb{Z}[\sqrt{-2}]$. Now using uniqueness of factorisation in $\mathbb{Z}[\sqrt{-2}]$ it follows that $x-\sqrt{-2}$ and $x+\sqrt{-2}$ are cubes* in $\mathbb{Z}[\sqrt{-2}]$. Setting $x-\sqrt{-2}=(a+b\sqrt{-2})^3$ for suitable $a,b\in\mathbb{Z}$ we find $x-\sqrt{-2}=a^3+3a^2b\sqrt{-2}+3ab^2(-2)-2b^3\sqrt{-2}$. It follows [why does it?] that $x=a^3-6ab^2=a(a^2-6b^2)$ and that $-1=3a^2b-2b^3=b(3a^2-2b^2)$. From this latter equality $b=\pm 1$ and hence $3a^2-2=\mp 1$. Then, necessarily, $b=-1$ and $a=\pm 1$. We deduce that $x=\pm 1\cdot(-5)$ so that x, being positive, is equal to 5, as required.

Exercises

1 Show that if $n=u^2+v^2$ is an odd integer then $n\equiv 1 \pmod 4$.

2 Show that for each positive prime p of the form $4k+1$, $p|1+\left\{\left(\frac{p-1}{2}\right)!\right\}^2$. (Use 2.5.4.)

3 Show that each positive prime in $\mathbb{Z}$ of the form $4k+3$ remains prime when regarded as an element of $\mathbb{Z}[\sqrt{-1}]$. Is the same true of any of the positive primes of the form $4k+1$?

4 For $\alpha=a+ib$ and $\beta=c+id$ in $\mathbb{Z}[\sqrt{-1}]$, find $N(\alpha\beta)$, $N(\alpha)$ and $N(\beta)$. Deduce that if x and y are integers each expressible as a sum of two squares then so is xy. Hence write 11 009 as a sum of two squares. [Hint: Factorise 11 009 as a product of primes.]

5 Show that (x_0, y_0, z_0) is a solution to $x^3+y^3+z^3=0$ iff $(x_0, y_0, -z_0)$ is a solution to $x^3+y^3=z^3$.

6 Using the fact that every solution of the equation $x^2+y^2=z^2$ is of the form given in exercise 3.5.2 prove that the equation $x^4+y^4=z^2$ has no solution

* Does it? What about the possibility that $x-\sqrt{-2}$ and $x+\sqrt{-2}$ are of the form $u\chi^3$ where $\chi\in\mathbb{Z}[\sqrt{-2}]$ and u is a unit in $\mathbb{Z}[\sqrt{-2}]$?

in integers (other than those for which $xyz = 0$) using Fermat's method of descent as follows.

Suppose $u^4 + v^4 = w^2$ where $u, v, w \in \mathbb{Z}$, $uvw \neq 0$ and w is as small as possible. Then

(i) $(u, v) = (v, w) = (w, u) = 1$ [otherwise a solution with smaller w can be found].

(ii) We may assume $u, v, w \in \mathbb{Z}^+$ and that not both u, v are odd [no square is of the form $4k + 2$].

(iii) Assuming u is odd, use exercise 3.5.2 on $(u^2)^2 + (v^2)^2 = w^2$ to deduce $u^2 = a^2 - b^2$, $v^2 = 2ab$, $w = a^2 + b^2$.

(iv) $(a, b) = 1$ and one of a, b is even, the other odd [w is odd].

(v) a must be odd [a even and b odd implies $a^2 - b^2 \equiv -1 \pmod 4$, whereas $u^2 \equiv 1 \pmod 4$, a contradiction].

(vi) Use exercise 3.5.2 on $u^2 + b^2 = a^2$ to get $u = l^2 - m^2, b = 2lm, a = l^2 + m^2$ where $(l, m) = 1$.

(vii) $v^2 = 2ab = 4lm(l^2 + m^2)$ and $(l, m) = (m, l^2 + m^2) = (l^2 + m^2, l) = 1$.

(viii) Hence each of $l, m, l^2 + m^2$ is a square, $l = r^2$, $m = s^2$, $l^2 + m^2 = t^2$, say.

(ix) $r^4 + s^4 = t^2$. But $rst \neq 0$ and $t < w$ so we have a new 'smaller' solution of the given equation, a contradiction.

Deduce the FC for $n = 4$. [Hint: rewrite $x^4 + y^4 = z^4$ as $x^4 + y^4 = (z^2)^2$.]

7 Following the line of argument given in exercise 6 try to prove *directly* that $x^4 + y^4 = z^4$ is impossible when $x, y, z \in \mathbb{Z}$ and $xyz \neq 0$. That is, start with a solution $u^4 + v^4 = w^4$ and try to obtain a smaller. [This is another example where a desired result is more easily obtained as a corollary of one stronger than that sought. An earlier example ocurs in exercise 1.2.18 which can be proved by establishing $1 + \frac{1}{2} + \cdots + \frac{1}{2^n} = 2 - \frac{1}{2^n}$ first.]

8 Show that there exist integers c, d as in the proof of 3.8.2 as follows.

Let $\beta = x + y\rho \in \mathbb{Z}[\rho]$. Then $\beta = u + v\pi$ where $u, v \in \mathbb{Z}$. Since $\pi \nmid \beta$ in $\mathbb{Z}[\rho]$ we see $3 \nmid u$ in $\mathbb{Z}$. Hence $u \equiv 1$ or $-1 \pmod 3$. Suppose $\beta \equiv 1 + v\pi \pmod 3$ in $\mathbb{Z}[\rho]$ (the other case being similar). Since $3 | \pi^2$, $\beta\rho^f \equiv (1 + v\pi)(1 - \pi)^f \equiv (1 + v\pi)(1 - f\pi) \equiv 1 + (v - f)\pi \pmod 3$ in $\mathbb{Z}[\rho]$. Taking $f = v$ does the trick. (Any problem if v is negative?)

9 Show that the only solution of $x^2 + 4 = y^3$ in which x and y are positive and x is odd is $x = 11$, $y = 5$ as follows: Write $x^2 + 4 = (x + 2i)(x - 2i)$. If $\pi | x + 2i$ and $\pi | x - 2i$ then $N(\pi) | x^2 + 4$ and $N(\pi) | N(4i) = 16$ (see 3.6.2). Thus $N(\pi) = 1$ and π is a unit. Thus $(x + 2i, x - 2i) = 1$. It follows that $x + 2i$ (and $x - 2i$) is a perfect cube*, $(a + bi)^3$, say. Then $x = a^3 - 3ab^2$, $2 = 3a^2b - b^3$.

10 In verifying Fermat's Conjecture for $n = 3$ Gauss considered the set $G(\rho) = \{a_0 + a_1\rho + a_2\rho^2 : a_0, a_1, a_2 \in \mathbb{Z}\}$ where $\rho = e^{2\pi i/3}$. Define the norm of $\alpha = a_0 + a_1\rho + a_2\rho^2$ by $\|\alpha\| = (a_0 + a_1\rho + a_2\rho^2)(a_0 + a_1\rho^2 + a_2\rho)$. Show that

* Might it not be, say, $i(a + bi)^3$? Cf. the proof of 3.8.2.

$G(\rho) = \mathbb{Z}[\rho] = \mathbb{Q}[\sqrt{-3}]$ (see 3.7.4 and exercise 3.2.14 for notation). Show that if $\alpha = a_0 + a_1\rho + a_2\rho^2 = b_0 + b_1\rho + b_2\rho^2 = \beta$ then $\|\alpha\| = \|\beta\|$ and that if

$$a_0 + a_1\rho + a_2\rho^2 = u + v\rho (\in \mathbb{Z}[\rho]) = \frac{l + m\sqrt{-3}}{2} (\in \mathbb{Q}[\sqrt{-3}])$$

then

$$\|\alpha\| = \delta(u + v\rho) = \left|\frac{l + m\sqrt{-3}}{2}\right|^2.$$

11. Euler's proof of the FC in the case $n = 3$ relied on his assertion that if $a, b \in \mathbb{Z}$ are such that $(a, b) = 1$ and if $a^2 + 3b^2$ is a cube in $\mathbb{Z}$ then $a + \sqrt{-3}b$ and $a - \sqrt{-3}b$ are both cubes in $\mathbb{Z}[\sqrt{-3}]$. Whilst this is in fact true ([46, p. 54]) it cannot be proved using uniqueness of factorisation in $\mathbb{Z}[\sqrt{-3}]$ since $\mathbb{Z}[\sqrt{-3}]$ is not a UFD! Give an example to show that if $a, b \in \mathbb{Z}$, if $(a, b) = 1$ and if $a^2 + 3b^2$ is a square in $\mathbb{Z}[\sqrt{-3}]$ then neither $a + \sqrt{-3}b$ nor $a - \sqrt{-3}b$ need be a square in $\mathbb{Z}[\sqrt{-3}]$ (see exercise 3.6.12).

Julius Wilhelm Richard Dedekind (*6 October 1831 – 12 February 1916*)
Richard Dedekind, born in Braunschweig (Brunswick), Germany, was the youngest of four children of a law professor. He studied with Gauss at Göttingen and took his PhD in 1852. After teaching in Göttingen and Zurich he moved in 1863 to the Technische Hochschule in Braunschweig where he remained until he retired in 1894. As a young man his first main interests were in chemistry and physics but he turned to mathematics because he appreciated its more precise arguments. Some of his early mathematical writings were in analysis and probability. In the late 1850s he became interested in algebraic number theory and was also disturbed by the lack of foundation to the real number system. The former interest found an outlet in his theory of ideals originally published in his supplements to Dirichlet's lectures on number theory. It was Dedekind who introduced the word *field* (*Körper*) and who gave the

first university course on Galois theory. Amongst his great work is his algebraic treatment, with H Weber, of algebraic functions. His interest in the real numbers arose in 1858 when, on teaching calculus, he saw there was little by way of solid basis to the various number systems necessary for a rigorous theory of limits. His solution to the problem, using 'cuts', was published in 1872. His methods made him a natural supporter of Cantor's set theory.

An accomplished musician—he even composed a small opera—he never married and lived with his sister until his death.

3.9 Unique factorisation reestablished. Prime and maximal ideals

In Section 3.5 we indicated why Lamé's 'proof' of the Fermat Conjecture was incomplete. In fact, the complex numbers appearing in Lamé's factorisation of x^p+y^p had already been investigated by Kummer* in his attempts to extend the reciprocity laws of Gauss beyond the quadratic, cubic and biquadratic cases. In more detail: Kummer had considered 'complex integers' of the form $f(\zeta)=a_0+a_1\zeta+\cdots+a_{p-2}\zeta^{p-2}$, p being an odd prime, $\zeta(\neq 1)$ a pth root of 1 and the $a_i \in \mathbb{Z}$. For a given ζ the set $\mathbb{Z}[\zeta]$ of all such integers forms an integral domain but, for $p=23$, not a UFD.

Now one can restore uniqueness of factorisation to the ring $\mathbb{Z}[\sqrt{-3}]$ by enlarging it to the UFD $\mathbb{Q}[\sqrt{-3}]$. Kummer tried the same kind of thing with the $\mathbb{Z}[\zeta]$ by introducing extra divisors in the form of his so-called *ideal* factors. In fact Kummer did not define the ideal factors explicitly but rather gave conditions under which a complex integer shall 'contain an ideal prime factor of the real prime p'.

We shall now outline very briefly the change in approach indicated by Dedekind around 1870.

Exercise 3.6.5 shows that the domain $D=\mathbb{Z}[\sqrt{-5}]=\{a+b\sqrt{-5}: a, b \in \mathbb{Z}\}$ is not a UFD; in particular $3\cdot 7=(1+2\theta)(1-2\theta)$ where $\theta=\sqrt{-5}$ and $3, 7, 1+2\theta, 1-2\theta$ are all irreducibles. Letting α_1 be the ideal factor common to 3 and $1+2\theta$ and letting $\alpha_2, \beta_1, \beta_2$ be defined analogously we see that we restore uniqueness of factorisation if we write $\alpha=3=\alpha_1\alpha_2$, $\beta=7=\beta_1\beta_2$, $\mu=1+2\theta=\alpha_1\beta_2$, $\nu=1-2\theta=\beta_1\alpha_2$. If $\omega=x+y\theta$ is any element of D it appears to follow that if ω is divisible by α_1 ($\alpha_2, \beta_1, \beta_2$ respectively) then $3|\nu\omega$ ($3|\mu\omega$, $7|\mu\omega$, $7|\nu\omega$ respectively). Now $\nu\omega=(1-2\theta)(x+y\theta)=(x+10y)+(y-2x)\theta$ whilst $\mu\omega=(x-10y)+(y+2x)\theta$ and so the above divisions yield $3|x+10y$ and hence $3|x+y$ ($3|x-y$, $7|x-3y$, $7|x+3y$ respectively). Thus from $\alpha_1|\omega=x+y\theta$ we find $3|x+y$, that is $x=3z-y$ for suitable $z\in\mathbb{Z}$. It follows that $\omega=3z+(\theta-1)y$.

Now Dedekind turns attention from the somewhat enigmatic ideal factor α_1 to the concrete subset $a_1=\{3z+(\theta-1)y: z, y\in\mathbb{Z}\}$ of elements of D. We

* Ernst Eduard Kummer (29 January 1810 – 14 May 1893).

denote a_1 more briefly by $\langle 3, \theta - 1\rangle$. Similarly $\alpha_2, \beta_1, \beta_2$ are replaced by $a_2 = \langle 3, \theta + 1\rangle$, $b_1 = \langle 7, \theta + 3\rangle$, $b_2 = \langle 7, \theta - 3\rangle$. One then naturally defines

$$a_1 a_2 = \langle 3, \theta - 1\rangle\langle 3, \theta + 1\rangle = \langle 9, 3\theta + 3, 3\theta - 3, -6\rangle$$
$$= \{9z_1 + (3\theta + 3)z_2 + (3\theta - 3)z_3 + (-6)z_4 \colon z_1, z_2, z_3, z_4 \in \mathbb{Z}\}$$

and this is easily seen to coincide with $\{3y_1 + 3\theta y_2 \colon y_1, y_2 \in \mathbb{Z}\} = 3\langle 1, \theta\rangle = 3D$. Thus $3D = a_1 a_2$. Similarly Dedekind finds $7D = b_1 b_2$, $(1 + 2\theta)D = a_1 b_2$, $(1 - 2\theta)D = b_1 a_2$, so that uniqueness of factorisation is restored via

$$a_1 a_2 \cdot b_1 b_2 = 3D \cdot 7D = 21D = (1 + 2\theta)D \cdot (1 - 2\theta)D$$
$$= a_1 b_2 \cdot b_1 a_2 \qquad (3.9.1)$$

Dedekind noted that for each pair of numbers $x, y \in a_1$ and for each $d \in D$ the numbers $x + y$, $x - y$ and xd again lie in a_1. That is a_1, and similarly a_2, b_1, b_2, αD, βD, μD, νD, satisfies the properties called for by 3.4.6. Because of their close relationship to Kummer's ideal numbers Dedekind called his subsets *ideals*.

To describe Dedekind's main result we first observe, given the ideal $[n]$ in $\mathbb{Z}$, where $n \neq 0, 1, -1$, that n is a prime in $\mathbb{Z}$ if and only if whenever $a, b \in \mathbb{Z}$ are such that $ab \in [n]$ then either $a \in [n]$ or $b \in [n]$ (exercise 1). This motivates

Definition 3.9.2 Let R be a commutative ring. An ideal $P(\neq R)$ is **prime** if and only if from $a, b \in R$ and $ab \in P$ we may always deduce $a \in P$ or $b \in P$.

Examples of prime ideals are given in the exercises. First, Dedekind's

Theorem 3.9.3 Let θ be an algebraic integer*, F the field† of all algebraic numbers of the form $r_0 + r_1\theta + \cdots + r_n\theta^n$ where $n \geqslant 0$ and $r_i \in \mathbb{Q}$, and R the ring† of all algebraic integers in F. Then each ideal $I(\neq \{0\}, \neq R)$ of R is expressible uniquely, except for the ordering of the factors, as a product‡ $P_1^{k_1} P_2^{k_2} \dots P_l^{k_l}$ of powers of prime ideals in R.

If, in particular, R is a PID and if $a \in R$ then we may write $[a] = P_1^{k_1} \dots P_l^{k_l}$ where each P_i is principal ideal, $[\pi_i]$ say. It follows that $[a] = [\pi_1]^{k_1} \dots [\pi_l]^{k_l} = [\pi_1^{k_1} \dots \pi_l^{k_l}]$ and hence (exercise 3) that, apart from units, a is uniquely expressible as a product of irreducible [?] elements in R. Since for rings of algebraic numbers the converse can also be proved we have

Theorem 3.9.4 ([49, vol. 2, p. 66]) Let R be as in 3.9.3. Then R is a UFD if and only if R is a PID.

* See exercise 3.2.14 for the definition.

† That the sets F, R are a field and a ring respectively is not obvious. See exercise 3.2.14 and Theorem 4.3.4.

‡ See exercise 3.4.17 for the definition of product.

To illustrate 3.9.3 we return to the above example in $D = \mathbb{Z}[\sqrt{-5}]$, where we have seen for example that $\langle 3, \theta - 1\rangle$ is the ideal $[3, \theta - 1]$ of D. Further, $3D$ is the principal ideal $[3]$ of D. Thus in terms of ideals (3.9.1) becomes

$$[3]\cdot[7]=[1+2\theta][1-2\theta]=[3,\theta-1][3,\theta+1][7,\theta+3][7,\theta-3]$$

Is this the required decomposition into prime ideals? One way to prove these last four ideals prime is to use the following definition and theorem.

Definition 3.9.5 Let R be a ring. An ideal $M \subsetneqq R$ is a **maximal** ideal of R if and only if R has no ideal I such that $M \subsetneqq I \subsetneqq R$.

Again, examples are given in the exercises.

Theorem 3.9.6 Let R be a commutative ring with unity. Each maximal ideal of R is a prime ideal.

Proof Suppose M is maximal in R but not prime, so there exist $a, b \in R$ such that $a \notin M$, $b \notin M$ but $ab \in M$.

Let S (respectively T) be the smallest ideal of R containing M and a (respectively M and b). Since $a \notin M$ we have $S = R$. Similarly $T = R$. Hence $S \cdot T = R \cdot R = R$ [why?]. However $S = \{m_1 + ar_1 \colon m_1 \in M, r_1 \in R\}$ whilst $T = \{m_2 + br_2 \colon m_2 \in M, r_2 \in R\}$ and by definition each element of ST is a sum of elements of the form $(m_1 + ar_1)(m_2 + br_2) = m_1m_2 + m_1br_2 + ar_1m_2 + abr_1r_2$, each of which is in M. Thus $ST \subseteq M \subsetneqq R$, a contradiction.

We leave the checking that $[3, \theta - 1]$ is a maximal ideal of D to exercise 7. We shall say a little more about prime and maximal ideals in exercises 4.3.

Exercises

1 Show that if $n \in \mathbb{Z}$ and if $n \neq 0, 1, -1$ then n is a prime integer iff $[n]$ is a prime ideal in $\mathbb{Z}$. Show that $[0]$ is a prime ideal, but not a maximal ideal in $\mathbb{Z}$.

2 Show that in $\mathbb{Z}[x, y, z]$ the ideals $[x] \subset [x, y] \subset [x, y, z]$ are all prime, but none is maximal. [Hint: For the first part work directly from the definition. For the second, consider the proper ideal $[x, y, z, 2]$.]

3 Let R be an integral domain and let $a, b \in R$. Show that $[a] = [b]$ if and only if a, b are associates. Deduce that if $[a]$ is a prime ideal then either $a = 0$ or a is irreducible.

4 Let R be the direct sum $\mathbb{Z} \oplus \mathbb{Z}$ (see exercise 3.2.9). Let a be the element $(1, 0)$. Show that $[a]$ is a prime ideal in R and yet $a = (1, 2)\cdot(1, 0)$ is the product of two non-unit elements. Doesn't this contradict the last assertion in exercise 3?

5 Show that in $\mathbb{Q}[x]$ a non-zero ideal is prime iff it is maximal. [Cf. 3.9.6. Corollary 3.7.16 might help.]

6 In $\mathbb{Z}[\sqrt{-5}]$ check:
(i) $3, 7, \sqrt{-5}-1$ are all irreducibles;
(ii) that $[3, \sqrt{-5}-1]$ is not a principal ideal. [Hint: suppose it is, $[\alpha]$, say. Then $3=\alpha\gamma$ and $\sqrt{-5}-1=\alpha\delta$ for suitable $\gamma, \delta \in \mathbb{Z}[\sqrt{-5}]$.];
(iii) setting $A=[3, \sqrt{-5}-1]$, $B=[3, \sqrt{-5}+1]$, $C=[7, \sqrt{-5}+3]$, $D=[7, \sqrt{-5}-3]$ it follows that $[3]=AB$, $[7]=CD$, $[1+2\sqrt{-5}]=AD$, $[1-2\sqrt{-5}]=BC$.

7 Show that $A=[3, \sqrt{-5}-1]$ is maximal (and hence prime) in $\mathbb{Z}[\sqrt{-5}]$ as follows.
(i) Every element of A is of the form $3(a+b\sqrt{-5})+(\sqrt{-5}-1)\cdot(c+d\sqrt{-5})=(3a-5d-c)+(3b-d+c)\sqrt{-5}=X+Y\sqrt{-5}$, say.
(ii) $3|X+Y$ in $\mathbb{Z}$.
(iii) Deduce $1 \notin A$ so that $A \neq \mathbb{Z}[\sqrt{-5}]$.
(iv) Show that if $x+y\sqrt{-5} \in \mathbb{Z}[\sqrt{-5}]\backslash A$ then X, Y can be chosen so that $Y=y$ and $X-x=1$ or 2.
(v) Deduce that any ideal T of $\mathbb{Z}[\sqrt{-5}]$ which contains A and $x+y\sqrt{-5}$ must contain 1 or else 2 (and hence $3-2=1$). Since $1 \in T$ we find $T=\mathbb{Z}[\sqrt{-5}]$.

8 Show that:
(i) in a UFD, if x is irreducible then $[x]$ is prime;
(ii) in a PID every prime ideal other than $[0]$ is maximal;
(iii) in a PID, if x is irreducible then $[x]$ is maximal;
(iv) in a PID x is a prime element if and only if $[x]$ is a non-zero prime ideal.

9 Show that in the ring of even integers the ideal $\{4z : z \in \mathbb{Z}\}$ is maximal but not prime. (Thus 3.9.6 may fail if the condition that R has a unity is omitted.)

10 Show that if I_1, I_2 are ideals in a ring R and if $I_1 \nsubseteq I_2$ and $I_2 \nsubseteq I_1$ then $I_1 \cap I_2$ is not prime in R.

11 Let A, B, P be ideals with P a prime. Show that: $(AB \subseteq P \Rightarrow A \subseteq P$ or $B \subseteq P)$. [This condition is actually equivalent to 3.9.2 in the commutative case. Furthermore it is taken as the definition of prime ideal in non-commutative rings where 3.9.2 is too restrictive.]

12 Let $\langle \mathbb{Z}, +, \square \rangle$ be the integers with the usual $+$, but with multiplication $a \square b=0$ for all $a, b \in \mathbb{Z}$. Show that the subset $2\mathbb{Z}$ is a maximal ideal which is not prime. Doesn't this contradict Theorem 3.9.6?

13 Show that $\mathbb{Z}$ has infinitely many distinct maximal ideals.

14 Theorem 3.9.6 says that in each $\mathbb{Z}_m$ each maximal ideal is prime. Do there exist integers m for which $\mathbb{Z}_m$ has primes which are not maximal?

15 Prove that each non-zero ideal in $\mathbb{Q}[[x]]$ has the form $[x^n]$ for some $n \in \mathbb{Z}^+ \cup \{0\}$. Deduce that $\mathbb{Q}[[x]]$ has just one maximal ideal.

3.10 Isomorphism. Fields of fractions. Prime subfields

Having finally confirmed in 3.7.13 that for each field F the polynomial ring $F[x]$ is a UFD, there is no need for apology if nothing more than intellectual curiosity leads one to ask if each polynomial ring $F[x_1, x_2, \ldots, x_n]$ on n letters is also a UFD. One feels instinctively that the answer must be 'yes' (how could it be otherwise?) but since, for $n \geqslant 2$, $F[x_1, x_2, \ldots, x_n]$ is not a PID (cf. exercise 3.4.22), there is no chance of applying 3.7.14 directly. On the other hand $\mathbb{Z}[x]$ is equally not a PID (exercise 3.4.22) and yet it can be shown to be a UFD (exercise 3.7.14) essentially by journeying into the UFD $\mathbb{Q}[x]$. We answer the question concerning $F[x_1, x_2, \ldots, x_n]$ in Section 3.11. In this section we set up the apparatus for the job.

Now we can think of $F[x_1, x_2, \ldots, x_n]$ as the polynomial ring $R[x_n]$ where $R = F[x_1, x_2, \ldots, x_{n-1}]$. For $n \geqslant 2$, R is, like $\mathbb{Z}$, an integral domain which is not a field. So we are led to ask: Can we enlarge R to some field $R°$, say, just as $\mathbb{Z}$ can be enlarged to $\mathbb{Q}$? We show that it can. To help us do this properly we introduce the following idea which is of fundamental importance throughout algebra.

Suppose $\langle R, +, \cdot \rangle$ is a ring with elements $a, b, \ldots$ and that $\langle R', \oplus, \odot \rangle$ is a ring with elements $a', b', \ldots$ in 1–1 correspondence (see 2.6.5) with those of R so that a corresponds to a', b corresponds to b', etc. It may be that for all $a, b \in R$ the elements $a + b$ and $a' \oplus b'$ and the elements $a \cdot b$ and $a' \odot b'$ also pair off under this correspondence. If so, then each calculation performed in R can be changed immediately to one in R' simply by 'dashing everything'! That is, $\langle R, +, \cdot \rangle$ and $\langle R', \oplus, \odot \rangle$ are structurally identical (in other words 'essentially the same'—see Remark (ii) in Section 1.8) even though their elements and their binary operations have different names.

Formally this idea of sameness becomes

Definition 3.10.1 Let $\langle R, +, \cdot \rangle$ and $\langle S, \oplus, \odot \rangle$ be rings. We say that R and S (or, more accurately, $\langle R, +, \cdot \rangle$ and $\langle S, \oplus, \odot \rangle$) are **isomorphic** iff there exists a $1-1$ map $\psi: R \to S$ from R onto S such that for all r_1, r_2 in R: (i) $(r_1 + r_2)\psi = r_1\psi \oplus r_2\psi$ and (ii) $(r_1 \cdot r_2)\psi = r_1\psi \odot r_2\psi$.

If R and S are isomorphic we write $R \cong S$. The map ψ is called an **isomorphism**.

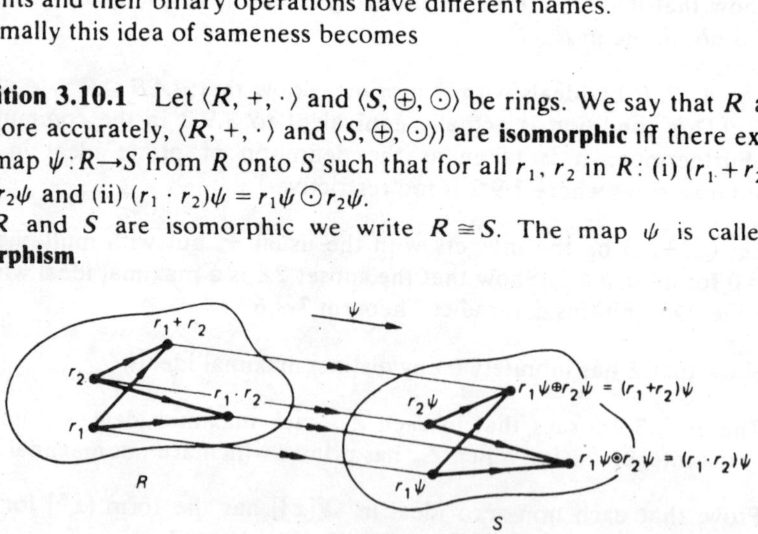

Fig. 3.2

It may help you to look at Fig. 3.2.

Thus ψ is a 1–1 and onto map which 'preserves' addition and multiplication.

Examples 3.10.2

(i) Let $N_2(\mathbb{R})$ be the ring [is it one?] of all matrices of the form $\begin{pmatrix} a & b \\ -b & a \end{pmatrix}$ where $a, b \in \mathbb{R}$. Then the map $\psi: \mathbb{C} \to N_2(\mathbb{R})$ given by $(x+iy)\psi = \begin{pmatrix} x & y \\ -y & x \end{pmatrix}$ is an isomorphism between the field of complex numbers and $N_2(\mathbb{R})$. In particular $N_2(\mathbb{R})$ is a field (exercise 7(v)).

(ii) The map θ from the ring $\mathbb{Z}_6$ to the direct sum $\mathbb{Z}_2 \oplus \mathbb{Z}_3$ (see exercise 3.2.9) given by* $\hat{m}_6\theta = (\hat{m}_2, \hat{m}_3)$ shows that the two rings mentioned are isomorphic.

For any algebraic system an investigation of the set (later, the *group*) of all the isomorphisms of the system with itself can be illuminating. Such an isomorphism, called an **automorphism**, exhibits a kind of internal symmetry of the system. As examples we mention

(iii) For any ring R the identity map $i: R \to R$ given by $ri = r$ for all $r \in R$ is clearly an automorphism of R. The fields $\mathbb{Q}$ and $\mathbb{R}$ possess no automorphisms apart from the identity (exercise 8) but the field $\mathbb{C}$ certainly has at least one other automorphism, namely that given by $(a+ib)\xi = a - ib$. [Can you think of any others?]

(iv) The map of $\mathbb{Q}[x_1, x_2, x_3, x_4]$ to itself in which each occurrence of x_1, x_2, x_3, x_4 is replaced by, say, x_2, x_4, x_1, x_3 respectively is clearly an automorphism of $\mathbb{Q}[x_1, x_2, x_3, x_4]$.

In this section the isomorphism concept is used in two ways: (i) to enlarge rings to bigger (but nicer) rings (Theorem 3.10.3); (ii) to elucidate the structure of special subrings of given rings (Theorem 3.10.9). These uses depend upon the following remarks.

(v) If T is a subring of a ring S and if $\theta: R \to T$ is an isomorphism we say that θ is an **embedding** of R in S. Of course R itself need not be contained in S, but something structurally identical to R, namely T, is. The algebraist tends to think of R as replacing the structurally identical subring T, often going so far as to rename each element $r\theta$ of T by calling it r. As an example, given any ring R, we can define the ring $R[x]$ of polynomials just as we did for $\mathbb{Q}$ in 1.6.1 and, using the map $\lambda: R \to R[x]$ given by $r\lambda = (r, 0, 0, \ldots)$, think of R itself as a subring of $R[x]$. In particular, we see how each of F, $F[x_1]$, $F[x_1, x_2], \ldots$ may be regarded as a subring of its successors and in particular of $F[x_1, x_2, \ldots, x_n]$.

(vi) Let S be the ring $R \times \mathbb{Z}$ of exercise 3.3.3. The map $\mu: R \to \{(k, 0): k \in R\} \subseteq S$ given by $r\mu = (r, 0)$ is an isomorphism. Thus can every ring be embedded in a ring with a unity element. Exercise 9 strengthens this statement slightly.

(vii) Without going into details the map $\theta: \mathbb{Z} \to 2\mathbb{Z}$ given by $z\theta = 2z$ is clearly not an isomorphism. Why not? What does $\mathbb{Z}$ have that $2\mathbb{Z}$ hasn't? (See exercise 7(ii).)

* Here $\hat{m}_k$ denotes the element $\hat{m}$ of $\mathbb{Z}_k$. (Cf. 2.4.2(i).)

Remarks It is clear that, given any concrete ring, an infinite number of distinct but structurally identical rings can be constructed merely by altering the names of the elements, but keeping their relationships intact. For this reason the algebraist, a person more interested in relationships between elements (i.e. in *structure*) than in the elements themselves, tends to collect all algebraic systems into heaps† of isomorphic ones, regarding all systems in the same heap as mutually indistinguishable. (One may consequently define‡ an algebraist to be a person who can't tell the difference between isomorphic systems!) The algebraist thus sees his task as the describing, in some manner, of a member out of each heap and the finding of a procedure which will determine, given two systems, whether or not they belong to the same heap. In particular we shall show in Section 3.12 that any two rings satisfying the axioms A1 through to I of Section 1.2 lie in the same heap whilst in Section 4.5 we find a very simple criterion for the isomorphism of two finite fields.

Some basic properties of isomorphisms are given in exercise 7.

We can now prove the main result of this section. Concrete examples are given in 3.10.6.

Theorem 3.10.3 Let D be an integral domain. Then there exists a field F_D containing a subring D^* isomorphic to D and such that every element of F_D is of the form uv^{-1} where $u, v \in D^*$ (and, of course, $v \neq 0$).

Note 3.10.4 In case $D = \mathbb{Z}$ the field F_D will be nothing more than $\mathbb{Q}$. Indeed, we model the proof of 3.10.3 on our understanding of how $\mathbb{Q}$ is obtained from $\mathbb{Z}$. Intuitively, elements of $\mathbb{Q}$ are of the form $\frac{r}{s}$ where $r, s \in \mathbb{Z}$ and $s \neq 0$. Of course $\frac{r_1}{s_1} = \frac{r_2}{s_2}$ is possible even though $r_1 \neq r_2$ and $s_1 \neq s_2$. But then $r_1 s_2 = s_1 r_2$. Thus a rational number $\frac{r}{s}$ is really a *representative* of all those $\frac{r_1}{s_1}$ for which $rs_1 = sr_1$. [If you do get a bit lost in the following abstraction, work through the entire proof assuming that $D = \mathbb{Z}$ so that the element $\{a, b\}$ introduced below may be thought of as $\frac{a}{b}$.]

Proof of 3.10.3 Consider the subset $P = \{(a, b): a, b \in D, b \neq 0_D\}$ of $D \times D$. Write $(a, b) \sim (c, d)$ iff $ad = cb$ in D. Then $\sim$ is an equivalence relation on P. [This needs proof!] We denote§ the equivalence class containing (a, b) by $\{a, b\}$ and the set of all such classes in P by F_D. On F_D we define $\oplus$ and $\odot$ by $\{a, b\} \oplus \{c, d\} = \{a \cdot d + b \cdot c, b \cdot d\}$ and $\{a, b\} \odot \{c, d\} = \{a \cdot c, b \cdot d\}$. [To what

† A better name for 'heap' is equivalence class! (See exercise 7(vii).)
‡ The definition of a topologist as one who can't tell the difference between a tea cup and a mint with a hole is much more fascinating!
§ Temporarily: Confusion with the set containing just a and b is unlikely.

do + and · refer?] An immediate problem is: Are $\oplus$ and $\odot$ well defined? (Cf. Section 2.4.) That is, if $\{a, b\}=\{A, B\}$ and if $\{c, d\}=\{C, D\}$, do we have $\{a\cdot d+b\cdot c, b\cdot d\}=\{A\cdot D+B\cdot C, B\cdot D\}$ and $\{a\cdot c, b\cdot d\}=\{A\cdot C, B\cdot D\}$? In fact everything turns out all right. [Prove it!]

We next check that $\langle F_D, \oplus, \odot\rangle$ is a field.

A1: Let $\{a, b\}, \{c, d\}\in F_D$. Then $\{a, b\}\oplus\{c, d\}=\{ad+bc, bd\}$ whilst $\{c, d\}\oplus\{a, b\}=\{cb+da, db\}$. [Why?] Since $ad+bc=cb+da$ and $bd=db$ in D [why?] we see that $\oplus$ is commutative as required.

M2: For $\{a, b\}$, $\{c, d\}$, $\{u, v\}\in F_D$ we have $(\{a, b\}\odot\{c, d\})\odot\{u, v\}=\{ac, bd\}\odot\{u, v\}=\{(ac)u, (bd)v\}=\{a(cu), b(dv)\}$ [why?] $=\{a, b\}\odot\{cu, dv\}=\{a, b\}\odot(\{c, d\}\odot\{u, v\})$ as required.

A3: For $\{a, b\}\in F_D$ we have $\{a, b\}\oplus\{0, 1\}=\{a1+b0, b1\}=\{a, b\}$. Similarly $\{0, 1\}\oplus\{a, b\}=\{a, b\}$ whence $\{0, 1\}$ is a zero element for F_D.

M4: Given $\{a, b\}$ $(\neq\{0, 1\})\in F_D$ we see that $\{a, b\}\odot\{b, a\}=\{ab, ba\}$, the element required by M3. Note that $\{b, a\}$ is in F_D since $a\neq 0_D$ [why not?].

M1, A2, M3, A4 and the distributive laws are left to you.

Now D itself is clearly not in F_D: the elements of F_D are equivalence classes $\{a, b\}$ and none of these belongs to D. The algebraist can however see D in F_D. First, the subset $D^*=\{\{d, 1\}: d\in D\}$ of F_D is easily seen to be a subring of F_D. Second, the map θ: $D\to D^*$ given by $d\theta=\{d, 1\}$ is an isomorphism. For, given $d_1, d_2\in D$ we have $(d_1+d_2)\theta=\{d_1+d_2, 1\}=\{d_1, 1\}\oplus\{d_2, 1\}$ [are you sure?] $=d_1\theta\oplus d_2\theta$. Similarly $(d_1\cdot d_2)\theta=\{d_1\cdot d_2, 1\}=\{d_1, 1\}\odot\{d_2, 1\}=d_1\theta\cdot d_2\theta$. Further θ, which is clearly onto, is also 1–1. For, if $d_1\theta=d_2\theta$ then $\{d_1, 1\}=\{d_2, 1\}$ which means [by definition] $(d_1, 1)\sim(d_2, 1)$ and hence $d_1 1=d_2 1$ in D. Thus $D\cong D^*$ as required.

Finally, given $\{a, b\}\in F_D$ we may write $\{a, b\}=\{a, 1\}\odot\{1, b\}=\{a, 1\}\odot\{b, 1\}^{-1}$ with both $\{a, 1\}$ and $\{b, 1\}\in D^*$ as required.

The observant reader will note that 3.10.3 does not strictly fulfill our promise to 'enlarge D to a bigger (but nicer) ring'. The algebraist is happy since he can't see any difference between D and D^*. Exercise 9 below should satisfy the uneasy reader.

Notes 3.10.5

(i) The field F_D is called the **field of fractions** or *quotient field* of the integral domain D. If we think of D rather than D^* as lying in F_D, we can then denote the element $\{a, b\}=\{a, 1\}\odot\{b, 1\}^{-1}$ of F_D by the symbol $a\odot b^{-1}$ or, more briefly, by ab^{-1}, or even by $\frac{a}{b}$.

(ii) If D_1 and D_2 are isomorphic domains then their corresponding fields of fractions F_{D_1} and F_{D_2} are isomorphic. (How could it be otherwise?) Further, if F is any field and if D is a subring of it then the subset $\bar{F}=\{ab^{-1}: a, b\in D, b\neq 0\}$ of F is a subfield of F such that $\bar{F}\cong F_D$. Summarising: a given domain D can always be embedded in a field F, say. If D cannot be embedded in any proper† subfield of F then $F\cong F_D$. F_D is unique (up to isomorphism).

† A subfield (subring) is called *proper* iff it is not the whole field (ring).

(iii) Even when applied to $\mathbb{Z}$ itself our construction is not devoid of interest. For in Section 3.12 we show that $\mathbb{Z}$ is the unique (up to isomorphism) well ordered integral domain. Thus (ii) will allow us to define $\mathbb{Q}$ as the (unique) field of fractions of the unique well ordered integral domain.
(iv) Dropping the commutativity of D it seems clear 3.10.3 will prove for us that every 'non-commutative domain' can be embedded in a 'non-commutative' field, that is, a division ring. In fact this assertion is *false* (exercise 15). Moreover there exist non-commutative domains which cannot be embedded in division rings no matter what method may be tried. The Russian mathematician A I Mal'cev* found such an example in 1937.
(v) Thinking of D as a subring of F_D leads one easily to think of $D[x]$ as a subring of $F_D[x]$.

Examples 3.10.6 According to 3.10.5(i) the fields of fractions of $\mathbb{Q}[x]$, $\mathbb{Q}[x, y]$ and $\mathbb{Z}[\sqrt{2}]$ comprise respectively the sets

$$\left\{\frac{u}{v}: u, v \in \mathbb{Q}[x], v \neq 0\right\},$$

$$\left\{\frac{u}{v}: u, v \in \mathbb{Q}[x, y], v \neq 0\right\} \quad \text{and} \quad \left\{\frac{u}{v}: u, v \in \mathbb{Z}[\sqrt{2}], v \neq 0\right\}.$$

In this last example we may write

$$\frac{a+b\sqrt{2}}{c+d\sqrt{2}} = \frac{(a+b\sqrt{2})(c-d\sqrt{2})}{c^2-2d^2} = r+s\sqrt{2}$$

for suitable $r, s \in \mathbb{Q}$. We thus see that, in the notation of exercise 3.2.14, the field of fractions of $\mathbb{Z}[\sqrt{2}]$ is the field $\mathbb{Q}(\sqrt{2})$. Notice that $\mathbb{Q}(\sqrt{2})$ is the smallest (sub)field (of $\mathbb{R}$) which contains $\mathbb{Q}$ and $\sqrt{2}$. Similarly one sees readily that the field of fractions of $\mathbb{Z}[\sqrt[3]{2}] = \{a + b\sqrt[3]{2} + c(\sqrt[3]{2})^2 : a, b, c \in \mathbb{Z}\}$ is the smallest (sub)field (of $\mathbb{R}$) containing $\mathbb{Q}$ and $\sqrt[3]{2}$. What is less obvious (but true, see 4.3.4) is that this field of fractions coincides with the set $\{u + v\sqrt[3]{2} + w(\sqrt[3]{2})^2 : u, v, w \in \mathbb{Q}\}$. The big surprise here is that multiplicative inverses of numbers of this form are again of this form. (Try writing $\{2 + \sqrt[3]{2} - \frac{1}{2}(\sqrt[3]{2})^2\}^{-1}$ in the form $u + v\sqrt[3]{2} + w(\sqrt[3]{2})^2$ where $u, v, w \in \mathbb{Q}$.)

Notation 3.10.7 Given fields $F \subseteq E$ and elements $\alpha, \beta, \gamma, \ldots \in E$ it is customary to denote the smallest subfield of E which contains $F, \alpha, \beta, \gamma, \ldots$ by $F(\alpha, \beta, \gamma, \ldots)$. In this notation the fields of fractions of $\mathbb{Q}[x]$, $\mathbb{Q}[x, y]$, $\mathbb{Z}[\sqrt{2}]$, $\mathbb{Z}[\sqrt[3]{2}]$ are $\mathbb{Q}(x)$, $\mathbb{Q}(x, y)$, $\mathbb{Q}(\sqrt{2})$, $\mathbb{Q}(\sqrt[3]{2})$, whilst the smallest subfield of $\mathbb{C}$ containing, for example, $\mathbb{Q}$, $\sqrt{2}$ and i is denoted by $\mathbb{Q}(\sqrt{2}, \mathrm{i})$.

A nice little application of a field of fractions to solve a problem posed solely in terms of domains is

* Anatoly Ivanovich Mal'cev (27 November 1909 – 7 July 1967).

Example 3.10.8 . Let D be an integral domain, a and b elements of D and $m, n \in \mathbb{Z}^+$, where $(m, n) = 1$. If $a^m = b^m$ and $a^n = b^n$ then $a = b$ in D. For: Since $(m, n) = 1$ there exist $s, t \in \mathbb{Z}$ such that $sm + tn = 1$. Form F_D. Assuming that $s < 0$ and $t > 0$ we find $(a^n)^t = (b^n)^t$ in $D \subseteq F_D$ and $(a^m)^s = (b^m)^s$ in F_D. But then $a = a^{nt}a^{ms} = b^{nt}b^{ms} = b$ in F_D. Hence $a = b$ in D.

Although we shall not need it until 4.5.7, now seems an appropriate time to present the following embedding theorem first proved in E Steinitz'* great paper of 1910. It shows that the vast totality of all fields is built around a rather restricted collection of 'basic' fields, namely the prime (sub)fields.

Theorem 3.10.9 Let F be any field and $p(F)$ the intersection of all subfields of F. Then $p(F)$, the **prime subfield** of F, is isomorphic to $\mathbb{Q}$ or to one of the finite fields $\mathbb{Z}_p$.

Sketch of proof Let $D = \{m 1_F : m \in \mathbb{Z}\}$. Since $1_F \in p(F)$ we see that $D \subseteq p(F)$. There are two possibilities: (i) for $m, n \in \mathbb{Z}$, $m \neq n \Rightarrow m 1_F \neq n 1_F$; (ii) for some pair $m, n \in \mathbb{Z}$, with $m \neq n$, we have $m 1_F = n 1_F$.

In case (ii) suppose WLOG that $m > n$. Since $(m - n)1_F = 0_F$ there exists a smallest positive integer r, say, such that $r 1_F = 0_F$. It follows that r is a prime in $\mathbb{Z}$. For, if not, then $r = st$ where $1 < s \leqslant t < r$ and we would have $(s 1_F)(t 1_F) = r 1_F = 0_F$. Since F is a field it would follow that either $s 1_F = 0_F$ or $t 1_F = 0_F$, both of which would contradict the choice of r. Consider then the subring $D = \{m 1_F : m \in \mathbb{Z}\} = \{k 1_F : 0 \leqslant k < r\}$ of F. It is not difficult to check—mind the well-definedness—that the map $\theta: \mathbb{Z}_r \to D$ given by $\hat{n}\theta = n 1_F$ is an isomorphism from $\mathbb{Z}_r$ onto D. Since r is a prime $\mathbb{Z}_r$, and hence D, are fields. Since $D \subseteq p(F)$ and since $p(F)$ cannot have any proper subfields [can you prove this?] we see that $p(F) = D \cong \mathbb{Z}_r$.

In case (i) it is not difficult to check—no well-definedness problem arises—that the map $\theta: \mathbb{Z} \to D$ given by $z\theta = z 1_F$ is an isomorphism between $\mathbb{Z}$ and D. Then, by 3.10.5(ii), the subfield $\bar{F} = \{(m 1_F)(n 1_F)^{-1} : m, n \in \mathbb{Z}, n \neq 0\}$ is a subfield of F isomorphic to $\mathbb{Z}$'s field of fractions, namely $\mathbb{Q}$. Further, since $p(F)$ is a field, we have $\bar{F} \subseteq p(F)$. Thus, as above, $p(F) = \bar{F} \cong \mathbb{Q}$ as required.

Exercises

1 Are the following maps isomorphisms between the given rings?

(i) $\theta: \mathbb{Q}[\sqrt{2}] \to \mathbb{Q}[\sqrt{2}]$ given by $(a + b\sqrt{2})\theta = a - b\sqrt{2}$;

(ii) $\theta: 2\mathbb{Z} \to 3\mathbb{Z}$ given by $(2n)\theta = 3n$;

(iii) $\theta: \mathbb{C} \to \mathbb{C}$ given by $(a + ib)\theta = b - ia$;

(iv) $\theta: K_5 \to K_5$ given by $(a_0 + a_1\lambda + a_2\lambda^2 + a_3\lambda^3)\theta = (a_0 - a_2) + (a_3 - a_2)\lambda + (a_1 - a_2)\lambda^2 - a_2\lambda^3 = a_0 + a_1\lambda^2 + a_2\lambda^4 + a_3\lambda$ where λ is a complex fifth root of 1 and the $a_i \in \mathbb{Z}$. [K is for Kummer—see Section 3.9];

* Ernst Steinitz (13 June 1871 – 29 September 1928). *Crelle's Journal*, Vol. 137, pp. 167–309.

(v) $\theta: \mathbb{Z}_{60} \to \mathbb{Z}_6 \oplus \mathbb{Z}_{10}$ given by† $\hat{m}_{60}\theta = (\hat{m}_6, \hat{m}_{10})$ where $0 \leq m < 60$;
(vi) Let u, σ be the real and one of the complex cube roots of 2 respectively. Define $\theta: \mathbb{Z}[u] \to \mathbb{Z}[\sigma]$ by $(a + bu + cu^2)\theta = a + b\sigma + c\sigma^2$.

2 On $\mathbb{Z}$ define $\oplus$ and $\odot$ by $a \oplus b = a + b + 1$, $a \odot b = ab + a + b$. Show $\langle \mathbb{Z}, \oplus, \odot \rangle$ is a ring with unity element. Is $\langle \mathbb{Z}, \oplus, \odot \rangle \cong \langle \mathbb{Z}, +, \cdot \rangle$?

3 Let $R_1 = M_2(\mathbb{Z})$ and $R_2 = M_2(R_1)$. Is $R_2 \cong M_4(\mathbb{Z})$? [$M_n$ denotes $n \times n$ matrices.]

4 Explain intuitively why $\mathbb{Z}[\sqrt{2}] \not\cong \mathbb{Z}[\sqrt{3}]$. Back your intuition with a proof. [Note: this example not only says that $a + b\sqrt{2} \xrightarrow{\theta} a + b\sqrt{3}$ is not an isomorphism. It says that *no isomorphism can be found at all*—no matter how clever a choice of mapping you might try to make.]

5 Let $H = \{(a, b, c, d): a, b, c, d \in \mathbb{R}\}$. Define $\boxplus$ componentwise and $\boxdot$ by $(a, b, c, d) \boxdot (A, B, C, D) = (aA - bB - cC - dD,\ aB + bA + cD - dC,\ aC - bD + cA + dB,\ aD + bC - cB + dA)$. Show that $\langle H, \boxplus, \boxdot \rangle$ is isomorphic to the ring of quaternions (Section 3.2). [H is, of course, for Hamilton.]

6 Show there exists $A = \begin{pmatrix} x & y \\ -y & x \end{pmatrix} \in M_2(\mathbb{R})$ such that $A \neq I_2$ and yet $A^{17} = I_2$. [Hint: think of 17th roots of 1 in $\mathbb{C}$ and use 3.10.2.]

7 Let $\phi: \langle R, +, \cdot \rangle \to \langle S, \oplus, \odot \rangle$ be an isomorphism. Show that
(i) $0_R\phi = 0_S$;
(ii) If R has a unity 1_R then $1_R\phi$ is a unity for S;
(iii) $(-a)\phi = \ominus(a\phi)$;
(iv) $(a^n)\phi = (a\phi)^n$;
(v) If R is commutative so is S; if R is a field so is S;
(vi) $\phi^{-1}: S \to R$ exists and is an isomorphism;
(vii) If $\psi: \langle S, \oplus, \odot \rangle \to \langle T, \boxplus, \boxdot \rangle$ is an isomorphism then so is $\phi \circ \psi: R \to T$.
Deduce that 'isomorphic to' is an equivalence relation on the class of all rings. [Hint: for (i) look at $0_R\phi = (0_R + 0_R)\phi = 0_R\phi \oplus 0_R\phi$ in S—or see 4.2.6.]

8 After noting that $\mathbb{Q} \subseteq \mathbb{Q}(\sqrt{2}) \subseteq \mathbb{R} \subseteq \mathbb{C}$ and that $\mathbb{Q}(\sqrt{2})$, $\mathbb{C}$ each possess non-identity automorphisms show that the only automorphisms of $\mathbb{Q}$ and of $\mathbb{R}$ are the identity maps. [Hint: By 7(ii) $1_\mathbb{Q} \xrightarrow{\phi} 1_\mathbb{Q}$. For $\mathbb{R}$: if $r \in \mathbb{R}^+$ then $r\phi = \sqrt{r}\phi \cdot \sqrt{r}\phi \in \mathbb{R}^+$. Hence, if $a < r < b$ with a, b in $\mathbb{Q}$ then $a = a\phi < r\phi < b\phi = b$ in $\mathbb{R}\phi = \mathbb{R}$.]

9 Let R, S be rings and $\theta: R \to S$ an embedding of R into S. Set $R\theta = \{r\theta : r \in R\} \subseteq S$. Let S^* be a set of symbols disjoint from R and in 1–1 correspondence with the elements of $S \backslash R\theta$. Let $R^* = S^* \cup R$. If $r \in R$, if $s^* \in S^*$ and if s^* corresponds with s, define $r \boxplus s^* = (r\theta \oplus s)^*$. Taking the hint from this show that R^* can be made into a ring isomorphic to S. Then R^* actually contains the subset R as a subring. [In particular for each domain we can

† See footnote p. 127.

find a field *actually containing it*, whilst for each ring R we can find a ring S with unity such that $R \subseteq S$.]

10 Describe other than in $\frac{u}{v}$ form the fields of fractions of the domains: $\mathbf{Z}[\sqrt{-1}]$; $\mathbf{Q}[[x]]$; $\mathbf{Q}_2 = \left\{\frac{m}{n} : m, n \in \mathbf{Z} \text{ and } 2 \nmid n\right\}$; $\{\hat{0}, \hat{2}, \hat{4}\}$ (mod 6).

11 Show that if $\theta : D \to D_1$ is an isomorphism then $F_D \cong F_{D_1}$. [Hint: try the map $\hat{\theta}$ defined by $\{d_1, d_2\}\hat{\theta} = \{d_1\theta, d_2\theta\}$.]

12 Show that between $\mathbf{Z}$ and $\mathbf{Q}$ there are infinitely many pairwise non-isomorphic rings each with the same field of fractions.

13 Is $\mathbf{Z}[x] \cong \mathbf{Q}[x]$? [Such an isomorphism would surely have to be a non-obvious one. Note that their fields of fractions coincide.]

14 Identify where in 3.10.3 commutativity and also the non-existence of non-trivial divisors of zero is used. Show that the existence of a unity element in D is not really required. What is the field of fractions determined by $2\mathbf{Z}$?

15 Obtain the result of 3.10.8 working entirely inside the integral domain D.

16 Show that the conclusion of 3.10.8 is false in $M_2(\mathbf{Z})$.

17 Let F be a field and S a subfield. Show that if $S \cong \mathbf{Q}$ or if $S \cong \mathbf{Z}_p$ then F contains no other subfield isomorphic to S.

18 By definition the **characteristic** of a ring R is the smallest positive integer n (if there is one) such that $na = 0_R$ for all $a \in R$. If there is no such integer we say R has characteristic 0. Show that if R has a 1 then n is also the smallest positive integer for which $n1 = 0_R$. Deduce that in an integral domain the characteristic is 0 or a prime p.

Show that, for all a, b in a domain D of characteristic p, $(a \pm b)^{p^k} = a^{p^k} \pm b^{p^k}$ for all $k \in \mathbf{Z}^+$. (Including the case $p = 2$, $k = 1$?) Deduce that the map $a \xrightarrow{\theta} a^p$ is a 1–1 homomorphism of D into D. Need θ be onto?

3.11 *U*[*x*] where *U* is a UFD

We now indicate how to prove that, for any field F, the ring $F[x_1, x_2, \ldots, x_n]$ is a UFD. Regarding $F[x_1, x_2, \ldots, x_n]$ as the polynomial ring $R[x_n]$ where $R = F[x_1, x_2, \ldots, x_{n-1}]$ the statement about $F[x_1, x_2, \ldots, x_n]$ will follow from the more general

Theorem 3.11.1 Let U be a UFD. Then $U[x]$ is a UFD.

One way to prove 3.11.1 is to copy, as far as possible, the corresponding proof for $\mathbf{Z}[x]$. Since the complete proof of that result is somewhat scattered, we shall identify the lemmas, theorems, etc. needed in the proof of 3.11.1 indicating the modifications required. (One advantage of this to the reader is that he will have the motivation to reread the appropriate proofs very carefully—to ensure the author isn't cheating!) An alternative proof is described in exercise 1.

We begin with proofs of parts of exercises 3.7.12 and 3.7.10.

Theorem 3.11.2 In a UFD an element π is irreducible iff π is prime.

Proof $\overset{\text{if}}{\Leftarrow}$: is true in any integral domain (exercise 3.6.12(a)). $\overset{\text{only if}}{\Rightarrow}$: let π be irreducible in the UFD U and suppose $\pi|ab$ where $a, b \in U$. Then $\pi c = ab$ for some c in U. Writing $a = \pi_1 \ldots \pi_r$, $b = \pi'_1 \ldots \pi'_s$, $c = \pi''_1 \ldots \pi''_t$, products of irreducibles, we find $\pi \cdot \pi''_1 \ldots \pi''_t = \pi_1 \ldots \pi_r \cdot \pi'_1 \ldots \pi'_s$. By definition of UFD, π is an associate of a π_i or a π'_j. Thus $\pi|a$ or $\pi|b$, as required.

[This proof needs a slight modification if a or b is a unit. Why can't *both* a and b be units?]

Theorem 3.11.3 In a UFD D every two elements a, b, not both zero, have a gcd unique except for units. So then does every set of n elements, not all zero.

Proof If one or both of a, b is a unit then 1 is the required gcd. Otherwise write $a = \pi_1 \ldots \pi_r$, $b = \pi'_1 \ldots \pi'_s$, where $r, s \geq 1$. Let $I = \{i_1, \ldots, i_m\}$ and $J = \{j_1, \ldots, j_m\}$ be subsets, as large as possible, of $M = \{1, \ldots, r\}$ and $N = \{1, \ldots, s\}$ respectively such that, for k $(1 \leq k \leq m)$, π_{i_k} is an associate $u_k \pi'_{j_k}$ of π'_{j_k}. Then $g = \pi_{i_1} \ldots \pi_{i_m}$ is a common divisor of a and b. Indeed it is a gcd of a and b. [This seems pretty obvious. Let us see what is required by way of proof.] If $g|\bar{g}$ where $\bar{g}|a$ and $\bar{g}|b$ then for suitable A, B and G in D we have $\bar{g} = Gg$, $a = A\bar{g} = AGg$, $b = B\bar{g} = BGg$. Comparing these factorisations of a and b with those above it quickly follows, by uniqueness of factorisation, that either G is a unit or $G = w_1 \pi_{\alpha_1} \ldots \pi_{\alpha_\lambda} = w_2 \pi'_{\beta_1} \ldots \pi'_{\beta_\mu}$ where $\lambda, \mu \geq 1$, w_1 and w_2 are units in D, the $\alpha_i \in M \backslash I$ and the $\beta_i \in N \backslash J$. Applying uniqueness of factorisation to G in this case we see that π_{α_1} is an associate of some π'_{β_i}. But this contradicts the definition of I and J. Thus G is a unit in D and g is a gcd, as required.

The uniqueness of g 'up to associates' and the extension of the result to sets of n elements are immediate.

Question In connection with the above proofs, why did we not just say 'Copy the proofs of 1.4.4 and 1.4.10'?

Notation 3.11.4 (cf. 1.9.7) For elements $z_0, z_1, \ldots, z_n \in U$ we denote their gcd, unique apart from occurrence of units (by 3.11.3), by $(z_0, z_1, \ldots, z_n)$.

Definition 3.11.5 (cf. 1.9.5) A polynomial $F = z_0 + z_1 x + \cdots + z_n x^n \in U[x]$ is **primitive** iff $(z_0, z_1, \ldots, z_n) = 1$.

Theorem 3.11.6 (cf. 1.9.10) If $F = y_0 + y_1x + \cdots + y_mx^m$ and $G = z_0 + z_1x + \cdots + z_nx^n \in U[x]$ are both primitive then so is FG.

Proof In 1.9.10 alter '(other than 1 and -1) in $\mathbb{Z}$' to '(other than units) in U' and note that every irreducible element in U is prime—by 3.11.2.

Definition 3.11.7 (cf. 1.9.11) Let $F = z_0 + z_1x + \cdots + z_nx^n \in U[x]$. Each gcd is called the **content** of F.

Theorem 3.11.8 (cf. 1.9.14) If $F \in U[x]$ and if $F = gh$ where $g, h \in F_U[x]$ then we can write $F = GH$ where $G, H \in U[x]$, $\deg G = \deg g$ and $\deg H = \deg h$.

Proof In 1.9.14 alter $\mathbb{Z}$ and $\mathbb{Z}^+$ to U [what has happened to U^+?] and deduce $ac = bdu$ where u is a unit in U.

Corollary 3.11.9 (cf. 1.9.15) If $F \in U[x]$ and if one cannot write F as a product of polynomials of smaller degree in $U[x]$, then one cannot write F as a product of polynomials of smaller degree in the bigger ring $F_U[x]$.

Theorem 3.11.10 (cf. the solution to Problem 3 in Section 1.9) Let $F \in U[x]$ be irreducible (in $U[x]$). Then F is prime (in $U[x]$).

Proof Replace $\mathbb{Z}$ and $\mathbb{Z}^+$ by U, $\mathbb{Q}[x]$ by $F_U[x]$ and '$a, b \in \mathbb{Z}^+$ with $(a, b) = 1$' by '$a, b \in U$ with $(a, b) = 1$'.

To prove 3.11.1 all we need, after 3.11.10 and exercise 3.7.12, is to show that every non-zero non-unit element of $U[x]$ is expressible as a product of irreducibles of $U[x]$. But this is easy. For if $F \in U[x]$ write $F = cG$ where c is the content of F and G is primitive* in $U[x]$. If c is not a unit in U then write it as a product of elements which are irreducibles in U and hence in $U[x]$ (exercise 3.6.2). By an argument on degree we can clearly factor G into a product of primitive irreducibles in $U[x]$. This does it.

Exercises

1 Prove 3.11.1 by copying the second proof of 1.5.1 as follows. Assuming $U[x]$ is not a UFD let $F = p_1p_2 \ldots p_r = q_1q_2 \ldots q_s$ be a polynomial of smallest degree with distinct factorisations into products of irreducibles in $U[x]$. Assuming $m = \deg p_1 \geqslant \cdots \geqslant \deg p_r$, $n = \deg q_1 \geqslant \cdots \geqslant \deg q_s$ and $m \leqslant n$, set $G = p_1x^{n-m}q_2 \ldots q_s$ and suppose a and b are the leading coefficient of p_1 and q_1 respectively. Show that $p_1 | aF - bG = (aq_1 - bp_1x^{n-m})q_2 \ldots q_s$. Now either $aF = bG$ or $\deg (aF - bG) < \deg F$; hence in either case $p_1 | aq_1$. Deduce [careful!] that $p_1 | q_1$. Thus p_1 and q_1 are associates, contradicting the obvious initial assumption that they are not.

* Is G *necessarily* primitive if the content of F is factored out? Can you prove this?

2 (i) Prove the converse of 3.11.1, namely: If U is a ring and if $U[x]$ is a UFD then U is a UFD.
(ii) Prove that if $U[x]$ is a PID then U is a field.
(iii) What can you say about U if $U[x]$ is Euclidean?
[Hint: for (ii) think first of $\mathbb{Z}[x]$.]

3 Let U be a UFD and let $f = a_0 + a_1x + \cdots + a_nx^n \in U[x]$. Show that if f is primitive and if p is a prime in U such that $p|a_0, p|a_1, \ldots, p \nmid a_n$ and $p^2 \nmid a_0$ then f is irreducible in $U[x]$. (That is, prove Eisenstein's test in $U[x]$.)

Deduce that $y^4 + 3x^2y^2 + 4x^7y + 2x$ is irreducible in $\mathbb{Q}[x, y]$. Is $y^6 + 3xy^4 + 3x^2y^2 + x^3$ irreducible in $\mathbb{Q}[x, y]$?

4 Use exercise 3 to show that $x^3 - 6x^2 + 4ix + 1 + 3i$ is irreducible in $\mathbb{Z}[\sqrt{-1}][x]$.

5 State the analogue for $U[x]$ of 1.11.6. Hence write $x^3 + yx^2 + (y - 2y^2)x - y^2$ as a product of a linear factor and a quadratic in x with coefficients in $\mathbb{Z}[y]$. Treating it as a polynomial in x with coefficients in $\mathbb{Z}[y]$ find the content C of $P = y^2x^3 + y^3x + y^3x^2 - y^4 - 2y^4x - x^3 - yx^2 - yx + 2y^2x + y^2$. Hence factorise P completely into irreducibles in $\mathbb{Z}[x, y]$.

6 Show that if $F_1, F_2 \in U[x]$ and if F_1, F_2 have a common divisor of degree greater than zero in $F_U[x]$ then they have a common divisor of degree greater than zero in $U[x]$.

7 Let f, g be polynomials in $\mathbb{R}[x, y]$ having no common factor except units. Show that there exist polynomials $s, t \in \mathbb{R}[x, y]$ and $u \in \mathbb{R}[x]$ such that $u = sf + tg$. [Hint: Let K be the field of fractions of $\mathbb{R}[x]$ and think of f, g as being in $K[y]$. By exercise 6, f, g have no common divisors other than units in $K[y]$. Thus there exist $v, w \in K[y]$ such that $1 = vf + wg$.]

Deduce that if $a, b \in \mathbb{R}$ are such that $f(a, b) = g(a, b) = 0$ then $u(a) = 0$. Show that there are at most finitely many pairs $(a, b) \in \mathbb{R} \times \mathbb{R}$ such that $f(a, b) = g(a, b) = 0$. (This result is used in algebraic geometry to determine the so-called irreducible varieties in the plane $\mathbb{R} \times \mathbb{R}$.)

8 Is $\mathbb{Q}[[x]]$ a UFD? What about $\mathbb{Q}[[x]][y]$ and $\mathbb{Q}[y][[x]]$? These latter two rings are isomorphic, aren't they?

3.12 Ordered domains. The uniqueness of $\mathbb{Z}$

In Section 1.2 we supposed that the set $\mathbb{Z}$ of integers, whatever they be, satisfies the axioms A1 through to I (excluding M4). Replacing axiom I by axiom M4 it is easy to give many distinct examples, including $\mathbb{Q}$ and $\mathbb{R}$, which satisfy axioms A1 through to P (including M4). Now $\mathbb{Q}$ and $\mathbb{R}$ are clearly not isomorphic [can you pinpoint one reason why not?] and so the question raised in Section 3.1 regarding the essential uniqueness of $\mathbb{Z}$ does seem to be somewhat less trivial than it at first appears.

We close this chapter by proving that $\mathbb{Z}$ is indeed unique up to isomorphism, thus fulfilling our promise of 3.10.5(iii).

A preliminary definition is helpful (cf. axiom P of Section 1.2).

Definition 3.12.1 Let $\langle R, +, \cdot \rangle$ be an integral domain. R is said to be **ordered** iff R contains a non-empty subset R^+ such that
(i) for all $a, b \in R^+$ we have $a + b \in R^+$ and $a \cdot b \in R^+$;
(ii) each element of R belongs to exactly one of the sets R^+, $\{0\}$, R^- where $R^- = \{-x : x \in R^+\}$.

R^+ is called a set of **positive elements** of R.

Examples 3.12.2
(i) According to Section 1.2, $\mathbb{Z}$ is an ordered domain. Intuitively, so are the fields $\mathbb{Q}$ and $\mathbb{R}$.
(ii) The ring $\mathbb{Z}[\sqrt{2}]$ of all $a + b\sqrt{2}$ where $a, b \in \mathbb{Z}$ sustains two distinct sets of positive elements as was noted in 1.2.3. So does the ring $\mathbb{Z}[x]$ (exercise 2 below).

Remark Each ordered domain must satisfy the analogue of property Z of Section 1.2 (exercise 6).

Now let R be any ordered (integral) domain. The generalisation to R of axiom I in Section 1.2 is easily written down. Do it by replacing N and 1 by R^+ and 1_R respectively. Furthermore if, just as we did for $\mathbb{Z}$ in 1.2.4, we write $a < b$ whenever $b - a \in R^+$ the generalisation to R of condition W in Section 1.2 reads
W: Every non-empty subset of R^+ contains a least member. That is, if $T \subseteq R^+$ and $T \neq \varnothing$ then T contains an element t such that $t < z$ for every other $z \in T$. Following the terminology of Section 1.2 we say that R^+ is **well ordered.**

It is not difficult to follow through the proof of 1.2.9 to show that these generalisations of I and W are logically equivalent. Thus it is reasonable to make

Definition 3.12.3 Let R be an ordered domain. If R satisfies axiom I (equivalently axiom W) we call R a **well ordered (integral) domain**.

We can now establish

Theorem 3.12.4 Let $\langle C, +, \cdot \rangle$ and $\langle D, +, \cdot \rangle$ be two well ordered domains*. Then $\langle C, +, \cdot \rangle \cong \langle D, +, \cdot \rangle$.

Proof Since C is a domain, C has a unity element 1_C such that $1_C \neq 0_C$. Since C is ordered, C contains a subset C^+ of positive elements. Hence either

* For simplicity we use + and · for each ring.

$1_C \in C^+$ or $-1_C \in C^+$. In this latter case $1_C = (-1_C)(-1_C) \in C^+$ so that in any event $1_C \in C^+$. Now set $U_C = \{n1_C : n \in \mathbb{Z}^+\}$. We have just seen that $1_C \in C^+$; furthermore C^+ is closed under addition. Thus $U_C \subseteq C^+$. Likewise, given $x = n1_C \in U_C$, we see that $x + 1_C = n1_C + 1_C = (n+1)1_C \in U_C$ for some $n \in \mathbb{Z}$. Since C satisfies axiom I we deduce that $U_C = C^+$. Since, by axiom P, $C = C^+ \cup \{0\} \cup C^-$, since $0_C = 01_C$ and since $-(n1_C) = (-n)1_C$ (exercise 3.3.2) we see that $C = \{m1_C : m \in \mathbb{Z}\}$.

Next suppose $m_1 > m_2$ in $\mathbb{Z}$ and set $t = m_1 1_C - m_2 1_C = (m_1 - m_2)1_C$. Because $m_1 - m_2 \in \mathbb{Z}^+$, $t \in U_C = C^+$ and hence $t \neq 0_C$. That is, $m_1 1_C \neq m_2 1_C$.

Now all remarks made so far apply equally to D. In particular $D = \{m1_D : m \in \mathbb{Z}\}$. Further, since $m_1 = m_2$ implies $m_1 1_C = m_2 1_C$ and $m_1 1_D = m_2 1_D$, the correspondence $\phi : C \to D$ given by $(m1_C)\phi = m1_D$ is indeed a map (i.e. it is well defined). Further it is 1–1 and onto D.

Finally, if $c_1, c_2 \in C$ then $c_1 = m_1 1_C$, $c_2 = m_2 1_C$ for suitable $m_1, m_2 \in \mathbb{Z}$ and we have

$$\begin{aligned}(c_1 + c_2)\phi &= (m_1 1_C + m_2 1_C)\phi = ((m_1 + m_2)1_C)\phi = (m_1 + m_2)1_D \\ &= m_1 1_D + m_2 1_D = (m_1 1_C)\phi + (m_2 1_C)\phi = c_1\phi + c_2\phi.\end{aligned}$$

Similarly

$$\begin{aligned}(c_1 c_2)\phi &= (m_1 1_C \cdot m_2 1_C)\phi = (m_1 m_2 1_C)\phi = m_1 m_2 1_D \\ &= m_1 1_D \cdot m_2 1_D = (m_1 1_C)\phi \cdot (m_2 1_C)\phi = c_1\phi \cdot c_2\phi.\end{aligned}$$

It follows that ϕ is an isomorphism, as required.

We thus see that whatever the integers are, there is, up to isomorphism, only one algebraic system satisfying precisely the axioms A1, A2, A3, A4, M1, M2, M3, D, P and I. A similar sort of result relating to the real numbers is given in 4.4.2 where it is far from obvious that each of two natural constructions of $\mathbb{R}$ based upon the rational numbers leads to the same result.

Problem 4 It is not difficult to check that the only automorphism of $\mathbb{Z}$ is the identity mapping. Suppose D is an ordered domain with just two units 1_D and -1_D and that the only automorphism of D is the identity. Is D necessarily isomorphic to the ring of integers?

Exercises

1 Is the (integral) domain of Gaussian integers (p. 108) an ordered domain?

2 Prove the assertion made in 3.12.2(ii) about $\mathbb{Z}[x]$. [Hint: Let $f = a_i x^i + \cdots + a_j x^j$ where $a_i \neq 0$, $a_j \neq 0$ and where $0 \leq i \leq j$ be a typical element of $\mathbb{Z}[x]$. Define subsets P_1, P_2 of $\mathbb{Z}[x]$ by $P_1 = \{f : a_i > 0 \text{ in } \mathbb{Z}\}$ and $P_2 = \{f : a_j > 0 \text{ in } \mathbb{Z}\}$. Show that P_1 and P_2 are subsets of positive elements of $\mathbb{Z}[x]$ in the sense of 3.12.1. Show that $P_1 \neq P_2$.] Can you find a third such subset distinct from P_1 and P_2?

3 Can $\mathbb{Z}_7$ be ordered? (See exercise 2.4.11.)

4 We know $\mathbb{Z}[\sqrt{2}]$ can be ordered. Can it be well ordered?

5 Show that 1_D is the smallest positive element in a well ordered domain D. Show that this assertion is generally false if D is only assumed to be ordered.

6 Let R be an ordered domain. Show that R has characteristic 0.

7 Can a well ordered domain sustain two distinct sets of positive elements?

8 Let D be an ordered domain and F_D its field of fractions. For $\frac{a}{b} \in F_D$ define $0 <_F \frac{a}{b}$ iff $0 <_D ab$. Show that the set $F_D^+ = \{x \in F_D : 0 <_F x\}$ is a set of positive elements as defined by 3.12.1. (A field which is ordered as a ring is called an **ordered field**. Setting $D = \mathbb{Z}$ we see that $\mathbb{Q}$ is an ordered field.). Write $\frac{c}{d} <_F \frac{a}{b}$ iff $0 <_F \frac{a}{b} - \frac{c}{d}$. Show $\frac{c}{d} <_F \frac{a}{b}$ iff $cdb^2 <_D abd^2$. Is it true that $\frac{c}{d} <_F \frac{a}{b}$ iff $bc <_D ad$?

9 $\mathbb{Z}$ can be ordered in only one way (1.2.3). Is the same true of $\mathbb{Q}$? (One way to order $\mathbb{Q}$ is given in exercise 8.)

10 An ordered domain D is called **Archimedean** iff for each $r \in D$ there exists $n \in \mathbb{Z}$ such that $r < n1_D$. Show that $\mathbb{Z}$ and $\mathbb{Q}$ are Archimedean but that, with P_2 as defined as in exercise 2 above, $\mathbb{Z}[x]$ is non-Archimedean. Deduce that its field of fractions $\mathbb{Z}(x) = F_{\mathbb{Z}[x]}$ is a non-Archimedean field (see exercise 8). Is there any way to order $\mathbb{Z}[x]$ (besides using P_1) to make it Archimedean?

11 Let D be an ordered domain in which there is no element t such that $0_D < t < 1_D$. Is $D \cong \mathbb{Z}$?

4

Factor rings and fields

4.1 Introduction

In Chapters 1, 2 and 3 we have given examples of fields. In this chapter we introduce the concept of ring homomorphism to help us make a deeper study of field structure. Our main objectives are:

(i) to describe the structure of all finite fields and then to complete the algebraist's dream as far as these fields are concerned by giving a simple criterion for telling any two of them apart (see 4.5.8);
(ii) to prove the impossibility of solving affirmatively the old Greek problems of angle trisection and cube duplication (see Section 4.6);
(iii) to give an almost totally algebraic proof of the Fundamental Theorem of Algebra (see 1.11.8 and 4.8.1).

On the way we shall take time off to show (Section 4.4) how one can construct the field $\mathbb{R}$ without making any assumptions other than that we already have to hand the field $\mathbb{Q}$ of rational numbers and how, in the same spirit but very much more easily, one can produce $\mathbb{C}$ using only known properties of $\mathbb{R}$. Having already indicated (Section 3.10) how to construct $\mathbb{Q}$ from $\mathbb{Z}$ we shall have built $\mathbb{C}$ on the same foundations as those of $\mathbb{Z}$. Further discussion can be found in [48] and [112].

4.2 Return to roots. Ring homomorphisms. Kronecker's theorem

Kronecker, whose attitude to mathematical 'existence' has been noted in the Prologue, had no difficulty in accepting the 'reality' of each individual integer. The creation of the integers he ascribed, in a much quoted statement,* to God. He could also accept the existence of each rational number since it can be created (by man*) in a finite way using only integers. But irrational numbers such as π, and even $\sqrt{2}$ when constructed by Dedekind's method (see Section 4.4), could not be said to have an existence since their definition required prior acceptance of infinite subsets of rational numbers. In fact Kronecker is reported to have asked Lindemann† after the latter had proved (1882) that π is not an algebraic number, 'Of what use is your beautiful investigation

* *Die ganzen Zahlen hat der liebe Gott gemacht, alles andere ist Menschenwerk.*
† C L F Lindemann (12 April 1852 – 6 March 1939).

regarding π? Why study such problems when irrational numbers do not exist?' Nevertheless Kronecker showed how to create irrationals like $\sqrt{2}$ in a way acceptable to him by employing the same method Cauchy had earlier used to introduce $\sqrt{-1}$ (see Section 4.4). The same ideas can be taken further and so the following theorem is usually credited to Kronecker.

Theorem 4.2.1 (Kronecker's Theorem) Let F be any field and let $f = a_0 + a_1x + \cdots + a_nx^n \in F[x]$ be such that f has no root in F. Then there exists a field E containing F as a subfield such that f has a root in E.

Terminology 4.2.2 E is a good letter to use for this new field since one may regard E as an **extension** (field) of F.

To obtain 4.2.1 in a modern way we introduce the concept of ring homomorphism.

Definition 4.2.3 Let $\langle R, +, \cdot\rangle$ and $\langle S, \oplus, \odot\rangle$ be rings and $\theta : R \to S$ a map. θ is called a **homomorphism** or **homomorphic mapping** from the ring $\langle R, +, \cdot\rangle$ (in)to the ring $\langle S, \oplus, \odot\rangle$ iff the following conditions hold for all $a, b \in R$:
(i) $(a+b)\theta = a\theta \oplus b\theta$;
(ii) $(a \cdot b)\theta = a\theta \odot b\theta$.

Remarks
(i) The requirements of a homomorphism are thus similar to those of an isomorphism (3.10.1) except that there is no insistence on θ being 1–1 nor onto.
(ii) Even if θ maps R onto S we do *not* say (cf. 3.10.1) that 'R and S are homomorphic' nor that 'S is homomorphic to R'. The symmetrical nature of the binary relation 'is isomorphic to' is missing in the case of homomorphisms.

Examples 4.2.4
(i) The map $\theta : \mathbf{Z} \to \mathbf{Z}_n$ given by $m\theta = \hat{m}$ defines a homomorphism from $\langle \mathbf{Z}, +, \cdot\rangle$ onto $\langle \mathbf{Z}_n, \oplus, \odot\rangle$. The only homomorphism from $\langle \mathbf{Z}_n, \oplus, \odot\rangle$ to $\langle \mathbf{Z}, +, \cdot\rangle$ is the *trivial homomorphism*, that is the map $\psi : \mathbf{Z}_n \to \mathbf{Z}$ given by $\hat{z}\psi = 0$.
(ii) The map $\theta : \mathbf{Z}_{18} \to \mathbf{Z}_3$ given by $\hat{m}_{18}\theta = \hat{m}_3$ (where $\hat{m}_t$ denotes, temporarily, the element $\hat{m}$ of $\mathbf{Z}_t$) defines a homomorphism from $\langle \mathbf{Z}_{18}, \oplus, \odot\rangle$ onto $\langle \mathbf{Z}_3, \oplus, \odot\rangle$.
(iii) The map $\theta : \mathbf{Z} \to \mathbf{Z}$ given by $z\theta = 2z$ is not a ring homomorphism from $\langle \mathbf{Z}, +, \cdot\rangle$ into itself. [Why not?]
(iv) The maps $\theta, \psi : \mathbf{Q}[x] \to \mathbf{Q}$ defined by $(a_0 + a_1x + \cdots + a_nx^n)\theta = a_0$ and $(a_0 + \cdots + a_nx^n)\psi = a_0 + a_1 + \cdots + a_n$ are homomorphisms. (In the notation of 1.11.1, ψ is the map given by $f\psi = f(1)$.)
(v) The map $\theta : \mathbf{Q}[x] \to \mathbf{R}$ given by $(a_0 + a_1x + \cdots + a_nx^n)\theta = a_0 + a_1\sqrt{2} + \cdots + a_n(\sqrt{2})^n$ is a homomorphism of $\langle \mathbf{Q}[x], +, \cdot\rangle$ onto $\langle \mathbf{Q}(\sqrt{2}), +, \cdot\rangle$.

(vi) The map $\theta:\mathbb{Z}\to\mathbb{Q}$ given by $z\theta=\left(\frac{z}{1}\right)$ is a homomorphism of $\langle\mathbb{Z},+,\cdot\rangle$ into $\langle\mathbb{Q},+,\cdot\rangle$.
(vii) The map $\theta:\mathbb{Q}[x,y]\to\mathbb{C}$ given by $(\Sigma a_{jk}x^jy^k)\theta=\Sigma a_{jk}(1+i)^j(\sqrt[3]{2})^k$ is a homomorphism.

The homomorphisms in (vi), (vii) are not onto and only that in (vi) is 1–1.

Notes 4.2.5
(i) $\theta:R\to S$ is above all a map from the *set* R to the *set* S. We shall however talk briefly of 'the homomorphism θ from the ring R to the ring S' or write at length $\theta:\langle R,+,\cdot\rangle\to\langle S,\oplus,\odot\rangle$, whichever seems more helpful.
(ii) The subset $R\theta=\{r\theta: r\in R\}$ of S is called the (**homomorphic**) **image of *R* under *θ***. $R\theta$ is easily shown to be a subring of S (exercise 4).
(iii) The reason for insisting on some 'preservation of structure' as in (i) and (ii) of 4.2.3 (rather than looking at, say, arbitrary mappings from R to S) is dealt with in the following Remarks and in exercise 7.
(iv) One can describe a homomorphism $\theta:R\to S$ pictorially as in Fig. 4.1.

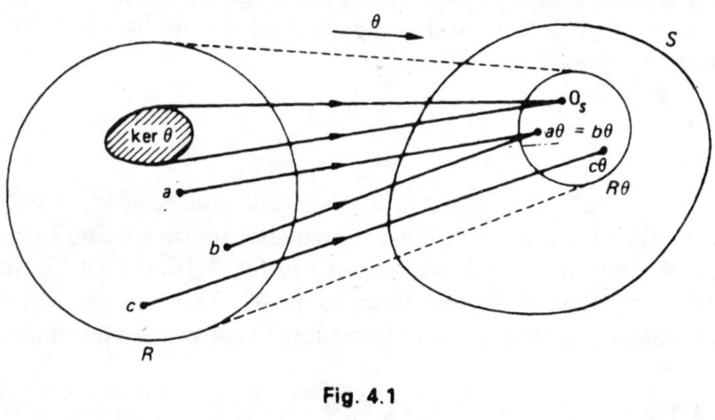

Fig. 4.1

Compare this with the pictorial description of isomorphism given in Section 3.10. The shaded region in R represents the *kernel* of θ (see 4.2.8).

Remarks
(i) Homomorphic images have well been likened to photographs ([18, p. 51], [37, p. 69]). Just as in a photograph of some object certain information present in the original—that round the back of the object—is lost, so in a homomorphic (as distinct from an isomorphic) image $R\theta$ of a ring R some information present in R (essentially that contained in ker θ—see 4.2.8) is lost. On the other hand, the information remaining in $R\theta$ might, if $R\theta$ is a 'simpler' ring than R, be more readily extracted than from the original R. Of course, to be useful, $R\theta$ will have to bear some structural similarity to R—hence the demands made for the preservation of structure in 4.2.3(i), (ii). The point

being made is that, for investigating a ring R, the investigation of a different but not totally unrelated ring will generally prove more beneficial than staring blankly at R. At the level of this text observations of this sort show up rather better in the investigation of finite groups but 4.2.10 below provides a relatively non-trivial ring-theoretic instance of the above remarks.
(ii) Some texts give the impression, by introducing them together, that the concepts of homomorphism and isomorphism play roles which are mild variants of one another. In fact they serve somewhat dissimilar purposes. The isomorphism concept brings some order to the plethora of algebraic systems (see Remarks in Section 3.10) by identifying 'algebraically identical' ones. (More fundamentally it *defines* the rather loose term in inverted commas!) Clearly the homomorphism concept is unsuitable for this purpose. As noted above, homomorphisms act in an investigative capacity. Although it may appear that isomorphisms cannot play such a role (surely any abstract system is just as difficult to penetrate as an isomorphic one?) an isomorphism between two systems, concrete or abstract, can often be revealing. See for example 5.9.5, 5.9.6 and 6.4.4.

Because of the similarity of 3.10.1 and 4.2.3 it is only to be expected that homomorphisms will have at least some of the properties possessed by isomorphic mappings. There is, for example (cf. exercise 3.10.7),

Theorem 4.2.6 Let $\theta:\langle R,+,\cdot\rangle\to\langle S,\oplus,\odot\rangle$ be a ring homomorphism. Then (i) $0_R\theta=0_S$; (ii) for each $a\in R$, $(-a)\theta=\ominus(a\theta)$; (iii) $(a-b)\theta=a\theta\ominus b\theta$.

Proof
(i) $0_R=0_R+0_R$. Hence $0_R\theta=(0_R+0_R)\theta=0_R\theta\oplus 0_R\theta$, in S. Thus, immediately, $0_R\theta=0_S$. [Why?]
(ii) $0_S=0_R\theta=(a+(-a))\theta=a\theta\oplus(-a)\theta$, in S. Thus, immediately, $(-a)\theta=\ominus(a\theta)$. [Why?]
(iii) Use part (ii).

At this point ideals (see Section 3.4) enter again. We see that they are intimately related to homomorphisms by proving

Theorem 4.2.7 Let $\theta:\langle R,+,\cdot\rangle\to\langle S,\oplus,\odot\rangle$ be a ring homomorphism. Then the subset $K=\{k\in R: k\theta=0_S\}$ is an ideal of R.

Proof Let $k_1,k_2\in K$. Then $(k_1-k_2)\theta=k_1\theta\ominus k_2\theta=0_S\ominus 0_S=0_S$. Thus $k_1-k_2\in K$. Next let $k\in K$ and $r\in R$. Then $(k\cdot r)\theta=k\theta\odot r\theta=0_S\odot r\theta=0_S$. Thus $k\cdot r\in K$. In a similar manner one proves $r\cdot k\in K$, so that K is an ideal, as asserted. [Question: Is K non-empty?]

Definition 4.2.8 The above ideal K is called the **kernel** of θ. We often denote it by $\ker\theta$.

Examples 4.2.9 The kernels of the homomorphisms given in 4.2.4 are

(i) For θ: the ideal $[n]=\{kn: k\in\mathbb{Z}\}$. For ψ: the whole ring $\mathbb{Z}_n$.

(ii) The ideal comprising $\{\hat{0}, \hat{3}, \hat{6}, \hat{9}, \widehat{12}, \widehat{15}\}$ in $\mathbb{Z}_{18}$.

(iv) For θ: the ideal $[x]=\{xf: f\in\mathbb{Q}[x]\}$ in $\mathbb{Q}[x]$. For ψ: the ideal comprising all f in $\mathbb{Q}[x]$ whose coefficients have sum zero, that is $\{(x-1)f: f\in\mathbb{Q}[x]\}=[x-1]$.

(v) The ideal $[x^2-2]$ in $\mathbb{Q}[x]$.

(vi) The subset $\{0\}$.

(vii) The least ideal of $\mathbb{Q}[x, y]$ containing (in other words, the ideal generated by) the polynomials x^2-2x+2 and y^3-2. [Can you see why?]

Applications 4.2.10

(i) A nice application of homomorphisms and kernels yields another proof of 1.9.10.

Suppose that $f=a_0+a_1x+\cdots+a_mx^m$ and $g=b_0+b_1x+\cdots+b_nx^n$ are primitive in $\mathbb{Z}[x]$ but that every coefficient in their product is divisible, in $\mathbb{Z}$, by p. The homomorphism $\theta:\mathbb{Z}\to\mathbb{Z}_p$ defined by $z\theta=\hat{z}$ clearly 'extends' to the homomorphism $\theta^*:\mathbb{Z}[x]\to\mathbb{Z}_p[x]$ defined by $(a_0+a_1x+\cdots+a_mx^m)\theta^*=\widehat{a_0}+\widehat{a_1}x+\cdots+\widehat{a_m}x^m$. Clearly ker θ^* comprises all polynomials in $\mathbb{Z}[x]$ whose every coefficient is divisible by p. Apply θ^* to the product fg. Since p divides all the coefficients of fg we see that $(fg)\theta^*=\hat{0}$ in $\mathbb{Z}_p[x]$. But $(fg)\theta^*=f\theta^*g\theta^*$. Since $\mathbb{Z}_p[x]$ is an integral domain (same proof as for $\mathbb{Q}[x]$; see p. 112) we deduce that $f\theta^*=\hat{0}$ or $g\theta^*=\hat{0}$. Thus f (or g) belongs to ker θ^*. That is, in $\mathbb{Z}$, each of the *a*s (or *b*s) is divisible by p, a contradiction.

(ii) $f=x^5+x^3+1$ is irreducible in $\mathbb{Z}[x]$. For if $f=gh$ in $\mathbb{Z}[x]$ then (notation as in (i)) $f\theta^*=g\theta^*h\theta^*$ in $\mathbb{Z}_2[x]$. Clearly $x^5+x^3+\hat{1}$ has no root in $\mathbb{Z}_2$ so $g\theta^*$, $h\theta^*$ must be of degrees 2, 3 respectively. But this is impossible by inspection. (Try it!) Hence $f\theta^*$ is irreducible in $\mathbb{Z}_2[x]$, whence so is f in $\mathbb{Z}[x]$. [Why is it better to map $\mathbb{Z}[x]$ onto $\mathbb{Z}_2[x]$ rather than onto $\mathbb{Z}_3[x]$?]

Problem 1 Can you prove Eisenstein's test using the same technique?

Remark For a quick proof of exercise 1.9.12 using the above methods see [38, p. 181].

Ideals are very useful in helping us construct new rings from old, as our proof of 4.2.1 will show. The following theorem is invaluable. The proof will take a while.

Theorem 4.2.11 Let $\langle R, +, \cdot\rangle$ be a ring and I an ideal of R. Then there exists a ring $\langle S, \oplus, \odot\rangle$ and a homomorphism $\theta: R\xrightarrow{\text{onto}} S$ such that ker $\theta=I$, exactly.

Thus we need (i) to construct S; (ii) to define θ; (iii) to check ker $\theta=I$. We first construct S.

Notation 4.2.12 Let $\langle R, +, \cdot \rangle$ be a ring, I an ideal and r_0 an element of R. We (rather naturally) denote by the symbol r_0+I the subset $\{r_0+i: i \in I\}$ of R.

Note 4.2.13 It is perfectly possible for $r_0 \neq r_1$ in R and yet $r_0+I=r_1+I$. Indeed $r_0+I=r_1+I$ if and only if $r_1-r_0 \in I$ (exercise 13). As an example, suppose I is the ideal $\{5k: k \in \mathbb{Z}\}$ in the ring $\mathbb{Z}$ of integers. Then $3+I = \{3+5k: k \in \mathbb{Z}\} = \{-22+5k: k \in \mathbb{Z}\}$ [can you see why?] $= -22+I$ whilst, of course, $3 \neq -22$. Note that $-22-3 \in I$.

You may recognise $0+I, 1+I, \ldots, 4+I$ as being new names for the elements $\hat{0}, \hat{1}, \ldots, \hat{4}$ of $\mathbb{Z}_5$. In fact the following construction of S from R is merely a generalisation of our construction of $\mathbb{Z}_n$ from $\mathbb{Z}$ (Section 2.4) using the notation $r+I$ instead of, say, $\hat{r}_I$ or just $\hat{r}$. In particular, replacing R and I by $\mathbb{Z}$ and $\{5k: k \in \mathbb{Z}\}$ in 4.2.14, 4.2.16 and 4.2.19 you'll retrieve familiar results about $\mathbb{Z}_5$.

Next we need

Lemma 4.2.14 Let $r_1, r_2 \in R$. Then either $r_1+I=r_2+I$ or else $(r_1+I) \cap (r_2+I) = \varnothing$, the empty set.

Proof Suppose that $(r_1+I) \cap (r_2+I) \neq \varnothing$. Then there exists an element t in both r_1+I and r_2+I. By definition this means that $t=r_1+i_1$ and that $t=r_2+i_2$ for suitable i_1 and i_2 in I. But then $r_1=r_2+i_2-i_1$ and hence, for each $i \in I$, $r_1+i=r_2+i_2-i_1+i$. Since $i_2, i_1, i \in I$ we see that $r_1+i \in r_2+I$. It follows that $r_1+I \subseteq r_2+I$. An identical argument shows that $r_2+I \subseteq r_1+I$ whence $r_1+I = r_2+I$, as required.

Notes 4.2.15

(i) r_1+I is called the **coset** of I in R determined by r_1. We shall meet a similar concept in the theory of groups in Section 5.7.

(ii) 4.2.14 says that any two cosets are either identically equal or else totally disjoint.

(iii) Since each $r \in R$ lies in the coset $r+I$ which it determines, we see that the various cosets of I in R entirely 'cover' R and, by (ii), do it in a pairwise non-overlapping manner. Does this remind you of something studied in Chapter 2? If not read Section 2.3 *now*. For an alternative way of presenting the coset concept see exercise 15.

Denoting by R/I the set of all (distinct) cosets of I in R we show below how to turn R/I into a ring in a natural way.

Problem 2 Why would the last sentence be redundant if its last four words were omitted?

Theorem 4.2.16 Let $\langle R, +, \cdot\rangle$ and R/I be as above. Then the set R/I can, by appropriately defining operations of 'addition' and 'multiplication', be made into a ring (later to be taken as the ring S in 4.2.11).

Proof [Proceeding somewhat naively] we attempt to define binary operations $\oplus$ and $\odot$ on R/I as follows. Let $a+I, b+I \in R/I$. Set

$$(a+I)\oplus(b+I)=(a+b)+I$$

and

$$(a+I)\odot(b+I)=a\cdot b+I$$

Clearly $(a+b)+I$ and $a\cdot b+I$ both lie in R/I [but once again there is the problem—see Section 2.4—of the well-definedness of $\oplus$ and $\odot$. The problem here is this: If $X=a+I=c+I$ and if $Y=b+I=d+I$ then according to the definition proposed above $X\odot Y=a\cdot b+I$ on the one hand and $c\cdot d+I$ on the other. Similar remarks apply to $X\oplus Y$. Consequently, we must check that, from the equalities $a+I=c+I$ and $b+I=d+I$, there follow the equalities $a\cdot b+I=c\cdot d+I$ and $(a+b)+I=(c+d)+I$.]

Now from $a+I=c+I$ we may deduce (exercise 13) that $a=c+j$ for suitable $j\in I$. Similarly from $b+I=d+I$ we deduce $b=d+k$ for suitable $k\in I$. Then $a\cdot b=(c+j)\cdot(d+k)=c\cdot d+j\cdot d+c\cdot k+j\cdot k\in c\cdot d+I$, since ... why?

It follows, as in the proof of 4.2.14, that $a\cdot b+I=c\cdot d+I$, as required.

The corresponding problem of showing that $(a+b)+I=(c+d)+I$ is left to the reader in exercise 14.

To check that $\langle R/I, \oplus, \odot\rangle$ is a ring we only have to check the six axioms A1, A2, A3, A4, M2, D as required by 3.2.2. We check only the axioms A1 and M2 leaving the others to exercise 14.

Axiom A1 For $a+I, b+I\in R/I$ we have
$(a+I)\oplus(b+I)=(a+b)+I=(b+a)+I=(b+I)\oplus(a+I)$, as required.

Axiom M2 For $a+I, b+I, c+I\in R/I$ we have
$\{(a+I)\odot(b+I)\}\odot(c+I)=\{ab+I\}\odot(c+I)=(ab)c+I=a(bc)+I=$
$(a+I)\odot\{bc+I\}=(a+I)\odot\{(b+I)\odot(c+I)\}$, as required.

Definition 4.2.17 The ring R/I is called the **factor ring*** of R by I.

Examples 4.2.18

(i) If $R=\mathbf{Q}[x]$ and if $I=[x^2-2]$ then every element of R/I can be written in the form $a+bx+I$ (where $a, b\in\mathbf{Q}$). Further, $\langle R/I, \oplus, \odot\rangle$ is clearly (?; exercise 20) isomorphic to the field of all real numbers of the form $a+b\sqrt{2}$ where $a, b\in\mathbf{Q}$.

* Some authors prefer the words *quotient ring*. We avoid this since it would otherwise be too easy to confuse this concept with that of quotient field. (3.10.5(i).) In any case quotient fields comprise quotients (i.e. fractions); factor rings, despite the (standard) notation R/I, do not.

(ii) If $R = \mathbb{Z}[x, y]$ and if $I = [x^2, y^2+1]$—the ideal of $\mathbb{Z}[x, y]$ generated (see 3.4.7(iv)) by x^2 and y^2+1—then every element of R/I has the form $a + bx + cy + dxy + I$. Here $\langle R/I, \boxplus, \boxdot \rangle$ is isomorphic to the ring of all quadruples (a, b, c, d) where $a, b, c, d \in \mathbb{Z}$, where addition $\boxplus$ is defined componentwise and multiplication by $(a, b, c, d) \boxdot (\alpha, \beta, \gamma, \delta) = (a\alpha - c\gamma, b\alpha + a\beta - d\gamma - c\delta, c\alpha + a\gamma, d\alpha + c\beta + b\gamma + a\delta)$.

Problem 3 Am I right here in (ii)? Can you see how I can be fairly sure I'm right even before I check anything. (At first glance it might not be apparent that the quadruples even form a ring under the given $\boxplus$ and $\boxdot$!)

As stated in 4.2.16 we shall take the ring S required in 4.2.11 to be the ring R/I constructed above. Thus we are left to prove

Lemma 4.2.19 Let $\langle R, +, \cdot \rangle$, I and $\langle R/I, \oplus, \odot \rangle$ be as in 4.2.16. The mapping $\theta: R \to R/I$ given by $r\theta = r + I$ is a homomorphism from R onto R/I and $\ker \theta = I$ exactly.

Proof Since every element of R/I has the form $a + I$ where $a \in R$ and since $a\theta = a + I$ by definition, it is clear that θ maps R *onto* R/I.

Now for $a, b \in R$ we have $(a \cdot b)\theta = (a \cdot b) + I = (a + I) \odot (b + I) = a\theta \odot b\theta$. One shows, similarly, that $(a + b)\theta = a\theta \oplus b\theta$. Thus θ is a homomorphism.

Finally, let $a \in \ker \theta$. Then, on the one hand, $a\theta = 0_R + I$ whilst, on the other, $a\theta = a + I$, by definition of θ. Consequently, $0_R + I = a + I$ and so $a \in I$ follows. Thus $\ker \theta \subseteq I$. Conversely if $i \in I$ then $i\theta = i + I = 0_R + I$, the zero element of R/I. Thus $I \subseteq \ker \theta$. It follows that $\ker \theta = I$, as required.

With the completion of 4.2.19, we also complete the proof of 4.2.11.

Remarks The passage from the ring $\langle R, +, \cdot \rangle$ to the ring $\langle R/I, \oplus, \odot \rangle$ (via the homomorphism θ) achieved by 'reducing I to zero' can be colourfully described as the process of **killing off I.** The elements of I, and they only, are sent to the zero element of R/I, but do note that this does *not* necessarily imply that if a and b are distinct elements of R which are not in I then $a\theta, b\theta$ are distinct in R/I (see Fig. 4.1 on p. 142). For it may be that a, b lie in the same coset of I in R in which case $a\theta = a + I = b + I = b\theta$.

We now give the

Proof of Theorem 4.2.1 [Although f has no root in F it may still factorise in $F[x]$ into (irreducible) factors (each of degree at least 2).] Suppose that $f = gg_2 \ldots g_r$ where the g_i are irreducible in $F[x]$. Let I be the (principal) ideal of $F[x]$ generated by g and set $S = \dfrac{F[x]}{I}$. [Thus the elements of S are cosets $p + I$ where $p \in F[x]$.]

Now S is a homomorphic image of a commutative ring with 1 (that is, with unity) and is thus itself a commutative ring with 1 (exercise 6). We show that S is a field.

Suppose $g = b_0 + b_1x + \cdots + b_mx^m$ and let $p + I$ be a non-zero element of S. Then $p \notin I$ and so p is not a (polynomial) multiple of g. Since the only divisors of g in $F[x]$ are its associates and the units of $F[x]$ we see that, in $F[x]$, $(p, g) = 1$. It follows from 3.7.5 that there exist, in $F[x]$, polynomials r, s such that

$$rp + sg = 1 \quad \text{in } F[x]$$

This implies $\quad rp + sg + I = 1 + I \quad$ in S. [Why?]

It follows that $\quad (rp + I) \oplus (sg + I) = 1 + I \quad$ in S, [Why?]

whence $\quad (r + I) \odot (p + I) = 1 + I \quad$ in S. [Why?]

Thus $r + I$ is a left (and hence two-sided [why?]) multiplicative inverse for $p + I$ in S which is thus proved to be a field.

Clearly F is not itself contained in S but an isomorphic copy is (see exercise 19). Hence, by exercise 3.10.9, we can construct a field $\bar{S} \overset{\psi}{\cong} S$ such that F itself *is actually a subset of* $\bar{S}$. Let* y denote the element of $\bar{S}$ which corresponds under ψ to the element $x + I$ of S. The element $b_0 + b_1y + \cdots + b_my^m$ in $\bar{S}$ then corresponds to the element $(b_0 + I) \oplus (b_1 + I) \odot (x + I) \oplus \cdots \oplus (b_m + I) \odot (x + I)^m = (b_0 + b_1x + \cdots + b_mx^m) + I = 0_S$. Thus $b_0 + b_1y + \cdots + b_my^m = 0_{\bar{S}}$, that is, y is a root, in $\bar{S}$, of the given polynomial g. It follows immediately that y is a root, in $\bar{S}$, of f.

Remark Part of the above result can be represented pictorially by Fig. 4.2.

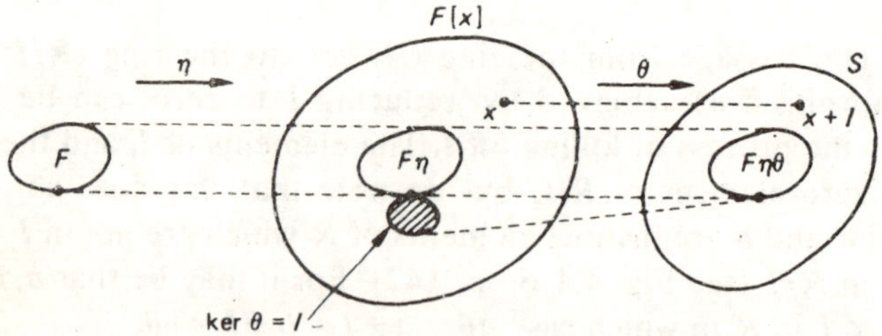

Fig. 4.2

The remarkable thing about the proof is not that we can make a structure S in which $b_0 + b_1x + \cdots + b_mx^m$ is zero—this is easily arranged by killing off $b_0 + \cdots + b_mx^m$—it is that, when one does this killing, S is a field which furthermore contains an undamaged replica (i.e. isomorphic copy) of F as a subfield.

* This use of $\bar{S}$ rather than S is not common, but the author feels it renders this particular proof more easily understood than that usually given.

Problem 4 $\mathbb{Q}$ has no roots of either x^2+1 or x^2+x+1. Let I be the ideal of $\mathbb{Q}[x]$ generated by x^2+1 and x^2+x+1. Then surely $\mathbb{Q}[x]/I$ is a field containing the required roots. Am I right?

Although we shall have no need of it until Section 4.5 it seems now appropriate to obtain the

Corollary 4.2.20 Let $f \in F[x]$, F any field. Then there exists a field E such that $E \supseteq F$ and in $E[x]$ f factorises completely into linear factors.

Proof (Outline) If $f = g_1 g_2 \ldots g_r$ expresses f as a product of polynomials irreducible over F, find E_1 as in 4.2.1 so that g_1 has a root α_1, say, in E_1. Then, in $E_1[x]$, f has at least one linear factor, $x-\alpha_1$. Factorise f into irreducible polynomials in $E_1[x]$ and repeat the process, with E_1 in place of F. After at most $\deg f$ steps we arrive at a field E with the desired properties.

Is the field E of 4.2.20 uniquely determined? By no means: take $f = x-1$ and $F = \mathbb{Q}$ whence E can be $\mathbb{Q}$ or $\mathbb{R}$ or $\mathbb{C}$.... But if we insist that E be as small as possible then we do indeed have

Theorem 4.2.21 Let $f \in F[x]$, F any field, and let E_1, E_2 be extension fields of F such that
(i) f factorises into linear factors* $(x-\alpha_1)(x-\alpha_2)\ldots(x-\alpha_n)$ in $E_1[x]$ and $(x-\beta_1)(x-\beta_2)\ldots(x-\beta_n)$ in $E_2[x]$;
(ii) E_1 (respectively E_2) is the smallest subfield of E_1 (respectively E_2) which contains F and all the roots α_i (respectively β_i).
Then $E_1 \cong E_2$.

Proof We ask the reader to be patient—until 7.2.5.

This (essentially unique) extension field E_1 ($\cong E_2$) is called the *splitting field* of f over F.

Exercises

Homomorphisms

1 (a) Let θ and ψ be maps $\mathbb{Q}[x] \to \mathbb{Q}[x]$ defined by

$$(a_0+a_1x+\cdots+a_nx^n)\theta = a_0+a_1(x+2)+\cdots+a_n(x+2)^n$$

and

$$(a_0+a_1x+\cdots+a_nx^n)\psi = a_0+a_1x^2+\cdots+a_nx^{2n}.$$

Determine whether or not θ, ψ are homomorphisms.
(b) Are the maps $z\theta = -z$ on $\mathbb{Z}$ and $(x+iy)\psi = x$ from $\mathbb{C}$ to $\mathbb{R}$ homomorphisms?

* We also say that f *splits in* $E_1[x]$ (or *over* E_1).

2 Describe all possible ring homomorphisms (i) from $\mathbb{Z}_{12}$ onto $\mathbb{Z}_4$; (ii) from $\mathbb{Z}_{12}$ into $\mathbb{Z}_8$; (iii) from $\mathbb{Z}_{12}$ into $\mathbb{Z}_5$; (iv) from $\mathbb{Z}$ into $\mathbb{Z}$. Can you, from (i), (ii), (iii), decide precisely when there exists a homomorphism from $\mathbb{Z}_m$ onto $\mathbb{Z}_n$?

3 Is the map θ from $\langle M_2(\mathbb{Z}), \oplus, \odot\rangle$ to $\langle \mathbb{Z}, +, \cdot\rangle$ given by $\begin{pmatrix} a & b \\ c & d\end{pmatrix} = ad - bc = \det\begin{pmatrix} a & b \\ c & d\end{pmatrix}$ a homomorphism?

4 Let $\theta : \langle R, +, \cdot\rangle \to \langle S, \oplus, \odot\rangle$ be a homomorphism.
(i) Prove that $R\theta$ is a subring of S. Give a concrete example where $R\theta$ is not an ideal of S.
(ii) Prove that, for $r_1, r_2 \in R$, $r_1\theta = r_2\theta$ iff $r_1 - r_2 \in \ker\theta$.

5 Let $\theta : \langle R, +, \cdot\rangle \to \langle S, \oplus, \odot\rangle$ and $\psi : \langle S, \oplus, \odot\rangle \to \langle T, \boxplus, \boxdot\rangle$ be homomorphisms. Prove that $\theta \circ \psi : \langle R, +, \cdot\rangle \to \langle T, \boxplus, \boxdot\rangle$ is a homomorphism.

6 Let $\langle R, +, \cdot\rangle$ be a commutative ring with a 1 and let θ be a homomorphism from R onto $\langle S, \oplus, \odot\rangle$. Prove that $\langle S, \oplus, \odot\rangle$ is a commutative ring with a 1. Is this proof still valid if θ is not assumed to be onto S?

7 Let $\langle R, +, \cdot\rangle$ be a ring and let $\langle S, \oplus, \odot\rangle$ be a ring which is *not* commutative and does *not* possess a 1. Given that there exists a mapping ζ from the set R onto the set S what can you say about $\langle R, +, \cdot\rangle$? Can you say any more if you are told that ζ is a homomorphism?

8 Let $L_2(\mathbb{Z})$ be the set of matrices of the form $\begin{pmatrix} a & 0 \\ b & c\end{pmatrix}$ where $a, b, c \in \mathbb{Z}$. Define $\theta : L_2(\mathbb{Z}) \to \mathbb{Z}$ by $\begin{pmatrix} a & 0 \\ b & c\end{pmatrix} = a$. Show that θ is a homomorphism onto $\mathbb{Z}$ and that $L_2(\mathbb{Z})$ is not a commutative ring.

9 Define $\theta : \mathbb{Q}[x] \to \mathbb{C}$ by

$$(a_0 + a_1x + \cdots + a_nx^n)\theta = a_0 + a_1(\tfrac{3}{7} + \tfrac{7}{3}i) + \cdots + a_n(\tfrac{3}{7} + \tfrac{7}{3}i)^n.$$

Find $\ker\theta$. If $\mathbb{Q}[x]$ is replaced by $\mathbb{Z}[x]$—so that all the $a_i \in \mathbb{Z}$—how is your answer changed?

10 Show that a homomorphism θ: $\langle R, +, \cdot\rangle \to \langle S, \oplus, \odot\rangle$ is $1-1$ if and only if $\ker\theta = \{0_R\}$.

11 Let F be a field. Use exercise 3.4.14 to describe all possible homomorphic images of F.

Cosets

12 Write the coset $9 + 2x + 2x^2 + 7x^3 + 4x^5 + I$ where $I = [x^2 + 5]$ in the form $a + bx + I$. Write down three more elements of this coset.

13 For the ideal I and the elements r_1, r_2 of the ring R show that $r_1 + I = r_2 + I$ iff $r_1 = r_2 + i$ for suitable $i \in I$. Deduce that if $a + bx + cx^2 + dx^3 + I =$

$A+Bx+Cx^2+Dx^3+I$ where $I=[x^4+2x+7]$ in $\mathbb{Q}[x]$ then $a=A, b=B, c=C$ and $d=D$ in $\mathbb{Q}$.

14 Complete the proof of 4.2.16.

15 Let I be an ideal in the ring R. For $r_1, r_2 \in R$ define $r_1 \sim r_2$ iff $r_1 - r_2 \in I$. Show directly that $\sim$ is an equivalence relation on R and that the cosets of I in R are the corresponding equivalence classes.

16 Let S be the subring of $\mathbb{Q}[x]$ [is it one?] comprising all polynomials whose coefficients of odd powers of x are all zero. For each $p \in \mathbb{Q}[x]$ denote the set $\{p+s : s \in S\}$ by $p+S$. Show that if $p_1+S \neq p_2+S$ then $(p_1+S) \cap (p_2+S) = \varnothing$. Attempt to make a ring out of these subsets of $\mathbb{Q}[x]$ by defining $(p_1+S) \oplus (p_2+S) = (p_1+p_2)+S$ and $(p_1+S) \odot (p_2+S) = p_1p_2+S$. Is your attempt successful? Demonstrate that it is—or explain why it is not. [Hint: $x+S=(1+x)+S$. Hence their squares are equal? Cf. exercise 2.4.12.]

Factor Rings

17 To which more familiar ring is R/I isomorphic when (i) $R=\mathbb{Q}[x], I=[x]$; (ii) $R=\mathbb{Q}[x], I=[x+2]$; (iii) $R=\mathbb{Z}[x], I=[x,4]$; (iv) $R=\mathbb{Z}[x], I=[x^2+1]$; (v) $R=\mathbb{Z}[x], I=[2x+1]$. Which of these factor rings are fields? [Hint: In (ii) think of $\mathbb{Q}[x]$ with $x+2$ 'put equal to 0'.]

18 Let $g \in F[x]$. Show that $\dfrac{F[x]}{[g]}$ is a field if *and only if* g is irreducible in $F[x]$. Is $\dfrac{\mathbb{Z}_5[x]}{[x^2+x+1]}$ a field? With how many elements?

19 Show that if $g \in F[x]$ and if g is irreducible in $F[x]$ then in the map $\theta : F[x] \to \dfrac{F[x]}{[g]}$ the subring F of $F[x]$ is mapped isomorphically into $\dfrac{F[x]}{[g]}$.

20 Establish the claim made concerning isomorphism in 4.2.18(i).

21 Show that x^3+x+1 is irreducible in $\mathbb{Q}[x]$. Find $r, s \in \mathbb{Q}[x]$ such that $r(x^3+x+1)+s(x^2+2x+2)=1$. Hence write in the form $a+bx+cx^2+I$ the multiplicative inverse of $2+2x+x^2+I$ in the field $\dfrac{\mathbb{Q}[x]}{I}$, I being the principal ideal $[x^3+x+1]$. [See the proof of 4.2.1.]

Extension Fields

22 Find inside $\mathbb{C}$ an extension of the field $\mathbb{Q}$ in which x^4+2x^2-4 has a root but over which x^4+2x^2-4 does not split into a product of four linear factors.

4.3 The isomorphism theorems

Because of the extreme importance of the homomorphism concept, not only for rings and groups (see Section 5.10) but for all algebraic systems, we offer the reader an opportunity to strengthen his understanding of the concept in the case of rings by proving three quite involved results concerning homomorphisms. Their collective name, 'The Isomorphism Theorems', indicates their standing. In Chapter 6 we shall make non-trivial use of the group-theoretic versions of these theorems. It must be admitted that, strictly speaking, we do not require 4.3.1, 4.3.2 and 4.3.3 in this text, the applications we give being obtainable just as naturally by other routes. On the other hand the application 4.3.4 of 4.3.1 is useful later.

We begin with

Theorem 4.3.1 (The First Isomorphism Theorem) Let $\theta:\langle R, +, \cdot\rangle \to \langle S, +, \cdot\rangle$ be a homomorphism from R onto S with kernel I. Then the ring $\langle S, +, \cdot\rangle$ and the factor ring $\langle R/I, \oplus, \odot\rangle$ are isomorphic.

Proof [There is an obvious mapping which, if the proposed assertion is true, one might expect to prove the theorem: namely the map $\psi: R/I \to S$ given by $(r+I)\psi = r\theta$. The first thing we must check is . . . what?*]

Suppose $r_1+I=r_2+I$. Then $r_1-r_2 \in I$. It follows [since $I = \ker\theta$, that $(r_1-r_2)\theta = 0_S$ and hence] that $r_1\theta = r_2\theta$. Thus the proposed mapping ψ *is* well-defined.

Now ψ is a homomorphism. For $\{(r_1+I)\oplus(r_2+I)\}\psi = \{(r_1+r_2)+I\}\psi = (r_1+r_2)\theta = r_1\theta + r_2\theta = (r_1+I)\psi + (r_2+I)\psi$, whilst $\{(r_1+I)\odot(r_2+I)\}\psi = \{r_1r_2+I\}\psi = (r_1r_2)\theta = r_1\theta \cdot r_2\theta = (r_1+I)\psi \cdot (r_2+I)\psi$.

Clearly ψ maps R/I onto S. Finally, if $(r_1+I)\psi = (r_2+I)\psi$ then $r_1\theta = r_2\theta$ whence $(r_1-r_2)\theta = 0_S$ so that $r_1-r_2 \in I$. That is, $r_1+I = r_2+I$. Thus R/I and S are isomorphic [via ψ] as required.

Remark This result tells us that all homomorphic images of a ring R are of a special kind—they are all (up to isomorphism) obtainable as factor rings R/I.

Next, let A be an ideal and B a subring of the ring R. Defining $A+B$ to be the subset $\{a+b: a \in A, b \in B\}$ of R one checks easily that $A+B$ is a subring of R, that A is an ideal of $A+B$ and that $A \cap B$ is an ideal of the ring B. We then have

Theorem 4.3.2 (The Second Isomorphism Theorem) Let A, B, R be as just described. Then the rings $(B+A)/A$ and $B/(A\cap B)$ are isomorphic.

Proof Define $\theta: B \to (B+A)/A$ by $b\theta = b+A$. Since $b+a+A = b+A$, it is clear that θ maps B onto $(B+A)/A$. We leave you to check that θ is a homomorphism. Suppose $x \in \ker\theta$. Then $x+A$ is the zero element $0+A$ of $(B+A)/A$. Thus $x \in A$. But $x \in B$, whence $x \in A \cap B$. It follows that $\ker\theta \subseteq A \cap B$.

* Well-definedness.

Conversely, if $x \in A \cap B$ then $x\theta = x + A = 0 + A$ since $x \in A$. Thus $\ker \theta = A \cap B$ and from 4.3.1 we deduce $B/(A \cap B) \cong (B+A)/A$.

Theorem 4.3.3 (The Third Isomorphism Theorem) Let I, K be ideals in the ring R such that $I \subseteq K$. Then I is an ideal in K, $K/I = \{k + I : k \in K\}$ is an ideal in R/I and $\dfrac{R/I}{K/I} \cong R/K$.

Proof We let the reader check the statement about I. Define the map $\theta : R/I \to R/K$ by $(r+I)\theta = r + K$. θ is a well-defined map since from $r_1 + I = r_2 + I$ we get $r_1 - r_2 \in I \subseteq K$ whence $r_1 + K = r_2 + K$. Clearly θ is onto. Again we leave you to check that θ is a homomorphism. From 4.3.1 it follows that $R/K \cong \dfrac{R/I}{\ker \theta}$. Now every element of R/I is of the form $r + I$ $(r \in R)$ and $(r+I) \in \ker \theta$ iff $r + K = 0 + K = 0_{R/K}$. But this is the case iff $r \in K$. Thus $\ker \theta = \{k + I : k \in K\} = K/I$. In particular K/I is an ideal of R/I (by 4.2.7) and $R/K \cong \dfrac{R/I}{K/I}$ by 4.3.1.

These three theorems can be represented pictorially as in Fig. 4.3.

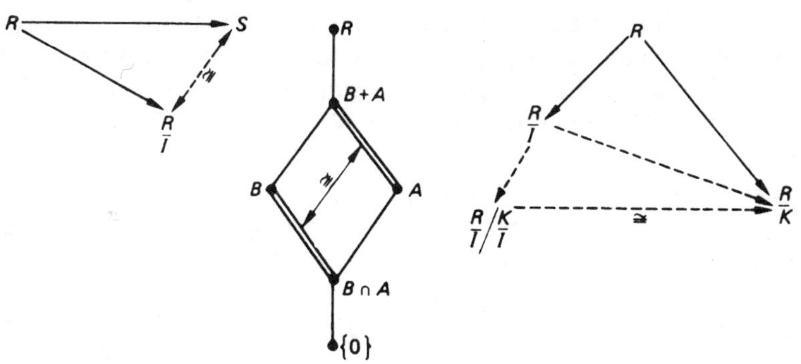

Fig. 4.3

Applications of the isomorphism theorems We have already applied 4.3.1 twice in proving 4.3.2 and 4.3.3. Other applications appear in Section 4.4 and the exercises below. Here we give just one, substantially extending the examples given in 3.10.6 and 4.2.18(i).

Let E and F be fields with $F \subseteq E$ and let $\alpha \in E$. α is called *algebraic over* F (cf. exercise 3.2.14) iff α is a root of some (monic) polynomial in $F[x]$. We let $F(\alpha)$ denote (cf. 3.10.7) the least subfield of E containing F and α. We then have

Theorem 4.3.4 Let E, F, α, and $F(\alpha)$ be as above and let $M_\alpha \in F[x]$ be a monic polynomial of least degree having α as a root. Then M_α is unique and is irreducible* in $F[x]$. If, further, $\deg M_\alpha = n$ then every element of $F(\alpha)$ is expressible uniquely in the form $f_0 + f_1\alpha + \cdots + f_{n-1}\alpha^{n-1}$ for suitable $f_i \in F$. Moreover $F(\alpha) \cong \dfrac{F[x]}{[M_\alpha]}$.

Proof If α were a root of two distinct monic polynomials M_α, N_α each of degree n then α would be a root of $M_\alpha - N_\alpha$ and hence of some (monic) polynomial of degree $\leqslant n-1$, contradicting the minimality of n. Next, M_α is irreducible. For, if $M_\alpha = R_\alpha S_\alpha$ where $\deg R_\alpha$, $\deg S_\alpha < \deg M_\alpha$ then we should have $0 = M_\alpha(\alpha) = R_\alpha(\alpha) \cdot S_\alpha(\alpha)$. Then at least one of $R_\alpha(\alpha)$, $S_\alpha(\alpha)$ would be 0 since E is a field. Hence α would be a root of at least one of R_α, S_α, a contradiction [why?].

Now consider the map $\theta : F[x] \to E$ given by $g\theta = g(\alpha)$ (cf. 4.2.4(iv)). Clearly θ is a homomorphism. What is its kernel? Let $k \in \ker\theta$. Use 1.10.1 to write $k = mM_\alpha + r$ where either $r = 0$ or $r \neq 0$ and $\deg r < \deg M_\alpha$. It follows that $0 = k\theta = m(\alpha)M_\alpha(\alpha) + r(\alpha) = r(\alpha)$ [why?]. But this is untenable unless r is the zero polynomial of $F[x]$ [why?]. Thus $r = 0$ and $k = mM_\alpha$, a polynomial multiple of M_α. This shows that $(F[x])\theta$, which clearly contains both F and α and is itself contained in the field $F(\alpha)$ is, by 4.3.1, isomorphic to $\dfrac{F[x]}{[M_\alpha]}$. But this factor ring is a field (exercise 4.2.18) since M_α is irreducible. Thus $(F[x])\theta = F(\alpha)$, the smallest field containing F and α. Using 1.10.1 again we see that every element of $(F[x])\theta$ is of the form $s(\alpha)$ for suitable s of degree $\leqslant n-1$ in $F[x]$ as claimed. Finally the uniqueness of this form is proved by an argument similar to that proving $M_\alpha = N_\alpha$.

Examples 4.3.5

(i) Taking $E = \mathbb{R}$, $F = \mathbb{Q}$ and $\alpha = \sqrt[3]{2}$ we see that $\mathbb{Q}(\sqrt[3]{2})$, the least subfield of $\mathbb{R}$ containing $\mathbb{Q}$ and $\sqrt[3]{2}$, comprises precisely all elements of the form $a + b(2)^{1/3} + c(2)^{2/3}$ where $a, b, c \in \mathbb{Q}$.

(ii) (See 3.10.7 for notation.) Taking $E = \mathbb{Q}(\sqrt{2}, \mathrm{i})$, $F = \mathbb{Q}(\sqrt{2})$ and $\alpha = \mathrm{i}$ one sees that $M_\alpha = x^2 + 1$. Thus every element of E is expressible uniquely in the form $f_0 + f_1\mathrm{i}$ where $f_0, f_1 \in F$. But similarly every element of F is expressible uniquely in the form $q_0 + q_1\sqrt{2}$ where $q_0, q_1 \in \mathbb{Q}$. Hence every element of E is expressible uniquely (uniquely?) in the form $r_0 + r_1\mathrm{i} + r_2\sqrt{2} + r_3\mathrm{i}\sqrt{2}$ where $r_0, r_1, r_2, r_3 \in \mathbb{Q}$.

As an application of 4.3.2 we reobtain the content of the Remark on p. 148.

Corollary 4.3.6 Let F be a field, p a polynomial of degree $\geqslant 1$ in $F[x]$ and let σ be the natural homomorphism from $F[x]$ onto $\dfrac{F[x]}{[p]} = (F[x])\sigma$. Then $F \cong F\sigma \subseteq (F[x])\sigma$.

* M_α is called the *minimum polynomial of α over F*.

Proof Set $I=[p]$. Then $F\sigma=\{\alpha\sigma:\alpha\in F\subseteq F[x]\}=\{\alpha+I:\alpha\in F\subseteq F[x]\}=\frac{(F+I)}{I}\subseteq\frac{F[x]}{I}$. By 4.3.2 $\frac{(F+I)}{I}\cong\frac{F}{(F\cap I)}$. But $F\cap I=[0]$ [since the non-zero elements of I have degree >0]. Hence $\frac{F}{(F\cap I)}=\frac{F}{[0]}\cong F$.

Genuine and significant applications of 4.3.3 at the level of this text seem less easy to find (although see exercise 6.2.12 and 6.5.8). One sees immediately from the statement of 4.3.3 that an ideal M of a ring R is a maximal ideal (see 3.9.5) iff R/M is a simple ring (exercise 3.4.20). But it really is more natural to deduce this assertion from exercise 4.

Exercises

1 Let $\theta:R\to S$ be a homomorphism not necessarily onto S. Show that $R\theta\cong\frac{R}{\ker\theta}$.

2 Let A_1, A_2 be ideals of the ring R. Show that the map* $\theta:R\to\frac{R}{A_1}\oplus\frac{R}{A_2}$ given by $r\theta=(r+A_1,r+A_2)$ is a homomorphism. Deduce that $\frac{R}{A_1}\oplus\frac{R}{A_2}$ has a subring isomorphic to $\frac{R}{(A_1\cap A_2)}$. Setting $R=\mathbb{Z}$, $A_1=[m]$ and $A_2=[n]$ where $(m,n)=1$, show that $\mathbb{Z}_{mn}\cong\frac{\mathbb{Z}}{([m]\cap[n])}\cong\frac{\mathbb{Z}}{[m]}\oplus\frac{\mathbb{Z}}{[n]}=\mathbb{Z}_m\oplus\mathbb{Z}_n$ (cf. 4.2.9(i)).

3 Let $\theta:R\to S$ be an onto homomorphism† with $\ker\theta=I$ and let $\psi:R\to R/K$ be the natural map where $K\subseteq I$. Show that there exists a homomorphism $\eta:R/K\to S$ such that $\theta=\psi\circ\eta$.

This result, which may be described by the following picture, generalises 4.3.1.

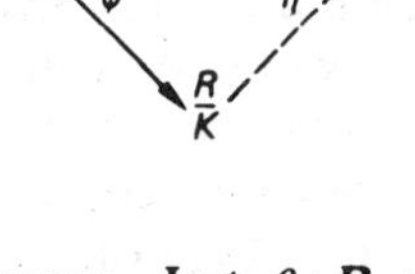

4 The Correspondence Theorem Let $\theta:R\to S$ be an epimorphism with $\ker\theta=I$. Let $U=\{A:A$ is a subring of R and $I\subseteq A\}$ and $V=\{B:B$ is a subring of $S\}$. Show that the map Θ: $U\to V$ defined by $A\Theta=A\theta$ is a 1–1 map of U onto

* $\oplus$ denotes the direct sum (exercise 3.2.9).
† Also called an *epimorphism*.

V. [Hint: 1–1ness is not too hard. To prove onto show that, given $B \subseteq S$, $B\theta^{-1}$ (see 2.6.5) is a subring of R containing I and that $(B\theta^{-1})\Theta = B$.] Show that, under this correspondence, ideals of R which contain I correspond to ideals of S.

Deduce that M is a maximal ideal of R iff R/M is a simple ring. If R is a commutative ring with unity show that M is a maximal ideal iff R/M is a field. [Exercise 3.4.14 might help.]

5 Let R be a commutative ring with unity. Show that an ideal P $(\neq R)$ is a prime ideal iff R/P is an integral domain.

6 Use exercises 4 and 5 to reprove Theorem 3.9.6.

7 (i) Let R be a commutative ring with unity. Show that if P is a prime ideal such that R/P is finite then P is a maximal ideal.
(ii) Show that in $F[x, y]$ the ideal $[x]$ is prime but not maximal. Is the ideal $[x, y]$ maximal in $F[x, y]$? [Cf. exercise 3.9.2.]

8 Let θ be a root of x^3+x+1 in $\mathbb{R}$. By 4.3.4 $\mathbb{Q}(\theta) \cong \dfrac{\mathbb{Q}[x]}{[x^3+x+1]}$. Use exercise 4.2.21 to show that $\dfrac{1}{2+2\theta+\theta^2} = \dfrac{1}{13}(7-5\theta+3\theta^2)$ in $\mathbb{Q}(\theta)$.

9 Let $F \subseteq E$ be fields and let $\alpha, \beta \in E$ be such that $M_\alpha = M_\beta$. Show that the fields $F(\alpha)$ and $F(\beta)$ are isomorphic.

4.4 Constructions of $\mathbb{R}$ from $\mathbb{Q}$ and of $\mathbb{C}$ from $\mathbb{R}$

We have seen in Section 3.10 how one can construct the field $\mathbb{Q}$ from the ring $\mathbb{Z}$ using only the properties assumed, in Section 1.2, to hold in $\mathbb{Z}$. Indeed $\mathbb{Q}$ can be formally defined as the unique field of fractions of the unique well ordered integral domain (see 3.10.5(iii)). Furthermore exercise 3.12.8 indicates how $\mathbb{Q}$ can be made into an ordered field, again using only the axiom P of Section 1.2. The question arises as to how we might similarly replace our intuitive and possibly vague ideas about real and complex numbers by formal constructions. Whilst employing some ideas from Sections 4.2 and 4.3, this section is meant to be easy reading—for information, if you like—and we shall only outline the formal procedures involved.

Informally we recognise that, whilst $\mathbb{R}$ is algebraically deficient ($x^2+1=0$ has no solution in $\mathbb{R}$), $\mathbb{Q}$ is even worse: it is analytically deficient too. For instance, in $\mathbb{Q}$ one can find infinite sequences (there is, for example, one whose first few terms are 14/10, 141/100, 1414/1000, 14142/10000, 141421/100000, . . .) which appear to 'home in'—technically *converge*—to non-rational quantities (in this instance $\sqrt{2}$). A geometrical interpretation of this is given by considering the (intuitive) real numbers stretched out along an infinite straight line. (Think of the x-axis in the plane.) Marking off the points on this straight line which correspond to rational numbers leaves infinitely many unmarked points on the line and it is possible, by alighting only on a sequence of rational points, to home in on the unmarked (irrational) point $\sqrt{2}$. Thus **Q** can be thought of informally as '**R**-with-gaps'—so that **Q** is analytically incomplete—with $\sqrt{2}$, $\sqrt{3}$, e, π, $\sqrt[8]{10}$ being amongst the infinitely many points at which such a gap is located.

The above intuitive picture leads one to propose *defining* $\sqrt{2}$ to *be* the above sequence. The elements of **Q** present no difficulty; we identify $\frac{a}{b}$ with the sequence $\left(\frac{a}{b}, \frac{a}{b}, \frac{a}{b}, \frac{a}{b}, \cdots\right)$. In fact this proposal won't quite suffice since, for instance, $\left(2\frac{1}{3}, 1, \frac{14}{10}, \frac{141}{100}, \cdots\right)$ and $\left(7, 3, \frac{6}{5}, \frac{a}{b}, \frac{a}{b}, \frac{a}{b}, \cdots\right)$ also seem to be candidates for $\sqrt{2}$ and $\frac{a}{b}$. To identify precisely which infinite sequences of rationals are of interest to us and to get round the insufficiency we proceed as follows.

The infinite sequence $(a_1, a_2, \ldots)$ of rational numbers is said to be **Cauchy convergent** if and only if, to each positive rational number $h > 0$, we can find a positive integer H (which will in general depend upon h) such that for all integers $i, j > H$ we have $|a_i - a_j| < h$. Two convergent infinite sequences $(a_1, a_2, \ldots)$, $(b_1, b_2, \ldots)$ are then said to be *equivalent* if and only if, given any rational $k > 0$, we can find a positive integer $K > 0$ such that $|a_s - b_s| < k$ for all $s > K$. [If these ideas are new to you, you may find you need to read them very carefully twenty times or more before they begin to stick.]

The set $\mathbb{S}$ of *real numbers* then comprises, by definition, the totality of all equivalence classes (see exercise 3) of Cauchy convergent sequences. Addition, $\oplus$, and multiplication, $\odot$, are defined as you would expect [and how is that?] and $\mathbb{S}$ can be shown to be a field containing (an isomorphic copy of) **Q** as a subfield. In $\mathbb{S}$ the analytic deficiencies apparent in **Q** are absent.

This method, using **Cauchy sequences,** is due to Cantor. Using the terminology of Section 4.2 we may proceed as follows. Under componentwise $\oplus$ and $\odot$ the set C of all Cauchy convergent sequences becomes a ring in which the subset of all those sequences which are equivalent to the zero sequence $(0, 0, 0, \ldots)$ form an ideal N (N for 'nought' or 'null'?). The above set $\mathbb{S}$ is nothing more than the factor ring C/N.

A second method of defining $\mathbb{R}$, this one due to Dedekind, is based on the informal observation that every real number r appears to split **Q** into two disjoint subsets: (i) all rationals less than or equal to r; (ii) all those greater than r (!!) This dissection is called a **Dedekind cut,** the subsets (i) and (ii) being called, respectively, *lower* and *upper sections.* Clearly r corresponds to the upper section of all those rationals greater than r but, of course, one cannot define r to be this upper section since such a definition (of r in terms of r) would be circular! Thus we have to find an honest way of defining the term upper section in **Q** solely in terms of **Q**. Assuming the familiar ordering which exercise 3.12.8 allows us to place on $\mathbb{Q}$ we make

Definition 4.4.1 An **upper section** in $\mathbb{Q}$ is a subset U of $\mathbb{Q}$ such that
(i) $U \neq \varnothing, U \neq \mathbb{Q}$;
(ii) if $x, y \in \mathbb{Q}$, if $x \in U$ and if $x \leqslant y$ then $y \in U$;
(iii) if $x \in U$ then there exists $y \in U$ such that $y < x$.
(To see what motivates this definition, try thinking pictorially of the upper sections corresponding to $\sqrt{2}$ and to, say, $5\frac{1}{4}$.)

By defining,* suitably, $\oplus$ and $\odot$ on the set $\mathbb{U}$ of all upper sections one can show that $\langle \mathbb{U}, \oplus, \odot \rangle$ is a field. (This is an arduous task, the proof that $\odot$ is associative requiring 9 subcases! See [50].) $\mathbb{U}$, like $\mathbb{S}$, can be shown to contain a subfield isomorphic to $\mathbb{Q}$ and, again like $\mathbb{S}$, be analytically complete. Before reading the next sentence the reader may care to ponder the question: Which of $\mathbb{S}$ and $\mathbb{U}$ more accurately reflects his intuitive conception of $\mathbb{R}$?

The answer is that neither should: for $\mathbb{S}$ and $\mathbb{U}$ are isomorphic!

Although we haven't properly defined the terms (see exercise 1) we quote, noting the close analogy with 3.12.4,

Theorem 4.4.2 Any two complete ordered fields ($\mathbb{S}$ and $\mathbb{U}$ being two examples of such) are isomorphic [50].

Any such field is called the **field of real numbers** and is denoted by $\mathbb{R}$.

One result we shall require in Section 4.8 is

Theorem 4.4.3 Let $f \in \mathbb{R}[x]$ be of odd degree. Then $f(c) = 0$ for some $c \in \mathbb{R}$.

Proof See exercise 7 below and then [50].

Problem 5 Why is it at this point unfair to say: 'Theorem 4.4.3 follows immediately from Theorem 1.11.9'?

* This is not as easy as in the Cauchy sequence case.

Augustin-Louis Cauchy *(21 August 1789 – 22 May 1857)*
Cauchy was the eldest child of a barrister. In 1807 he entered the Ecole des Ponts et Chausees intending to become a civil engineer. In 1812 he proved Fermat's assertion that every positive integer is a sum of n n-gonal integers. By 1816 he was teaching at the Ecole Polytechnique and a member of the Académie des Sciences. After the July revolution of 1830 Cauchy, an ardent Royalist, refused to take an oath of allegiance to Louis Philippe. He left for Turin where it appears he taught Latin and Italian, later returning to Paris in 1838. The records on his subsequent career are astonishingly confused for a person of his eminence.

Cauchy was a prolific author, producing about 800 papers ranging over many areas including optics, elasticity, ordinary and partial differential equations, mechanics, determinants, theory of permutations and probability. He wrote the classic textbooks in which analysis was made more rigorous. The first developments in complex-variable theory are all due to him.

In his undergraduate career the reader can hardly fail to meet Cauchy's root test, ratio test, convergence criterion, inequality, integral theorem, integral formula as well as the Cauchy–Riemann equations.

We now come to the formal construction of the field ℂ of complex numbers. As in the construction of ℝ from ℚ, we offer two approaches (each based on the assumption that ℝ is already 'known'). Above we recognised that ℝ (even though it contains substantially more elements* than ℚ) is still not sufficiently rich to contain all roots of all polynomials with real coefficients, x^2+1 providing the simplest example. Informal introduction of the missing roots was made by A Girard as early as c. 1620, but it was not until Gauss in 1831 (unpublished) and Hamilton in 1833 put them on an apparently firmer (i.e. arithmetic) footing that the complex numbers met with a fair degree of approval. Gauss himself said (c. 1825) that 'the true metaphysics of $\sqrt{-1}$ is hard' whilst Cauchy (1847) said: 'We repudiate the sign $\sqrt{-1}$ completely because one does not know what the sign signifies, nor what meaning one should attribute to it'. Simply introducing the symbol i and asserting that $\mathrm{i}^2=-1$ and that i behaves as do real numbers raises problems (see exercise 16). Euler, in his book *Complete Introduction to Algebra* (1768) had said that 'Since all conceivable numbers are greater than, less than or equal to zero, square roots of negative numbers cannot be included amongst the possible numbers'. Now if i is *not* a number of previous acquaintance, on what grounds can one assert that $2\cdot\mathrm{i}=\mathrm{i}\cdot 2$? Gauss and Hamilton got round the problem by noting that each complex number $a+\mathrm{i}b$ determines and is determined by the (ordered) pair (a, b) of real numbers. Noting that we would ideally like $a+\mathrm{i}b=c+\mathrm{i}d$ if and only if $a=c$ and $b=d$, $(a+\mathrm{i}b)+(c+\mathrm{i}d)=(a+c)+\mathrm{i}(b+d)$ and $(a+\mathrm{i}b)\times(c+\mathrm{i}d)=(ac-bd)+\mathrm{i}(bc+ad)$ they *defined* a complex number to be a pair,

* In technical terms, ℚ is countable, ℝ is uncountable (exercise 2.6.15).

(a, b) of real numbers and *defined* equality, addition $\oplus$ and multiplication $\odot$ of complex numbers by:

(i) $(a, b) = (c, d)$ if and only if $a = c$ and $b = d$;
(ii) $(a, b) \oplus (c, d) = (a + c, b + d)$;
(iii) $(a, b) \odot (c, d) = (ac - bd, bc + ad)$.

It is not difficult to check (exercise 8) that the set $\mathbb{C}$ of all complex numbers (a, b) forms a field with respect to $\oplus$ and $\odot$. One observes that the mapping $a \to (a, 0)$ shows that $\mathbb{C}$ contains (an isomorphic copy of) $\mathbb{R}$ as a subfield. Denoting by i the pair $(0, 1)$ one finds $\mathrm{i}^2 = (0, 1) \odot (0, 1) = (-1, 0)$. Finally $(a, b) = (a, 0) \oplus \{(0, 1) \odot (b, 0)\}$ as is easily checked. Thus, identifying each $r \in \mathbb{R}$ with $(r, 0)$ in $\mathbb{C}$, the ordered pair (a, b) can be more recognisably written as $a \oplus \{\mathrm{i} \odot b\}$ or $a + ib$ if we resort to the more usual notations for addition and multiplication.

Note 4.4.4 The reason for returning to the $a + ib$ notation is one of convenience only. It seems easier to work with than does the ordered pair notation, simply because most of us have been used to working with it for so long (recall exercise 1.6.4 and see exercise 10 below).

An alternative way of constructing $\mathbb{C}$ is to follow Cauchy and extend Gauss' notion of congruence from $\mathbb{Z}$ to $\mathbb{R}[x]$. Given $f, g, h \in \mathbb{R}[x]$ Cauchy wrote $f \equiv g \pmod{h}$ iff $h \mid f - g$ in $\mathbb{R}[x]$. He then replaced each x in f and g by a new symbol i and wrote $f(\mathrm{i}) = g(\mathrm{i})$ rather than $f \equiv g \pmod{h}$. In particular when $h = x^2 + 1$ we find, for example, that

$$2x^4 + 3x^3 + 3x^2 - x + 4 \equiv -4x + 3 \pmod{x^2 + 1}$$

which Cauchy wrote as

$$2\mathrm{i}^4 + 3\mathrm{i}^3 + 3\mathrm{i}^2 - \mathrm{i} + 4 = -4\mathrm{i} + 3$$

Of course from $x^2 + 1 \equiv 0 \pmod{x^2 + 1}$ it follows that $\mathrm{i}^2 + 1 = 0$. In modern terminology this amounts to nothing more than forming the factor ring $\mathbb{D} = \dfrac{\mathbb{R}[x]}{[x^2 + 1]}$. (Kronecker's method mentioned in Section 4.2 amounts to forming factor rings of $\mathbb{Q}[x]$ in a similar way.)

Since $x^2 + 1$ is irreducible over $\mathbb{R}$ the proof of 4.2.1 tells us that $\mathbb{D}$ is a field with typical element $a + bx + I$ where I is the ideal $[x^2 + 1]$ of $\mathbb{R}[x]$. We leave it to the reader to show (exercise 9) that $\mathbb{C}$ and $\mathbb{D}$ are isomorphic fields each containing a root of the polynomial $x^2 + 1$.

$\mathbb{C}$ was introduced in order to produce a solution to the equation $x^2 + 1 = 0$. But there are many other equations [how many?] with real coefficients which do not possess any real roots. Even worse, we are now morally obliged to consider polynomials with complex coefficients. Surely to find roots for such polynomials we shall need to construct some sort of super-complex numbers? That indeed we do *not* is the content of the Fundamental Theorem of Algebra (Theorem 4.8.1).

Exercises

1 Based on our intuitive picture of **Q** as the real line **R** with gaps we define, formally, for any ordered field $\langle F, +, \cdot, <\rangle$ a **gap** to be a pair of subsets A, B of F such that (i) $A \neq \varnothing$, $B \neq \varnothing$; (ii) $A \cup B = F$; (iii) $r \in A$, $s \in B \Rightarrow r < s$; (iv) $x \in A$, $y \in B \Rightarrow$ there exist $u \in A$, $v \in B$ such that $x < u < v < y$ (i.e. A has no maximum element and B no minimum element). Prove that if B is the set of all positive rationals t such that $t^2 > 2$ then the pair A, B (where $A = \mathbf{Q} \backslash B$) is a gap in **Q**. (An ordered field without gaps is called **complete.**)

2 (i) Show that between any two rational numbers there exists one and hence infinitely many others.
(ii) Show (using the less cumbersome notation but remembering what is meant and hence avoiding proofs by intuition) that if $0 < \frac{x}{y}$ and $0 < \frac{r}{s}$ then there exists $n \in \mathbf{Z}^+$ such that $\frac{r}{s} < n \cdot \frac{x}{y}$. (Cf. exercise 3.12.10.)

3 (i) Show that the relation of equivalence on the collection of all Cauchy convergent sequences is indeed an equivalence relation.
(ii) How should one define $\oplus$ and $\odot$ on the set $\mathbb{S}$ in order that the result is a field modelling our intuitive grasp of $\mathbb{R}$?
(iii) Let $s = (x_1, x_2, \ldots)$ be the sequence defined by $x_1 = 1$ and, for each $n \geqslant 1$, $x_{n+1} = \frac{x_n}{2} + \frac{1}{x_n}$. Show that s is a Cauchy sequence. Which real number is represented by the equivalence class containing s?

4 As stated in the text, the set C of all Cauchy convergent sequences in $\mathbb{Q}$ can be turned into a ring by defining $\oplus$ and $\odot$ componentwise. Show that the subset N of C comprising all sequences equivalent to the zero sequence forms an ideal. Show that the factor ring C/N is a field, by showing each non-zero element of C/N has a multiplicative inverse in C/N.

5 Let V, W be two upper sections in **Q**. Define $V + W$ to be $\{v + w : v \in V, w \in W\}$. Show that $V + W$ is an upper section in **Q**. In order to make a field out of this set of upper sections we shall need to identify a zero element and an additive inverse for each upper section V. Can you do that? [Be careful over $\ominus V$. Keep upper sections corresponding to $\sqrt{2}$ and to $-\sqrt{2}$ in mind.]

6 Working informally show that (i) $\mathbb{Q}(\sqrt{2})$ can sustain two sets of positive elements (cf. 1.2.3); and that (ii) $\mathbb{Q}$ and $\mathbb{R}$ can each sustain only one set of positive elements. [Hint: All of **Q** is determined by the fact that $1 \in \mathbf{Q}^+$. In $\mathbb{R}$, $r \in \mathbb{R}^+$ iff r is non-zero and a square.]

7 Show that if $f = x^n + a_{n-1}x^{n-1} + \cdots + a_0 \in \mathbb{R}[x]$ and n is odd then

$$f(s) > 0 \text{ if } s > \text{maximum of } \{1, |a_0| + |a_1| + \cdots + |a_{n-1}|\}$$

$$f(s) < 0 \text{ if } s < \text{minimum of } \{-1, -|a_0| - |a_1| - \cdots - |a_{n-1}|\}$$

Hence deduce that 4.4.3 follows at once if only we can prove: If g is a function continuous on the closed interval $[a, b]$ and if $g(a)<0$ and $g(b)>0$ then there exists c such that $a<c<b$ and such that $g(c)=0$. [How might you prove this result? Bolzano* (1817) found he didn't know enough about the real numbers to prove this to his satisfaction.]

8 Check that with respect to $\oplus$ and $\odot$ as defined above the set $\mathbb{C}$ of ordered pairs (a, b) of reals do form a field.

9 Show that the field of exercise 8 is isomorphic to the field $\left\langle \dfrac{\mathbb{R}[x]}{[x^2+1]}, \oplus, \odot \right\rangle$.

10 Solve for x, y; (i) $(3, 5) \odot (x, y) = (1, 1)$ formally; (ii) $(3+5\mathrm{i})(x+\mathrm{i}y) = 1+\mathrm{i}$ informally. Which way is easier? Why?

11 On the set S of all ordered pairs (a, b) of reals define $\oplus$ and $\odot$ by: $(a, b)\oplus(c, d) = (a+c, b+d)$, $(a, b)\odot(c, d) = (ac, bd)$. Is $\langle S, \oplus, \odot\rangle$ a field?

12 With S as in exercise 11 define $\oplus$ as in exercise 11 and $\odot$ by $(a, b)\odot(c, d) = (ac+bd, bc-ad)$. Is $\langle S, \oplus, \odot\rangle$ a field?

13 Apply the definitions of $\oplus$ and $\odot$ of the text to ordered pairs (a, b) where a, b lie in the fields (i) $\mathbb{C}$; (ii) $\mathbb{Z}_3$; (iii) $\mathbb{Z}_5$. Is the resulting system always a field?

14 Does the map $\theta: \mathbb{R} \to \mathbb{C}$ defined by $a\theta = (0, a)$ embed $\mathbb{R}$ in $\mathbb{C}$? (See 3.10.2(v).)

15 Let $f = ax^2+bx+c \in \mathbb{R}[x]$ be irreducible. Then $\dfrac{\mathbb{R}[x]}{[f]}$ is necessarily a field. Must $\dfrac{\mathbb{R}[x]}{[f]} \cong \mathbb{C}$?

16 Show that $\mathbb{C}$ cannot be ordered. [Hint: Assume it can. Then i^2 will be in $\mathbb{C}^+$.]

17 Let $\mathbb{N}$ denote $\mathbb{Z}^+ \cup \{0\}$. On the set $\mathbb{N}\times\mathbb{N}$ define the binary relation $\sim$ by $(m_1, n_1) \sim (m_2, n_2)$ iff $m_1+n_2 = m_2+n_1$ in $\mathbb{N}$. Show that $\sim$ is an equivalence relation on $\mathbb{N}$. Denote the equivalence class containing (m, n) by $\langle m, n\rangle$ and the set of all classes by S. On S define $\oplus$ by $\langle a, b\rangle \oplus \langle u, v\rangle = \langle a+u, b+v\rangle$. Show that under $\oplus$ the subset $S_1 = \{\langle m, n\rangle : n = 0\}$ is structurally identical to $\mathbb{N}$ under $+$. (Technically $\langle S_1, \oplus\rangle$ and $\langle \mathbb{N}, +\rangle$ are isomorphic semigroups.) Denoting $\langle m, 0\rangle$ by $\bar{m}$ show that for all $m, n \in \mathbb{N}$ the equation $x \oplus \bar{m} = \bar{n}$ has a solution in S (whereas, of course, $x+m = n$ may not have a solution in $\mathbb{N}$). Can you see what we are doing here? See Section 0.1.

4.5 Finite fields

Having indicated in Section 4.4 how one can use 4.2.1 to generate $\mathbb{C}$ from $\mathbb{R}$ we now use its corollary (4.2.20) to prove, as part of the first objective posed in Section 4.1,

* Bernard Bolzano (5 October 1781 – 18 December 1848).

Theorem 4.5.1 Let p be a positive prime and n a positive integer. Then there exists a field with exactly p^n elements.

Proof In 4.2.20 take $F = \mathbb{Z}_p$, $f = x^{p^n} - x$ and E to be any field containing $\mathbb{Z}_p$ and in which $x^{p^n} - x$ factorises into p^n linear factors, $x^{p^n} - x = x(x - t_1) \ldots (x - t_{p^n - 1})$, say. Now all the t_i are distinct. For, as in exercise 1.11.2, we can argue that if $t_i = t_j$ where $i \neq j$ then t_i is a root of the formal derivative $f' = p^n x^{p^n - 1} - 1$ of f. But then $f' = -1$ since $\mathbb{Z}_p$, E and hence $E[x]$ have characteristic p (see exercise 3.10.18). Thus t_i cannot possibly be a root of f'.

Let T be the above set of roots (including 0). For $t_i, t_j \in T$ we see easily (exercise 3.10.18) that $(t_i - t_j)^{p^n} = t_i^{p^n} - t_j^{p^n} = t_i - t_j$ and that $(t_i t_j^{-1})^{p^n} = t_i^{p^n} t_j^{-p^n} = t_i t_j^{-1}$. Hence $t_i - t_j$ and $t_i t_j^{-1} \in T$ and thus T is a subfield of E (by 3.4.2(F)) containing exactly p^n elements.

Theorem 4.5.1 raises more questions than it settles. For example

Question 1 Do there exist fields with other than a prime power number of elements?

Noting that if f is a polynomial of degree n which is irreducible in $\mathbb{Z}_p[x]$ then $\mathbb{Z}_p[x]/[f]$ is (see the proof of 4.2.1) a field with exactly p^n elements (in which f has a root), we naturally ask

Question 2 Is every field with p^n elements obtainable in this way?

Or at least

Question 3 Is there for each prime p and $n \in \mathbb{Z}^+$ a polynomial f of degree n which is irreducible in $\mathbb{Z}_p[x]$?

Finally

Question 4 What relationship, if any, is there between two fields with the same number of elements?

To help answer these questions we introduce some concepts which are used time and again throughout the whole of mathematics and its applications to problems in the real world; for example in differential equations, statistics, linear programming and even in modelling an economy (see [67]). The proper mathematical setting for these concepts is the subject of Linear Algebra, of which they, along with matrices, constitute the life blood. For reasons of space we take the development only as far as required in this text. Since, however, the machinery to be developed will also find application in Section 4.6, we save effort by working, even here, in terms of general fields rather than with just finite ones. We begin with

Definition 4.5.2 Let E be a field and F a subfield of E.
(i) The set $u_1, u_2, \ldots, u_r$ of elements of E is said to **span** E over F iff every element e of E can be expressed in the form $e = f_1u_1 + \cdots + f_ru_r$, that is, iff e is a **linear combination** of the u_i using suitable $f_1, \ldots, f_r$ in F.
(ii) The set $v_1, \ldots, v_s$ of elements of E is said to be **linearly independent** over F iff the only way of writing 0_E in the form $f_1v_1 + \cdots + f_sv_s$, where the $f_i \in F$, is by taking each f_i to be 0_F.
(iii) The set $w_1, \ldots, w_t$ of elements of E is said to constitute a **basis** for E over F iff it spans E over F and is linearly independent over F.

Notes 4.5.3
(i) It is clear that if E is a finite field and if F is any subfield then a spanning set (as in (i)) for E over F certainly exists; the set of all elements of E constitutes such a set.
(ii) $\{1, i\}$ forms a spanning set—indeed a basis—for $\mathbb{C}$ over $\mathbb{R}$; $\{1, \sqrt{2}\}$ forms a basis for $\mathbb{Q}(\sqrt{2})$ over $\mathbb{Q}$ (see 3.10.6); $\{1, i, \sqrt{2}, i\sqrt{2}\}$ forms a basis for $\mathbb{Q}(i, \sqrt{2})$ over $\mathbb{Q}$ (see 4.3.5); $\{1+[f], x+[f], \ldots, x^{n-1}+[f]\}$ forms a basis for the field $\dfrac{\mathbb{Q}[x]}{[f]}$ over $\mathbb{Q}$, if f is irreducible of degree n in $\mathbb{Q}[x]$. (See exercise 4.2.18 and the proof of 4.2.1.)
(iii) Let $w_1, \ldots, w_t$ be a basis as in 4.5.2(iii). From 4.5.2(ii) it follows immediately that, for $f_1, \ldots, f_t, g_1, \ldots, g_t \in F$ we have

$$f_1w_1 + \cdots + f_tw_t = g_1w_1 + \cdots + g_tw_t$$

if and only if $f_1 = g_1, f_2 = g_2, \ldots, f_t = g_t$ (exercise 2). (Roughly stated: in terms of a given basis different *looking* elements *are* different.)

Problem 6 Does $\mathbb{R}$ have a finite basis over $\mathbb{Q}$?

Our first result relates the respective sizes of spanning sets and independent sets.

Theorem 4.5.4 If $\{u_1, \ldots, u_r\}$ spans E over F and if $\{v_1, \ldots, v_s\}$ is linearly independent over F then $s \leq r$.

Proof We may suppose that no proper subset $\bar{U}$ of $U = \{u_1, \ldots, u_r\}$ spans E. (If so, replace U by a smallest possible such U at the outset.) Now suppose $r < s$. Since $\{u_1, \ldots, u_r\}$ spans E over F there exist $f_{11}, f_{12}, \ldots, f_{1r}$ in F such that $v_1 = f_{11}u_1 + \cdots + f_{1r}u_r$. Since $v_1 \neq 0_E$ [why not?] $f_{1i} \neq 0_F$ for at least one i. WLOG we can assume $i = 1$. Then we have $u_1 = \frac{1}{f_{11}}(v_1 - f_{12}u_2 - \cdots - f_{1r}u_r)$. It follows [prove it *now*] that $\{v_1, u_2, \ldots, u_r\}$ spans E over F and hence that there are $f_{21}, f_{22}, \ldots, f_{2r}$ in F such that $v_2 = f_{21}v_1 + f_{22}u_2 + \cdots + f_{2r}u_r$. Here, at least one of the f_{2j} with $j \geq 2$ is non-zero. (Otherwise we should have $v_2 - f_{21}v_1 = 0$ which is impossible whether or not $f_{21} = 0_F$ [why?].) WLOG, we suppose $f_{22} \neq 0_F$. Then we may write $u_2 = \frac{1}{f_{22}}\{-f_{21}v_1 + v_2 - f_{23}u_3 - \cdots - f_{2r}u_r\}$ so that $\{v_1, v_2, u_3, \ldots, u_r\}$ spans E over F. One can continue in this manner replacing, at each step, a u_i by a v_i [why never a $v_k (k < i)$ by a v_i?] until all the u_i have been replaced by $v_1, v_2, \ldots, v_r$, thus showing that $\{v_1, \ldots, v_r\}$ spans E over F. There then exist $f_{s1}, \ldots, f_{sr}$ in F such that $v_s = f_{s1}v_1 + \cdots + f_{sr}v_r$. But this equality can be rewritten $f_{s1}v_1 + \cdots + f_{sr}v_r + 0v_{r+1} + \cdots + 0v_{s-1} + (-1)v_s = 0$, which is an impossibility since the $v_1, \ldots, v_s$ were supposed to be independent over F. This contradiction shows that $r < s$ is untenable and we deduce that $s \leq r$ as required.

Corollary 4.5.5 Let $F \subseteq E$ be fields such that E has a (finite) spanning set over F. Then E has a (finite) basis over F and any two such bases contain the same number of elements.

Proof Let $S = \{u_1, u_2, \ldots, u_r\}$ be a spanning set for E over F. We define a subset B of S as follows. For each k $(1 \leq k \leq r)$ we take $u_k \in B$ if and only if u_k cannot be expressed in the form $f_{k1}u_1 + \cdots + f_{k,k-1}u_{k-1}$. (For $k = 1$ this is interpreted as: $u_1 \in B$ if and only if $u_1 \neq 0$.) [Intuitively we are relieving S of redundant elements.] It is easy to check (exercise 2(ii)) that the subset B still spans E over F and that it is linearly independent over F. That is, B is a basis for E over F.

If B_1 is another basis and if B, B_1 have t, t_1 elements respectively then, thinking of B as an independent set and B_1 as a spanning set, $t \leq t_1$. Similarly $t_1 \leq t$ [why?]. Thus $t = t_1$ as asserted.

Notation and Definition 4.5.6 This unique number of elements in any (finite) basis for E over F is denoted by $[E:F]$. We call $[E:F]$ the **dimension** of E over F and say that E is a **finite-dimensional** extension of F.

It is now an easy matter to answer Question 1 above.

Theorem 4.5.7 Let E be a finite field. Then E has p^t elements where p is some prime and t some positive integer.

Proof Since E is finite the prime subfield of E must be $\mathbf{Z}_p$ for some prime p (3.10.9). By 4.5.3(i) and 4.5.5, E has a basis $w_1, \ldots, w_t$, say, over $\mathbf{Z}_p$. Then

4.5.3(iii) tells us that the p elements $f_1w_1 + \cdots + f_tw_t$ where the f_i are in $\mathbb{Z}_p$ are pairwise distinct.

To answer (affirmatively) Questions 2 and 3 raised above we first observe (exercise 4) that in any field E with p^n elements every non-zero element satisfies the equation $x^{p^n-1} - 1 = 0$. Using exercise 5 we deduce that the set of all non-zero elements of E is exactly the set $\{a^k : 0 \leq k < p^n - 1\}$ for some suitable $a \in E$. Let M_a be the (unique irreducible) minimum polynomial (see 4.3.4) of a over $\mathbb{Z}_p$ ($\subseteq E$). [How do you know M_a exists?] Suppose $\deg M_a = m$. Then, by 4.3.4, the smallest subfield $\mathbb{Z}_p(a)$ of E containing $\mathbb{Z}_p$ and a, namely E itself, contains p^m elements. It follows that (i) $p^n = p^m$, whence $n = m$ and $\deg M_a = n$ and (ii) $E = \mathbb{Z}_p(a) = \dfrac{\mathbb{Z}_p[x]}{[M_a]}$.

This still leaves Question 4. The answer is quite amazing: *Every two fields with precisely p^n elements are isomorphic*! Given 4.2.21 this is not difficult: for each field with p^n elements is easily seen (exercise 4) to be a splitting field for the polynomial $x^{p^n} - x$ over $\mathbb{Z}_p$ and the uniqueness then follows from 4.2.21.

To give a specific example we note that $f = x^3 + x + 1$ is irreducible over $\mathbb{Z}_2$. (Being a cubic it suffices to check that f has no root in $\mathbb{Z}_2$ (cf. Remark after 1.11.7).)

Thus $\dfrac{\mathbb{Z}_2[x]}{[f]}$ is a field whose elements are of the form $a_0 + a_1x + a_2x^2 + [f]$ where $a_0, a_1, a_2 \in \mathbb{Z}_2$.* Denoting this element briefly by $a_0a_1a_2$ we find that the multiplication table for $\dfrac{\mathbb{Z}_2[x]}{[f]}$ is [can you complete it?] as follows.

	000	100	010	001	110	101	011	111
000	000	000	000		000			
100	000	100	010	001		101		
010	000	010	001					
001	000	001	110					
110			011	111				
101				010	001		110	
011					100	110		
111						011	001	

The entry in 'row' 101 and 'column' 011 is 110 since $(1 + 0x + x^2 + [f]) \cdot (0 + x + x^2 + [f]) = x + x^2 + x^3 + x^4 + [f] = 1 + x + [f]$.

Finite fields were introduced by Galois in 1830 during number-theoretic investigations. Wishing to solve congruences $f(x) \equiv 0 \pmod p$ where p is a

* We denote the elements of $\mathbb{Z}_2$ by 0 and 1.

Picture by courtesy of the American Mathematical Society

Eliakim Hastings Moore *(26 January 1862–30 December 1932)*
Moore was born at Marietta, Ohio, into a family with strong university connections. His interest in mathematics was strengthened by his spending a summer, whilst still at school, as assistant to the astronomer Ormond Stone.

After Moore had completed his PhD in 1885, the Yale professor, H A Newton, financed Moore's visit to Göttingen and Berlin where he made contact with Weierstrass, Kronecker and Klein. After holding positions at Northwestern and Yale, Moore was appointed acting head of the mathematics department at the University of Chicago when it opened in 1892. By persuading the university to appoint Bolza and Maschke to the department, Moore began a pursuit of excellence which was one of his characteristics. Of the twenty nine students whose doctoral theses were supervised by Moore, three, L E Dickson, O Veblen and G D Birkhoff, all mathematicians of the very top rank (the first two coming from Iowa) followed Moore as President of the American Mathematical Society.

Moore's teaching methods were also revolutionary. Instead of lectures of fixed length he ran classes as 'laboratories' whose durations were independent of the clock and meal times. Sessions ended when he felt the subject of the day was exhausted.

Moore's research, characterised by rigour and generality, took in algebra, number theory, integral equations and analysis. His algebraic work included the results mentioned on finite fields.

Moore's crusade to promote precision in mathematical argument won him at least seven honorary doctorates. At one ceremony he met an old friend, the physicist Carl Barus, who had instructed Moore in his student days in the art of organ pumping. 'Who knows', asked Barus later, 'what my instruction in organ pumping may have done for Professor Moore?'

---

prime and f is irreducible of degree n (over $\mathbb{Z}_p$) he introduced, in analogy with the introduction of i to 'solve' the equation $x^2+1=0$, a hypothetical symbol j to act as a root for f and considered the p^n formally distinct expressions $z_0+z_1\mathrm{j}+\cdots+z_{n-1}\mathrm{j}^{n-1}$ where z_i are integers such that $0 \leq z_i < p$. With the obvious definitions of addition and multiplication these expressions can be turned into a field—a Galois field, $GF(p^n)$. (Thus $GF(p) \cong \mathbb{Z}_p$.)

The objection to this kind of introduction of j is the same as that raised over the introduction of i (Section 1.6). The above construction of $GF(p^n)$ is replaced in present day mathematics by forming the factor ring $\dfrac{\mathbb{Z}_p[x]}{[f]}$.

The answer to Question 4 was first obtained in 1893 by the American E H Moore. Together with the answers to Questions 2 and 3, part of the algebraist's dream of classifying all algebraic systems comes true. We have

Theorem 4.5.8

(i) Every finite field is of the form $\dfrac{\mathbb{Z}_p[x]}{[f]}$ where f is a polynomial irreducible in $\mathbb{Z}_p[x]$. In particular every finite field has p^n elements for some prime p and $n \in \mathbb{Z}^+$.
(ii) To each such p and n there exists a field with p^n elements.
(iii) Two finite fields are isomorphic iff they have the same number of elements.

For applications of finite fields see, for example, [7], [61].

Exercises

1 Show that the set $\{1, \mathrm{i}, \sqrt{3}, \mathrm{i}\sqrt{3}\}$ is linearly independent over $\mathbf{Q}$. Do the same for $\{1, \sqrt{2}, \sqrt{3}, \sqrt{5}\}$. Does one retain independence on extending this set to include the various products $\sqrt{6}, \sqrt{10}, \sqrt{15}, \sqrt{30}$?

2 (i) Prove 4.5.3(iii).
(ii) Prove that if $\{u_1, \ldots, u_m\}$ spans E over F and if u_m is a linear combination of $u_1, \ldots, u_{m-1}$ then $\{u_1, \ldots, u_{m-1}\}$ spans E over F.

3 Prove that $[\mathbf{Q}(\sqrt[3]{2}) : \mathbf{Q}] = 3$. (See 4.3.5(i).)

4 Prove that in a finite field with p^n elements each non-zero element satisfies the equation $x^{p^n-1} - 1 = 0$. [Hint: Follow the method of proof of 2.5.3.] Deduce that each such field is a splitting field for $x^{p^n} - x$ over $\mathbb{Z}_p$.

5 Prove that the multiplicative group of non-zero elements in a finite field is cyclic as follows. Let F be a finite field with $q = p^n$ elements and let $\alpha \in F$ where $\alpha \neq 0$. We know (exercise 4) that $\alpha^{q-1} = 1$. Let k be the smallest positive integer such that $\alpha^k = 1$. Show that $k | q-1$. (Exercise 2.5.7 might help.) We call k the *order* of α. Show that if α, β have coprime orders k, l then $\alpha\beta$ has order kl (exactly). Now suppose F has no element of order $q-1$. Let γ be an element of largest order e, say, in F and let δ be any element of F, of order m, say. Suppose $m \nmid e$. Then, for some prime r, $m = r^b m_1$ and $e = r^c e_1$ where $b > c$, $r \nmid m_1$

and $r \nmid e_1$. Deduce that $\delta^{m_1}\gamma^{r^c}$ has order $r^b e_1 (> e)$, a contradiction. Thus $m|e$. It follows that $\delta^e = 1$. Thus the equation $x^e = 1$ has $q-1$ solutions in the field F, a contradiction (see exercise 2.4.4). Thus F has an element a, say, of order $q-1$. Hence $F = \{a^t : 0 \leqslant t < q-1\} \cup \{0_F\}$.

6 Show that $x^3 - x + 1$ and $x^3 - x - 1$ are irreducible over $\mathbb{Z}_3$. Is $\dfrac{\mathbb{Z}_3[x]}{[x^3-x+1]} \cong \dfrac{\mathbb{Z}_3[x]}{[x^3-x-1]}$?

7 Show that $P = x^4 + x + 1$ and $Q = x^4 + x^3 + x^2 + x + 1$ are both irreducible in $\mathbb{Z}_2[x]$. Show that $\dfrac{\mathbb{Z}_2[x]}{[P]} \cong \dfrac{\mathbb{Z}_2[x]}{[Q]}$ but that $x + [P]$ and $x + [Q]$ have multiplicative orders 15 and 5 respectively. Deduce that the 'obvious' mapping between $\dfrac{\mathbb{Z}_2[x]}{[P]}$ and $\dfrac{\mathbb{Z}_2[x]}{[Q]}$, namely $f + [P] \to f + [Q]$, is not an isomorphism.

8 $H = \dfrac{\mathbb{Z}_3[x]}{[x^2+1]}$ has 9 elements. Find $a \in H$ such that $H = \{a^k : 0 \leqslant k \leqslant 7\} \cup \{0\}$.

9 (i) Show that if $m | p^n - 1$ then $x^m - 1 = 0$ has exactly m distinct roots in $GF(p^n)$.
(ii) If $t|n$ show that $GF(p^n)$ has exactly one subfield with p^t elements. [Hint: $t|n$ implies $p^t - 1 | p^n - 1$.]

10 Find all monic irreducible polynomials of degree 2 over $GF(3)$. Show that their product divides $x^8 - 1$.

11 Is $GF(4)$ (isomorphic to) a subfield of $GF(8)$?

12 Let f be irreducible of degree t in $\mathbb{Z}_p[x]$. Show that $f | x^{p^t} - x$ in $\mathbb{Z}_p[x]$. [Hint: let $g = (f, x^{p^t} - x)$ in $\mathbb{Z}_p[x] \subseteq GF(p^t)[x]$. There exists $u \in GF(p^t)$ such that $x - u | f$ and $x - u | x^{p^t} - x$ in $GF(p^t)[x]$. Thus $g \neq 1$. Therefore $f | x^{p^t} - x$ in $\mathbb{Z}_p[x]$.]

13 Try to convince yourself of the following. Let $\mathbb{Z}_p[t]$ be the ring of polynomials in the letter t and let $F = \mathbb{Z}_p(t)$ denote its field of fractions (see Section 3.10). Consider the polynomial $f = x^p - t$ in $F[x]$. Let E be a splitting field for f over F.

First claim: f is irreducible in $F[x]$. Second claim: Since $f' = px^{p-1} = 0$, f has repeated roots in E. Thus: whilst it is true that irreducible polynomials cannot have repeated roots in their splitting fields E when the base field F is finite or of characteristic zero they can have repeated roots if F is infinite of prime characteristic.

4.6 Constructions with compass and straightedge

Turning now to the second of the objectives listed at the start of the chapter, we show how the algebra we have developed can help in discussing problems of a purely geometrical nature, namely in deciding whether or not certain constructions can be carried out (theoretically) using only a compass and straightedge, a straightedge being an ungraduated ruler.

Some initial remarks are in order. Most readers will recall from schooldays how to bisect a given angle using only a ruler and a compass (no use being made of the markings on the ruler, the often-used term 'ruler and compass construction' thus being erroneous). However, few will have given much thought to the apparently similar problem of *trisecting* a given angle. In the 5th Century BC the Greeks were studying this problem, although the precise reason appears unknown ([84, p. 92]). Other famous construction problems were also under consideration at this time. Two of them were: using only straightedge and compass construct (i) a square with area equal to that of a given circle; (ii) a line segment equal in length to each edge of a cube of volume 2 units given the length of an edge of a cube of unit volume. It does not seem clear how or why problem (i) arose ([80, p. 70]), but there are stories relating to the origin of problem (ii) ([84, p. 90]). Now, it is the case that these problems can be solved if one is allowed to introduce fixed curves other than circles and straight lines (Archimedes' spiral $r = a\theta$, for instance) or a marked straightedge ([84, p. 93]). The restriction to use of compass and straightedge only might well reflect a supposed primitiveness of the straight line and circle. Whatever the reason, repeated attempts to solve the above problems led to the introduction of an immense amount of good (and bad!) mathematics (just as did attempts to establish Fermat's Conjecture). One other point. The problems posed are wholly theoretical ones requiring exact answers. The physical drawing is to be seen merely as an aid to understanding. Further, whilst we shall prove that all three problems above are insoluble, it is easy enough, even theoretically, to solve them to within any required degree of approximation (example 11).

We first recall how to bisect a given angle $\widehat{AOB}$. Figure 4.4 should suffice (but see Problem 8).

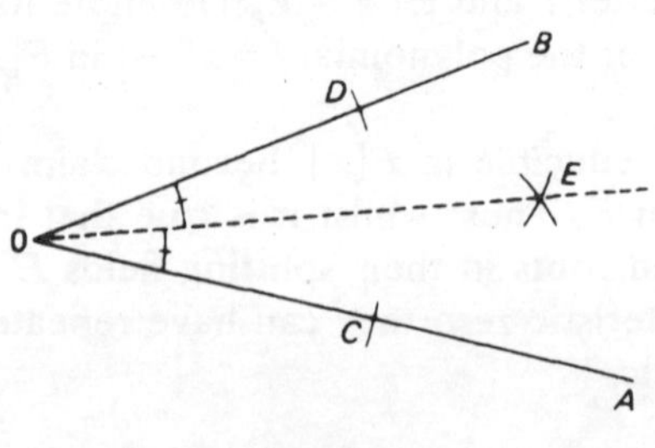

Fig. 4.4

In any construction problem of the above type we are given points, lines, circles, etc. and are required to construct (or show the impossibility of constructing) further such objects. To prove that a construction can be performed one only needs to describe a method for doing it. To prove it cannot requires us first to list the complete 'rules of the game'. We adopt the following rules.

Let a finite set P_1 of points be given in the plane. Joining all pairs of points in P_1 by straight lines of arbitrary length and drawing each circle which has a point of P_1 at its centre and a point of P_1 on its circumference, we augment P_1 with the (finitely many) points of intersection of all these straight lines and circles. Calling this new (finite) set P_2 we repeat the procedure to obtain sets $P_1 \subset P_2 \subset P_3 \subset \cdots$ of points in the plane. The totality P of points so obtained we call *the set of points constructible from* P_1.

In the above problems we are given an angle, a circle and a unit line segment and we are required to construct, respectively, another angle, a square and another line segment. Each of these configurations determines and is determined by a finite set of points and so it is seen that problems concerning geometrical constructions are equivalent to problems concerning constructible points.

Problem 7 If basing our work on sets of points rather than on complete line segments, complete circles, etc. seems pedantic, contemplate construction (ii) above where a 'unit line segment' is given. The question arises: Given a line segment are we also given all points on it? If not, which exactly *are* we given?

Problem 8 According to the rules we have adopted, the only circles we can draw are those whose centres lie at and which pass through points already given or constructed. We do not allow for drawing circles of arbitrary radius nor for transferring radii using fixed-opening compass, from one centre to another. Both these 'moves' were used above to bisect a given angle, so shouldn't we allow them here? Does it make any difference?

Since in every construction problem we are given at least two points we can take these to be the points 0, 1 on the straight line (the 'x-axis') joining them. We then easily construct (exercise 1; see Fig. 4.5) a 'y-axis' through 0

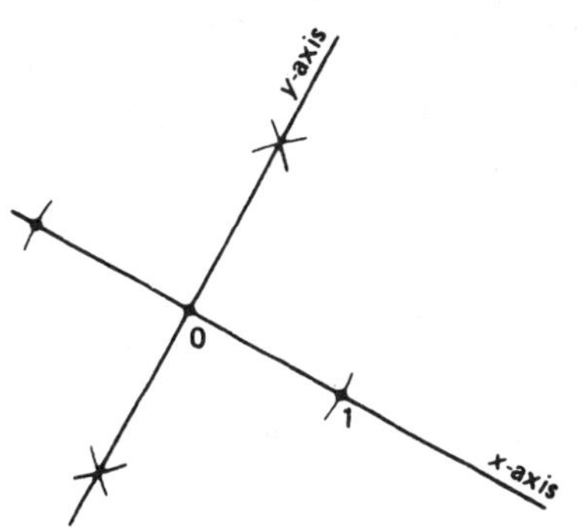

Fig. 4.5

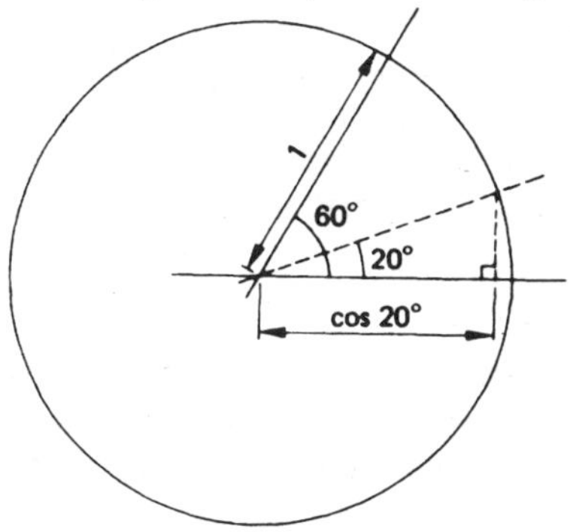

Fig. 4.6

perpendicular to the x-axis. This permits us to give each constructible point coordinates (a, b) determined by the unit of length from 0 to 1.

It is easily seen (exercise 1) that all points (u, v) where $u, v \in \mathbb{Q}$ are constructible and that problems (i) and (ii) above become: (i) construct the point $(\sqrt{\pi}, 0)$; (ii) construct the point $(\sqrt[3]{2}, 0)$. Further, from Fig. 4.6 the problems of trisecting the specific angle of 60° is clearly equivalent to: construct $(\cos 20°, 0)$.

Before we can complete the solutions to the three classical problems, we require one more technical result, namely

Theorem 4.6.1 Let $F \subseteq E \subseteq K$ be fields such that* $[K:E] = r$ and $[E:F] = s$ are both finite. Then $[K:F]$ is finite. Indeed $[K:F] = rs = [K:E][E:F]$.

Proof Let $u_1, \ldots, u_r$ and $v_1, \ldots, v_s$ be bases for K over E and E over F respectively. Then each element k of K is expressible in the form $k = e_1u_1 + \cdots + e_ru_r$ for suitable $e_i \in E$. But each e_i $(1 \leqslant i \leqslant r)$ can be written as $e_i = f_{i1}v_1 + \cdots + f_{is}v_s$. It follows easily, on substituting for the e_i in k, that the v_iu_j certainly span K over F.

To prove that the v_iu_j are also linearly independent over F one essentially reverses the above procedure. The details are left to exercise 2.

In the notation of 3.10.7 (see also 4.3.5(ii)) we have

Examples 4.6.2

(i) $[\mathbb{Q}(i, \sqrt{2}): \mathbb{Q}] = [\mathbb{Q}(i, \sqrt{2}): \mathbb{Q}(\sqrt{2})][\mathbb{Q}(\sqrt{2}): \mathbb{Q}] = 2 \cdot 2 = 4$.

(ii) $[\mathbb{Q}(\sqrt[3]{2}, \sqrt[2]{3}): \mathbb{Q}] = [\mathbb{Q}(\sqrt[3]{2}, \sqrt[2]{3}): \mathbb{Q}(\sqrt[3]{2})][\mathbb{Q}(\sqrt[3]{2}): \mathbb{Q}] = 2 \cdot 3 = 6$.

Returning to constructible points, suppose we have labelled two points in P_1 as $p_1 = (0, 0)$ and $p_2 = (1, 0)$ and suppose $p_{n+1} = (a_{n+1}, b_{n+1})$ is the $(n+1)$th point of a sequence constructible from P_1— including the given points $p_1, \ldots, p_r$ of P_1.

The straight line joining p_i and p_j $(1 \leqslant i < j \leqslant n)$ has equation $(a_j - a_i)(y - b_i) = (b_j - b_i)(x - a_i)$ whilst the circle centre p_k and passing through p_l has equation $(x - a_k)^2 + (y - b_k)^2 = (a_l - a_k)^2 + (b_l - b_k)^2$. These equations can be rewritten $ax + by + c = 0$ and $x^2 + y^2 + 2fx + 2gy + h = 0$ where $a, b, c, f, g, h \in \mathbb{Q}_n =$† $\mathbb{Q}(a_3, b_3, \ldots, a_n, b_n)$. Now it is easy to check (exercise 3) that if p_{n+1} is obtained as the intersection of two distinct lines then $a_{n+1}, b_{n+1} \in \mathbb{Q}_n$ whilst if p_{n+1} is obtained as the intersection of a line and a circle, or of two distinct circles, then either $a_{n+1}, b_{n+1} \in \mathbb{Q}_n$ or—at the worst—$a_{n+1}, b_{n+1} \in \mathbb{Q}_n(\sqrt{d})$ where $d \in \mathbb{Q}_n$. Thus for each $n \geqslant r$ we have $[\mathbb{Q}_{n+1}:\mathbb{Q}_n] = 1$ or 2. There then follows

Theorem 4.6.3 If $p_t = (a_t, b_t)$ is a point constructible from $P_1 = \{p_1, \ldots, p_r\}$, as above, then there exists a sequence $\mathbb{Q}_r \subseteq \mathbb{Q}_{r+1} \subseteq \cdots \subseteq \mathbb{Q}_t$ of subfields of $\mathbb{R}$ such that $a_t, b_t \in \mathbb{Q}_t$ and $[\mathbb{Q}_{i+1}:\mathbb{Q}_i] \leqslant 2$ for each i $(r \leqslant i \leqslant t-1)$.

* For notation see 4.5.6.

† See 3.10.7 for notation: note that $\mathbb{Q}_2 = \mathbb{Q}$.

We can now prove the impossibility of solving the three classical problems of the Greeks.

(A) **Duplication of the cube** Here we are given a line segment of unit length. Equivalently we are given $P_1 = \{(0, 0), (1, 0)\}$. We are asked to construct the point $p = (\sqrt[3]{2}, 0)$. If p is constructible then according to 4.6.3 we see that $\sqrt[3]{2}$ lies in some subfield $\mathbb{Q}_t$ of $\mathbb{R}$ where $[\mathbb{Q}_t:\mathbb{Q}] = 2^m$ for some m. But by 4.6.1 $2^m = [\mathbb{Q}_t:\mathbb{Q}] = [\mathbb{Q}_t: \mathbb{Q}(\sqrt[3]{2})][\mathbb{Q}(\sqrt[3]{2}): \mathbb{Q}]$ whilst $[\mathbb{Q}(\sqrt[3]{2}):\mathbb{Q}] = 3$ (exercise 4.5.3). This contradiction shows that p is not constructible from $(0, 0)$, $(1, 0)$ alone.

By similar means we investigate

(B) **Angle trisection** Of course some angles *can* be trisected, 90° and 27° for example.* The point is that, if we exhibit just one angle which cannot be trisected, this suffices to quash the assertion that *every* angle can be trisected by straightedge and compass. We claim that 60° is not trisectible. Thus (see Fig. 4.7) we are given three points $(0, 0)$, $(1, 0)$ and one other which we may take as $\left(\frac{1}{2}, \frac{\sqrt{3}}{2}\right)$. [Why may we? Cf. Problem 9.]

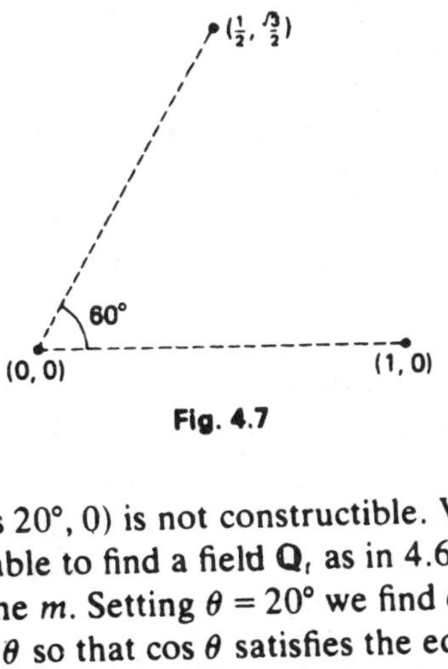

Fig. 4.7

We show that $p = (\cos 20°, 0)$ is not constructible. We argue as above that, if it were, we would be able to find a field $\mathbb{Q}_t$ as in 4.6.3 such that $\cos 20° \in \mathbb{Q}_t$ and $[\mathbb{Q}_t:\mathbb{Q}] = 2^m$ for some m. Setting $\theta = 20°$ we find $\cos 3\theta = \cos 60° = \frac{1}{2}$. But $\cos 3\theta = 4\cos^3\theta - 3\cos\theta$ so that $\cos\theta$ satisfies the equation $8x^3 - 6x - 1 = 0$. Since this is irreducible over $\mathbb{Q}$ (exercise 1.11.7(d)) we see that $[\mathbb{Q}(\cos 20°):\mathbb{Q}] = 3$. Thus, from 4.6.3, we deduce that p is not constructible from $(0, 0)$, $(1, 0)$ and $\left(\frac{1}{2}, \frac{\sqrt{3}}{2}\right)$ (see exercise 4.)

Problem 9 Is $\cos 20°$ constructible if we are given the points $(0, 0)$, $(1, 0)$, $(\cos 20°, \sqrt{3}\cos 20°)$? What if we are given this latter point—but not told its coordinates? Does this mean we can trisect 60°?

* 27°? I bet you can't prove it yet!

(C) **Squaring the circle** To show that $p = (\sqrt{\pi}, 0)$ is not constructible (given $P_1 = \{(0, 0), (1, 0)\}$) we argue that if it *were* constructible so would be $(\pi, 0)$. [Why? See Fig. 4.8.] We now cheat by stating Lindemann's result (see p. 140) that π satisfies no polynomial equation at all with rational coefficients. As a consequence π cannot lie in any of the fields $\mathbb{Q}_t$. (For a proof see [38, p. 74].)

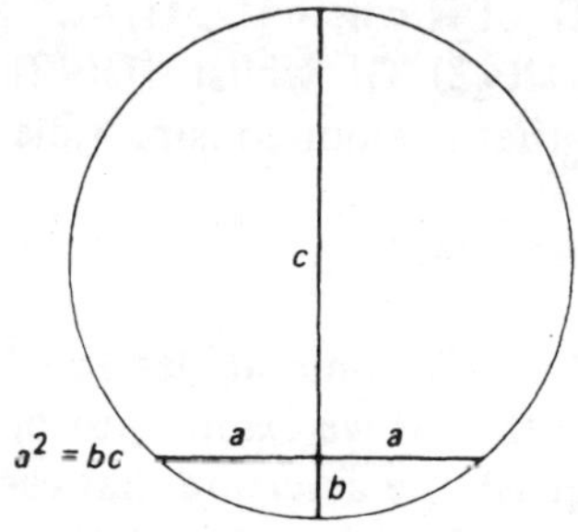

Fig. 4.8

A fourth straightedge and compass construction which occupied the Greeks, partial positive solutions to which appear in Euclid's Book 4, was the problem of constructing regular n-sided polygons. It was known how to construct polygons of 2^α, $2^\alpha \cdot 3$, $2^\alpha \cdot 5$, $2^\alpha \cdot 15$ sides and a construction for a regular heptagon ($n = 7$) was sought. In 1796 Gauss constructed a regular polygon with 17 sides, the first new such polygon in 2000 years! (It is said that this discovery made him decide to devote himself to mathematics rather than linguistics.) In his famous *Disquisitiones Arithmeticae* of 1801 he shows that polygons with p sides are constructible if the prime p has the form $2^{2^n} + 1$. (Fermat conjectured that *all* numbers of the form $2^{2^n} + 1$ are primes. See exercise 5.) He also warns the reader that for no other prime p can a regular p-sided polygon be constructed—but offers no proof. He further asserts that an n-sided polygon is constructible iff $n = 2^\alpha p_1 \ldots p_r$ where the p_i are distinct Fermat primes. The necessity of this condition was not proved by Gauss. It was first proved in 1837 by Wantzel* in a paper in which he also showed the impossibility of the trisection and duplication problems by the methods indicated above (viz. every constructible quantity satisfies an equation of degree 2^t for some t). (See exercise 8(b). For the sufficiency use exercise 8(a) and 7.6.6).

For details about alternative construction problems, for example those using a 'rusty' compass (!) or no compass at all see, for example, [84].

Exercises

1 (a) Given points $(0, 0)$ and $(1, 0)$ show how to construct $(0, 1)$. Use ideas from Fig. 4.8 to construct $\left(\frac{1}{b}, 0\right)$ and then $\left(\frac{a}{b}, \frac{c}{d}\right)$ where $\frac{1}{b}, \frac{a}{b}, \frac{c}{d} \in \mathbb{Q}$.

* There is an article on Wantzel in the *Bulletin of the American Math. Soc.*, Vol. 24, 1917–18, p. 339.

(b) Show that (l, m) is constructible from $\{(0,0),(1,0)\}$ where

$$l=\sqrt{1+\sqrt{2+\sqrt{3}}} \quad \text{and} \quad m=\frac{1+\sqrt{5}}{4}.$$

(c) Prove that the set $\{a: a\in \mathbb{R} \text{ and } (a,b) \text{ is constructible from } \{(0,0),(1,0)\}\}$ is a subfield of $\mathbb{R}$.

2 Complete the proof of Theorem 4.6.1.

3 Complete the proof of Theorem 4.6.3. State and prove its converse.

4 In problem (B) shouldn't we be looking at $[\mathbb{Q}(\cos 20°): \mathbb{Q}(\sqrt{3}/2)]$—rather than $[\mathbb{Q}(\cos 20°):\mathbb{Q}]$—in order to apply 4.6.3?

5 (See exercise 2.5.2.) Show that, if 2^k+1 is a prime integer, then k must be a power of 2. [Hint: Assume $k=2^n v$, where v is odd. Now use: $x+1 \mid x^v+1$ in $\mathbb{Z}[x]$ and put $x=2^{2^n}$.]

6 Construct a regular pentagon. [Hint: Use the formula $2\cos\theta=e^{i\theta}+e^{-i\theta}$ and the fact that $e^{2\pi i/5}$ is a root of $y^4+y^3+y^2+y+1$ to deduce that $2\cos(2\pi/5)$ is a root of x^2+x-1.]

7 Using the same technique as in exercise 6, show that the regular heptagon is *not* constructible. [Hint: Write $y^6+y^5+\cdots+y+1=y^3(x^3+x^2-2x-1)=y^3 f(x)$ and show that $f(x)$ is irreducible in $\mathbb{Q}[x]$.]

8 (a) Show that, if a (regular) m-gon and a (regular) n-gon are constructible and if $(m,n)=1$, then a (regular) mn-gon is constructible.
(b) Let p be an odd prime. Show that: (i) no p^2-gon is constructible; (ii) if a p-gon is constructible then $p=2^{2^n}+1$ for some n. [Hint: Use the technique of exercises 6, 7 to deduce that constructibility implies: (i) $\frac{p(p-1)}{2}$; (ii) $\frac{p-1}{2}$ must both be powers of 2. See 1.9.17 and exercise 1.9.12.]

9 Construct, or show that it is impossible to construct,
(a) Regular polygons with 8 sides, 9 sides, 10 sides. (Do this directly—don't use Wantzel's result.)
(b) Angles of 15°, 2°, 3°, 9°.

10 Construct, or prove it is impossible to construct,
(a) A square whose area is the sum of the area of 2 given squares.
(b) A cube whose volume is the sum of the volumes of 2 given cubes.

11 Using the equality $\frac{1}{2}-\frac{1}{4}+\frac{1}{8}-\frac{1}{16}+\cdots=\frac{\frac{1}{2}}{1-(-\frac{1}{2})}=\frac{1}{3}$ to show that any angle can be trisected to within any required degree of accuracy.

4.7 Symmetric polynomials

Here we prove for use in Section 4.8 a basic result concerning certain special polynomials in several letters which arise naturally in the problem of factorising polynomials in one letter into linear factors.

Let F be a field, $f=x^n+a_{n-1}x^{n-1}+\cdots+a_1x+a_0$ a polynomial in $F[x]$, and E a field (containing F) in which f factorises, $f=(x-\alpha_1)\ldots(x-\alpha_n)$, into a product of distinct linear factors. From the two expressions for f one sees that

$$a_{n-1}=-(\alpha_1+\alpha_2+\cdots+\alpha_n)=-\sum_{i=1}^{n}\alpha_i$$

$$a_{n-2}=\alpha_1\alpha_2+\alpha_1\alpha_3+\cdots+\alpha_1\alpha_n$$
$$+\alpha_2\alpha_3+\cdots+\alpha_2\alpha_n$$
$$\vdots$$
$$+\alpha_{n-1}\alpha_n=\sum_{1\le i<j\le n}\alpha_i\alpha_j$$
$$\vdots$$
$$a_0=(-1)^n\alpha_1\alpha_2\ldots\alpha_n$$

Now consider the ring $F[x_1,x_2,\ldots,x_n]$ of polynomials in the (commuting) letters $x_1,\ldots,x_n$. Setting

$$s_1=\sum_{i=1}^{n}x_i$$

$$s_2=\sum_{1\le i<j\le n}x_ix_j$$

$$s_3=\sum_{1\le i<j<k\le n}x_ix_jx_k$$

$$\vdots$$

$$s_n=x_1x_2\ldots x_n$$

we call s_i **the *i*th elementary symmetric polynomial in $F[x_1,\ldots,x_n]$.**

By a **symmetric polynomial** in $F[x_1,\ldots,x_n]$ we understand any element f of $F[x_1,\ldots,x_n]$ which is left unaltered by every permutation* of the set $\{x_1,\ldots,x_n\}$. In particular each of the s_i (indeed every polynomial in the s_i) is a symmetric polynomial in $F[x_1,\ldots,x_n]$. In $\mathbb{Q}[x_1,x_2,x_3]$, $x_1x_2^7+x_2x_3^7+x_3x_1^7+x_2x_1^7+x_3x_2^7+x_1x_3^7-4x_1-4x_2-4x_3$ is a symmetric polynomial whilst $x_1x_2+x_1x_3$, $x_1^2+x_2^2+x_3$, $x_1x_2^7+x_2x_3^7+x_3x_1^7$ are not.

The basic result referred to above is

Theorem 4.7.1.† (The Fundamental Theorem on Symmetric Polynomials) Let P be a symmetric polynomial in $F[x_1,\ldots,x_n]$. Then $P\in F[s_1,\ldots,s_n]$.

* See 2.6.5.

† This theorem also holds if F is just an integral domain.

That is, P can be written as a polynomial, with coefficients from F, in the elementary symmetric polynomials $s_1, \ldots, s_n$.

Proof P is a sum of monomials $\alpha x_1^{a_1} x_2^{a_2} \ldots x_n^{a_n}$ where $\alpha \in F$ and the a_i are non-negative integers. We can order the monomials of $F[x_1, \ldots, x_n]$ in a manner reminiscent of 'dictionary ordering' by writing $\alpha x_1^{a_1} \ldots x_n^{a_n} < \beta x_1^{b_1} \ldots x_n^{b_n}$ iff $a_s < b_s$ where s is the least integer for which $a_t \neq b_t$. Thus, for example, in $\mathbb{Q}[x_1, x_2, x_3, x_4]$ we would have

$$7x_1^6 x_2^0 x_3^3 x_4^{11} < -3x_1^6 x_2^0 x_3^4 x_4^2$$

where zero exponents are used in the obvious way.

Suppose that, with respect to this ordering, $\alpha x_1^{a_1} \ldots x_n^{a_n}$ is the largest monomial in P. Let σ be a permutation of $\{1, 2, \ldots, n\}$. Since P is symmetric, the monomial $\alpha x_{1\sigma}^{a_1} x_{2\sigma}^{a_2} \ldots x_{n\sigma}^{a_n}$ is a summand of P. It follows that $a_1 \geqslant a_2 \geqslant \cdots \geqslant a_n$. Now $\alpha s_1^{a_1-a_2} s_2^{a_2-a_3} \ldots s_n^{a_n} = g_1$, say, is symmetric in $x_1, \ldots, x_n$. Writing g_1 as a sum of monomials in the x_i, the largest of these is evidently $\alpha x_1^{a_1-a_2}(x_1x_2)^{a_2-a_3} \ldots (x_1 \ldots x_n)^{a_n}$—that is, it is precisely the largest monomial in P. Consider the (obviously symmetric [why?]) polynomial $P_1 = P - g_1$. P_1 possibly involves more monomials than P but the largest monomial of P_1 is smaller than that of P. Clearly this gives scope for a formal induction proof (which we ask the reader to supply for himself). Informally: we repeat the above process with P_1 replacing P to obtain a new symmetric polynomial P_2 whose largest monomial is smaller than that of P_1. This process clearly stops after at most N steps where $N \leqslant n \cdot a_1$ [is this estimate correct?] when we arrive at $P - g_1 - g_2 - \cdots - g_N = 0$. Then $P = g_1 + g_2 + \cdots + g_N$, with each g_i similar in form to g_1, as required.

Note 4.7.2 Clearly this procedure is one which can be put to practical use. Alternative methods are to be found in the exercises.

Example 4.7.3 In $f = \sum_{i \neq j} x_i^3 x_j \in F[x_1, \ldots, x_n]$, the largest monomial is $x_1^3 x_2$.

Then (for $n \geqslant 4$: what happens if $n = 2, 3$?)

$$f - s_1^2 s_2 = \sum_{i \neq j} x_i^3 x_j - \left(\sum_{i=1}^{n} x_i\right)^2 \left(\sum_{1 \leqslant i < j \leqslant n} x_i x_j\right) = -\sum_{i \neq j \neq k \neq i} x_i x_j x_k^2 - 2s_2^2$$

Next $\sum_{i \neq j \neq k \neq i} x_i x_j x_k^2 - s_1 s_3 = -4s_4$. Hence $f = s_1^2 s_2 - 2s_2^2 - s_1 s_3 + 4s_4$.

Exercises

1 Express, in terms of elementary symmetric polynomials, the following symmetric polynomials: (Assume those in (a), (d) $\in \mathbb{Z}[x_1, \ldots, x_n]$ with n big enough.)

(a) $\sum_{i \neq j} x_i^2 x_j$; (b) $x_1^3 + x_2^3 + x_3^3$; (c) $\sum_{1 \leqslant i < j \leqslant 4} x_i^2 x_j^2$; (d) $\sum_{i \neq j \neq k \neq i} x_i^2 x_j^2 x_k$.

2 Given that the equation $x^3 + 3x^2 + 7x - 2$ has roots a, b, c find the polynomial with roots
(i) a^2, b^2, c^2; (ii) $a + b, b + c, c + a$; (iii) $1/a, 1/b, 1/c$.

3 Prove that the s_i are algebraically independent, that is: For each $n \geqslant 1$, if $f(y_1, \ldots, y_n)$ is a non-zero polynomial in $\mathbb{Q}[x]$, then $f(s_1, \ldots, s_n)$ is also non-zero. [Hint: Assume false for some $h(y_1, \ldots, y_k) = h_m(y_1, \ldots, y_{k-1})y_k^m + \cdots + h_0(y_1, \ldots, y_{k-1}) \neq 0$ with k as small as possible. Assuming, for this k, that m is also chosen as small as possible, prove: (i) $h_0(y_1, \ldots, y_{k-1}) \neq 0$ and, putting $x_k = 0$, (ii) $h_0(\bar{s}_1, \ldots, \bar{s}_{k-1}) \neq 0$ where the $\bar{s}_i$ are (elementary) symmetric in $x_1, \ldots, x_{k-1}$.] Deduce that the polynomial in the s_i mentioned in 4.7.1 is unique.

4 Express $f = \sum_{1 \leqslant i,j \leqslant n} x_i^2 x_j$ in terms of elementary symmetric polynomials as follows:

(i) Show that f is expressible as a sum $As_1^3 + Bs_1s_2 + Cs_3$ where A, B, C are constants.

(ii) Find A, B, C by choosing (α) $x_1 = 1$, $x_2 = x_3 = 0$; (β) $x_1 = x_2 = 1$, $x_3 = 0$, etc.

5 For each $k \in \mathbb{Z}^+ \cup \{0\}$ put $\tau_k = \sum_{i=1}^{n} x_i^k$. Use the equality $(t - x_1)\ldots(t - x_n) = t^n - s_1t^{n-1} + \cdots + (-1)^n s_n$ in $\mathbb{Z}[x_1, \ldots, x_n, t]$ to show that, for $1 \leqslant i \leqslant n$, $x_i^n - s_1x_i^{n-1} + \cdots + (-1)^n s_n = 0$. Deduce that, for $r \geqslant n$, $\tau_r - s_1\tau_{r-1} + \cdots + (-1)^n s_n \tau_{r-n} = 0$. For $1 \leqslant r < n$, put $U_r(x_1, \ldots, x_n) = \tau_r - s_1\tau_{r-1} + \cdots + (-1)^r s_r r$. From the above conclude $U_r(x_1, \ldots, x_r, 0, \ldots, 0) = 0$. Now, using ideas from exercise 3, try to deduce

$$U_r(x_1, \ldots, x_r, \ldots, x_{r+t}, 0, \ldots, 0) = 0$$

implies

$$U_r(x_1, \ldots, x_r, \ldots, x_{r+t+1}, 0, \ldots, 0) = 0$$

Use these results to find $\tau_0, \tau_1, \tau_2, \tau_3, \tau_4$, where x_1, x_2, x_3 are the roots of $z^3 + 3z^2 + 7z - 4$.

4.8 The fundamental theorem of algebra

Here, using the minimum of analysis and quite a bit of our earlier developments, we prove the third result mentioned at the start of the chapter, the Fundamental Theorem of Algebra. The first reasonable proof of this was given by Gauss in 1799 (where others, including Euler and Lagrange, had failed). In a second proof in 1815 Gauss assumed a result we assume here, namely 4.4.3. It is interesting to note ([85, p. 598]) that proofs of this fact then given would not pass today's more rigorous demands.

Theorem 4.8.1 Let* $f(x) = a_0 + a_1x + \cdots + a_nx^n \in \mathbb{C}[x]$. There exists in $\mathbb{C}$ a number α such that $f(\alpha) = 0$.

Remark The proof is quite long but is not so difficult if you take it steadily.

Proof Clearly we may assume that $f(x)$ is irreducible over $\mathbb{C}$ since if $f(x)$ were reducible we should only need to prove the theorem for one of $f(x)$'s irreducible factors.

* For the reader's convenience we write $f(x)$ rather than f throughout 4.8.1.

Let us suppose initially that n is odd. Then if all the a_i are real numbers, $f(x)$ has a real, and hence complex, root by 4.4.3. If at least one of the a_i is not real we introduce the polynomial $\bar{f}(x)=\bar{a}_0+\bar{a}_1x+\cdots+\bar{a}_nx^n$ and form the polynomial $h(x)=f(x)\bar{f}(x)$ of degree $2n$. It is easily seen that $\bar{f}(x)$ is irreducible in $\mathbb{C}[x]$ [since $f(x)$ is] and that $h(x)\in\mathbb{R}[x]$ [since $\bar{h}(x)=f(x)\bar{\bar{f}}(x)=\bar{f}(x)f(x)=h(x)$]. If $h(x)$ factorises in $\mathbb{R}[x]$, $h(x)=g_1(x)g_2(x)$, say, where $\deg g_1<\deg h$ and $\deg g_2<\deg h$, then since $h(x)=f(x)\bar{f}(x)=g_1(x)g_2(x)$ in $\mathbb{C}[x]$ and since $f(x)$, $\bar{f}(x)$ are irreducible and hence prime in $\mathbb{C}[x]$ we must conclude that $g_1(x)=c_1f(x)$ and $g_2(x)=c_2\bar{f}(x)$ (or $g_1(x)=c_1\bar{f}(x)$ and $g_2(x)=c_2f(x)$) for suitable units c_1, $c_2\in\mathbb{C}[x]$. Thus c_1, $c_2\in\mathbb{C}$. In particular $\deg g_1=\deg g_2=n$. Since n is odd we can infer that $g_1(x)$, and hence $f(x)$, has a real root as required. Thus we may suppose that $h(x)$ is irreducible in $\mathbb{R}[x]$.

Now, using 4.2.20, we construct a splitting field E for $h(x)$ over $\mathbb{R}$. In E, $h(x)$ factorises into a product of $2n$ distinct [why?] linear factors, $h(x)=a_n\bar{a}_n(x-t_1)\ldots(x-t_{2n})$. Consider the polynomial $k(x)=\prod_{1\leq i<j\leq 2n}(x-t_i-t_j)$ in $E[x]$. Clearly $k(x)$ has degree $\dfrac{2n(2n-1)}{2}$, an odd integer. Further one sees easily that any permutation of the t_i $(1\leq i\leq 2n)$ simply rearranges the factors of $k(x)$. Hence, on multiplying out the product, the coefficients of the various powers of x are seen to be symmetric polynomials in the t_i. Consequently each of these coefficients is expressible as a polynomial (with integer coefficients) in the elementary symmetric polynomials on the t_i. But each of *these* being a coefficient (or its negative) in $h(x)$ is a real number. It follows that each coefficient in $k(x)$ is a real number, that is $k(x)\in\mathbb{R}[x]\subseteq E[x]$. Using 4.4.3 again we deduce that $k(x)$ has a real root: $r=t_i+t_j$, say.

Now consider the polynomials $h_1(x)=h\left(x+\dfrac{r}{2}\right)$, $h_2(x)=h\left(-x+\dfrac{r}{2}\right)$. Clearly $h_1(x)$, $h_2(x)\in\mathbb{R}[x]$ and are irreducible in $\mathbb{R}[x]$ (cf. example 1.9.17(ii)). If $(h_1(x), h_2(x))=1$ in $\mathbb{R}[x]$ then there exist* $u(x)$, $v(x)\in\mathbb{R}[x]$ such that $u(x)h_1(x)+v(x)h_2(x)=1$ in $\mathbb{R}[x]$, an identity which holds equally well in $E[x]$.

Evaluating each side at the element $\dfrac{t_i-t_j}{2}$ in E we obtain, on noting that $h_1\left(\dfrac{t_i-t_j}{2}\right)=h(t_i)=0$ and $h_2\left(\dfrac{t_i-t_j}{2}\right)=h(t_j)=0$, the equality $0=1$ in E. Thus $(h_1(x), h_2(x))\neq 1$ in $\mathbb{R}[x]$. Since $h_1(x)$, $h_2(x)$ are both irreducible in $\mathbb{R}[x]$ and since both have the same leading coefficient $(a_n\bar{a}_n)$ we deduce $h_1(x)=h_2(x)$.

From this it follows that $h_1(x)=h\left(x+\dfrac{r}{2}\right)=h\left(-x+\dfrac{r}{2}\right)$ is a polynomial $m(x^2)$ of degree n in x^2. Since n is odd it follows from 4.4.3 that $m(\beta)=0$ for some $\beta\in\mathbb{R}$. Hence $h\left(\sqrt{\beta}+\dfrac{r}{2}\right)=h_1(\sqrt{\beta})=m(\beta)=0$ where $\sqrt{\beta}$ lies in $\mathbb{C}$. Thus

* See 3.7.5.

$\sqrt{\beta}+\frac{r}{2}$ is a root of $h(x)$ and hence of $f(x)$ or $\bar{f}(x)$. This completes the proof of the theorem in the case where n is odd.

So assume $n = 2^{s+1}w$ where w is odd and $s \geqslant 0$ and let $t_1, \ldots, t_n$ be the distinct roots of $f(x)$ in the splitting field F for $f(x)$ over $\mathbb{C}$. We proceed by induction on s. Forming $k(x)$ as before we have $\deg k(x) = \dfrac{2^{s+1}w(2^{s+1}w-1)}{2}$ and $k(x) \in \mathbb{C}[x]$. Consequently, by induction hypothesis, there exists $c \in \mathbb{C}$ such that $t_i + t_j = c$ for some i, j.

Form $f_1(x) = f\left(x + \frac{c}{2}\right)$ and $f_2(x) = f\left(-x + \frac{c}{2}\right)$. Then $f_1(x), f_2(x)$ are irreducible over $\mathbb{C}$ and as above we can deduce [can we?] that $(f_1(x), f_2(x)) \neq 1$ in $\mathbb{C}[x]$. It follows that $f_1(x) = ef_2(x)$ where $e \in \mathbb{C}$. Thus $f\left(x + \frac{c}{2}\right) = ef\left(-x + \frac{c}{2}\right)$ and comparing leading coefficients we find $a_n = (-1)^n e a_n$, that is $e = 1$ (since n is even). Thus again, as above, $f\left(x + \frac{c}{2}\right) = M(x^2)$ is a polynomial in x^2 in $\mathbb{C}[x]$. Since $\deg M = 2^s w$ we can conclude that $M(\gamma) = 0$ for some $\gamma \in \mathbb{C}$. But then $\sqrt{\gamma} \in \mathbb{C}$ and $f\left(\sqrt{\gamma} + \frac{c}{2}\right) = 0$, as required.

Notes 4.8.2

(i) The complex numbers play a somewhat lesser role in algebra today than in their heyday (namely the majority of the 19th Century) and in the algebraists' armoury the fundamental theorem has been replaced by the result that to every field F one can find an algebraically closed extension field E. To say that a field E is *algebraically closed* is to say that each polynomial $f \in E[x]$ has a root in E. (This is equivalent to saying that each polynomial in $E[x]$ factorises into linear factors in $E[x]$.) Thus 4.8.1 says that $\mathbb{C}$ *is algebraically closed*.

(ii) $\mathbb{C}$ is not the smallest subfield of $\mathbb{C}$ which is algebraically closed. In fact one can prove [49, Vol. 2, p. 40] that the set of all algebraic numbers (exercise 3.2.14) forms a field which is algebraically closed. Of course $\mathbb{C}$ is the smallest algebraically closed field containing $\mathbb{R}$.

(iii) The above algebraic proof of 4.8.1 is quite long. There are short (once the machinery has been set up!) proofs using complex analysis. For a nice heuristic argument see [5, p. 107]. See also [101].

Exercises

None! If you've worked hard on this section you deserve a break.

5

Basic group theory

5.1 Introduction

This chapter introduces, and Chapter 6 examines more deeply, the third of the three main types of algebraic systems that we study in this text, namely that of group. We begin by outlining how Lagrange's investigations into the problem of finding an algebraic formula yielding the roots of the general polynomial of degree n led, naturally, to the introduction of what we now call permutations. The culmination of these investigations, namely the remarkable result of the 20-year-old Evariste Galois, which explains precisely when a polynomial equation is soluble by radicals, is obtained in Theorem 7.10.5.

Following Section 5.2 we first introduce to the reader many concrete examples of groups and then the basic tools and concepts of group theory. Several of these concepts are the analogues of concepts already introduced for rings and fields. This is not really so surprising. Homomorphisms, kernels, cosets, subsystems, etc. are concepts of a general algebraic nature: we are merely meeting their specialisations to rings, fields and now groups. We close Chapter 5 with a rather sketchy description of the kinds of ideas involved in the first application of group theory to a problem from outside mathematics.

Within the next two chapters there are the occasional references to Chapters 3 and 4. The reader who has not studied Chapters 3 and 4 will lose nothing if he simply ignores references to earlier definitions and takes on trust the very few references to earlier theorems. On the other hand the reader is assumed to be acquainted with the concept of binary operation (see 2.7.1). In addition, reference to Section 3.1 and Remarks in Section 3.10 may help the reader to understand the algebraist's general philosophy.

5.2 Beginnings

We give here a brief account of the ideas involved in Lagrange's explanation of why, by 1770, the quest for a formula for the roots of the general quadratic, cubic and quartic equations had proved successful whilst that for the quintic had not.

In the introduction to his paper *Réflexions sur la Résolution Algébrique des Equations* (published in two parts in the years 1770 and 1771) Lagrange

Joseph Louis Lagrange *(25 January 1736 – 10 April 1813)*
Lagrange was born in Turin, the eldest of eleven children, into a family of French origin. On entering the University of Turin his initial interests were in physics and geometry but he soon dropped the former in favour of analysis. At about the time of his appointment at the Royal Artillery School in Turin he began researches which made him the founder of the Calculus of Variations.

In 1766 Euler left Berlin for St Petersburg. Frederick the Great urged Lagrange to fill the vacancy with the words 'The greatest king in Europe wishes to have at his court the greatest mathematician in Europe'.

Between 1764 and 1788 Lagrange several times won the biennial prizes offered by the French Académie des Sciences for solutions to problems in celestial mechanics. From 1787 onwards he resided in Paris and in 1788 he wrote his famous *Méchanique analytique*, a beautiful work providing a unified treatment of mechanics and which, he proudly announced, 'contains no pictures'.

He devoted much time to number theory. He was the first to prove that
(i) for each square-free $a \in \mathbb{Z}^+$ the so-called Pell equation $x^2 - ay^2 = 1$ has infinitely many solutions (1768);
(ii) every positive integer is the sum of 4 integer squares (1770);
(iii) Wilson's Theorem (1771) (see (2.5.4)).

Apart from his work in algebraic equations (Section 5.2) he did sterling work in numerical equations.

---

writes (*R206)

> 'I propose in this memoir to examine the different methods that have been found for the algebraic solution of equations, to reduce these to general principles, and to show *a priori* why these methods succeed for the third and fourth degree and fail for higher degrees.'†

* *Rxyz* indicates page *xyz* of Volume 3 of Lagrange's *Oeuvres* edited by T-A Serret.
† This doesn't mean that the quintic can't be solved, merely that it can't be done this way.

Lagrange found that all the methods proposed by del Ferro, Tartaglia, Ferrari, Descartes,* Tschirnhaus, Euler and Bezout found previously for the solution of the cubic and quartic equations depended essentially on the same principle. As he wrote at the beginning of the fourth section of his paper (R355)

> '... all the methods are reducible to the same general principle, knowing how to find functions of the roots of the proposed equation, which are such: 1° that the equation or the equations by which they are given, that is of which they are the roots (equations which are commonly called the reduites), have themselves degree less than that of the proposed equation or are at least decomposable into other equations of a degree less than that ...'

Let us, by following Lagrange's discussion of two methods of solving the general cubic, see how these reduites and the consequent appearance of permutations of roots arose.

First note that the substitution $y = x + M/3$ reduces the equation $x^3 + Mx^2 + Nx + P = 0$ to $y^3 + ny + p = 0$ for suitable n and p. The method of solution proposed by Hudde (R207) tells us to put $y = z + t$ whence $y^3 + ny + p = 0$ becomes $z^3 + t^3 + p + (3zt + n)(z + t) = 0$ and this will certainly be satisfied if we arrange for $z^3 + t^3 + p = 0$ and $3zt + n = 0$. From these two equations we get $z^3 - \left(\frac{n}{3z}\right)^3 + p = 0$, that is $z^6 + pz^3 - \frac{n^3}{27} = 0$. This equation, the reduite of sixth degree (R213) is a quadratic in z^3 with solution $z^3 = -\frac{p}{2} \pm \sqrt{q}$ where we have put $q = \frac{p^2}{4} + \frac{n^3}{27}$ for brevity. Thus z can take any one of six possible values, namely α, $\omega\alpha$, $\omega^2\alpha$, β, $\omega\beta$ and $\omega^2\beta$ where α is one of the cube roots $\sqrt[3]{\frac{-p}{2} + \sqrt{q}}$, β is one of the cube roots $\sqrt[3]{\frac{-p}{2} - \sqrt{q}}$ and ω is a complex cube root of 1. Thus $y = z + t = z - \frac{3n}{z}$ appears to be six-valued whilst being a root of a cubic equation! (R208) However, Lagrange notes that if we choose α, β so that $\alpha\beta = \sqrt[3]{\frac{-p}{2} + \sqrt{q}} \cdot \sqrt[3]{\frac{-p}{2} - \sqrt{q}} = \sqrt[3]{\frac{p^2}{4} - q} = \sqrt[3]{\frac{-n^3}{27}} = -\frac{n}{3}$ then we have

$$y = y_1 = \alpha - \frac{n}{3\alpha} = \beta - \frac{n}{3\beta} = \alpha + \beta \tag{i}$$

or
$$y = y_2 = \omega\alpha - \frac{n}{3\omega\alpha} = \omega^2\beta - \frac{n}{3\omega^2\beta} = \omega\alpha + \omega^2\beta \tag{ii}$$

* René du Perron Descartes (31 March 1596 – 11 February 1650).

$$\text{or} \quad y = y_3 = \omega^2\alpha - \frac{n}{3\omega^2\alpha} = \omega\beta - \frac{n}{3\omega\beta} = \omega^2\alpha + \omega\beta \qquad \text{(iii)}$$

so that y is 3-valued, after all.

Lagrange goes on to observe that just as we have written the roots y_1, y_2, y_3 of the given cubic in terms of the 6 roots of the reduite so the roots of the reduite can be expressed in terms of the roots y_1, y_2, y_3. Indeed from (i), (ii) and (iii) above we soon find

$$\alpha = \tfrac{1}{3}(y_1 + \omega^2 y_2 + \omega y_3)$$
$$\omega\alpha = \tfrac{1}{3}(y_2 + \omega^2 y_3 + \omega y_1)$$
$$\omega^2\alpha = \tfrac{1}{3}(y_3 + \omega^2 y_1 + \omega y_2)$$
$$\beta = \tfrac{1}{3}(y_1 + \omega^2 y_3 + \omega y_2)$$
$$\omega\beta = \tfrac{1}{3}(y_3 + \omega^2 y_2 + \omega y_1)$$
$$\omega^2\beta = \tfrac{1}{3}(y_2 + \omega^2 y_1 + \omega y_3)$$

Thus one sees why (R215) the reduite is necessarily of degree 6. (There is one root corresponding to each of the six permutations of the 3 'letters' y_1, y_2, y_3.) Further, one sees that the reason the reduite is a quadratic in z^3 is that the expression $(y_1 + \omega^2 y_2 + \omega y_3)^3$ takes on only two distinct values when the roots y_1, y_2, y_3 are permuted in all six possible ways.

A second method, in which we meet another 2-valued function, is that due to Tschirnhaus (R222). The idea here is to determine constants a and b such that, on solving the equation $x^2 = bx + a + y$ simultaneously with the given equation $x^3 + mx^2 + nx + p = 0$, we are reduced to an equation of the form $y^3 + C = 0$. Then y can be found directly and x from the quadratic above. Making the substitution required it is a few lines work to check that a and b have to satisfy the equations

$$3a - mb - m^2 + 2n = 0 \qquad \text{(i)}$$

and

$$3a^2 - 2a(mb + m^2 - 2n) + nb^2 + (mn - 3p)b + n^2 - 2mp = 0 \qquad \text{(ii)}$$

Substituting for a from (i) into (ii) we find a quadratic which must be satisfied by b (whence the necessary value of a is immediate from (i)).

We suppose that the roots of $y^3 + C = 0$ are labelled y_1, y_2, y_3, where $y_1 = -\sqrt[3]{C}$, $y_2 = -\omega \cdot \sqrt[3]{C}$, $y_3 = -\omega^2 \cdot \sqrt[3]{C}$. If the corresponding xs are x_1, x_2, x_3 we have

$$x_1^2 = bx_1 + a - \sqrt[3]{C}$$
$$x_2^2 = bx_2 + a - \omega\sqrt[3]{C}$$
$$x_3^2 = bx_3 + a - \omega^2 \cdot \sqrt[3]{C}$$

whence it follows that $b=\frac{x_1^2+\omega x_2^2+\omega^2 x_3^2}{x_1+\omega x_2+\omega^2 x_3}$ so that a, too, is found. The six apparent values of b (all of which are equally valid since b is surely independent of the order in which we choose the roots of the given equation!) reduce, in fact, to two.

Thus we see once again how the solution of the cubic depends upon being able to find functions of the roots x_1, x_2, x_3 (of the given equation) which yield only two distinct values when the x_1, x_2, x_3 are permuted in all $3! = 6$ ways.

As regards the quartic equation $x^4+mx^3+nx^2+px+r=0$, it is soluble (by radicals) essentially because there exists a 3-valued function of four variables, namely $x_1x_2+x_3x_4$. Under the $4! = 24$ different permutations of x_1, x_2, x_3, x_4 the only values this function takes are $g_1 = x_1x_2+x_3x_4$, $g_2 = x_1x_3+x_2x_4$, $g_3 = x_1x_4+x_2x_3$. For then the cubic $(y-g_1)(y-g_2)(y-g_3)$ is a cubic with coefficients which can be determined from the coefficients of the given equation (exercise 1).

If we then set $z_1 = x_1+x_2-x_3-x_4$ we see that

$$\begin{aligned} z_1^2 &= (x_1+x_2+x_3+x_4)^2 - 4(x_1x_2+x_1x_3+x_1x_4+x_2x_3+x_2x_4+x_3x_4) + 4(x_1x_2+x_3x_4) \\ &= (-m)^2 \qquad\qquad -4n \qquad\qquad +4g_1 \end{aligned}$$

Hence the two possible values of z_1 can be found. In a similar manner one can find the two possible values of $z_2 = x_1-x_2+x_3-x_4$ and $z_3 = x_1-x_2-x_3+x_4$. Recalling, finally, that $-m = x_1+x_2+x_3+x_4$ we can deduce

$$x_1 = -\tfrac{1}{4}m+\tfrac{1}{4}(z_1+z_2+z_3)$$

$$x_2 = -\tfrac{1}{4}m+\tfrac{1}{4}(z_1-z_2-z_3)$$

$$x_3 = -\tfrac{1}{4}m+\tfrac{1}{4}(-z_1+z_2-z_3)$$

$$x_4 = -\tfrac{1}{4}m+\tfrac{1}{4}(-z_1-z_2+z_3)$$

Each x_i has $2\cdot 2\cdot 2 = 8$ different values but once again it can soon be determined which the correct four are. The point to be observed is that solving the given quartic reduces to solving a cubic (because there is a 3-valued function of x_1, x_2, x_3, x_4 and hence a cubic whose coefficients are obtainable from the coefficients of the given equation) and three quadratics, whose coefficients involve (i) the coefficients of the original equation (m and n above) and (ii) a quantity g_1 which is expressible, using radicals, in terms of the coefficients of the original equation since it is a root of the above cubic.

Thus Lagrange saw that to obtain, in like manner, the solution (by radicals*) of the general quintic, he would have to begin by finding a function (equivalent to g_1 above in the case of the quartic) of x_1, x_2, x_3, x_4, x_5 which under the

* That is, by means of a formula involving $+, \cdot, -, \div$ and various $\sqrt[n]{\ }$ and the coefficients of the given equation. See 7.7.2 for an 'official' definition.

$5! = 120$ permutations of $x_1, \ldots, x_5$ yields at most four formally distinct values. Having found such a function he'd then need to find the equivalent of the zs above. Maybe a third (or more) step would be needed in order finally to arrive, as above, at the required general formula.

Lagrange remarked (R403):

> 'Here, if I mistake not, are the true principles of the solution of equations and the most suitable analysis to guide us to it; all is reduced as one can see to a kind of calculus of combinations, by which one finds *a priori* the results one might expect. It would be appropriate to apply these to equations of the fifth degree and higher degrees whose solution is at present unknown; but this application requires too large a number of trials and combinations, whose success is as yet much in doubt, for us to be able to spend all our time on it now; we hope to be able to return to it at another time and we content ourselves here with having laid the foundations of a theory which seems to us new and general.'

In fact Lagrange never returned to the study of the general equation. He may even have suspected the general quintic to be insoluble by radicals for he succeeded in finding a 6-valued function of five variables but not a 4-valued one. Indeed such a function cannot exist as was proved by Ruffini in 1799 not only for $n = 5$ but for all $n \geq 5$.

For further details relating to Lagrange's work the reader might profitably consult [85], [103], [108], [114], and, of course, Lagrange's paper itself!

Exercises

1 Given that x_1, x_2, x_3, x_4 are the roots of $x^4 + mx^3 + nx^2 + px + r = 0$ find the coefficients of the equation whose roots are $x_1x_2 + x_3x_4, x_1x_3 + x_2x_4, x_1x_4 + x_2x_3$.

2 Show that under the 24 permutations of x_1, x_2, x_3, x_4 each of the functions $(x_1 + x_2)(x_3 + x_4)$ and $(x_1 - x_2 + x_3 - x_4)^2$ is three-valued.

Find, in each case, the corresponding cubic equation.

5.3 Axioms and examples

Having seen in Section 5.2 a little of the mathematics which essentially gave rise to the theory of (permutation) groups, we now begin an account in modern style of the rudiments of that theory. Although in practice an abstract concept tends to emerge only when there are sufficiently many interesting examples to warrant it, we reverse this ordering here and begin with

Definition 5.3.1 A **group** is an (ordered) pair $\langle G, \circ \rangle$ where G is a non-empty set and $\circ$ is a binary operation on G satisfying the following axioms:

(A) **The associative law**: that is, for all $a, b, c \in G$ we have $(a \circ b) \circ c = a \circ (b \circ c)$;
(N) Existence of **neutral** or **identity element**: that is, there exists in G an element e, say, such that for all $a \in G$, $e \circ a = a \circ e = a$;
(I) Existence of **inverses**: that is, to each $a \in G$ there exists in G an element denoted by a^{-1} such that $a \circ a^{-1} = a^{-1} \circ a = e$.

Remarks
(i) (cf. 3.2.3(i)) It is common practice to talk of *the group G* rather than the group $\langle G, \circ \rangle$. We shall follow this practice except when seeking extra clarity.
(ii) (cf. 3.2.3(ii)) The letter e (an abbreviation for the German word *einheit*) is used here rather than the symbol '1' for the identity element to remind the reader that the most important applications of group theory are to systems of elements which are not numbers.
(iii) According to our definition of binary operation on G (see 2.7.1) it is automatically the case that, if $x, y \in G$ then $x \circ y \in G$. One often emphasises this fact, however, by saying that G is **closed** under $\circ$. If one denotes this property by C (for closure) a mnemonic for recalling the group axioms is given by the name CAIN.* This is particularly appropriate in view of

Definition 5.3.2 If for all x, y in G we have $x \circ y = y \circ x$, the group G is called **commutative** or **abelian** (after N H Abel who showed the importance of such groups in the theory of equations).

Notation and Terminology 5.3.3 (cf. 3.3.3) It is usual, in dealing with groups in general, to replace $\circ$ by $\cdot$, or even to omit the dot altogether and simply juxtapose the elements being combined. The method of combination is called **multiplication** with $a \circ b$ (or $a \cdot b$ or ab) being called the **product** of a and b. When the group one is dealing with is known to be abelian one sometimes (but not always) replaces $\circ$ by $+$, and refers to the binary operation as **addition** and to $a+b$ as the **sum** of a and b. In this latter case we also write 0 rather than e, $-a$ rather than a^{-1} and, of course, $(a+b)+c$ rather than $(a \cdot b) \cdot c$ or $(ab)c$.

To show that there do exist objects to which the theory we are going to develop does apply, we now give an extensive list of specific groups of different kinds. The examples given can be used by the reader as test cases on which to try out the theorems we prove as well as the various questions which might naturally occur to him.

Examples 5.3.4
(a) $\langle \mathbb{Z}, + \rangle$, where $+$ indicates the usual addition on $\mathbb{Z}$, is an abelian group. The identity element is, of course, 0 and the inverse of, say, -3 is $+3$.
(b) $\langle \mathbb{R}, + \rangle$ and $\langle \mathbb{R}^+, \cdot \rangle$, where $\cdot$ is ordinary multiplication, are abelian groups.

* This is taken from the book [65, p. 29].

Remark (a), (b) are rather trivial examples and there would be no need of a theory of groups if these were the only examples. Indeed we learn nothing we didn't know already about these three groups from our theory, although there is an interesting structural similarity, which we shall uncover in 5.9.5 between the two groups in (b) which is perhaps not apparent at first glance.

Examples 5.3.4

(c) (Generalising (a) and (b).) Let $\langle R, +, \cdot\rangle$ be a ring and $\langle F, +, \cdot\rangle$ be a field. Then $\langle R, +\rangle$, $\langle F, +\rangle$ and $\langle F^{\times}, \cdot\rangle$ are all abelian groups. (For any field F the set of non-zero elements of F is usually denoted by $F^{\times}$.)

It might be wise, before going further, to give examples which fail to be groups.

(d) $\langle \mathbb{Z}^+, -\rangle$. In fact it's a bit unfair even to write down $\langle \mathbb{Z}^+, -\rangle$ since $-$ is not a binary operation on $\mathbb{Z}^+$. For instance, $3-5 \notin \mathbb{Z}^+$.

(e) $\langle \mathbb{Z}, -\rangle$. This time $-$ *is* a binary operation on the given set $\mathbb{Z}$ but axiom A fails. [What about N and I? Do they hold or not?]

(f) $\langle \mathbb{Z}^+, +\rangle$. Here N fails since the only conceivable candidate, namely 0, is not in $\mathbb{Z}^+$.

(g) $\langle \mathbb{Z}^+, \cdot\rangle$. Here I fails. For instance $2 \in \mathbb{Z}^+$ but there is no $x \in \mathbb{Z}^+$ such that $2 \cdot x = 1$ (1 being the clear candidate to satisfy N).

(h) Let $n \in \mathbb{Z}^+$ and let $M(n)$ denote the set of all equivalence classes, modulo n, of integers which are coprime to n. Then $\langle M(n), \odot\rangle$ is an abelian group with $\phi(n)$ elements. As we saw in Section 2.4, $\hat{1}$ acts as the identity element for $\odot$ and, if $n = 42$, say, then $\hat{5}^{-1} = \widehat{17}$ (since $\hat{5} \odot \widehat{17} = \widehat{85} = \hat{1}$).

Remark We do not use the additive notation with this abelian group. The multiplicative notation is much more suggestive and certainly less confusing.

In Section 5.1 we talked of permutations. We recall from 2.6.5 the

Definition 5.3.5 Let X be any non-empty set. A **permutation** on X is a function $f: X \to X$ such that f is 1–1 and onto.

Reminder If f, g are two 1–1 functions from X onto X then their composition $f \circ g$ is defined by $x(f \circ g) = (xf)g$ for all $x \in X$ and is itself 1–1 and onto. (See 2.6.7 and exercise 2.6.9(ii).)

Example 5.3.4

(j) If $P(X)$ denotes the set of all permutations on X then $\langle P(X), \circ\rangle$ is a group. For, we have just remarked that $\circ$ is a binary operation on $P(X)$; composition of functions is associative (see 2.7.6); the identity function, $I: X \to X$ given by $xI = x$, for all $x \in X$, acts as the identity element of $P(X)$ with respect to $\circ$ and, finally, given $f \in P(X)$ the function f^{-1} acts as an inverse to f (see 2.6.5 and exercises 2.6.11, 2.6.12).

Remarks

(i) When X is a finite set with n elements the set $P(X)$ is often denoted by S_n. The group S_n (more accurately, the group $\langle S_n, \circ\rangle$) is called the **symmetric group** on n symbols since S_n leaves fixed each of the n (formal) elementary symmetric polynomials on the elements of X (see Section 4.7). Clearly S_n contains exactly $n!$ elements.

(ii) (h) gives the first (specific) example in this chapter of a **finite group**, that is a group with only a finite number of elements. ((a), (b) give examples of groups with infinitely many elements, otherwise called **infinite groups**.) (j) gives, in the case when X has more than two elements, our first example of a **non-abelian group**. For example, if $X=\{a, b, c\}$ and if

$$af=c \quad ag=b$$

$$bf=a \quad bg=a$$

$$cf=b \quad cg=c$$

then $a(f\circ g)=(af)g=cg=c$ whereas $a(g\circ f)=(ag)f=bf=a$. Thus $f\circ g$ and $g\circ f$ have different effects on $a\in X$; hence $f\circ g\neq g\circ f$, immediately.

For practice, and to assist with 5.3.6, you might care to confirm that $b(f\circ g)=b$ whilst $c(f\circ g)=a$.

Notation 5.3.6 If we express f and g in the form suggested by Cauchy, namely $f=\begin{pmatrix} a & b & c \\ c & a & b\end{pmatrix}$ and $g=\begin{pmatrix} a & b & c \\ b & a & c\end{pmatrix}$, then their product $f\circ g$ is, according to the above calculations, expressible as $\begin{pmatrix} a & b & c \\ c & b & a\end{pmatrix}$. $f\circ g$ can be calculated quite quickly according to the following self-explanatory scheme

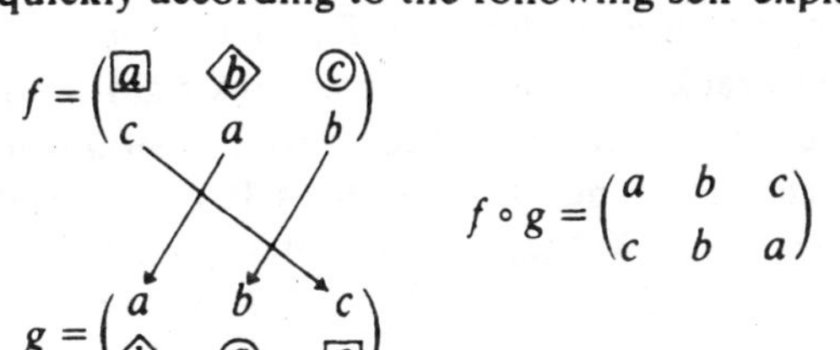

Note that $f^{-1}=\begin{pmatrix} c & a & b \\ a & b & c\end{pmatrix}$ which may be expressed as $\begin{pmatrix} a & b & c \\ b & c & a\end{pmatrix}$ if desired. (A still more economical way of working with permutations will be developed in Section 5.5.)

Definition 5.3.7 The number of elements in the group $\langle G, \circ\rangle$ is denoted by $|G|$ and is called its **order**.

Thus we can talk of 'a group of order 6' (for instance the group S_3 just dealt with) or 'a group of infinite order' (as for example in 5.3.4(a)).

To show that there exist groups of every finite order n ($n \in \mathbb{Z}^+$) we note

Examples 5.3.4

(k) Let C_n denote the set $\left\{\cos\frac{2\pi k}{n} + i \sin\frac{2\pi k}{n} : k = 0, 1, \ldots, n-1\right\}$ of the n complex nth roots of 1. With respect to multiplication C_n forms a group of order n. [This is essentially de Moivre's* theorem.]

To get hold of some more examples of non-abelian groups which are group-theoretically significant and to connect group theory, in albeit a rather trivial way, with the symmetry of geometrical figures we look at

(l) The **group of isometries** of the plane. This is the set E of all those permutations f of the points in the real plane $\mathbb{R} \times \mathbb{R}$ such that for all $x, y \in \mathbb{R} \times \mathbb{R}$ the distance between x and y is equal to that between xf and yf. Such permutations include rotations, translations and reflections (about a line or point). Given a geometrical figure in the plane the number of such motions which leave the figure 'unchanged' is clearly a measure of the figure's symmetry, larger numbers corresponding to greater symmetry.

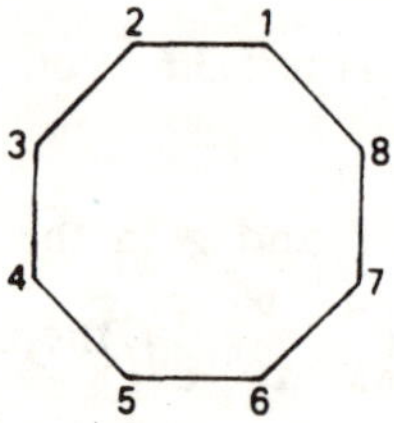

Fig. 5.1

Consider, for instance, the regular n-gon (drawn in Fig. 5.1 for $n = 8$). We label the vertices solely to keep track of them under various rigid motions. Now any isometry of the plane which preserves the overall position of the n-gon clearly must send the vertex 1 to one of the vertices $1, 2, \ldots, n$. If we suppose that 1 is mapped to the vertex K then the isometry must, since it is distance preserving, send 2 and n to $K-1$ and $K+1$ (or $K+1$ and $K-1$) respectively. Whichever of these two possibilities occurs the positions of the remaining vertices (and indeed all points in the plane†) are then automatically determined. Thus there are exactly $2n$ distinct isometries of the plane which 'preserve' the regular n-gon. This group of isometries is called the **dihedral group** of order $2n$. We denote it by D_n.

We look a little more closely at this example in the case of $n = 8$. Let a denote the isometry determined by the permutation $\begin{pmatrix} 1 & 2 & 3 & 4 & 5 & 6 & 7 & 8 \\ 2 & 3 & 4 & 5 & 6 & 7 & 8 & 1 \end{pmatrix}$ so

* Abraham de Moivre (26 May 1667 – 27 November 1754).

† Since an isometry is completely determined once the images of three non-collinear points are known ([68, p. 140]).

that a is an anticlockwise rotation through $\pi/4$ and let b denote the reflection in the vertical line through the mid-point of the sides 1–2 and 5–6. Thus b is completely determined by the permutation $\begin{pmatrix} 1 & 2 & 3 & 4 & 5 & 6 & 7 & 8 \\ 2 & 1 & 8 & 7 & 6 & 5 & 4 & 3 \end{pmatrix}$. Using the same letters for the permutations as for the isometries they correspond to, we see, using the notation introduced in 2.7.9, that

$$a^2 = a \circ a = \begin{pmatrix} 1 & 2 & 3 & 4 & 5 & 6 & 7 & 8 \\ 3 & 4 & 5 & 6 & 7 & 8 & 1 & 2 \end{pmatrix}, \qquad a^3 = \begin{pmatrix} 1 & 2 & 3 & 4 & 5 & 6 & 7 & 8 \\ 4 & 5 & 6 & 7 & 8 & 1 & 2 & 3 \end{pmatrix}, \text{ etc.,}$$

whilst a^8 is the least positive power of a equal to the identity permutation (or isometry). It is not difficult to check that the 16 $(=2n)$ symmetries of the regular octagon can be expressed (omitting the $\circ$ signs) as

$$I, a, a^2, \ldots, a^7, b, ba, \ldots, ba^7$$

where I is the identity permutation and ba^3, for example, denotes (the isometry determined by) the permutation

$$\begin{pmatrix} 1 & 2 & 3 & 4 & 5 & 6 & 7 & 8 \\ 2 & 1 & 8 & 7 & 6 & 5 & 4 & 3 \end{pmatrix}\begin{pmatrix} 1 & 2 & 3 & 4 & 5 & 6 & 7 & 8 \\ 4 & 5 & 6 & 7 & 8 & 1 & 2 & 3 \end{pmatrix} = \begin{pmatrix} 1 & 2 & 3 & 4 & 5 & 6 & 7 & 8 \\ 5 & 4 & 3 & 2 & 1 & 8 & 7 & 6 \end{pmatrix}$$

One final remark: It is not difficult to check—see the demonstration in Fig. 5.2, but also check it from the permutations given above—that $b^{-1}ab = a^{-1} = a^7 \neq a$. It follows that $ab \neq ba$ so that the group of symmetries of the regular octagon is not abelian.

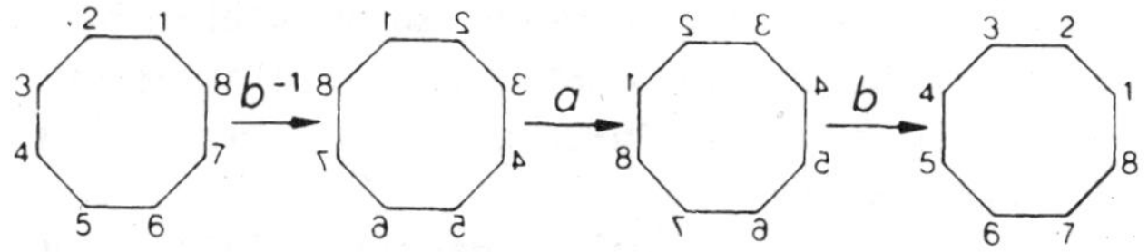

Fig. 5.2

As a final example on permutations

(m) Let $V = \left\{ \begin{pmatrix} x & y & z & t \\ x & y & z & t \end{pmatrix}, \begin{pmatrix} x & y & z & t \\ y & x & z & t \end{pmatrix}, \begin{pmatrix} x & y & z & t \\ x & y & t & z \end{pmatrix}, \begin{pmatrix} x & y & z & t \\ y & x & t & z \end{pmatrix} \right\}$.

Then each element of V gives rise in an obvious manner to an automorphism of the ring $\mathbb{Z}[x, y, z, t]$ (see 3.10.2(iv)). We leave the reader to check that $\langle V, \circ \rangle$, where $\circ$ denotes composition, is a group.

Applying these four permutations to the polynomial $xz + y$ yields the four distinct 'values' $xz + y$, $yz + x$, $xt + y$, $yt + x$. Thus, under the group V, $xy + z$ is a 4-valued polynomial which is fixed only by the trivial subgroup of V. Similarly $xy + zt$ is a one-valued function whilst $xy + z$ is two-valued. As we have seen in Section 5.2 these kinds of considerations were of interest to Lagrange.

Problem 1 Does V give rise to a 3-valued function on x, y, z, t?

Next we present three more specific groups of order 4.

Examples 5.3.4

(n) (i) Let $V_1 = \{1, -1, i, -i\}$ and let $\circ$ denote (ordinary) multiplication of complex numbers.

(ii) Let $V_2 = \left\{\begin{pmatrix}1 & 0\\0 & 1\end{pmatrix}, \begin{pmatrix}-1 & 0\\0 & -1\end{pmatrix}, \begin{pmatrix}0 & 1\\-1 & 0\end{pmatrix}, \begin{pmatrix}0 & -1\\1 & 0\end{pmatrix}\right\}$ with $\circ$ taken as matrix multiplication.

(iii) Let $V_3 = \left\{\begin{pmatrix}1 & 0\\n & 1\end{pmatrix} : n \in \mathbf{Z}_4\right\}$ with $\circ$ taken as matrix multiplication.

(p)* The set $GL_n(X)$ of all *invertible* (i.e. non-singular) $n \times n$ matrices with coefficients in the set X, where $X = \mathbb{Z}$, $\mathbb{Q}$, $\mathbb{R}$, $\mathbb{C}$ or $\mathbb{Z}_m$, forms a group with respect to matrix multiplication. (By definition, the element A of $M_n(X)$ is invertible iff there exists $B \in M_n(X)$ such that $AB = BA = I_n$, the identity matrix of $M_n(X)$. Thus $A \in M_n(\mathbf{Z})$ is invertible iff $\det A = 1$ or -1, and $A \in M_n(\mathbf{Z}_m)$ is invertible iff $\det A$ is a unit in $\mathbf{Z}_m$. For $X = \mathbf{Q}$ or $\mathbf{R}$ or $\mathbb{C}$, $A \in M_n(X)$ is invertible iff ... what?)

(q)* For each X above the subset $SL_n(X)$ of $M_n(X)$ comprising those matrices with determinant $+1$ forms a group with respect to matrix multiplication.

Groups of this kind are, for $X = \mathbf{R}$ or $X = \mathbb{C}$, of interest to physicists and chemists. They are special instances of so called **Lie groups**. These are essentially groups L in which (i) the elements can be labelled by r-tuples of continuously varying real number parameters so that (ii) some sort of 'nearness' condition (a 'near' c in L and b 'near' d in L is to imply ab is 'near' cd in L) holds. As another example

(r) The set $T = \{T_{a,b} : a, b \in \mathbf{R}, a \neq 0\}$ of all mappings of $\mathbf{R}$ to $\mathbf{R}$ of the form $T_{a,b}(x) = ax + b$ constitutes a group with respect to composition of mappings. Each real number pair (a, b) with $a \neq 0$ provides a group element. The identity element is $T_{1,0}$. For products and inverses see exercise 10.

Exercises

1 Are the following sets, together with the given multiplication, groups? If not, list the first of the axioms C, A, N, I which fails.

(a) $M_2(\mathbf{Z}_5)$: matrix multiplication;
(b) all elements of $M_2(\mathbf{Z}_6)$ with determinant $\hat{1}$: matrix multiplication;
(c) $\langle \mathbf{Z}, \circ \rangle$ where $a \circ b = a + b - 37$;

* If anything in this example is unfamiliar to you look it up *now*. For $M_n(X)$, cf. 3.2.5. *GL* and *SL* stand for *general linear* (group) and *special linear* (group) respectively.

(d) $\langle \mathbf{R}, \circ \rangle$ where $a \circ b = a + b - ab$;
(e) $\langle \{\hat{2}, \hat{4}, \hat{6}, \hat{8}\}, \odot \rangle$ where $\odot$ is multiplication mod 10;
(f) the set of all rotations about the origin of 3-dimensional space which permute the unit vectors $\pm\mathbf{i}$, $\pm\mathbf{j}$, $\pm\mathbf{k}$: composition of rotations;
(g) the set of all isometries of 3-dimensional space leaving a given cube 'unchanged': multiplication of isometries of 3-space (cf. 5.3.4 (1));
(h) the set of all permutations on $\{1, 2, 3, 4\}$ which send 1 to 3 or send 2 to 3: composition;
(i) the set of all vectors in 3-dimensional space: vector product;
(j) $\{a + b\sqrt{2}: a, b \in \mathbf{Q}\}$: ordinary addition;
(k) the non-zero elements of exercise (j): ordinary multiplication;
(l) the six functions $f_1(x) = x$, $f_2(x) = \dfrac{1}{1-x}$, $f_3(x) = \dfrac{x-1}{x}$, $f_4(x) = \dfrac{1}{x}$, $f_5(x) = 1 - x$, $f_6(x) = \dfrac{x}{x-1}$, defined on the set of all real numbers other than 0 and 1: composition of functions.

2 For those parts of exercise 1 which are groups state which are abelian and which have finite order.

3 Let $\langle G, \circ \rangle$ and $\langle H, * \rangle$ be groups. Define $\cdot$ on $G \times H$ by $(g_1, h_1) \cdot (g_2, h_2) = (g_1 \circ g_2, h_1 * h_2)$. Show that $\langle G \times H, \cdot \rangle$ is a group. (Cf. exercise 3.2.9.)

4 Let P be your favourite pop record and H your favourite hammer. Show how to make the set $\{P, H\}$ into a group by defining 'multiplication' suitably. Deduce that there exist infinitely many distinct groups of order 2.

5 Let $\langle G, \circ \rangle$ be a group. Define $*$ on G by $a * b = b \circ a$. Is $\langle G, * \rangle$ a group?

6 An **affine geometry** comprises a set X whose elements are called *points* together with various subsets of X called *lines* such that
(i) Each pair of distinct points is contained in exactly one line.
(ii) Each pair of distinct lines has at most one point in common.
(iii) Given a line l and a point P not on it there exists exactly one line n which contains P and has no point in common with l.

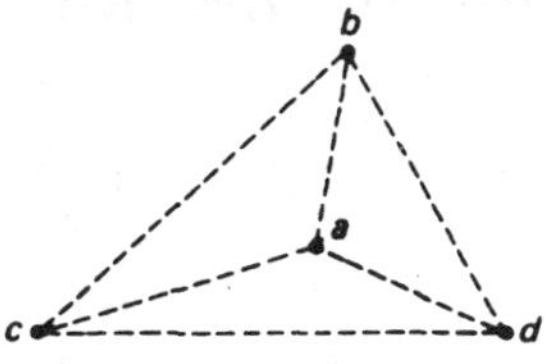

Fig. 5.3

(iv) There are at least two lines. Figure 5.3 gives a pictorial representation of an affine geometry with 4 points and 6 lines.

A **collineation** of an affine geometry is a permutation of the points of X which maps lines to lines. Show that the set of all collineations forms a group under composition. What is the order of the collineation group of the above 4-element affine geometry?

7 (An example for those who have read Section 3.10) An **automorphism** of a field $\langle F, +, \cdot \rangle$ is a $1-1$ mapping ψ from F onto itself such that, for all $a, b \in F$, $(a+b)\psi = a\psi + b\psi$ and $(a \cdot b)\psi = a\psi \cdot b\psi$. Show that, under composition, the set of all automorphisms of F forms a group G, say. Let K be a subfield of F. Show that the set of all automorphisms τ of F which are such that $k\tau = k$ for all $k \in K$ is also a group. It is called the **Galois group of F over K**.

8 Let P be the set of all positive rationals of the form $\frac{a^2+b^2}{c^2+d^2}$ where $a, b, c, d \in \mathbb{Z}$. Is $\langle P, \cdot \rangle$ a group, being ordinary multiplication? (Exercise 3.8.4 might help.)

9 Find the inverse of $\hat{5}$ in $\langle M(11), \odot \rangle$ and in $\langle M(12), \odot \rangle$.

10 Show that, in example 5.3.4(r), $T_{a,b} \circ T_{c,d} = T_{ac,ad+b}$. What is the inverse in T of $T_{a,b}$?

11 How many elements has the group of isometries of the plane leaving unchanged (i) the regular 27-gon?; (ii) the circle of unit radius?

12 Let $x_1x_2^2 + x_2x_3^2 + x_3x_1^2 \in \mathbb{Z}[x_1, x_2, x_3]$. List all the elements of S_3 which leave this polynomial fixed.

13 Let S be any set with more than one element and define for all $x, y \in S$, $x \circ y = x$. Show that $\circ$ is an associative binary operation on S. Is $\langle S, \circ \rangle$ a group? (A pair $\langle S, \circ \rangle$ where S is a set and $\circ$ is an associative binary operation on S is called a **semigroup**.)

14 Let S be a non-empty set and $\circ$ an associative binary operation on S, such that, for every $a, b \in S$ the equation $a \circ x = b$ has a unique solution for x in S. Is $\langle S, \circ \rangle$ necessarily a group?

5.4 Deductions from the axioms

We now parallel the development of Section 3.3 by making some logical deductions from the axioms listed in 5.3.1. As in the case of rings, no assumptions are made concerning the nature of the elements involved. Consequently any conclusions we draw will be applicable to any of the specific groups mentioned in Section 5.3—and to any other groups we come across.

Our first observation is that the axioms of 5.3.1 conceal the fact that the elements given to exist in (N) and (I) are unique. They also conceal some redundancies, as exercise 9 shows.

Theorem 5.4.1 (cf. 3.3.1) In any group G there is exactly one identity element. Further, to each $a \in G$ there corresponds exactly one inverse.

Proof Axiom N of 5.3.1 assures us of at least one element e such that $ea = ae = a$ for all a in G. If $f \in G$ also satisfies axiom N we have $fb = bf = b$ for all b in G. Taking a and b to be the particular elements f and e we get $ef = fe = f$ and $fe = ef = e$. Combining these we see that $e = ef = f$ thus proving the uniqueness of the identity element.

Axiom I assures us that to each $a \in G$ there corresponds at least one inverse element, namely a^{-1}. Let $b \in G$ be any element such that $ab = ba = e$. Then $(a^{-1}a)b = eb = b$ whereas $a^{-1}(ab) = a^{-1}e = a^{-1}$. Since $(a^{-1}a)b = a^{-1}(ab)$ we have $b = a^{-1}$ and uniqueness is proved.

An immediate consequence is that *the inverse of the inverse is the original.*

Theorem 5.4.2 (cf. 3.3.2(ii)) For any group G and any $a \in G$, $(a^{-1})^{-1} = a$.

Proof We know $a \cdot a^{-1} = a^{-1} \cdot a = e$. From this it follows immediately that a is *an* inverse for a^{-1}. But $(a^{-1})^{-1}$ is the notation for the (*unique*) inverse of a^{-1}. Thus $(a^{-1})^{-1} = a$ as required.

Another consequence shows that, unlike the case in the theories of rings (exercise 3.2.10) and of semigroups (exercise 7) cancellation is always possible in a group.

Theorem 5.4.3 If $a, b, c \in G$ are such that $ab = ac$, then $b = c$.

Proof From $ab = ac$ we deduce $a^{-1}(ab) = a^{-1}(ac)$ [why?]

whence $(a^{-1}a)b = (a^{-1}a)c$ [why?]

so that $eb = ec$ [why?]

that is $b = c$ [why?]

A final consequence is a slight surprise (unless you've done exercise 3.3.9(a)).

Theorem 5.4.4 Let $a, b \in G$. Then $(ab)^{-1} = b^{-1}a^{-1}$ (and *not*, in general, $a^{-1}b^{-1}$).

Proof $(ab) \cdot (b^{-1}a^{-1}) = ((ab)b^{-1})a^{-1} = (a(bb^{-1}))a^{-1} = (ae)a = aa^{-1} = e$. Similarly $(b^{-1}a^{-1}) \cdot (ab) = e$ and $b^{-1}a^{-1} = (ab)^{-1}$ follows.

We confirm this in one special case.

Example 5.4.5 Let $A = \begin{pmatrix} 3 & 2 \\ 1 & \frac{1}{2} \end{pmatrix}$, $B = \begin{pmatrix} -2 & 1 \\ 3 & -5 \end{pmatrix}$; then $A, B \in GL_2(\mathbf{Q})$.

$$A^{-1} = \begin{pmatrix} -1 & 4 \\ 2 & -6 \end{pmatrix}, \quad B^{-1} = \begin{pmatrix} -\frac{5}{7} & -\frac{1}{7} \\ -\frac{3}{7} & -\frac{2}{7} \end{pmatrix},$$

$$(AB)^{-1} = \begin{pmatrix} 0 & -7 \\ -\frac{1}{2} & -\frac{3}{2} \end{pmatrix}^{-1} = \begin{pmatrix} \frac{3}{7} & -2 \\ -\frac{1}{7} & 0 \end{pmatrix} = B^{-1}A^{-1}$$

whereas

$$A^{-1}B^{-1}=\begin{pmatrix}-1 & -1\\ \frac{8}{7} & \frac{10}{7}\end{pmatrix}$$

Remarks

(i) The fact that, in general, $(ab)^{-1}\neq a^{-1}b^{-1}$ is just a fact of life that the beginner must learn to live with. In particular the reader should refrain from using the notation $\frac{1}{a}$ in preference to a^{-1}, as many beginners are then tempted to write $\frac{1}{a\cdot b}=\frac{1}{a}\cdot\frac{1}{b}$—a quite natural reaction since the only occasions on which most readers will have used this equality is in the case of rational, real or complex numbers where of course the equality *is* valid.

(ii) The equality $(ab)^{-1}=a^{-1}b^{-1}$ is universally true in a given group G *when and only when* the group is abelian (see exercise 2). That is, this property characterises abelian groups within the class of all groups.

The brackets were left in the proof of 5.4.4 in order to exhibit clearly exactly which axioms were being employed at each stage. They could have been left out because (see 2.7.7) the generalised associative law follows from axiom A of 5.3.1.

For completeness and amusement we note, without proof,

Theorem 5.4.6 (cf. 2.7.7) Let $a_1, a_2, \ldots, a_n$ be elements of G. There are $\frac{(2n-2)!}{n!(n-1)!}$ ways of evaluating the product of $a_1, a_2, \ldots, a_n$ *in that order* (see exercise 1.2.1 and [20, Vol. 1, p. 19]). All these ways yield the same group element which can thus be denoted unambiguously by $a_1a_2\ldots a_n$.

Exercises

1 Show that if G is a group and if $x\in G$ is such that $x^2=x$ then $x=e$.

2 Show that a group G is abelian iff for all $a, b\in G$ we have $(ab)^{-1}=a^{-1}b^{-1}$

3 Let G be a group such that $x^2=e$ for all $x\in G$. Prove that G is abelian.

4 Given a, b, c elements of the group G, show that there exists in G a unique element x such that $axb=c$.

5 Show that if the finite group G comprises the elements $g_1, g_2, \ldots, g_n$ and if g is one of these, then the list $g_1g, g_2g, \ldots, g_ng$ contains each of the elements of G exactly once.

6 Let G be a group, a an element of G and m and n elements of $\mathbb{Z}^+$. Defining $a^0=e$, and $a^{-n}=(a^{-1})^n$ show that $a^{-n}=(a^n)^{-1}$ and that the usual laws of exponents hold; that is, for all $u, v \in \mathbb{Z}$, $a^u a^v = a^{u+v}$ and $(a^u)^v = a^{uv}$. Show that if G is abelian and if $a, b \in G$ then $(ab)^t = a^t b^t$ for all $t \in \mathbb{Z}$.

7 Give a specific example of a semigroup S and elements a, b, $c \in S$ such that $ab = ac$ but $b \neq c$ [You've had examples already in this chapter!]

8 Try to prove the formula in 5.4.6.

9 Let G be a set and $\circ$ an associative binary operation on G.

(i) Consider the axioms

N_R: There exists in G an element e such that, for all $a \in A$, $a \circ e = a$.

I_R: To each $a \in G$ there exists in G an element a^{-1} such that $a \circ a^{-1} = e$.

Show that $\langle G, \circ \rangle$ is a group.

(ii) Defining N_L in an analogous manner, show, by exhibiting an example, that if $*$ is an associative binary operation on G and if $*$ satisfies N_L and I_R then $\langle G, * \rangle$ might fail to be a group. [You have already seen a suitable example in this chapter.]

5.5 The symmetric and the alternating groups

In this section we look more closely at the groups S_n introduced in 5.3.4(j). There are three main reasons for this. The least important is that we thereby offer the reader even more concrete groups on which to test our theorems and his theories—indeed, in a sense, we offer him all groups! (See 5.9.6.) The other two reasons are: (i) with each S_n we shall associate a group A_n called the alternating group on n letters which, for $n \geq 5$, will prove to be a basic building block in the theory of finite groups (see 6.5.9 and 6.6.1); (ii) the problem concerning the existence of a radical formula for the roots of the general quintic equation is resolved using S_5 (Section 7.9).

We begin by replacing the 'double row' notation of 5.3.6 by something easier.

It will probably help the reader if he reads through the following three definitions and theorem with 5.5.5, below, in mind.

Definition 5.5.1 Let f be a permutation on the set $X = \{1, 2, \ldots, n\}$ and let $x_1, x_2, \ldots, x_r$ $(1 \leq r \leq n)$ be distinct elements of X. If for $1 \leq i < r$ we have $x_i f = x_{i+1}$, if $x_r f = x_1$ and if $yf = y$ for every other element of X, then f is called a **cyclic permutation** or **cycle** and is denoted briefly by $(x_1, x_2, \ldots, x_r)$. r is called the **length** of the cycle.

Next, let f be any permutation on X and let $\alpha \in X$ be chosen arbitrarily. Writing f^k for the composition of f with itself k times $(k > 0)$ in $\langle S_n, \circ \rangle$ consider the subset

$$A = \{\alpha, \alpha f, \alpha f^2, \ldots, \alpha f^n\}$$

of X. As there are only n elements in X not all of these $n+1$ elements of X can be distinct. Reading the above list from the left, suppose αf^j is the first element of X to be repeated. Interpreting αf^0 as α we can then say that there exists $i \in \mathbb{Z}$ such that $0 \leq i < j$ and $\alpha f^i = \alpha f^j$. Applying the permutation f^{-i} to this equality* we find that $(\alpha f^i)f^{-i} = (\alpha f^j)f^{-i}$, that is, $\alpha = \alpha f^{j-i}$. By the minimality of j we see that $i = 0$. It follows (see exercise 7) that A comprises precisely the set of distinct elements $\alpha, \alpha f, \ldots, \alpha f^{j-1}$. We make

Definition 5.5.2 The set A described above is called the **orbit** of f on X determined by α.

Now suppose that the elements $\alpha, \alpha f, \ldots, \alpha f^{j-1}$ do not exhaust X. Then there exists an element β, say, in X but not in A. As we found A so we find B, say, the orbit of f determined by β. Clearly $A \cap B = \varnothing$ (see exercise 8). Continuing to obtain orbits in this fashion until all the elements of X are exhausted, we can put all this information, and more, into a theorem if we first make

Definition 5.5.3 To each orbit A $(B, \ldots, \text{etc.})$ define the function f_A $(f_B, \ldots, \text{etc.})$ by

$$\left.\begin{array}{ll} xf_A = xf & \text{if } x \in A \\ xf_A = x & \text{if } x \notin A \end{array}\right\} \quad \text{(and similarly for } f_B, \ldots, \text{etc.)}$$

We then have the promised

Theorem 5.5.4 Let f be a permutation on the finite set X and let $A, B, \ldots, T$ be the finitely many distinct, pairwise disjoint orbits of f on X. Then each of $f_A, f_B, \ldots, f_T$ is a cyclic permutation and f is equal to their product, taken in any order. That is, f is expressible as a product of disjoint cycles.

Example 5.5.5 If

$$f = \begin{pmatrix} 1 & 2 & 3 & 4 & 5 & 6 & 7 & 8 & 9 & 10 & 11 & 12 & 13 & 14 & 15 \\ 7 & 10 & 13 & 4 & 5 & 12 & 3 & 15 & 2 & 9 & 11 & 6 & 1 & 14 & 8 \end{pmatrix}$$

then, since f maps 1 to 7, 7 to 3, 3 to 13 and 13 back to 1; 2 to 10, 10 to 9 and 9 back to 2; 4 to 4; 5 to 5; 6 to 12 and 12 back to 6; 8 to 15 and 15 back to 8; 11 to 11 and finally 14 to 14, we see that the various orbits are $\{1, 7, 3, 13\}$, $\{2, 10, 9\}$, $\{4\}$, $\{5\}$, $\{6, 12\}$, $\{8, 15\}$, $\{11\}$, $\{14\}$ and in the notation of 5.5.1 we can write (omitting the $\circ$ symbols)

$$f = (1, 7, 3, 13)(2, 10, 9)(4)(5)(6, 12)(8, 15)(11)(14)$$

According to the theorem we can express f in many other ways, one of which is

$$f = (2, 10, 9)(8, 15)(4)(6, 12)(11)(1, 7, 3, 13)(14)(5)$$

* Or using the 1–1ness of f^i on the equality $\alpha f^i = (\alpha f^{j-i})^i$.

On the other hand, except for variants of this type, and those arising from noting that, for example, $(1, 7, 3, 13) = (7, 3, 13, 1) = (3, 13, 1, 7) = (13, 1, 7, 3)$, the way of representing f is clearly unique.

Notation 5.5.6

(i) It is usual to omit cyclic permutations involving only one element. Thus, one more way of writing f above is as follows:

$$f = (6, 12)(8, 15)(2, 10, 9)(1, 7, 3, 13)$$

(ii) Commas too are often omitted within the cycles.

To introduce the important concept of alternating group we first make

Definition 5.5.7 If the permutation f on the set X is a cycle which interchanges just two elements of X, leaving all the others fixed, then f is called a **transposition**.

We have immediately

Theorem 5.5.8 Every cyclic permutation (and hence every permutation) can be expressed as a product of transpositions.

Proof One only has to check that

$$(x_1x_2)\cdot(x_1x_3)\cdot\ldots\cdot(x_1x_n) = (x_1, x_2, \ldots, x_n).$$

Note that the way of writing f as a product of transpositions is by no means unique. For instance, $(1\,2\,3\,4\,5) = (12)(13)(14)(15) = (24)(45)(35)(12)(14)(45)$.

On the other hand we do at least have

Theorem 5.5.9 Let the permutation f on the set $X = \{1, 2, \ldots, n\}$ be expressible in some way as an even (respectively odd) number of transpositions. Then *every* way of expressing f as a product of transpositions requires an even (respectively odd) number of transpositions.

*Proof** Consider, in $\mathbb{Z}[x_1, x_2, \ldots, x_n]$, the polynomial

$$\begin{aligned} P = \prod_{1\le i<j\le n}(x_i - x_j) = {} & (x_1-x_2)(x_1-x_3)\ldots(x_1-x_n) \\ & (x_2-x_3)\ldots(x_2-x_n) \\ & \vdots \\ & (x_{n-1}-x_n) \end{aligned}$$

* For a more direct proof see exercise 11.

Now each transposition (ij), say, in* S_X gives rise to a mapping (indeed an automorphism†) of $\mathbb{Z}[x_1, x_2, \ldots, x_n]$ onto itself, namely that determined by interchanging x_i and x_j and leaving the remaining x_k fixed. Such a mapping changes P to $-P$. Likewise each permutation f of S_X gives rise to a mapping of $\mathbb{Z}[x_1, x_2, \ldots, x_n]$ onto itself and similarly maps P either to P or to $-P$. Clearly if f can be expressed as a product of an even number of transpositions then f maps P to P. It is then immediate that f cannot be written in any way as a product of an odd number of transpositions, for if it could f would simultaneously map P to $-P$, an obvious contradiction. An identical argument holds if f is given to be expressible in at least one way as a product of an odd number of transpositions.

Because of this last theorem the following definition is unambiguous.

Definition 5.5.10 Let f be a permutation on the finite set X. f is an **even** (respectively **odd**) **permutation** iff f can be expressed as a product of an even (respectively odd) number of transpositions.

Finally, since the product of two even permutations is now clearly even, since the identity permutation is even and since the inverse of an even permutation is even, we have

Theorem 5.5.11 Let $X = \{1, 2, \ldots, n\}$. The set of all even permutations on X forms a group under composition of functions. This group, called the **alternating group** on n symbols, is denoted by A_X or A_n and has $\frac{n!}{2}$ elements.

Proof The only part left to be proved is that $|A_n| = \frac{n!}{2}$. Now every permutation on X is either even or odd. Let $p_1, p_2, \ldots, p_r$ be the set of all even permutations and $q_1, \ldots, q_s$ be the set of all odd permutations [so that $r + s = n!$]. The permutations $(12)p_1, (12)p_2, \ldots, (12)p_r$ are all odd, are pairwise unequal [why?] and there are clearly r of them. Thus $r \leq s$. Similarly the permutations $(12)q_1, \ldots, (12)q_s$ are all even, pairwise unequal and there are s of them. Thus $s \leq r$. Consequently $r = s = \frac{n!}{2}$.

Exercises

1 The set S of all functions from $X = \{1, 2, 3\}$ to itself is, with respect to composition, a semigroup (see exercise 5.3.13) with 27 elements. Show that it is possible to find $\alpha \in X$ and $f \in S$ such that $\alpha f^i = \alpha f^j$ with $i < j$ but $\alpha f^{j-i} \neq \alpha$. [Hint: Look for α, f such that α, αf, αf^2 are distinct but $\alpha f^i = \alpha f^j$ if $i, j \geq 2$.]

* S_X is used as an alternative to $P(X)$.
† See 3.10.2(iv).

2 Express as a product of disjoint cycles:

$(\alpha)\ \begin{pmatrix}1&2&3&4&5&6&7&8\\2&4&1&6&8&3&7&5\end{pmatrix}$; $(\beta)\ \begin{pmatrix}1&2&3&4&5&6&7\\1&4&2&7&5&3&6\end{pmatrix}$.

3 Express as a product of disjoint cycles:

(a) $\begin{pmatrix}1&2&3&4&5&6&7\\1&6&4&7&2&5&3\end{pmatrix}\cdot\begin{pmatrix}1&2&3&4&5\\5&4&3&1&2\end{pmatrix}$; (b) $(265)\cdot(347)\cdot(1524)$; (c) $(265)\cdot(1524)\cdot(347)$.

4 Let

$$X=\begin{pmatrix}1&2&3&4&5&6&7\\5&2&7&1&3&6&4\end{pmatrix},\qquad Y=\begin{pmatrix}1&2&3&4&5&6\\2&5&4&6&3&1\end{pmatrix}.$$

Using *only* the double row notation find XY, YX, $Y^{-1}XY$. Can you see any relationship at all between X and $Y^{-1}XY$?

5 Repeat exercise 4 using the single row notation.

6 (i) Prove that if f is any permutation on X and if $x_1, x_2, \ldots, x_r$ are distinct elements of X then $f^{-1}(x_1, x_2, \ldots, x_r)f = (x_1f, x_2f, \ldots, x_rf)$.
(ii) Show conversely that if u and v are cyclic permutations of the same length on X then $f^{-1}uf = v$ for some $f \in S_X$.

7 (i) Show that the orbit of the permutation f on X determined by α is equally well determined by αf, by αf^2, etc.
(ii) Prove that if k is the smallest positive integer such that $\alpha f^k = \alpha$ then for each $l \in \mathbb{Z}$ such that $l > k$, $\alpha f^l \in \{\alpha, \alpha f, \ldots, \alpha f^{k-1}\}$. [Hint: use the division algorithm on l and k.]

8 Show that distinct orbits determined by any permutation are necessarily disjoint.

9 Is a cycle of odd length an odd or an even permutation? Determine which elements of S_4 are even and which are odd.

10 Find two distinct even permutations on $X = \{1, 2, 3, \ldots, 9\}$, each of which has just one orbit with 4 elements and moves 1 to 6, 2 to 3 and 7 to 9.

11 Show that $(\ldots xay\ldots)(ab) = (\ldots xbay\ldots)$

that $(\ldots xaby\ldots)(ab) = (\ldots xby\ldots)$

that $(\ldots xay\ldots zbt\ldots)(ab) = (\ldots xbt\ldots)(\ldots zay\ldots)$

and that $(\ldots xay\ldots)(\ldots zbt\ldots)(ab) = (\ldots xbt\ldots zay\ldots)$

Use these results to prove by induction on m: Let $t_1, \ldots, t_m$ be transpositions, and let $t_1t_2\ldots t_m = c_1\ldots c_r$ where $c_1, \ldots, c_r$ are disjoint cycles. Then $m \equiv \sum_{i=1}^{r} (l(c_i)-1) \pmod 2$, where $l(c_i)$ is the length of c_i. Deduce that a product of an even number of transpositions can never be equal to the product of an odd number of transpositions.

12 Show that the permutation

$$\begin{pmatrix} 1 & 2 & 3 & 4 & 5 & 6 & 7 & 8 & 9 & 10 & 11 & 12 & 13 & 14 & 15 \\ 15 & 14 & 13 & 12 & 11 & 10 & 9 & 8 & 7 & 6 & 5 & 4 & 3 & 2 & 1 \end{pmatrix}$$

is odd. It is this which lies behind the fact that the numbers $1, 2, \ldots, 15$ in the well-known '15-puzzle' cannot (without cheating!) be rearranged in the reverse order.

In the 15-puzzle can the position

14	11	7	1
12	2		9
6	5	10	4
13	8	3	15

(i)

be changed (legally!) to

15	12	7	13
4	11	3	14
6	9	8	10
5		2	1

(ii)

5.6 Subgroups. The order of an element

In studying rings and fields certain subsets, namely subrings and subfields, arose in a natural way. A similar situation exists in group theory where, amongst all the subsets of a group G, those which are groups in their own right, the so-called subgroups of G, stand out as worthy of consideration. A special type of subgroups, normal subgroups, plays a role analogous to that played by ideals in the theory of rings at least as far as their connections with homomorphisms are concerned. On the other hand it seems fair to say that the concept of subgroup is more important in group theory than is the concept of subring in ring theory.*

* Of course in field theory subfields are more important than ideals (see exercise 3.4.14 and Theorem 3.10.9).

We begin with

Definition 5.6.1 (cf. 3.4.1) A non-empty subset S of the group $\langle G, \circ\rangle$ is called a **subgroup** of G iff (α) the restriction $\bar{\circ}$ of $\circ$ to $S \times S$ is a binary operation on S, and (β) $\langle S, \bar{\circ}\rangle$ is a group.

If S is a subgroup of G we shall write $S \leqslant G$. If we know (and if we care) that $S \neq G$ we can denote this fact by $S < G$.

Remarks (cf. 2.7.3 (ii) and 2.7.1)
(i) Although it is technically incorrect to do so, no danger will result if we replace $\bar{\circ}$ by $\circ$. Requirement (α) above can then be restated as: S is closed under $\circ$.
(ii) Thus, informally, a subgroup is a subset which is itself a group *with respect to the binary operation induced by* (or passed down from) *the group.*
(iii) We emphasise the requirements in italics above (just as we did for rings; see 3.4.4(ii)), by noting that whilst $\langle \mathbb{R}, +\rangle$ and $\langle \mathbb{Q}^+, \cdot\rangle$ are both (abelian) groups, and whilst $\mathbb{Q}^+ \subset \mathbb{R}$, $\langle \mathbb{Q}^+, \cdot\rangle$ cannot be regarded as a subgroup of $\langle \mathbb{R}, +\rangle$ since the method of combination $(\cdot)$ for the elements of $\mathbb{Q}^+$ is not that $(+)$ handed down from the containing group $\langle \mathbb{R}, +\rangle$.

Examples 5.6.2
(i) Consider the group S_4 of all permutations on the set $X = \{1, 2, 3, 4\}$. Thus S_4 comprises the $4! = 24$ elements $\begin{pmatrix} 1 & 2 & 3 & 4 \\ a & b & c & d \end{pmatrix}$ where a, b, c, d are the integers 1, 2, 3, 4 in some order. (The method of combination, composition of permutations, is, from now on, understood.) According to example 5.3.4(m) the subset V comprising the four permutations, I (the identity), (12), (34), (12)(34), is itself a group with respect to composition of permutations. Thus $V < S_4$.

The subsets $U = \{s : s \in S_4 \text{ and } d = 4\}$ and $W = \{s : s \in S_4 \text{ and } \{b, c\} = \{2, 3\}\}$ are subgroups of orders 6 and . . . what? . . . of S_4. A_4 is a subgroup of S_4 of order 12.
(ii) In the group $M = GL_n(\mathbb{C})$ the subset S of all those $n \times n$ matrices with determinant $+1$ or -1 and the subset T of all matrices with determinant $+1$ are subgroups. Thus $S < M$ and $T < M$. Further, regarding S as a group in its own right, $T < S$. This raises the question: Is each subgroup T of each (sub)group S of a group R necessarily a subgroup of R? We leave this to exercise 7.
(iii) Let $\langle G, \circ\rangle$ be a group. The subsets G and $\{e\}$ are two (extreme) subgroups of G. $\{e\}$ is called the **trivial subgroup**, any subgroup $S \neq \{e\}$ is called a **non-trivial subgroup** of G and any subgroup S other than G is called a **proper subgroup** of G.
(iv) Let $\langle G, \circ\rangle$ be a group. Let $x \in G$. Let $C = \{x^k : k \in \mathbb{Z}\}$. It is not difficult to check that C is a subgroup of G. It is called the **cyclic subgroup** of G generated by x.

Note that we may have $x^r = x^s$ for some $r, s \in \mathbb{Z}$ with $r > s$ (as, for example, when G is a finite group). In this case $x^{r-s} = x^r x^{-s} = x_x^{s-s} = e$. That is, some positive power of x coincides with the identity element. This leads us naturally to

Definition 5.6.3 Let G be a group and let $a \in G$. If there exists a positive integer m such that $a^m = e$ then the smallest such positive integer is called the **order** of a. If no such integer m exists then a is said to be of **infinite order**.

Note that this use of the word *order* does not clash with that of 5.3.7. For, clearly, the element x of G has finite order n iff the cyclic subgroup $\{x^k : k \in \mathbb{Z}\}$ of G has finite order n.

Examples 5.6.4

(i) In $\langle \mathbb{C}^{\times}, \cdot \rangle$, -1 has order 2, i has order 4, and de Moivre's Theorem tells us that $\cos \frac{2\pi}{n} + \mathrm{i} \sin \frac{2\pi}{n}$ has order n. 2 has infinite order.

(ii) In $GL_2(\mathbb{Z})$, $\begin{pmatrix} 0 & -1 \\ 1 & 0 \end{pmatrix}$, $\begin{pmatrix} 1 & -1 \\ 1 & 0 \end{pmatrix}$ and $\begin{pmatrix} 1 & 1 \\ 0 & 1 \end{pmatrix}$ have orders 4, 6 and infinity respectively.

(iii) 5 has infinite order in $\langle \mathbb{Z}, + \rangle$; $\hat{5}$ has order 2 in $\langle \mathbb{Z}_{10}, \oplus \rangle$ and order 11 in $\langle \mathbb{Z}_{11}, \oplus \rangle$.

(iv) In a finite group every element has finite order.

(v) The order of $(1\ 3\ 2\ 6\ 7)(4\ 5)$ in S_7 is 10.

Let us return to the groups in examples 5.6.2. The reader who has carefully checked all the group axioms for each of the parts of 5.6.2 will surely know the group axioms like the back of his hand! And if he had read 3.4.2 and 3.4.2(F) he should prepare to kick himself! For, in those sections were given theorems which showed how to minimise the work of establishing that a subset of a given ring (or field) was a subring (subfield). The reader should expect a similar result to hold in the case of any algebraic system, in particular in the case of groups. Thus we prove

Theorem 5.6.5 (cf. 3.4.2 and 3.4.2(F)) Let $\langle G, \circ \rangle$ be a group and S a non-empty subset of G. Then $S \leq G$ iff for all $a, b \in S$ we have both

(i) $a \circ b \in S$

and

(ii) $a^{-1} \in S$.

Proof If S is a subgroup of G, so that $\langle S, \circ \rangle$ is a group, then clearly, from $a, b \in S$ we deduce $a \circ b \in S$, since S is closed under $\circ$. [It is not quite so straightforward to prove (ii).] Since S is a group it has an identity element f, say, which is such that $f \circ f = f$. It follows from exercise 5.4.1 that f coincides with e, the identity element of $\langle G, \circ\rangle$. Finally, a^{-1} is the unique inverse of a with respect to e *in* G. But, since S is a group, a has an inverse $\bar{a}$, say, with respect to f $(=e)$ *in* S. It follows that $a^{-1} = \bar{a} \in S$ [why?], as required.

Conversely, let S be a non-empty subset of G such that (i) and (ii) hold. [We must check that $\langle S, \circ\rangle$ is a group.] From (i) S is closed under $\circ$. If a, b, $c \in S$ then $a, b, c \in G$ and so $(a \circ b) \circ c = a \circ (b \circ c)$ automatically. Given $a \in S$ we know $a^{-1} \in S$ [by (ii)] and then $e = a \circ a^{-1} \in S$ [by (i)]. Then, trivially, $e \circ s = s \circ e = s$ for all $s \in S$ [why?]. Thus e is in S and acts as an identity element there. Finally, given $a \in S$ there exists [by (ii)] an element (namely a^{-1}) in S which is such that $a \circ a^{-1} = a^{-1} \circ a = e$. As we have just shown CAIN for $\langle S, \circ\rangle$ we know that $\langle S, \circ\rangle$ is a group; that is $S \leqslant G$.

Remarks

(i) We ask the reader to show in exercise 2 that if G is known to be a finite group then even 5.6.5(ii) can be dispensed with since it is then a consequence of 5.6.5(i). That is, *a non-empty subset S of a finite group G is a subgroup of G iff S is closed under the operation on G.*

(ii) Some texts *define* a subgroup of a group G to be a non-empty subset for which conditions 5.6.5(i), (ii) both hold. Since 5.6.5 is an if and only if theorem such a definition is equivalent to 5.6.1.

We leave the reader to re-do 5.6.2 in the light of 5.6.5 and pass on to several of its consequences. First we invite the reader to prove, with the aid of 5.6.5, the group-theoretic analogue of exercise 3.4.5 and Theorem 3.4.5(F), namely

Theorem 5.6.6 Let $\{S_\lambda : \lambda \in \Lambda\}$ be any set of subgroups of a group G. Then the set-theoretic intersection $\bigcap_{\lambda \in \Lambda} S_\lambda$ is also a subgroup of G.

(The reader who has not read Chapter 3 will lose little by taking this result on trust. Equally it will not harm him to try to prove it!)

Example 5.6.7 Let $GL_n(\mathbb{Z})$ and $SL_n(\mathbb{R})$ have the same meaning as in 5.3.4 (p) and (q). Then both are subgroups of $GL_n(\mathbb{R})$ and their intersection is the subgroup of $GL_n(\mathbb{R})$ comprising all $n \times n$ matrices with integer entries *and* determinant $+1$ (i.e. $SL_n(\mathbb{Z})$).

Remark We leave it to the reader to show (exercise 3) that the set-theoretic union $S_1 \cup S_2$ of two subgroups of a group G is itself a subgroup only under special circumstances.

In each group G there exist subgroups of such general importance that they bear special names. Definition 5.6.8(ii) introduces one such.

Definition 5.6.8
(i) Let $a, b \in G$ be such that $ab = ba$. We then say that a and b **commute**.
(ii) Put $\zeta(G) = \{x : x \in G \text{ and } xg = gx \text{ for all } g \in G\}$. $\zeta(G)$ is called the **centre** of G. In words, $\zeta(G)$ is the subset of G comprising those elements which commute with every element of G.

Theorem 5.6.9 $\zeta(G)$ is an abelian subgroup of G.

Proof Clearly $e \in \zeta(G)$ so that $\zeta(G)$ is not empty. Let $a, b \in \zeta(G)$ and let $g \in G$. [Then $ag = ga$ and $bg = gb$.] It follows that $(ab)g = a(bg) = a(gb) = (ag)b = (ga)b = g(ab)$, that is $ab \in \zeta(G)$. Also $a^{-1}(ag)a^{-1} = a^{-1}(ga)a^{-1}$; that is $ga^{-1} = a^{-1}g$. Thus $a^{-1} \in \zeta(G)$ so that $\zeta(G)$ is a subgroup of G. Finally, $\zeta(G)$ is abelian. For, let a, b be any two elements of $\zeta(G)$. Then $ag = ga$ for all $g \in G$; in particular when $g = b$.

Examples 5.6.10
(i) The group with 16 elements in example 5.3.4(l) has centre comprising the two elements a^4 and e. For, as observed there, every element of the group can be expressed in one of the 16 forms a^i or ba^i $(i = 0, 1, \ldots, 7)$. Now $b^{-1}ab = a^{-1}$ and so $b^{-1}a^tb = a^{-t}$ for all integers t (see exercise 14). This means that $a^tb = ba^{-t} \neq ba^t$ unless $a^{-t} = a^t$, that is unless $a^{2t} = e$. But a has order 8 exactly, so $8|2t$, that is $4|t$. Thus the only (distinct) powers of a which could *possibly* lie in the centre are a^4 and $a^8 = e$: and these *do* lie in the centre, as is readily verified. We leave it to the reader to check that no element of the form ba^i can possibly be in the centre.
(ii) If A is an abelian group then $A = \zeta(A)$.
At the other extreme:
(iii) $\zeta(S_3) = \{e\}$. The inequality $(12)(123) \neq (123)(12)$ shows that neither (12) nor (123) is central in S_3. Other elements can be dealt with similarly.

Any group whose centre is no bigger than the trivial subgroup is called a **group with trivial centre** or even a **group with no centre!**

Another important deduction from 5.6.5 can be given after

Definition 5.6.11 Let G be a group and let $U = \{a, b, c \ldots\}$ be a non-empty (possibly infinite) set of elements of G. We denote by $\langle U \rangle$ or $\langle a, b, c, \ldots \rangle$ the set of all elements g of G which can be expressed as a product, $g = x_{i_1}^{\varepsilon_1} x_{i_2}^{\varepsilon_2} \ldots x_{i_r}^{\varepsilon_r}$ where $r \in \mathbb{Z}^+$, where each $x_i \in U$ and each $\varepsilon_i \in \{-1, 1\}$. We call $\langle U \rangle$ the **subgroup* of G generated by $a, b, c, \ldots$ (or by U)** and $\{a, b, c, \ldots\}$ a **set of generators** for $\langle U \rangle$. If $\langle U \rangle = G$ then $\{a, b, c, \ldots\}$ is described as a set of generators for G. If the set U is finite and $G = \langle U \rangle$ then G is called a **finitely generated group**.

* See 5.6.12.

Remark Every subgroup S of G (including G itself) certainly has a set of generators: just take for U all the elements of S. In general G will have many different sets of generators; one often tries to find as convenient a set as possible (for example, some infinite groups do possess finite generating sets).

Our first task is to prove

Theorem 5.6.12 $\langle U\rangle$ is a subgroup of G.

Proof Since U is non-empty certainly $\langle U\rangle$ is not empty. Suppose $u_1 = x_{i_1}^{\varepsilon_1}\ldots x_{i_r}^{\varepsilon_r}$ and $u_2 = y_{j_1}^{\eta_1}\ldots y_{j_s}^{\eta_s}\in\langle U\rangle$. Then, rather trivially, $u_1u_2 = x_{i_1}^{\varepsilon_1}\ldots x_{i_r}^{\varepsilon_r}y_{j_1}^{\eta_1}\ldots y_{j_s}^{\eta_s}\in\langle U\rangle$ and $u_1^{-1} = x_{i_r}^{-\varepsilon_r}\ldots x_{i_1}^{-\varepsilon_1}\in\langle U\rangle$, This does it.

Examples 5.6.13
(i) $\langle \mathbb{Z}, +\rangle$ has, amongst infinitely many others, the generating sets $\{1\}$, $\{-1\}$, $\{2, 3\}$, $\{1, 2, 3, 4\}$. The first two are the only one-generator subsets; the third has two elements, neither of which is redundant.
(ii) $\langle \mathbb{Q}^+, \cdot\rangle$ has, as one of its generating sets, the set of all (positive) primes. $\langle \mathbb{Q}^+, \cdot\rangle$ has no generating sets with only finitely many elements (see exercise 22).

A characterisation of $\langle U\rangle$ which is often taken as the definition is given by

Theorem 5.6.14 Let $\{S_\lambda : \lambda\in\Lambda\}$ be the set of all subgroups of G which contain the subset U. Then $\langle U\rangle = \bigcap_{\lambda\in\Lambda} S_\lambda$.

Proof By 5.6.6 $\bigcap_{\lambda\in\Lambda} S_\lambda$ is a subgroup of G containing U and hence $\langle U\rangle$. The remaining details are left to you (see exercise 24).

Remark We see that 'the cyclic subgroup generated by x' of 5.6.2(iv) has, according to 5.6.11, a generating set comprising just $\{x\}$. Thus 5.6.11 generalises the notion of generator for a cyclic (sub)group.

Exercises

1 Let S be a non-empty subset of a group G. Show that S is a subgroup of G iff for all $a, b\in S$ we have $ab^{-1}\in S$ (cf. exercise 3.4.4).

2 Let S be a non-empty subset of a finite group G. Show that S is a subgroup of G iff for all $a, b\in S$ we have $ab\in S$. Show, using the simplest infinite group you know, that this result is in general false for subsets of infinite groups.

3 Let $A\leqslant G$, $B\leqslant G$. Show that $A\cup B\leqslant G$ iff $A\subseteq B$ or $B\subseteq A$. Give a specific example to show that $G = A\cup B\cup C$ where $A<G$, $B<G$, $C<G$ is possible. [Hint: 5.3.4(m) has three suitable subgroups of order 2.]

4 Show that in an abelian group G the set of all elements of finite order forms a subgroup. (This conclusion may not be valid if G is not abelian; see exercise 12.)

5 Show that the set of all elements of S_{15} which fix the symbols 3 and 7 and permute 4, 9, 13 amongst themselves forms a subgroup T of S_{15}. Find $|T|$.

6 Let S be the subset of all elements of S_{15} which map 5 to 7, 7 to 13 and 13 to 5. Is S a subgroup of G?

7 Given that $\langle S, \bar{\circ}\rangle \leqslant \langle G, \circ\rangle$ and given that $\langle T, \bar{\bar{\circ}}\rangle \leqslant \langle S, \bar{\circ}\rangle$ (where $\bar{\bar{\circ}}$ is the restriction of $\bar{\circ}$ to $T \times T$) show that $\langle T, \bar{\bar{\circ}}\rangle \leqslant \langle G, \circ\rangle$. (This proves that *a subgroup of a subgroup of G is a subgroup of G*.)

8 (a) List all the subgroups of the group of order 16 in example 5.3.4(l). [Lagrange's theorem in Section 5.7 would help a lot here!]
(b) List all the subgroups of order 4 in S_4 and all those of order 5 in S_5.

9 Show that S_7 has an element of order 12, but S_6 does not. [Hint: $12 = 3 \cdot 4$. See 5.6.4(v).]

10 (i) Show that if for all $a, b \in G$ we know $a^2b^2 = (ab)^2$ then G is abelian. [Hint: $aabb = abab$ is given.]
(ii) Find elements $a, b \in S_3$ such that $a^2b^2 \neq (ab)^2$.

11 Let G be an abelian group and let $a, b \in G$ be such that a has order m and b has order n. What is the order of ab? [Note: $a^l b^l = (ab)^l$ for all $l \in \mathbb{Z}$.]

12 Let $A = \begin{pmatrix} 0 & -1 \\ 1 & 0 \end{pmatrix}$, $B = \begin{pmatrix} 0 & -1 \\ 1 & -1 \end{pmatrix} \in GL_2(\mathbb{Z})$. Show that A has order 4, B has order 3 and AB has infinite order.

13 Show that in a finite group, every element has finite order.

14 Show that for any elements x, y, a, b in a group G we have
(i) x, x^{-1} and $y^{-1}xy$ have the same (finite or infinite) order. [Note: $(y^{-1}xy)^r = y^{-1}x^r y$ (prove it!).]
(ii) ab and ba have the same order.
If $c \in G$ do abc and cba necessarily have the same order? [Hint: try taking a, b, c out of S_3.]

15 Show that if G is a group with an even number of elements then G necessarily contains an element of order 2. [Hint: Show that $a^2 = e$ iff $a = a^{-1}$ Try pairing off elements of G with their inverses, beginning with e.]

16 Let $\mathbb{Q}/\mathbb{Z}$ denote the set of all rational numbers x such that $0 \leqslant x < 1$. Define $\oplus$ on $\mathbb{Q}/\mathbb{Z}$ by

$$x \oplus y = x + y \qquad \text{if } 0 \leqslant x + y < 1$$

$$x \oplus y = x + y - 1 \quad \text{if } 1 \leqslant x + y$$

Show that $\langle \mathbb{Q}/\mathbb{Z}, \oplus\rangle$ is an infinite group.

Show that every element of this group is of finite order and that for any $n \in \mathbb{Z}^+$ there exists an element of order exactly n.

17 Show that for each $n \geqslant 3$, $\zeta(S_n) = \{e\} = \langle e \rangle$.

18 Find the centre of $GL_2(\mathbb{R})$. $\left[\text{Hint: If } g \text{ is in the centre it commutes with } \begin{pmatrix} 1 & 1 \\ 0 & 1 \end{pmatrix} \text{ and } \begin{pmatrix} 1 & 0 \\ 1 & 1 \end{pmatrix}.\right]$

19 Find the order of

(i) the subgroup of S_3 generated by the set $\{(12), (13)\}$;

(ii) the subgroup of S_5 generated by (12) and (13)(45).

20 Find four distinct generating sets each with two elements for the group of order 16 in 5.3.4(l).

21 For each $n \in \mathbb{Z}^+$ find a set of n generators for $\langle \mathbb{Z}, + \rangle$ such that no subset of $n-1$ of these generates the whole group. [Hint: for $n=2$ see 5.6.13(ii). Then try to generalise the idea of that case.]

22 Show that $\langle \mathbb{Q}^+, \cdot \rangle$ is not finitely generated. [Hint: Suppose it is. Get a contradiction.]

23 Prove that the group of exercise 5.3.1(l) can be generated by two elements.

24 Complete the proof of 5.6.14. [Hint: $\langle U \rangle$ is a subgroup which contains U and is contained in every S_λ.]

25 Show that S_n is generated by (12) and $(12 \ldots n)$.

5.7 Cosets of subgroups. Lagrange's theorem

We now come to a result, Lagrange's Theorem, described in [31] as 'the most important theorem in finite group theory'. It relates the number of elements in a subgroup of a finite group G to the number of elements in G. As most of current interest in ring theory (as distinct from field theory) concerns rings with infinitely many elements there is not much use in the theory of rings for an analogue of this theorem. (Note, however, the dimension equality 4.6.1 in the theory of field extensions.)

In its original form Lagrange's Theorem related two functions $\phi(x_1, x_2, \ldots, x_n)$ and $\psi(x_1, x_2, \ldots, x_n)$ of the roots $x_1, x_2, \ldots, x_n$ of a given equation when the group of all permutations on the x_i which fix ϕ (say) is also known to fix ψ (R373*). As part of the proof of this Lagrange needed to observe that (in modern terminology) the order of a subgroup of S_n divides n!

* See first footnote on p. 182.

We begin with

Definition 5.7.1 (cf. 4.2.15(i)) Let $H \leqslant G$ and let g be a fixed element of G. We denote by gH the subset $\{gh : h \in H\}$ of G and call this subset the **left coset of H in G determined by g.** One defines analogously the concept of **right coset of H in G** determined by g. Denoted by Hg, this is the subset $\{hg : h \in H\}$.

Remark (cf. 4.2.15(iii)) A more sophisticated way of introducing cosets is given in exercise 5.

Examples 5.7.2
(i) The group S_3 comprising the $3! = 6$ permutations on the set $X = \{1, 2, 3\}$ contains the subset $H = \{I, (12)\}$ as a subgroup. One easily checks that there are three distinct left cosets of H in S_3, namely:

$$H = IH = (12)H; \quad (23)H = (123)H; \quad (13)H = (132)H$$

and three distinct right cosets

$$H = HI = H(12); \quad H(23) = H(132); \quad H(13) = H(123)$$

(ii) In our construction of factor rings we had occasion to consider subsets of the form $a + I = \{a + i : i \in I\}$ where I is an ideal in a ring R and a is a fixed element of R. Note that in this case $a + I = I + a$ since the group $\langle R, + \rangle$ is abelian.
(iii) Let $\mathbb{R}^3$ denote the set of all triples (x, y, z) of real numbers. Then $\mathbb{R}^3$ forms a group under $\oplus$ if $\oplus$ is defined by $(x_1, y_1, z_1) \oplus (x_2, y_2, z_2) = (x_1 + x_2, y_1 + y_2, z_1 + z_2)$. If one considers the subset $H = \{(x, y, z) : 2\pi x + 7y - \frac{5}{3}z = 0\}$ one soon sees that H is a subgroup of G. In geometrical terms H is a plane through the origin in 3-dimensional space. The coset $\left(3, 5e, \frac{1}{4\sqrt{2}}\right) \oplus H$ is, in geometrical terms, nothing more than the plane K passing through the point $\left(3, 5e, \frac{1}{4\sqrt{2}}\right)$ and parallel in $\mathbb{R}^3$ to H.

Remark Certain general facts about cosets are reflected in (i) above. There we have:

(A) $(23)H = \{(23), (123)\} \neq \{(23), (132)\} = H(23)$;
(B) $H(23) = H(132)$ even though $(23) \neq (132)$;
(C) all the cosets—on right and left—contain the same number of elements (in this case 2);
(D) H is always one of the left cosets and always one of the right cosets.

That the behaviour discussed in (C) occurs in every group is shown by

Lemma 5.7.3 Let $H \leqslant G$ and let $g \in G$. The mapping τ from H to gH defined by $h\tau = gh$ is a 1–1 map of H onto gH.

Proof Clearly τ *is* a mapping. Next if $h_1\tau = h_2\tau$ then $gh_1 = gh_2$, whence $h_1 = h_2$ [why?]. Thus τ is 1–1. Now let $x \in gH$. Then $x = gh^*$ for some suitable $h^* \in H$. Trivially $h^*\tau = gh^*$ so that τ is onto.

To obtain Lagrange's Theorem we need one more observation.

Lemma 5.7.4 (cf. 4.2.14) Let $H \leqslant G$ and let $g_1, g_2 \in G$. Then either $g_1H = g_2H$ or $g_1H \cap g_2H = \varnothing$. (That is, any two left cosets of H in G are either identical or miss each other completely.)

Proof Suppose $g_1H \cap g_2H$ is not empty and let c be an element common to both cosets. In particular $c \in g_1H$ and so $c = g_1h^*$ for suitable $h^* \in H$. Then $cH = \{ch : h \in H\} = \{g_1h^*h : h \in H\} = \{g_1\bar{h} : \bar{h} \in H\}$ (see exercise 3) $= g_1H$. In a like manner, since $c \in g_2H$ we deduce that $cH = g_2H$. Thus $g_1H = cH = g_2H$, as claimed.

Now it is trivial to note that each element of G lies in *some* left coset of H in G; indeed $g = ge \in gH$. Thus we see that G is the set-theoretic union of a number of left cosets of H. From 5.7.4 and 5.7.3 we see that distinct cosets are mutually disjoint and contain the same number, namely $|H|$, of elements. Thus if G is a finite group and if G is the union of r distinct left cosets of H in G we see (by counting!) that $|G| = r|H|$. An identical argument for right cosets shows that if G is the union of s distinct right cosets of H in G then $|G| = s|H|$. It follows that $r = s$. This leads to

Definition 5.7.5 Let $H \leqslant G$ with $|G|$ finite. The number of right cosets of H in G (which is equal to the number of left cosets of H in G) is called the **index** of H in G. It is denoted by $|G:H|$.

We then have

Lagrange's Theorem Let $H \leqslant G$, G a finite group. Then $|G| = |G:H| \cdot |H|$.

Remark Lagrange's Theorem is often stated in the form: *The order of a subgroup divides the order of the group.* Combining this with the observation following 5.6.3 we deduce that: *The order of an element in a finite group G divides the order of G.*

Problem 2 Here is a converse to Lagrange's Theorem: Let G be a finite group and let $n \mid |G|$. Then G has at least one subgroup of order n. Do you think this assertion is true? If so, try and prove it; if not, try to find a counterexample. [As the answer to the problem will be given later, don't spend more than an hour at it. You might begin by looking at a few specific examples—in theory you have infinitely many permutation groups at your fingertips!]

Problem 3 Can you say anything significant about a group G which is known to have order 107?

Exercises

1 List the left and right cosets of the subgroup $S = \{I, (123), (132)\}$ in A_4.

2 List the left and right cosets of the subgroup $V = \{I, (12)(34), (13)(24), (14)(23)\}$ in A_4.

3 Let $H \leqslant G$ and let $g \in G$ and $h^* \in H$. Show that $gh^*H = \{gh^*h : h \in H\} = \{gh : h \in H\} = gH$. Deduce that $h^*H = H = Hh^*$.

4 Show that, for $c, d \in G$, $cH = dH$ iff $d^{-1}c \in H$. Deduce that $cH = H$ iff $c \in H$.

5 Given $H \leqslant G$ define a binary relation on G by putting $a \sim b$ iff $a^{-1}b \in H$. Show that $\sim$ is an equivalence relation and that the corresponding equivalence classes are just the left cosets of H in G.

Given $a, b, c \in G$ and $a \sim b$ is it necessarily true that (i) $a^{-1} \sim b^{-1}$; (ii) $ca \sim cb$; (iii) $ac \sim bc$?

6 Let $A \leqslant G$ and $B \leqslant G$. Show that for $x, y \in G$ either $xA \cap yB = \varnothing$ or else $xA \cap yB$ is a left coset of $A \cap B$. Deduce (even if G be infinite) that if $H \leqslant G$, if $K \leqslant G$ and if $|G:H|$ and $|G:K|$ are finite, then $|G:H \cap K| \leqslant |G:H| \cdot |G:K|$ and is finite. Can you describe any circumstances under which we have equality here?

7 Show that if $H \leqslant K$ and $K \leqslant G$ then $|G:H| = |G:K||K:H|$.

8 Suppose $|G:H| = n$ and $a \in G$. Show that there exists $t \in \mathbb{Z}^+$ such that $t \leqslant n$ and $a^t \in H$. Need $t|n$? [Hint: look at (12) and $\{I, (13)\}$ in S_3.]

9 Let $g_1, g_2, \ldots, g_r$ be a set of elements, one from each of the r left cosets of H in G. Show that $g_1^{-1}, g_2^{-1}, \ldots, g_r^{-1}$ is a set of elements one from each of the r right cosets of H in G.

10 Let G be a finite group. Show that for all $g \in G$ one has $g^{|G|} = e$. [Hint: the order of an element . . . .]

11 A finite group G contains elements of every finite order up to and including 12. Find the least possible value of $|G|$.

5.8 Cyclic groups

Since every group contains, along with each of its elements x, the whole of the cyclic subgroup generated by x we see that (i) every group is built up from (specifically, is the set-theoretic union of) its cyclic subgroups, and that (ii) the simplest possible types of groups are those which comprise the distinct powers of some one element. This section contains a few easy observations about the family of groups described by (ii), the so-called **cyclic groups**. Although it is in general very difficult to relate the structure of a group to its cyclic structure,* we shall see in Section 6.4 how this is possible in the case of abelian groups.

The reason for the terminology *cyclic* is that, at least in the finite case, the prototype amongst cyclic groups of order n is the multiplicative group of the n complex nth roots of 1. In the case $n = 6$ these roots can be pictured as in Fig. 5.4. Figure 5.5 then presents itself naturally as a pictorial representation of the abstract cyclic group of order 6 with generator x.

In the case of infinite cyclic groups the prototype is the group $\langle \mathbb{Z}, +\rangle$.

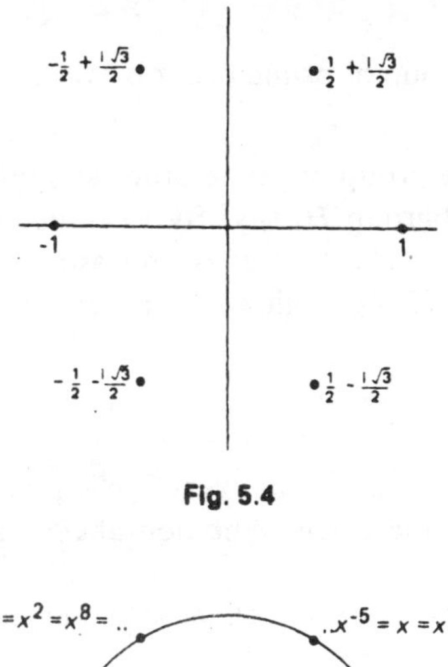

Fig. 5.4

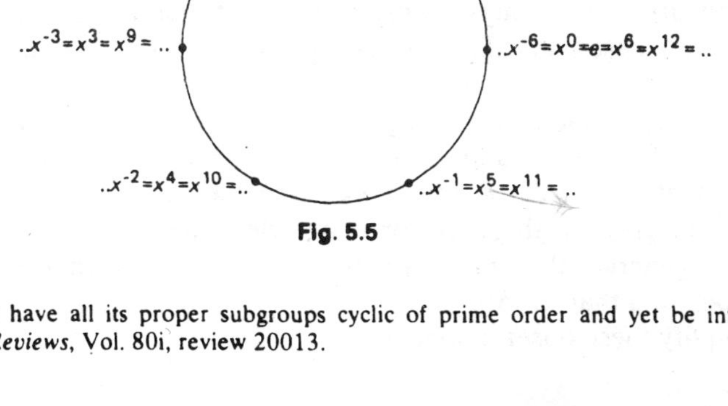

Fig. 5.5

* A group may have all its proper subgroups cyclic of prime order and yet be infinite. See *Mathematical Reviews*, Vol. 80i, review 20013.

Remark We show in 5.9.3 that the above prototypes are essentially the only examples of cyclic groups of orders 6 and infinity respectively.

Other concrete examples of finite cyclic groups are

(i) $P_4 = \{I, (1234), (13)(24), (1432)\}$ (usual multiplication).

(ii) All the groups $\langle M(n), \odot\rangle$ of equivalence classes of integers coprime to n (see 5.3.4(h)) for which $n=2$ or 4 or p^α or $2p^\alpha$ where p is an odd prime and $\alpha \in \mathbb{Z}^+$. (Actually, for all other positive integers n, the groups $\langle M(n), \odot\rangle$ are not cyclic. You can easily check a few individual instances, but the general result just stated is a quite difficult and very important result in number theory. In those cases when $M(n)$ is cyclic any generating element for $M(n)$ is called a *primitive root* modulo n.)

We know (by 5.3.4(k)) that for each $n \in \mathbb{Z}^+$ there exists at least one cyclic group of order n. We illustrate the power of Lagrange's Theorem by showing that, if n is a prime, then *all* groups of order n are cyclic. (In particular we answer problem 3 above.)

Theorem 5.8.1 Any group of prime order is cyclic.

Proof Let G be a finite group of prime order and select x $(x \neq e)$ in G. Now x generates a cyclic subgroup H, say. By Lagrange's Theorem $|H|\,|\,|G| = p$. Hence $|H| = 1$ or $|H| = p$. The former is impossible since H contains e and x. Thus $|H| = p$ whence $H = G$ follows. Consequently G is cyclic! [Isn't that a neat proof?]

Remarks

(i) Since it is clear (exercise 1) that every cyclic group is necessarily abelian we see immediately that there can be no non-abelian groups of order n when n is a prime.

(ii) In a group of prime order any element, other than e, can be taken as a generator.

Not only are cyclic groups simply defined but their subgroup structure can be completely described. (Such a description is out of the question for most groups.)

If G is a finite cyclic group with generator x and if $|G| = j \cdot k$ where $j, k \in \mathbb{Z}^+$ then it should be reasonably clear that the element x^j generates a (cyclic) subgroup S of G of order k and index j (see exercise 2). Similarly, if G is an infinite cyclic group with generator x, it is clear that, for $j \in \mathbb{Z}^+$, the elements x^j and x^{-j} generate the same (cyclic) subgroup S of G, that S is an infinite cyclic group and that $|G:S| = j$.

We amplify these observations in

Theorem 5.8.2
(i) Each subgroup S of a finite cyclic group G is a (finite) cyclic group whose index in G divides $|G|$. Further, given any $j \in \mathbb{Z}^+$ such that $j \mid |G|$ there exists *exactly one* subgroup of G with index j.
(ii) Each subgroup S (other than $S = \{e\}$) of an infinite cyclic group G is an infinite cyclic group of finite index in G. Further, given any $j \in \mathbb{Z}^+$ there exists *exactly one* subgroup of G with index j.

Proof Let S be a subgroup of the [finite or infinite] cyclic group $G = \langle x \rangle$. If $S = \langle e \rangle$ then S is certainly cyclic, generated by e. Assuming $S \neq \langle e \rangle$ choose α to be the smallest positive integer such that $x^\alpha \in S$. [Why does such an α exist?] Now suppose that x^β $(\beta \in \mathbb{Z})$ also lies in S. By the division algorithm there exist integers $m, r \in \mathbb{Z}$ such that $\beta = m\alpha + r$ where $0 \leqslant r < \alpha$. Since $x^\beta \in S$ and since $(x^\alpha)^m \in S$ we see that $x^r = x^\beta x^{-m\alpha} \in S$. By the choice of α, we see that $r = 0$ whence $\beta = m\alpha$ and $x^\beta = (x^\alpha)^m$. Thus S is the cyclic subgroup of G generated by x^α. Further use of the division algorithm* shows that in either case G is the disjoint set-theoretic union of cosets $S, xS, \ldots, x^{\alpha-1}S$. Thus $|G:S| = \alpha$ is finite and, in case $|G|$ is finite, $|G:S| \mid |G|$.

To complete the proof we note that the discussion preceding this theorem shows that in each case there is certainly at least one subgroup of the desired kind. Suppose now that S, T are (in case (i) or in case (ii)) subgroups of the same index in G and let σ, τ denote respectively the smallest positive integers such that $x^\sigma \in S$ and $x^\tau \in T$. Then, whether G be finite or infinite, $\sigma = |G:S| = |G:T| = \tau$ follows, as required.

Remark In the infinite cyclic group generated by x the single subgroup of index 12 can be generated only by x^{12} or x^{-12} whereas in the cyclic group of order 120 the single subgroup of index 12 can be generated by x^{12}, by x^{36}, by x^{84} and by x^{108}. [See exercise 3 if you don't understand why.]

Another way of phrasing part of 5.8.2 is to say: *Each subgroup of a group which can be generated by a single element can itself be generated by a single element.*

The reader may care, after considering a few specific examples, to try and settle

Problem 4 Let G be a group which can be generated by two elements. Is it necessarily the case that every subgroup of G can be generated by (at most) two elements? [Exercise 5.6.25 looks helpful.]

Exercises

In these exercises C_n will denote the cyclic group of order n with generator x.

1 Prove that every cyclic group is abelian. Give an example of an abelian group which isn't cyclic. [Hint: search examples 5.3.4.]

* Do you recognize this argument? See 3.7.15, 1.4.4, 4.3.4.

2 Given that $n = kl$ where $k, l \in \mathbb{Z}^+$ show that C_n has a subgroup of order k.

3 If $n = 10$ show that C_n is also generated by x^3, by x^7 and by x^9. Deduce that if S is the subgroup of index 12 in C_{120} then S can be generated by x^{12}, by x^{36}, by x^{84} and by x^{108}

4 Find all elements $y \in C_{15}$ such that $C_{15} = \langle y \rangle$.

5 How many elements $y \in C_n$ are there such that $C_n = \langle y \rangle$? Find all n for which there are (i) just 2 such y; (ii) just 3 such y.

6 Let S be the subgroup of C_{154} generated by x^{28} and x^{88}. Find n such that $S = \langle x^n \rangle$.

7 Show that if the group $G \neq \langle e \rangle$ has no subgroups apart from $\langle e \rangle$ and G then G is finite and has prime order. (You are not meant to assume G is finite. Prove that *first*!)

8 List all subgroups of C_{120}. (To see relative difficulty now try to list all the subgroups of S_5.)

9 Give an example of a finite group G such that all proper subgroups of G are cyclic but G is not even abelian and, hence, certainly not cyclic. (You do know of such an example.)

10 Is $\langle \mathbb{Q}, + \rangle$ cyclic? Let $a, b \in \mathbb{Q}$. Is the subgroup $\langle a, b \rangle$ of $\mathbb{Q}$ cyclic?

11 Is $\langle \{a + b\sqrt{2}: a, b \in \mathbb{Q}\}, + \rangle$ cyclic?

12 Is $\langle \{a + b\sqrt{2}: a, b \in \mathbb{Q}; a, b \text{ not both zero}\}, \cdot \rangle$ cyclic?

13 It has been conjectured* that 2 is a primitive root of infinitely many primes. Show that $2^{20} \equiv 1 \pmod{41}$ and deduce that 2 is not a primitive root of 41.

14 Let $P = \{x : x \in \mathbb{C}$ and for some $n \in \mathbb{Z}^+, x^{p^n} = 1\}$. Show that $\langle P, \cdot \rangle$ is an infinite group. Show that P is not cyclic but that every proper subgroup of P is a finite cyclic group of order p^m for some m.

15 What can you say about a group with $391\,581 \times 2^{216\,193} - 1$ elements?

5.9 Isomorphism. Group tables

In Section 3.10 we made precise the idea of two rings being 'essentially the same' by introducing the concept of ring isomorphism. Under the heading 'Remarks' we indicated not only ring-theoretic uses for the concept but also that isomorphism is a basic concept for all of algebra. It is therefore no surprise that we shall have occasion to call on

* This conjecture is attributed to Emil Artin (3 March 1898 – 20 December 1962).

Definition 5.9.1 (cf. 3.10.1)
(i) Let $\langle G, \circ\rangle$ and $\langle H, *\rangle$ be groups. A 1–1 mapping $\psi: G \to H$ from the set G onto the set H is called an **isomorphism** iff for all $a, b \in G$ we have

$$(a \circ b)\psi = (a\psi) * (b\psi)$$

(ii) If $\langle G, \circ\rangle$, $\langle H, *\rangle$ are groups such that *at least one isomorphism* can be found between them then we say (loosely) that G and H are **isomorphic** (or that G and H are the same, *up to isomorphism*) and we write $G \cong H$.

Elementary consequences of 5.9.1, analogous to those for rings in exercise 3.10.7, are left to 5.10.3 and exercise 5.10.3. Figure 5.6, illustrating this concept pictorially, is analogous to Fig. 3.2 in Section 3.10.

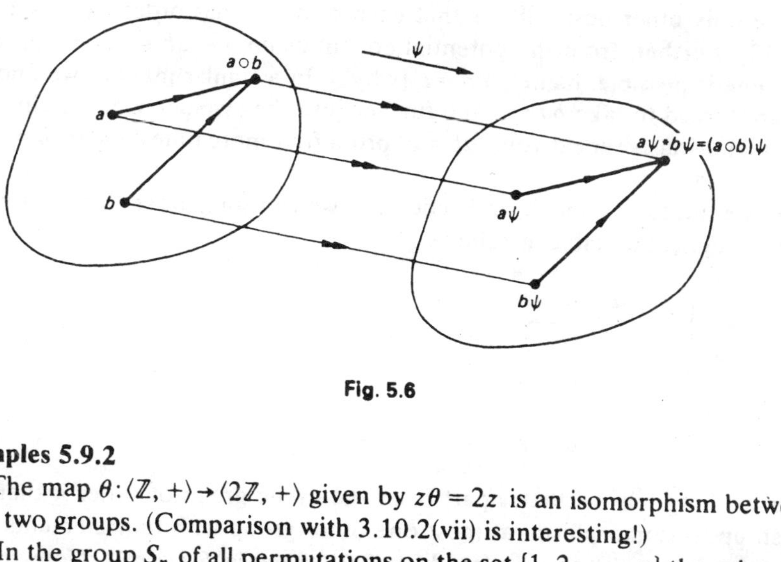

Fig. 5.6

Examples 5.9.2
(i) The map $\theta: \langle \mathbb{Z}, +\rangle \to \langle 2\mathbb{Z}, +\rangle$ given by $z\theta = 2z$ is an isomorphism between these two groups. (Comparison with 3.10.2(vii) is interesting!)
(ii) In the group S_n of all permutations on the set $\{1, 2, \dots, n\}$ the subgroup comprising all permutations fixing '1' is a (sub)group isomorphic to S_{n-1}.
(iii) The group of 8 symmetries of a square is isomorphic to the subgroup of S_4 generated by (12)(34) and (1234) but not to the subgroup of S_6 generated by (12), (34) and (56), the latter being abelian (of order 8).
(iv) The group of matrices generated by $\begin{pmatrix} 0 & 1 \\ -1 & 0 \end{pmatrix}$ and $\begin{pmatrix} 0 & i \\ i & 0 \end{pmatrix}$ with respect to multiplication is a group of order 8. It is called the *group of quaternions* (cf. Section 3.2). It is not isomorphic to the group of symmetries of the square given in (iii).
(v) $\langle \mathbb{Q}^+, \cdot\rangle \cong \langle \mathbb{Z}[x], +\rangle$. For $r = 2^{\alpha_0}3^{\alpha_1}5^{\alpha_2}\dots p_{s+1}^{\alpha_s} \in \mathbb{Q}$ define $r\theta = \alpha_0 + \alpha_1 x + \alpha_2 x^2 + \dots + \alpha_s x^s$. Then θ is an isomorphism.
(vi) $\langle \mathbb{Z}, +\rangle \not\cong \langle \mathbb{Q}, +\rangle$; $\langle \mathbb{Q}, +\rangle \not\cong \langle \mathbb{R}, +\rangle$. The former is the case since the equation $nx = a$ $(n \in \mathbb{Z})$ is always soluble within $\mathbb{Q}$ but not always, indeed rarely, soluble in $\mathbb{Z}$. The last two groups are not isomorphic since there is no 1–1

map between $\mathbb{Q}$ and $\mathbb{R}$. (Recall exercise 2.6.15: $\mathbb{Q}$ is countable whereas $\mathbb{R}$ is not.)

(vii) $\langle \mathbb{Z}_6, \oplus \rangle \cong \langle M(7), \odot \rangle$ (see 5.3.4(h)). The maps $\hat{n}\theta = \widehat{3^n}$ and $\hat{n}\psi = \widehat{5^n}$ both establish this. The map $\hat{n}\sigma = \widehat{2^n}$ does not. [Why not?]

We have already exhibited in 5.3.4(m), (n) distinct concrete examples of groups with four elements, but before attempting to see if any two of them are essentially the same let us do a little abstract working.

Let us suppose that G is a group with four elements e, a, b, c of unspecified nature where e denotes the identity element. By Lagrange's Theorem the only possible orders for elements of G are 1, 2 and 4. If G has an element of order 4 then the 4 distinct powers of that element already account for all the elements of G. Hence G is cyclic.

The only other possibility is that each of a, b, c has order exactly 2 [why not 1?]. Further, from the potential equalities $ab = e$, $ab = a$, $ab = b$, $ab = c$ only one is possible, namely $ab = c$ [why?]. In a similar manner we find that we are forced to take $ba = c$ and further [use the symmetry of the situation; don't laboriously repeat your '$ab = c$' proof four more times!] $bc = cb = a$ and $ca = ac = b$.

As we did for $\mathbb{Z}_6$ and $\mathbb{Z}_7$ in Section 2.4 we can summarise this information in a multiplication table as follows:

	e	a	b	c
e	e	a	b	c
a	a	e	c	b
b	b	c	e	a
c	c	b	a	e

Our working tells us that, apart from the cyclic group of order 4 there can exist, up to isomorphism, at most one other group with four elements. Such a group does exist as we already know (example 5.3.4(m)). Note that, in particular, *every* group with four elements must be abelian.

We leave the reader to check (exercise 14) that there is, up to isomorphism, only one group of order 3 and likewise that there are only two groups of order 6. (And you already know of a concrete example of each type.)

This sort of analysis can be carried out on groups of higher order: the reader might care to try to show there are (up to isomorphism) five distinct groups of order 8, three of which are abelian.

The usefulness of multiplication tables, other than for purposes of exposition, is just about nil! You can convince yourself of this by writing out the 256 entries in the multiplication table of the group in 5.3.4(l) and then observing that in fact the group appears to be summarised by the assertions (i) that group is generated by two elements, a and b; (ii) all relationships between the various elements can be deduced from three basic ones, viz, $a^8 = e$, $b^2 = e$, $b^{-1}ab = a^{-1}$. This description of the group, presented in the form $\langle a, b; a^8 = e, b^2 = e, b^{-1}ab = a^{-1}\rangle$, is clearly much more economical than a 16×16 multiplication table. Such 'presentations' have obvious advantages in the case of infinite groups; furthermore groups tend to make their appearances in terms of presentations in certain branches of mathematics, in particular in topology.

We now give three instances of the isomorphism concept in action. Firstly we confirm the essential uniqueness of cyclic groups of a given order.

Theorem 5.9.3
(i) Every cyclic group of finite order n is isomorphic to the multiplicative group of all complex nth roots of 1.
(ii) Every infinite cyclic group is isomorphic to the group $\langle \mathbb{Z}, +\rangle$.

Proof We shall prove part (i) only (leaving (ii) to exercise 1), using the additive notation for the given cyclic group.

Thus let $\langle G, +\rangle$ be a cyclic group of order n with generator a [say, and consequently elements $0, a, 2a, \ldots, (n-1)a$]. Let $\langle C_n, \cdot\rangle$ denote the group of nth roots of 1 under multiplication. We know the element $\cos \frac{2\pi}{n} + i \sin \frac{2\pi}{n}$ $\left(\text{which we abbreviate to cis } \frac{2\pi}{n}\right)$ generates C_n. Clearly the mapping $\psi: G \to C_n$ defined by $(ka)\psi = \text{cis } \frac{2\pi k}{n}$ $(0 \leqslant k < n)$ is a 1–1 map of G onto C_n. To prove it is a homomorphism we take $ua, va \in G$ $(0 \leqslant u,\ v < n)$. By the division algorithm there exist unique integers m, r such that $u + v = mn + r$ where $0 \leqslant r < n$. Thus we have

$$\begin{aligned}(ua + va)\psi &= ((u+v)a)\psi = ((mn+r)a)\psi = (ra)\psi \\ &= \text{cis } \frac{2\pi r}{n} = \text{cis } \frac{2\pi(mn+r)}{n} \\ &= \text{cis } \frac{2\pi(u+v)}{n} = \text{cis } \frac{2\pi u}{n} \cdot \text{cis } \frac{2\pi v}{n} = (ua)\psi \cdot (va)\psi\end{aligned}$$

thus showing that ψ is a homomorphism, as required.

Questions
(i) Can you supply the reason for each equality asserted above?
(ii) Explain why it is not enough to say $(ua + va)\psi = ((u+v)a)\psi = \text{cis } \frac{2\pi(u+v)}{n} = \text{cis } \frac{2\pi u}{n} \cdot \text{cis } \frac{2\pi v}{n} = (ua)\psi \cdot (va)\psi$.

It is now a small step to prove that *every* two cyclic groups of order n (n finite or infinite) are isomorphic to each other. Clearly we need

Lemma 5.9.4 If $\langle G, \circ\rangle \cong \langle H, *\rangle$ and $\langle G, \circ\rangle \cong \langle K, \cdot\rangle$ then $\langle H, *\rangle \cong \langle K, \cdot\rangle$.

Proof Let ρ, σ be the 1–1 mappings from G onto H and from G onto K which show that $G \cong H$ and $G \cong K$. Since ρ is 1–1 and onto there exists the inverse mapping $\rho^{-1}: H \to G$ and it is easy to check ρ^{-1} is an isomorphism between H and G (exercise 2).

Now define the map $\tau: H \to K$ by $\tau = \rho^{-1} \circ \sigma$ (where $\circ$ denotes composition of functions). We already know that τ is 1–1 and onto K (exercise 2.6.9(ii)) and we leave the checking of the remaining property to exercise 2.

The second example of isomorphisms in action establishes a result which is a surprise—until you see the proof!

Theorem 5.9.5 $\langle \mathbb{R}, + \rangle \cong \langle \mathbb{R}^+, \cdot \rangle$.

Proof One easily checks that the map $\psi: \mathbb{R} \to \mathbb{R}^+$ given by $x\psi = e^x$ is an isomorphism.

(Exercise 12 asks the obvious follow-up question!)

Remark The above proof was kept to a minimum in order to impress the reader with its brevity. You might argue that I ought to show that ψ is 1–1 and onto, but these facts together with the third requirement, which is nothing more than the fact that for all $x, y \in \mathbb{R}$ we have $e^{x+y} = e^x \cdot e^y$, I feel I can leave to you solely because the proposed mapping is such a familiar one. In the case of a less familiar, or more awkwardly defined, mapping between two groups I would probably feel obliged to go into more specific details in order to convince you of my claim.

Arthur Cayley *(16 August 1821 – 26 January 1895)*
Arthur Cayley was the second son of a Yorkshire businessman who lived for some time in St Petersburg. Although born in England he spent the first eight years of his life in Russia. In 1842 he was elected a Fellow of Trinity College, Cambridge; but he left Cambridge in

1846 rather than take the Holy Orders necessary for prolonging his stay. He turned to the Law and during 14 years at the bar produced almost 300 mathematical papers. Whilst at Lincoln's Inn he became friendly with J J Sylvester who, like Cayley, was pursuing mathematics and law simultaneously. Together they essentially founded the theory of invariants. They were so prolific in this area that they were given the name 'Invariant Twins'. In 1863 Cayley took the Sadlerian chair at Cambridge where he remained until his death. Cayley, who was a good linguist with a keen interest in painting and mountaineering, wrote 966 papers. Only one book* (on elliptic functions) bears his name; but he willingly helped in the writing of several others.

Cayley contributed greatly to the geometry of curves and surfaces. In 1854 he introduced the concept of abstract group. He also wrote on matrices, determinants, quaternions, the theory of equations, dynamics, and astronomy.

---

* He did write a book on the Principles of Double Entry Bookkeeping!

As an example of the last point let us produce another surprise.

Theorem 5.9.6 (Cayley's Theorem) Let $\langle G, *\rangle$ be any group. Then $\langle G, *\rangle$ is isomorphic to a group of permutations on the set G.

Proof To each element $a \in G$ define a map $\rho_a : G \to G$ by $g\rho_a = g * a$ for all $g \in G$. Clearly ρ_a *is* a map from G to G. Further, ρ_a is a permutation on the set G. For: given any $h \in G$ one notes that $(h * a^{-1})\rho_a = h$ whence ρ_a is clearly onto G; further, from $g_1\rho_a = g_2\rho_a$ we get $g_1 * a = g_2 * a$ whence $g_1 = g_2$ [why?] so that ρ_a is 1–1.

Next, the subset $S = \{\rho_a : a \in G\}$ is a subgroup of the group of all permutations on the set G. For, given $\rho_a, \rho_b \in S$ we see that for all $g \in G$, $g(\rho_a \circ \rho_b) = (g\rho_a)\rho_b = (g * a)\rho_b = (g * a) * b = g * (a * b)$. Thus $\rho_a \circ \rho_b = \rho_{a*b}$. Similarly we find that $(\rho_a)^{-1} = \rho_{a^{-1}} \in S$. Thus $\langle S, \circ\rangle$ is a group.

Finally we claim that the mapping $\theta : G \to S$ given by $a\theta = \rho_a$ establishes the isomorphism of G and S. Briefly:

(i) θ is clearly onto.

(ii) θ is 1–1. [For: if $a\theta = b\theta$ then $\rho_a = \rho_b$. In particular we'd have $e\rho_a = e\rho_b$, i.e. $a = b$.]

(iii) Given $a, b \in G$, $(a * b)\theta = \rho_{a*b} = \rho_a \circ \rho_b = (a\theta) \circ (b\theta)$.

Remark This theorem, proved by Cayley in 1854, showed that the new 'abstract' theory of groups was subsumed under the theory of permutation groups from which the abstract theory was emerging. Since the theorem says that every finite group of order n is (isomorphically) contained in S_n it is sometimes remarked that the study of finite group theory is reduced to the study of the subgroups of every S_n. But can one really hope to find out very much about the five groups of order 8, for example, by looking at the subgroup structure of the group S_8 of order 40 320? Interestingly enough, however, 5.9.6 does give us the idea of attempting to 'represent' a given group by other groups in which more or easier calculations might be possible and a generalisation of Cayley's theorem (see 5.10.2(iii)) is useful in 6.6.3 and 6.6.4.

Exercises

1 Show that each infinite cyclic group is isomorphic to $\langle \mathbb{Z}, + \rangle$. Deduce that all infinite cyclic groups are pairwise isomorphic.

2 Show (i) that for each group G the identity map $i: G \to G$ (see 2.6.5) is an isomorphism; (ii) that if $\rho: G \to H$ is an isomorphism from G onto H then $\rho^{-1}: H \to G$ (see 2.6.5) is an isomorphism from H onto G; (iii) that if $\sigma: G \to H$ and $\tau: H \to K$ are isomorphisms then $\sigma \circ \tau: G \to K$ is an isomorphism. Deduce that 'is isomorphic to' is an equivalence relation on the class of all groups.

3 Is $A_6 \cong S_5$? (A proof, or a reason why not, is required.)

4 Show that S_6 has an abelian subgroup and also a non-abelian subgroup each of order 6. Are these subgroups isomorphic?

5 Is the map $x \to 3x$ an isomorphism (i) of $\langle \mathbb{Q}^+, \cdot \rangle$ with itself; (ii) of $\langle \mathbb{Q}, + \rangle$ with itself?

6 Is $\langle \{\hat{2}, \hat{4}, \hat{6}, \hat{8}\}, \odot \rangle \cong \langle \{\hat{1}, \hat{3}, \hat{5}, \hat{7}\}, \odot \rangle$ – numbers mod 10 and mod 8 respectively?

7 Is $\langle M(14), \odot \rangle \cong \langle \mathbb{Z}_6, \oplus \rangle$?

8 Is $\langle \mathbb{Z}, \circ \rangle \cong \langle \mathbb{Z}, + \rangle$ where $a \circ b = a + b - 7$?

9 Show that the set of all non-zero matrices of the form $\begin{pmatrix} a & b \\ -b & a \end{pmatrix}$ in $M_2(\mathbb{R})$ forms, under matrix multiplication, a group isomorphic to $\langle \mathbb{C}^\times, \cdot \rangle$.

10 Show that for every $a, b \in \mathbb{R}$ and every $n \in \mathbb{Z}^+$ there exist $x, y \in \mathbb{R}$ such that $\begin{pmatrix} x & y \\ -y & x \end{pmatrix}^n = \begin{pmatrix} a & b \\ -b & a \end{pmatrix}$.

11 (a) Is $\langle \{a + b\sqrt{2}: a, b \in \mathbb{Q}\}, + \rangle \cong \langle \{a + b\sqrt{3}: a, b \in \mathbb{Q}\}, + \rangle$?
(b) Is $\langle \{a + b\sqrt{2}: a, b \in \mathbb{Q};\ a, b \text{ not both zero}\}, \cdot \rangle \cong$
$\langle \{a + b\sqrt{3}: a, b \in \mathbb{Q};\ a, b \text{ not both zero}\}, \cdot \rangle$?

12 Is $\langle \mathbb{R}^+, \cdot \rangle \cong \langle \mathbb{Q}[x], + \rangle$? (Cf. 5.9.2(v).) Is $\langle \mathbb{Q}, + \rangle \cong \langle \mathbb{Q}^+, \cdot \rangle$? (Cf. 5.9.5.)*

13 Let ω be a complex cube root of unity and $\sqrt[3]{2}$ the real cube root of 2. Is $\langle \mathbb{Q}(\sqrt[3]{2}) \backslash \{0\}, \cdot \rangle \cong \langle \mathbb{Q}(\omega\sqrt[3]{2}) \backslash \{0\}, \cdot \rangle$? (4.3.4 may help.)

14 Prove that there is, up to isomorphism, only one group of order 3 and just two groups of order 6. Write out their multiplication tables.

* It's not enough to say $r\psi = e'$ is not an isomorphism (since e' may not belong to $\mathbb{Q}$). You must allow the possibility that a very peculiarly defined map may establish the isomorphism.

15 Do the following multiplication tables yield groups? If not, why not?

(i)

	e	*a*	*b*	*c*	*d*	*f*
e	*e*	*a*	*b*	*c*	*d*	*f*
a	*a*	*e*	*c*	*f*	*b*	*d*
b	*b*	*c*	*e*	*d*	*f*	*a*
c	*c*	*d*	*f*	*e*	*a*	*b*
d	*d*	*f*	*a*	*b*	*e*	*c*
f	*f*	*b*	*d*	*a*	*c*	*e*

(ii)

	a	*b*	*c*	*R*	*S*	*T*
a	*c*	*a*	*b*	*T*	*R*	*S*
b	*a*	*b*	*c*	*R*	*S*	*T*
c	*b*	*c*	*a*	*S*	*T*	*R*
R	*T*	*R*	*S*	*b*	*c*	*a*
S	*R*	*S*	*T*	*a*	*b*	*c*
T	*S*	*T*	*R*	*c*	*a*	*b*

(iii)

	L	*M*	*N*	*O*	*P*
L	*N*	*P*	*M*	*L*	*O*
M	*P*	*L*	*O*	*M*	*N*
N	*M*	*O*	*P*	*N*	*L*
O	*L*	*M*	*N*	*O*	*P*
P	*O*	*N*	*L*	*P*	*M*

(Note that (i), (ii) and (iii) certainly do have identity elements.)

16 Let $\langle G, \circ\rangle$ and $\langle H, *\rangle$ be groups and ψ a 1–1 map from G onto H such that for all $g_1, g_2 \in G$ we have $(g_1 \circ g_2)\psi = g_2\psi * g_1\psi$. Show that $G \cong H$.

17 Show that the group of exercise 5.3.1(f) is isomorphic to a subgroup of index 2 of the group of exercise 5.3.1(g).

18 An isomorphism $\theta: G \to G$ is called an **automorphism** of G. (Cf. exercise 5.3.7.) Show that the set of all automorphisms of G forms a group with respect to composition of functions. We call this group the **automorphism group** of G and denote it by Aut (G).

For each $g \in G$ define $\psi_g: G \to G$ by $x\psi_g = g^{-1}xg$ for all $x \in G$. Show that ψ_g is, for each g, an automorphism of G. Such an automorphism is called an **inner automorphism** of G. Show that the set Inn (G) of all inner automorphisms forms a subgroup of Aut (G).

19 Try to formulate the concept of 'isomorphism of semigroups' (see exercise 5.3.13). Let S be a semigroup with a multiplicative identity e. For each $a \in S$ define ρ_a: $S \to S$ by $s\rho_a = s * a$ for each $s \in S$. Show that $\{\rho_a: a \in S\}$ forms a semigroup with respect to composition of mappings and that this semigroup is isomorphic to S. Show that ρ_a need not be a permutation on S. What happens if S has no multiplicative identity?

20 Show that the map, $\psi: S_n \to A_{n+2}$ given by $s\psi = s$ if s is even, $s\psi = s \cdot (n+1, n+2)$ if s is odd, is an isomorphism between S_n and a subgroup of A_{n+2}. Deduce that every finite group is isomorphic to a finite group of *even* permutations.

21 Show that to each integer $n \in \mathbb{Z}^+$ there exist only a finite number of pairwise non-isomorphic groups of order n. [The present section suggests two ways in which this can be proved.]

5.10 Homomorphisms. Normal subgroups

In Section 4.2 we defined the concept of homomorphism from one ring to another. One reason was to help construct new rings from old (4.2.1, 4.5.1); another was to investigate given rings by looking at homomorphic images in which, loosely speaking, calculation is easier (4.2.10). In this section we introduce the same concept into the theory of groups where the idea is again of vital importance, as we shall see. Apart from uses within group theory of the kind just described for the case of rings, many of the applications of group theory to physics and chemistry involve 'representing' groups by mapping them homomorphically onto groups of matrices (see, for example, [62], [72]). With no more motivation we give

Definition 5.10.1 (cf. 4.2.3) Let $\langle G, \circ\rangle$ and $\langle H, *\rangle$ be groups. A mapping $\psi: G \to H$ from the set G into the set H is called a **homomorphism** iff for all $a, b \in G$ we have $(a \circ b)\psi = (a\psi) * (b\psi)$.

Remarks

(i) As with 4.2.3 we do not insist that ψ is 1–1, nor that it is onto. Of course, if it is both 1–1 and onto then ψ is an isomorphism between G and H.

(ii) The subset $G\psi = \{g\psi : g \in G\} \subseteq H$ is easily shown to be a subgroup of H. [Show it!] We refer to the group $G\psi$ as a **homomorphic image of** G. Because 'homomorphic image of' is not a symmetrical binary relation on the class of all groups we do not use expressions such as '$G\psi$ is homomorphic to G'.

(iii) You may picture a homomorphism as in Fig. 4.1.

Examples 5.10.2

(i) The map $\tau: \mathbb{Z} \to \mathbb{Z}_n$ given by $z\tau = \hat{z}$ is a rather simple example of a homomorphism from $\langle \mathbb{Z}, +\rangle$ onto $\langle \mathbb{Z}_n, \oplus\rangle$. τ is clearly not 1–1.

(ii) The map $\tau: GL_n(\mathbb{R}) \to \mathbb{R}^\times$ given by $M\tau = \det M$ is a homomorphism which is onto and, for $n \geqslant 2$, not 1–1.

(iii) The following generalises Cayley's theorem, 5.9.6. Let G be any group and H any subgroup. Let $C_0(H)$ be the set of all right cosets of H in G. For each $a \in G$ define ρ_a to be the permutation of $C_0(H)$ defined by $(Hg)\rho_a = Hga$. Letting S denote the group of all permutations on $C_0(H)$, the map $\tau: G \to S$ defined by $a\tau \to \rho_a$ is a homomorphism (exercise 19).

(iv) The map τ from the group with 16 elements in 5.3.4(l) to the group comprising the four matrices

$$\begin{pmatrix}1 & 0\\ 0 & 1\end{pmatrix}, \begin{pmatrix}1 & -2\\ 0 & -1\end{pmatrix}, \begin{pmatrix}-1 & 2\\ 0 & 1\end{pmatrix}, \begin{pmatrix}-1 & 0\\ 0 & -1\end{pmatrix}$$

(*do* these form a group?) given by

$$x\tau = \begin{pmatrix}1 & 0\\ 0 & 1\end{pmatrix} \qquad \text{if } x \in \{I, a^2, a^4, a^6\}$$

$$x\tau = \begin{pmatrix}1 & -2\\ 0 & -1\end{pmatrix} \qquad \text{if } x \in \{a, a^3, a^5, a^7\}$$

$$x\tau = \begin{pmatrix}-1 & 2\\ 0 & 1\end{pmatrix} \qquad \text{if } x \in \{b, ba^2, ba^4, ba^6\}$$

$$x\tau = \begin{pmatrix} -1 & 0 \\ 0 & -1 \end{pmatrix} \quad \text{if } x \in \{ba, ba^3, ba^5, ba^7\}$$

is a homomorphism.

In order to use the homomorphism concept we need some immediate consequences of the definition. Of course these consequences hold if, in particular, ψ is an isomorphism between G and H.

Theorem 5.10.3 (cf. 4.2.6) Let ψ be a homomorphism from the group G into the group H. Then (i) $e_G\psi = e_H$; (ii) For all $g \in G$, $g^{-1}\psi = (g\psi)^{-1}$. (We use the notation e_G, e_H for the identity elements of G and H in order to distinguish them more readily.)

Proof
(i) $e_G = e_G e_G$. Hence $e_G\psi = (e_G e_G)\psi = (e_G\psi)(e_G\psi)$. From exercise 5.4.1 it follows immediately that $e_G\psi = e_H$.
(ii) $g \cdot g^{-1} = e_G$. Hence $g\psi \cdot g^{-1}\psi = (gg^{-1})\psi = e_G\psi = e_H$. Thus $g^{-1}\psi = (g\psi)^{-1}$, by 5.4.1.

As in the case of rings the concept of homomorphism leads to

Definition 5.10.4 Let $\psi: G \to H$ be a homomorphism from G into H. The subset ker ψ of G defined by ker $\psi = \{g: g\psi = e_H\}$ is called the **kernel** of ψ.

The reader should have little difficulty in proving that ker ψ is a subgroup of G (exercise 10).

To qualify as a kernel of some homomorphism a subgroup must have special qualities. We introduce these in

Definition 5.10.5 Let $N \leqslant G$. If, for all $g \in G$, we have $N = g^{-1}Ng$ (where $g^{-1}Ng$ is defined to be the subset $\{g^{-1}ng: n \in N\}$ of G) then N is called a **normal subgroup*** of G. We write this briefly as $N \lhd G$.

Theorem 5.10.6 Let $\psi: G \to H$ be a homomorphism. Then ker ψ is a normal subgroup of G.

Proof Ker ψ is a subgroup, by exercise 10. Suppose $n \in \ker \psi$ and let $g \in G$. Then $(g^{-1}ng)\psi = (g^{-1}\psi) \cdot (n\psi) \cdot (g\psi) = (g\psi)^{-1} \cdot e_H \cdot (g\psi) = e_H$. Thus $g^{-1}ng \in$ ker ψ for all $n \in \ker \psi$. Thus for each $g \in G$ we have $g^{-1}Ng \subseteq N$. In particular, for each $g \in G$, $(g^{-1})^{-1}Ng^{-1} \subseteq N$ whence $gNg^{-1} \subseteq N$, that is, $N \subseteq g^{-1}Ng$ (see exercise 11). Thus $g^{-1}Ng = N$ for each $g \in G$, as required.

* This concept was introduced by Galois.

Examples 5.10.7
(i) If ψ is an isomorphism, or more generally if ψ is 1–1, then $\ker\psi = \langle e_G\rangle$. Clearly $\langle e_G\rangle \lhd G$.
(ii) If ψ is defined by $g\psi = e_H$ for all $g \in G$ then $\ker\psi = G$. Clearly $G \lhd G$.
(iii) The kernels of the homomorphisms described in 5.10.2 are respectively $\{zn : z \in \mathbb{Z}\}$; the set of all $M \in GL_n(\mathbb{R})$ such that $\det M = +1$; $\bigcap_{g\in G} g^{-1}Hg$ (which one can show directly is the largest normal subgroup of G which is contained in H: see exercise 19); the subgroup comprising I, a^2, a^4, a^6 (see exercise 13).
(iv) Every subgroup of an abelian group is a normal subgroup (see exercise 12).

We have shown that every homomorphic image of a group G gives rise to a normal subgroup of G. In the next section we establish the converse assertion (see 5.11.3). This 1–1 correspondence between normal subgroups and homomorphic images shows the importance attached to identifying, whenever possible, the normal subgroups of a given group. We have, in 5,10.7 (iii) and exercise 14, indicated two ways of picking up normal subgroups in any group. Another is given by

Theorem 5.10.8 Let G be a group and $N < G$ such that $|G:N| = 2$. Then $N \lhd G$.

Proof Let g be any element of G which is not in N. Since there are just two left cosets of N in G we conclude that $G = N \dot{\cup} gN$ (the dot indicating disjoint union). But also $G = N \dot{\cup} Ng$ for the same reason. Thus the subsets gN and Ng of G are identical. That is $Ng = gN$, whence (by exercise 11) $g^{-1}Ng = N$. On the other hand if $g \in N$ then $gN = N = Ng$ (exercise 5.7.3) so that $g^{-1}Ng = N$ in that case too.

The most commonly quoted instance of 5.10.8 is

Example 5.10.9 For each $n \in \mathbb{Z}^+$, $A_n \lhd S_n$.

Proof We have already seen in 5.5.11 that $|A_n| = \dfrac{n!}{2}$. Hence $|S_n : A_n| = 2$.

We finish this section by proving a result we shall call upon from time to time,

Theorem 5.10.10 Let ψ be a homomorphism from the group G into the group H. (i) if $S \leqslant G$ then $S\psi \leqslant H$; (ii) if $T \leqslant H$ then* $T\psi^{-1} \leqslant G$. Further, (iii) if $T \lhd H$ then $T\psi^{-1} \lhd G$; (iv) if $S \lhd G$ *and* ψ *is onto* H then $S\psi \lhd H$.

* Recall $T\psi^{-1} = \{g : g \in G \text{ and } g\psi \in T\}$ (2.6.5).

Remark The conclusion in (iv) is not necessarily true if ψ is not onto H (see exercise 7).

Proof of 5.10.10
(i) First note that $S\psi \neq \varnothing$ since $S \neq \varnothing$. Now let $h_1, h_2 \in S\psi$. Then there exist elements $g_1, g_2 \in S$ such that $g_1\psi = h_1$ and $g_2\psi = h_2$. Then $h_1h_2 = g_1\psi \cdot g_2\psi = (g_1g_2)\psi \in S\psi$ since $g_1, g_2 \in S$ and $S \leqslant G$. Also $h_1^{-1} = (g_1\psi)^{-1} = g_1^{-1}\psi \in S\psi$ since $g_1 \in S$ and $S \leqslant G$. Thus $S\psi \leqslant H$ by 5.6.5.
(ii) First note that $T\psi^{-1} \neq \varnothing$ since $e_G \in T\psi^{-1}$. Now if $g_1, g_2 \in T\psi^{-1}$ then there exist $h_1, h_2 \in T$ such that $g_1\psi = h_1$ and $g_2\psi = h_2$. But then $(g_1g_2)\psi = g_1\psi \cdot g_2\psi = h_1h_2 \in T$ since $h_1, h_2 \in T$ and $T \leqslant H$. Thus $g_1g_2 \in T\psi^{-1}$. Also $g_1^{-1}\psi = (g_1\psi)^{-1} = h_1^{-1} \in T$ since $h_1 \in T$ and $T \leqslant G$. Thus $g_1^{-1} \in T\psi^{-1}$ and so $T\psi^{-1} \leqslant G$ by 5.6.5.
(iii) Let $g \in G$ and let $k \in T\psi^{-1}$. Then $g\psi \in H$ and $k\psi \in T$. Thus $(g^{-1}kg)\psi = (g\psi)^{-1}k\psi(g\psi) \in T$ (since $T \lhd H$). Thus $g^{-1}kg \in T\psi^{-1}$ and hence $T\psi^{-1} \lhd G$.
(iv) We leave to the reader.

Exercises

1 Show that the maps defined in example 5.10.2(i) and (ii) are homomorphisms.

2 Let $\langle A, +\rangle$ be an abelian group. Show that for each $m \in \mathbb{Z}$ the map θ_m defined by $a\theta_m = ma$, for all $a \in A$, is a homomorphism of A to itself. If $m > 1$ can θ_m still map A *onto* itself?

3 Show that every homomorphic image of an abelian group is abelian and of a cyclic group is cyclic. Show that under a homomorphism an element of finite order n maps to an element of order m *dividing* n. Give a specific instance where $m \neq n$.

4 How many pairwise non-isomorphic groups can you find which are homomorphic images of S_3?

5 Find, or prove the impossibility of finding, homomorphisms (i) from $\langle \mathbb{Z}_{12}, \oplus\rangle$ onto $\langle \mathbb{Z}_6, \oplus\rangle$; (ii) from $\langle \mathbb{Z}_{12}, \oplus\rangle$ onto $\langle \mathbb{Z}_5, \oplus\rangle$; (iii) from $\langle \mathbb{Z}, +\rangle$ onto $\langle M(14), \odot\rangle$.

6 Let $\mathbb{C}^\infty$ be the complex plane augmented by an extra point '∞'. Let $a, b, c, d \in \mathbb{C}$ be such that $ad - bc \neq 0$. Define $f: \mathbb{C}^\infty \to \mathbb{C}^\infty$ by:

$$\left.\begin{aligned} f(z) &= \frac{az+b}{cz+d} \quad \text{if } z \neq \infty,\ z \neq -\frac{d}{c} \\ f\left(-\frac{d}{c}\right) &= \infty \\ f(\infty) &= \frac{a}{c} \end{aligned}\right\} \text{if } c \neq 0$$

and by

$$\left.\begin{aligned} f(z) &= \frac{az+b}{d} \quad \text{if } z \neq \infty \\ f(\infty) &= \infty \end{aligned}\right\} \text{if } c = 0.$$

Show that (i) under composition these mappings form a group G; and (ii) G is a homomorphic image of $GL_2(\mathbb{C})$. $\left[\text{Hint: } \begin{pmatrix} a & b \\ c & d \end{pmatrix} \to f.\right]$ Is this homomorphism an isomorphism?

7 Find groups G and H, a subgroup S of G and a homomorphism $\psi: G \to H$ such that $S \lhd G$ but $S\psi \not\!\lhd H$.

8 Give an example of a homomorphism from $\langle S_n, \cdot \rangle$ onto $\langle \{-1, 1\}, \cdot \rangle$.

9 Let U be a set of generators for a group G. Let $\psi: G \to H$ be a homomorphism. Explain why, for each $x \in G$, $x\psi$ is completely determined once $u\psi$ is known for each $u \in U$.

10 Show that the kernel of any homomorphism is a subgroup.

11 Show that if $M \subseteq G$ and $N \subseteq G$ and if $g \in G$ then (i) $gN = Ng$ iff $g^{-1}Ng = N$; (ii) $g^{-1}Mg \subseteq N$ iff $M \subseteq gNg^{-1}$.

12 Show that in an abelian group all subgroups are normal. Find an example of a non-abelian group G in which all subgroups are normal. [Hint: example 5.9.2(iv).]

13 (a) Show that the subset I, a^2, a^4, a^6 forms a normal subgroup of the group of 5.3.4(l).

(b) Show that $\langle(12)\rangle$ is not a normal subgroup of S_3. Thus $H < G$ and $|G:H| = 3$ are together not enough to imply $H \lhd G$. (Cf. 5.10.8.)

14 Show that every subgroup of $\zeta(G)$ is normal in G.

15 Show that if G has just one subgroup S of order 20 then $S \lhd G$. What significance has the integer 20?

16 Show that if $H \lhd G$ and $K \lhd G$ then $H \cap K \lhd G$, and $\langle H \cup K \rangle \lhd G$.

17 Show by example that $H \lhd K \leq G$ and $K \lhd G$ do not necessarily imply $H \lhd G$. [Hint: A_4.]

18 Show that for every group G we have $\operatorname{Inn}(G) \lhd \operatorname{Aut}(G)$. (See exercise 5.9.18.)

19 Let $H \leq G$ with $|G:H| = r$ and let $Hg_1, Hg_2, \ldots, Hg_r$ be the distinct right cosets of H in G. For each $g \in G$ let ρ_g be the map sending each Hg_i to Hg_ig. Show that each ρ_g is a permutation on the set $X = \{Hg_1, \ldots, Hg_r\}$. Show that the map $\psi: G \to S_X$ defined by $g\psi = \rho_g$ is a homomorphism whose kernel is

$\bigcap_{g\in G} g^{-1}Hg$. Deduce that $\bigcap_{g\in G} g^{-1}Hg$ is the largest normal subgroup of G to be found in H.

20 Use exercise 19 to show that if a finite group G with more than 24 elements has a subgroup of index 4 then G has a normal subgroup N such that $\langle e\rangle < N < G$. (We shall use this idea later in the case when $|G| = 36$. See 6.6.3.)

21 Let $\psi: G \twoheadrightarrow M$ be a homomorphism of the group G onto a (multiplicative) group of matrices. Given $g, h \in G$ show that det $(h\psi) =$ det $((g^{-1}hg)\psi)$ and that trace $(h\psi) =$ trace $((g^{-1}hg)\psi)$.

5.11 Factor groups. The first isomorphism theorem

In this section we construct to each normal subgroup N of a group G a homomorphic image of G with kernel N. This construction and consequent theorems are so important that we do not apologise for essentially repeating the statements and proofs of 4.2.16 and 4.2.19. Besides its theoretical use in characterising all homomorphic images of G (see 5.11.6) we also immediately illustrate the practical use of 5.11.1 and 5.11.3 by answering Problem 2 for finite abelian groups.

Theorem 5.11.1 Let G be a group and N a subgroup of G. The set of all left cosets of N in G forms a group with respect to the 'natural' multiplication, namely $aN \cdot bN = abN$, if and only if $N \lhd G$.

Proof ($\overset{\text{if}}{\Leftarrow}$) Suppose $N \lhd G$ and denote the set of (left) cosets of N in G by the symbol G/N. In attempting to define a binary operation on G/N by the above formula we first have to check that this 'product' is well defined. (See Section 2.4 and 4.2.16.) That is, we must check: if $aN = cN$ and if $bN = dN$ then $abN = cdN$. The given equalities show (see exercise 5.7.4) that $a^{-1}c \in N$ and $b^{-1}d \in N$. Since $N \lhd G$ we see that $b^{-1}(a^{-1}c)b \in N$ and hence that $b^{-1}(a^{-1}c)b \cdot b^{-1}d = b^{-1}a^{-1} \cdot cd \in N$. That is, $abN = cdN$. Thus the proposed method of multiplication on the elements of G/N is a well-defined binary operation and we only have the group axioms A, N, I left to check.

(A) Given $aN, bN, cN \in G/N$ we have $(aN \cdot bN) \cdot cN = abN \cdot cN = (ab)cN = a(bc)N = aN \cdot (bcN) = aN \cdot (bN \cdot cN)$;

(N) $N(=eN)$ is clearly such that $eN \cdot aN = aN \cdot eN = aN$ for all $aN \in G/N$;

(I) Given $aN \in G/N$ we see that $a^{-1}N \cdot aN = aN \cdot a^{-1}N = eN = N$ so that $a^{-1}N = (aN)^{-1}$, as required.

For the 'only if' ($\Rightarrow$) part of this theorem see exercise 1 below.

Definition 5.11.2 The group just constructed is called the **factor group*** (or *quotient group*) of G with respect to N.

* See the footnote to 4.2.17.

Remark The order of G/N is equal to the index of N in G: $|G/N| = |G:N|$.

We thus see the possibility of attempting proofs by mathematical induction. Theorem 5.11.4 is an example of this. First, however, we show that the group constructed in 5.11.1 fulfills the requirements of the first sentence of this section.

Theorem 5.11.3 Let G and $N \lhd G$ be given. Then there exists a homomorphism $\theta: G \to G/N$ such that $\ker \theta = N$ exactly.

Proof We define θ (clearly it *is* a map) from G to G/N by $g\theta = gN$, for all $g \in G$. For all $g, h \in G$ we have $(gh)\theta = ghN = gN \cdot hN = g\theta \cdot h\theta$ as required. Clearly $g\theta = e_{G/N}$ if and only if $gN = N$, that is, if and only if $g \in N$ (exercise 5.7.4). Thus $\ker \theta = N$. (θ is called the *natural homomorphism* from G onto G/N.)

Problem 5 Will *every* homomorphism from G onto G/N have kernel *exactly* equal to N?

As an application of 5.11.3 we show that, for abelian groups at least, Lagrange's Theorem has a converse. First we prove

Theorem 5.11.4 Let G be a finite abelian group and let the prime $p \in \mathbb{Z}^+$ be such that $p \,|\, |G|$. Then G has an element of order p.

Proof If the theorem is false then one can choose a counterexample C, say, of smallest possible order. Clearly $|C| \neq p$. Now let x be any non-identity element of C and suppose the order of x is n. Then $p \nmid n$ (or else $x^{n/p}$ is an element of C of order p).

Set $N = \langle x \rangle$. Then $N \lhd C$ (since C is abelian) and so we can form* $\frac{C}{N}$. Since $p \,|\, |C| = |C:N||N|$ we see that $p \,|\, |C:N| = \left|\frac{C}{N}\right| < |C|$. By the choice of C we see that $\frac{C}{N}$ *does* contain an element yN, say, of order p. Thus $(yN)^p = e_{C/N} = N$. It follows that $y^p \in N$ (exercise 3) and hence that $y^p = x^r$ for some r. Consequently $y^{np} = x^{nr} = e$. Thus y^n has order p or order 1. But C has no elements of order p and so $y^n = e$. But then yN has order dividing n (exercise 5.10.3) so that $p|n$ after all. This contradiction shows no such counterexample C can exist and so the theorem is proved.

And now for the converse of Lagrange's Theorem for finite abelian groups. (See Problem 2.)

Theorem 5.11.5 Let G be a finite abelian group. For each $m \in \mathbb{Z}^+$ such that $m \,|\, |G|$, G has a subgroup of order m.

* We occasionally write $\frac{C}{N}$ as an alternative to C/N.

Proof Suppose that the theorem is false and let C be a counterexample of smallest possible order. Suppose that $k \,|\, |C|$ but C has no subgroup of order k. Let $p \in \mathbb{Z}^+$ be a prime such that $p|k$. By 5.11.4, C has a (normal) cyclic subgroup $N = \langle x \rangle$, say, of order p. Now $\left|\frac{C}{N}\right| < |C|$ and $\frac{k}{p} \,\Big|\, \left|\frac{C}{N}\right|$. By choice of C, $\frac{C}{N}$ has a subgroup M, say, of order $\frac{k}{p}$. It is not difficult to check (see 5.10.10(ii) and exercise 8) that under the natural map $\psi: C \to \frac{C}{N}$ the inverse image $M\psi^{-1}$ of M is a subgroup of order $\frac{k}{p} \cdot p = k$. Thus no such counterexample C can exist and the theorem is proved.

Actually Problem 2 has, in general, a negative answer. For suppose N were a subgroup of order 6 in A_4. Then $|A_4 \colon N| = 2$ since $|A_4| = 12$. It follows (see 5.10.8) that $N \lhd A_4$. Now $(123) \in A_4$ hence $(123)^2 \in N$ (since $\left|\frac{A_4}{N}\right| = 2$). Thus $(123) = (123)^4 \in N$. Consequently N contains I, (123), $(123)^2 = (132)$ and similarly (124), (142), (134) and (143) which is already too many elements. It follows that Lagrange's Theorem has a converse for finite abelian groups but not for finite groups in general.

Remark We shall improve substantially upon 5.11.4 in the direction of non-abelian groups in 6.2.8. (See also Theorem H of Section 6.5.)

We have seen in 5.11.3 that every group of the form G/N, where $N \lhd G$, is a homomorphic image of G. We now show that, conversely, the set of all homomorphic images of G is essentially restricted to groups of this type. Applications of this theorem are given in exercise 6 below, and in exercises 6.2.17, 6.3.16 and 6.3.17.

Theorem 5.11.6 (cf. 4.3.1) **(The First Isomorphism Theorem)** Let G be a group and ψ a homomorphism of G onto H with kernel K. Then $H \cong G/K$.

Proof Define a map $\theta: H \to G/K$ by: If $h = g\psi$ set $h\theta = gK$. We must first check that θ is indeed a map.* Thus suppose $h = g_1\psi = g_2\psi$. Then $(g_1^{-1}g_2)\psi = e_H$ so that $g_1^{-1}g_2 \in \ker \psi = K$. Consequently $g_1K = g_2K$ so that θ is well defined. [Do you understand that? Are you sure?]

Clearly θ is onto G/K. [If $gK \in G/K$ then $h\theta = gK$ if we choose $h = g\psi$.]

θ is 1–1. [If $h_1\theta = h_2\theta$ where $h_1 = g_1\psi$ and $h_2 = g_2\psi$, then $g_1K = h_1\theta = h_2\theta = g_2K$. Thus $g_1^{-1}g_2 \in K = \ker \psi$. Therefore $g_1\psi = g_2\psi$; that is $h_1 = h_2$.]

θ is a homomorphism. [With notation as above $h_1\theta \cdot h_2\theta = g_1K \cdot g_2K = g_1g_2K = ((g_1g_2)\psi)\theta = (g_1\psi \cdot g_2\psi)\theta = (h_1h_2)\theta$, as required.]

* That is, we must check θ is well defined.

Exercises

1 Show that if one is to make a group out of the left cosets of the subgroup N of the group G by defining, for $a, b \in G$, $aN \cdot bN = abN$, then it is necessary that N be normal in G. [Hint: $aN = cN$, $bN = dN$ imply $a^{-1}c \in N$ and $b^{-1}d \in N$. Hence if $b^{-1}a^{-1}cd = b^{-1}(a^{-1}c)b \cdot b^{-1}d \in N$ we must have $b^{-1}(a^{-1}c)b \in N$. Let a be any element of N, $c = 1$ and b any element of G. Cf. exercise 4.2.16.]

2 Prove that $V = \{I, (12)(34), (13)(24), (14)(23)\}$ is a normal subgroup of S_4. To which group, with which we are already familiar, is S_4/V isomorphic? [Hint: What is $|S_4/V|$?]

3 Let $N \lhd G$. Show that if gN has order n in G/N then $g^m \in N$ iff $n|m$.

4 Let T be the set of all elements of finite order in the (possibly infinite) abelian group A. Show that T is a (normal) subgroup of A and prove that A/T has no elements of finite order (other than the identity element).

5 Let $N \lhd G$ and $H \leq G$ with $N \leq H$. Define $\frac{H}{N}$ to be the subset $\{hN : h \in H\}$ of $\frac{G}{N}$. Show that $\frac{H}{N} \leq \frac{G}{N}$ and that $\frac{H}{N} \lhd \frac{G}{N}$ iff $H \lhd G$. Show that if $S \leq \frac{G}{N}$ then there exists $K \leq G$ such that $N \leq K$ and $S = \frac{K}{N} = \{kN : k \in K\}$.

6 Show that, for each group G, Inn $(G) \cong G/\zeta(G)$. [Hint: map $G \to \text{Inn } G$ by $g \to \psi_g$ (see exercise 5.9.18) and show that the kernel is $\zeta(G)$. Now use 5.11.6.]

7 (a) If $N_1 \lhd G$ and $N_2 \lhd G$ and if $N_1 \cong N_2$ is $\frac{G}{N_1} \cong \frac{G}{N_2}$?

(b) If $N_1 \lhd G$ and $N_2 \lhd G$ and if $\frac{G}{N_1} \cong \frac{G}{N_2}$ is $N_1 \cong N_2$? [These may prove considerably easier when the reader has assimilated Section 6.3.]

8 Let $\psi : G \to H$ be a homomorphism with kernel K of G onto H, both finite groups. If $T \leq H$ show that $|T\psi^{-1}| = |T| \cdot |K|$.

9 Prove the **Third* Isomorphism Theorem** (cf 4.3.3). Let $N \lhd G$, $K \lhd G$ with $N \leq K$. Show that $\theta : \frac{G}{N} \to \frac{G}{K}$ given by $(gN)\theta = gK$ is well defined and that it is a homomorphism. Show that ker $\theta = \frac{K}{N}$ (cf. exercise 5) and deduce that $\frac{G}{K} \cong \frac{G/N}{K/N}$.

* The Second Isomorphism Theorem is proved in Section 6.5.

5.12 Space groups and plane symmetry groups

According to [132, p. 290], the first direct application of group theory to an important problem in the natural sciences was in the determining, by the Russian crystallographer Fedorov,* in 1890, of all the possible symmetry groups of 3-dimensional repeating patterns. The interest in such patterns arose from the then widely held view, not confirmed until 1912, that crystals were assemblages of very small basic units of matter repeated periodically in 3 dimensions. The connection with group theory is that each pattern can be characterised by its symmetry group. (One must take a little care. See the exercise at the end of this section.) It turns out that there are only 230 of these so-called *crystallographic space groups* amongst which are 22 which crystallographers prefer to regard as distinct but which, from an abstract point of view, form 11 pairs of isomorphic groups. Thus the space groups fall into 219 isomorphism classes [70]. The enumeration of these space groups is built upon the 14 lattices determined by Bravais in 1848 and the 32 crystallographic point groups (which themselves fall into 18 distinct isomorphism classes [71]) already known to Hessel in 1830. Since this enumeration is quite complicated ([63]) we here look at some of the corresponding ideas involved in the analogous 2-dimensional problem where 17 groups, no two of which are isomorphic ([10]), arise. We leave the reader to find the not too difficult details in, for example, [69] and [73]. (This section is just for fun. You are left to make what you will of most of the provocative statements in the text. They are left to you as exercises.)

First recall that an *isometry* of the plane $\mathbb{R}^2$ is a distance preserving mapping τ, say, of $\mathbb{R}^2$ onto itself. Amongst such isometries are translations, rotations, reflections (in lines) and glide reflections. Figure 5.7 adequately describes these movements, a glide reflection being the result of an ordinary reflection in some line l followed by a translation parallel to l. Translations and rotations preserve orientation, reflections and glide reflections clearly reverse it; note the arrowheads on the circles in Fig. 5.7(c), (d).

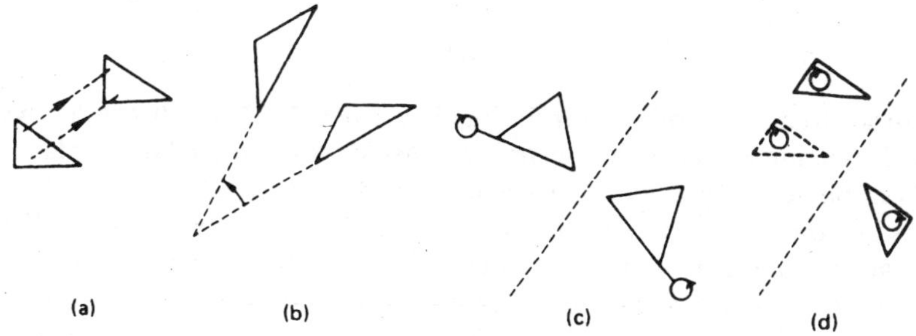

Fig. 5.7

* Western authors tend to credit A Schönflies and W Barlow with their independent discovery. In fact it appears both Fedorov and Schönflies made errors which were eliminated by their corresponding with one another.

It is not difficult to prove ([69, p. 98]).

Theorem 5.12.1 Every proper (i.e. sense preserving) isometry of $\mathbb{R}^2$ is of type (a) or (b); every improper (i.e. sense reversing) isometry of $\mathbb{R}^2$ is of type (c) or (d).

With an appropriate definition of 'product' the set E of all isometries of $\mathbb{R}^2$ forms a group (5.3.4(l)). As is easily checked, the product of two proper, or of two improper, isometries is a proper isometry whilst the product, in either order, of a proper and an improper isometry is an improper isometry ([69, p. 98]).

The 2-dimensional repeating patterns we consider are commonly called *wallpaper patterns* (guess why!). By definition these are patterns in which it is possible to find a *basic pattern unit* repeated periodically but not 'continuously' in each of two non-parallel directions. Thus we are interested in patterns such as that in Fig. 5.8(i) and not those in Fig. 5.8(ii), (iii).

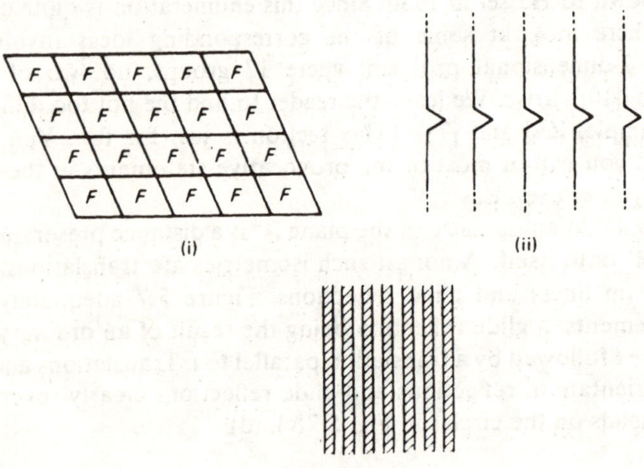

Fig. 5.8

In order to accommodate translational symmetries in our discussion we are obliged to assume that the patterns we consider completely fill the plane.

It is implicit in Fig. 5.8(i) that we are dealing with what are called discrete groups of isometries of $\mathbb{R}^2$. Formally: A group G of isometries of $\mathbb{R}^2$ is **discrete** iff to each point $p \in \mathbb{R}^2$ there exists a circle C_p with centre p such that for each $g \in G$ either p is not moved by g or it is moved to a point, pg, outside C_p. Thus the symmetry group of the circle $x^2 + y^2 = 1$ is not discrete since it contains arbitrarily small (rotational) isometries.

Now the complete symmetry group of a wallpaper pattern is a special kind of discrete group known as a plane symmetry group.* By definition, a **plane**

* Also called a plane, or 2-dimensional, crystallographic group.

symmetry group is a discrete subgroup of E which contains translations s, t in non-parallel directions. Let G be such a group and let $p \in \mathbf{R}^2$. Applying the isometries $s^i t^j$ (each such is again a translation) to p produces an infinite lattice of points (Fig. 5.9).

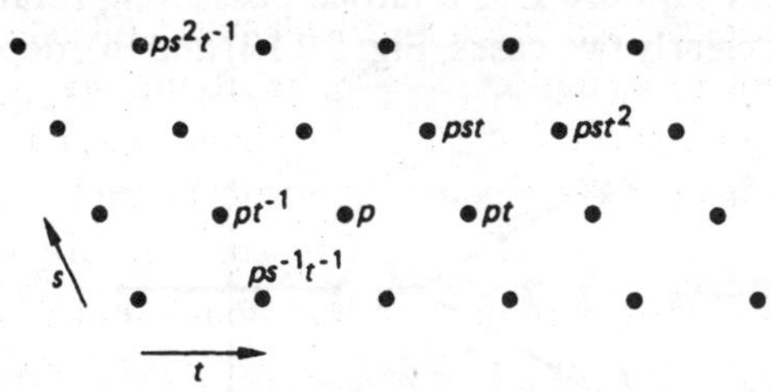

Fig. 5.9

If we take s, t to be non-parallel translations of smallest possible magnitudes [how do we know that such 'smallest' translations exist?] it is not difficult to check that every translation in G is of the form $s^i t^j$ ([69, p. 114]). That is, the subgroup of all translations of G requires no more than two generators.

Now there do exist (infinite) wallpaper patterns which possess only translational symmetry (Fig. 5.8(i)). On the other hand some patterns also exhibit rotational or reflectional symmetry (Fig. 5.13(ii), (iii) for example).

Suppose, then, that G contains rotations too. Let t be the smallest translation in G and let r be a rotation (centre c_0) in G of smallest (positive) angle $\theta = \dfrac{2\pi}{n}$, say. [Why does such a smallest exist; why does it take the form $\dfrac{2\pi}{n}$?

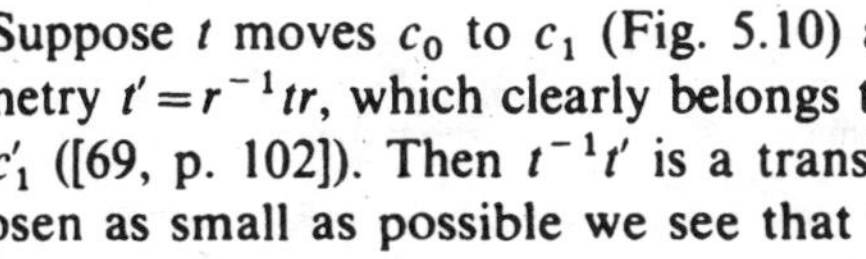

Fig. 5.10

See [132, p. 286].] Suppose t moves c_0 to c_1 (Fig. 5.10) and that r moves c_1 to c_1'. Now the isometry $t' = r^{-1}tr$, which clearly belongs to G, is a translation which sends c_0 to c_1' ([69, p. 102]). Then $t^{-1}t'$ is a translation sending c_1 to c_1'. Since t was chosen as small as possible we see that the distance c_1c_1' is no shorter than c_0c_1. It follows that angle $c_1c_0c_1'$ is at least 60°. Thus $\dfrac{2\pi}{n} \geqslant \dfrac{\pi}{3}$ whence $n \leqslant 6$. The reader is invited to prove that the case $n = 5$ is also impossible ([69, p. 111]). We thus have

Theorem 5.12.2 (The Crystallographic Restriction) Let G be a plane symmetry group. Each rotation of G necessarily has order 1, 2, 3, 4 or 6.

A consequence of 5.12.2 is that there are only 5 basic types of lattice which can underlie a plane symmetry group. Let us first consider what possible symmetries, apart from translations, a lattice of points as in Fig. 5.9 may have. Clearly every lattice has rotational symmetry through angle π (so-called diad rotation). Next suppose L is a lattice possessing rotational symmetry of order 3. There are clearly two cases, Fig. 5.11(i) and (ii), depending on whether

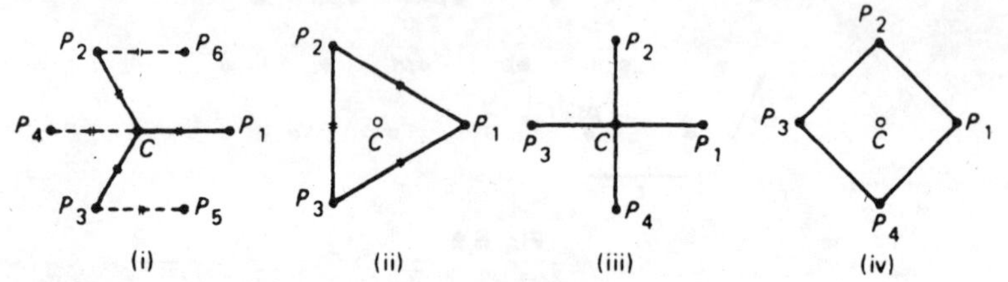

Fig. 5.11

or not the rotation centre C is or is not a lattice point. In fact the two cases give rise to identical lattices* comprising equilateral triangles. Clearly this lattice has 6-fold rotational symmetry about each of its lattice points—and in fact any 'hexad' lattice must be of this type ([69, p. 112]). A similar analysis shows that a lattice with 4-fold rotational symmetry must be a square lattice (Fig. 5.11(iii) and (iv)).

Lattices possessing reflectional symmetry in a line (which need not contain any lattice point) must be made up of rectangles or rhombuses ([69, p. 112]). In this latter case the lattice is often thought of as 'centred rectangular' (Fig. 5.12). Finally, if a lattice has glide-reflectional symmetry it is necessarily of centred rectangular type ([69, p. 113]).

Thus there are only 5 kinds of lattices when we distinguish them by their symmetries, namely those in Fig. 5.12. These are the 2-dimensional analogues of the 14 Bravais lattices mentioned earlier.

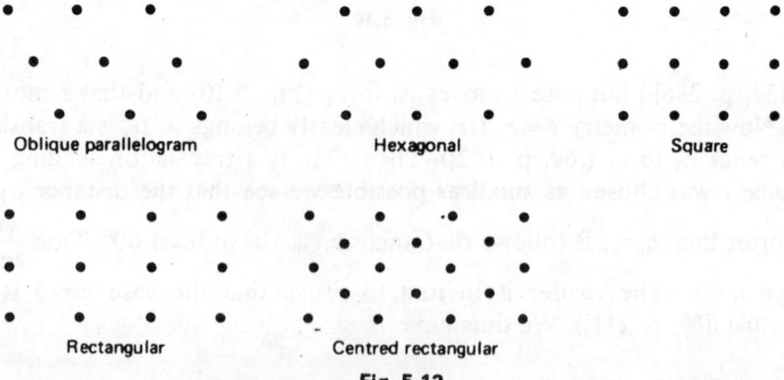

Fig. 5.12

* In (i) P_4, P_5, P_6 arise from P_1, P_2, P_3 under translations.

We now make

Definition 5.12.3 A 2-dimensional crystallographic **point group** K is a group of isometries of $\mathbb{R}^2$ which fixes a point p and maps a 2-dimensional lattice containing p into itself.

In any such group there can be neither translations nor glide reflections (why not?). Consequently either all the elements of K are rotations or one half of them are rotations and the other half reflections (cf. the proof of 5.5.11). It follows from 5.12.2 that K is (isomorphic to) one of the cyclic groups C_n or one of the dihedral* groups D_n where $n = 1, 2, 3, 4$ or 6.

We now show how each plane group G determines a crystallographic point group as a homomorphic image. We choose some point O, say, in the plane. Note that if ρ is any isometry of $\mathbb{R}^2$ which moves the point O to the point a and if t is the translation of $\mathbb{R}^2$ moving O to a then $s = \rho t^{-1}$ moves O to O, that is, s is either a rotation about O or a reflection in a line through O. Further we may write $\rho = st$. Moreover the subgroup T of translations of $\mathbb{R}^2$ is a normal subgroup of E (cf. the proof of 5.12.2). It is then immediate that $H = T \cap G$, that is, the subgroup of translations in G is a normal subgroup of G.

Now suppose that $g_1 = s_1t_1$ and $g_2 = s_2t_2$ are elements of G written in the above form. (Note: it isn't asserted that either s_1 or t_1 lies in G.) Then $g_1g_2 = s_1s_2 \cdot s_2^{-1}t_1s_2 \cdot t_2$. Here $s_2^{-1}t_1s_2$ is a translation (cf. the proof of 5.12.2), whilst s_1s_2 fixes O (since both s_1 and s_2 do). In a similar manner $g_1^{-1} = (s_1t_1)^{-1}$ may be written $s_1^{-1} \cdot s_1t_1^{-1}s_1^{-1}$, s_1^{-1} being a rotation about O, $s_1t_1s_1^{-1}$ being a translation. We thus see that the set of all s which appear as 'rotation-reflection components' of elements of G forms a group K fixing the point O. Further the map $\theta: G \to K$ given by $g_1\theta = (s_1t_1)\theta = s_1$ is a homomorphism of G onto K with $\ker \theta = H$. Finally let the elements of H be applied to the point O to produce the lattice L. Suppose $a \in L$ so that $a = Ot$ for suitable $t \in H$. Let $s \in K$. Then there is $g \in G$ such that $g = st_1$ for suitable $t_1 \in T$. Then $as = Ots = Otgt_1^{-1} = Og \cdot g^{-1}tg \cdot t_1^{-1} = Ost_1 \cdot g^{-1}tg \cdot t_1^{-1}$. But s fixes O and $g^{-1}tg$ translates the lattice into itself. Thus $as = Og^{-1}tg$ is a point of L. Thus $K \cong G/H$ is a point group mapping the lattice determined by H and O to itself.

The above analysis shows how we should be able to find systematically all possible plane groups starting from the five lattices and the ten point groups. The method is indicated in [70] and [120]. Bearing in mind the remark in [120] that 'the passive contemplation of wallpaper patterns ... is not mathematics' we invite the reader whom we have managed to interest to read [73] or [120] and look at the 3-dimensional case in [70]. A more geometrical approach is given in [69] (see also [63]).

There follows a description of the 17 plane groups in pictorial terms and also in terms of generators and relations (see [10]).

* D_n is, for each $n \in \mathbb{Z}^+$, the group of $2n$ symmetries of a regular n-gon (see 5.3.4(l)). The 1-gon and 2-gon are pictured respectively as and .

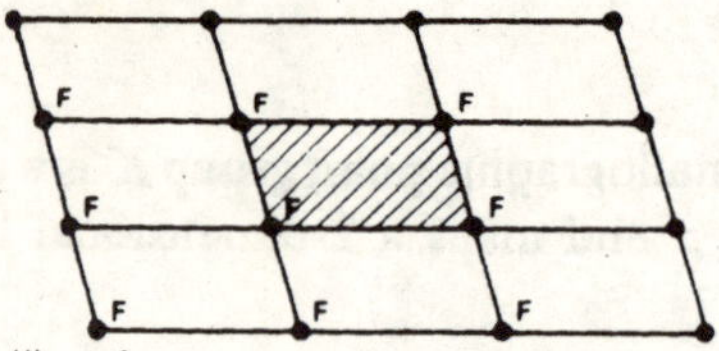

(i) $p1$

generated by two translations

$\langle x,y : xy = yx \rangle$

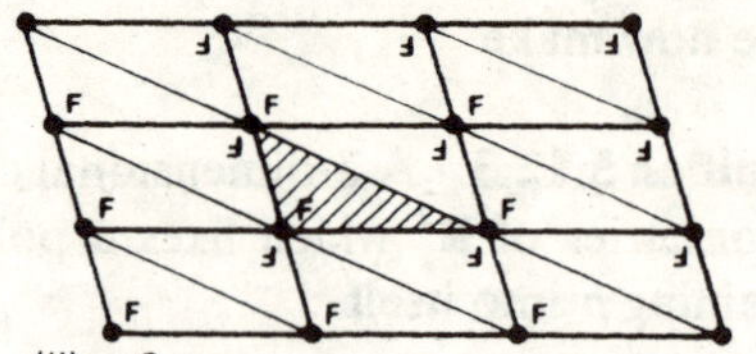

(ii) $p2$

generated by three half turns

$\langle t_1,t_2,t_3 : t_1^2 = t_2^2 = t_3^2 = (t_1t_2t_3)^2 = 1 \rangle$

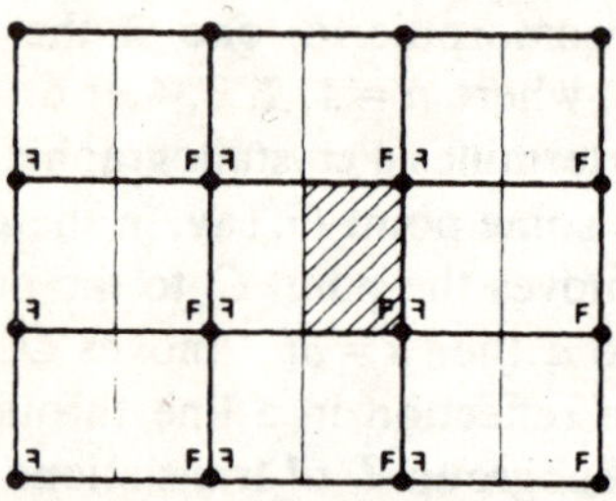

(iii) pm

generated by two reflections and a translation

$\langle r_1, r_2, y : r_1^2=r_2^2 = 1, r_1y = yr_1, r_2y = yr_2 \rangle$

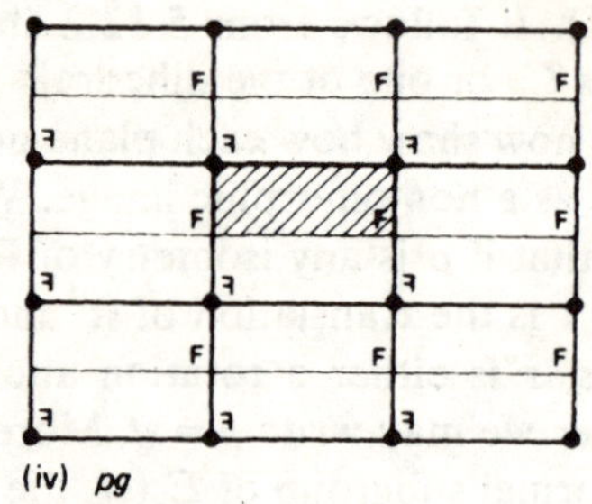

(iv) pg

generated by two parallel glide reflections

$\langle p,q : p^2 = q^2 \rangle$

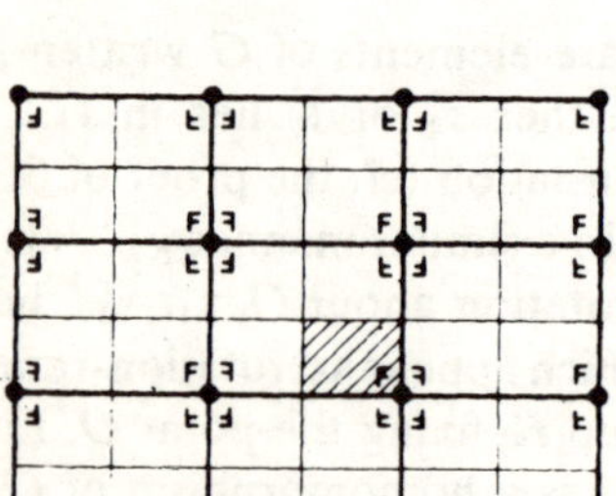

(v) pmm

generated by four reflections

$\langle r_1, r_2, r_3, r_4 : r_1^2 = r_2^2 = r_3^2 = r_4^2 = (r_1r_2)^2 = (r_2r_3)^2=(r_3r_4)^2=(r_4r_1)^2 = 1 \rangle$

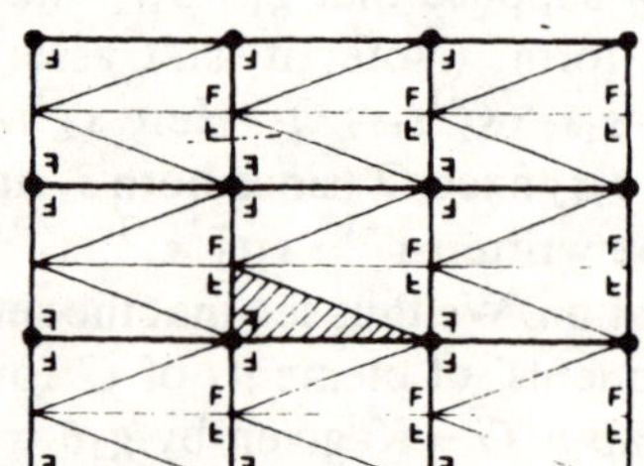

(vi) pmg

generated by a reflection and two half turns

$\langle r, t_1, t_2 : r^2 = t_1^2 = t_2^2 = 1, t_1rt_1 = t_2rt_2 \rangle$

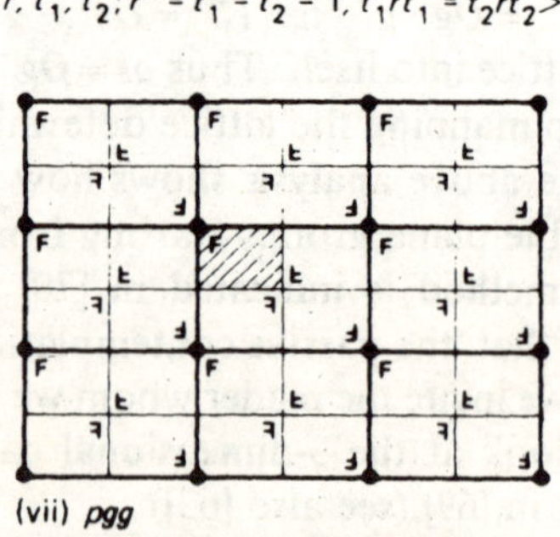

(vii) pgg

generated by two perpendicular glide reflections

$\langle p,q : (pq)^2 = (p^{-1}q)^2 = 1 \rangle$

Fig. 5.13

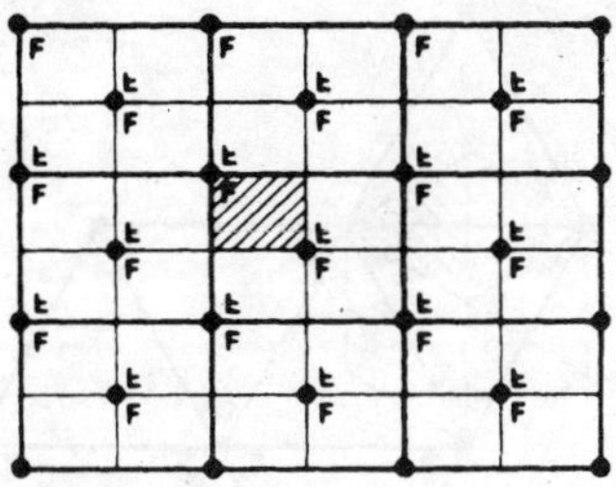

(viii) *cm*
generated by a reflection and a parallel glide reflection
$\langle r, p : r^2 = 1, rp^2 = p^2 r \rangle$

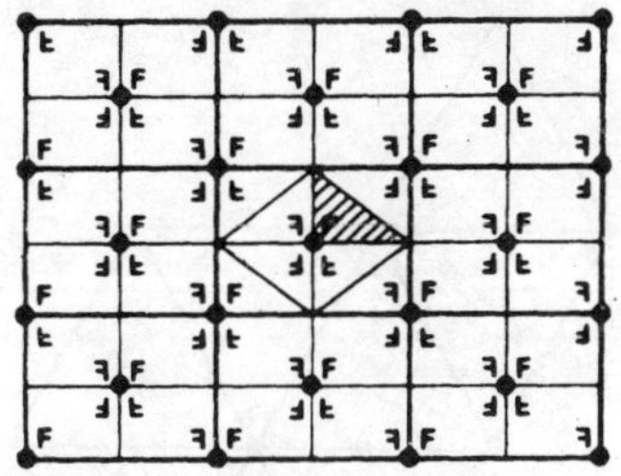

(ix) *cmm*
generated by two perpendicular reflections and a half turn
$\langle r_1, r_2, t ; r_1^2 = r_2^2 = t^2 = (r_1 r_2)^2 = (r_1 t r_2 t)^2 = 1 \rangle$

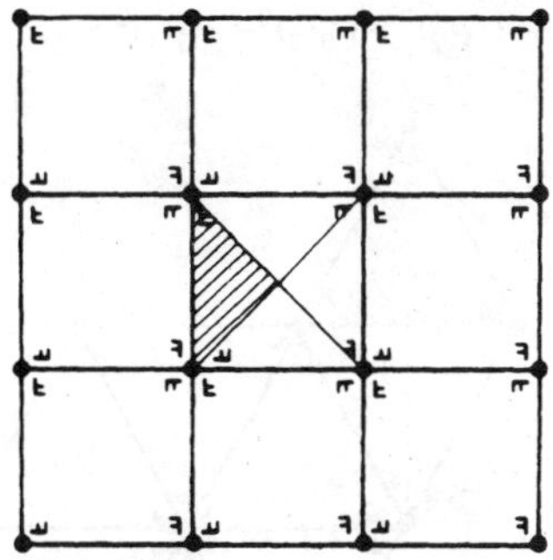

(x) *p*4
generated by a half turn and a quarter turn
$\langle s, t ; s^4 = t^2 = (st)^4 = 1 \rangle$

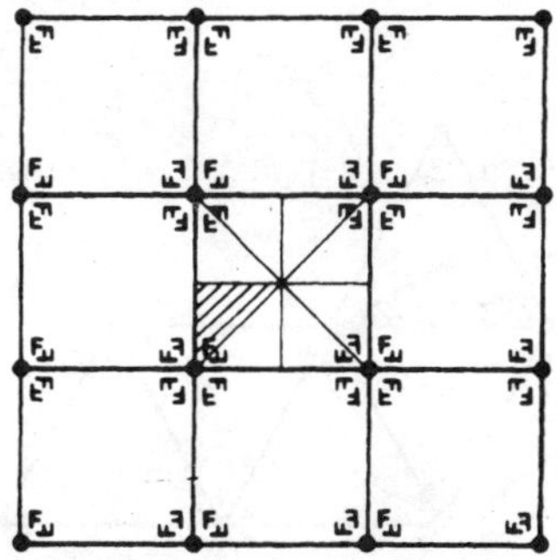

(xi) *p*4*m*
generated by reflections in the sides of a (45°, 45°, 90°) triangle
$\langle r_1, r_2, r_3 ; r_1^2 = r_2^2 = r_3^2 = (r_1 r_2)^4 = (r_2 r_3)^4 = (r_3 r_1)^2 = 1 \rangle$

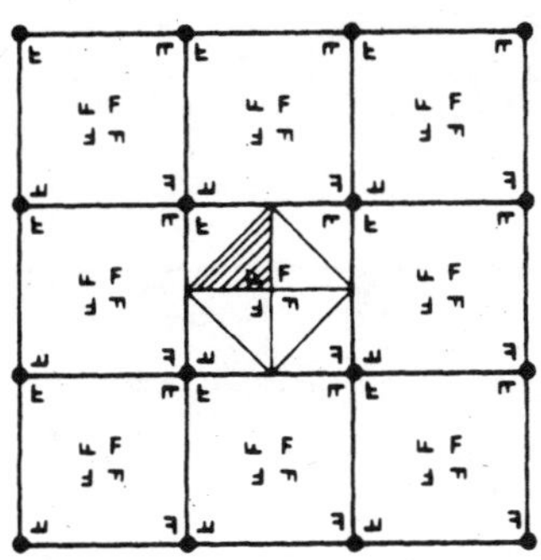

(xii) *p*4*g*
generated by a reflection and a quarter turn
$\langle r, s ; r^2 = s^4 = (s^{-1} r s r)^2 = 1 \rangle$

Fig. 5.13 (*cont.*)

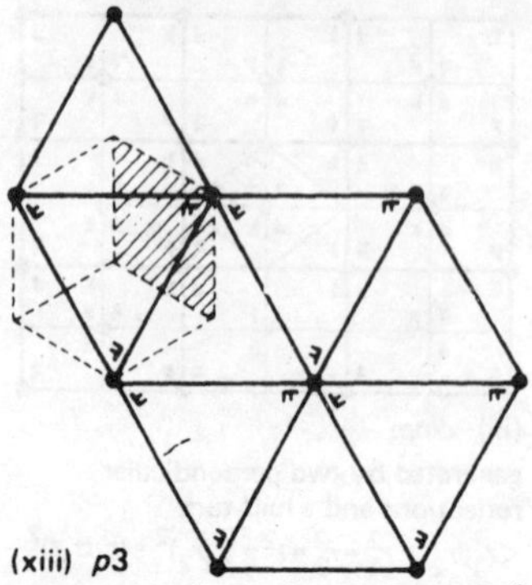

(xiii) *p*3
generated by 3 rotations through 120°
$< s_1, s_2, s_3 : s_1^3 = s_2^3 = s_3^3 = s_1 s_2 s_3 = 1 >$

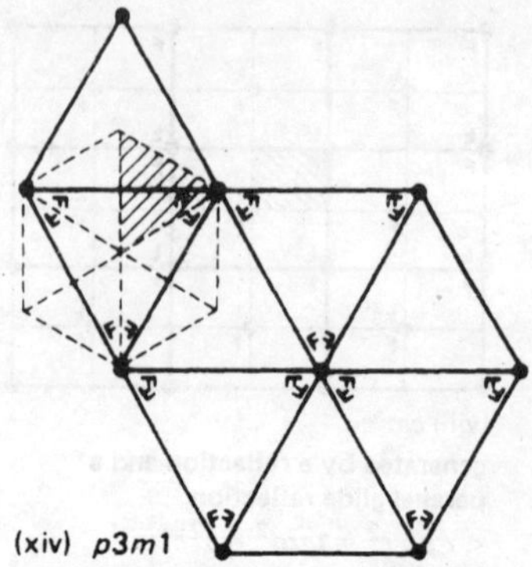

(xiv) *p*3*m*1
generated by three reflections in the sides of an equilateral triangle
$< r_1, r_2, r_3 : r_1^2 = r_2^2 = r_3^2 = (r_1 r_2)^3 = (r_2 r_3)^3 = (r_3 r_1)^3 = 1 >$

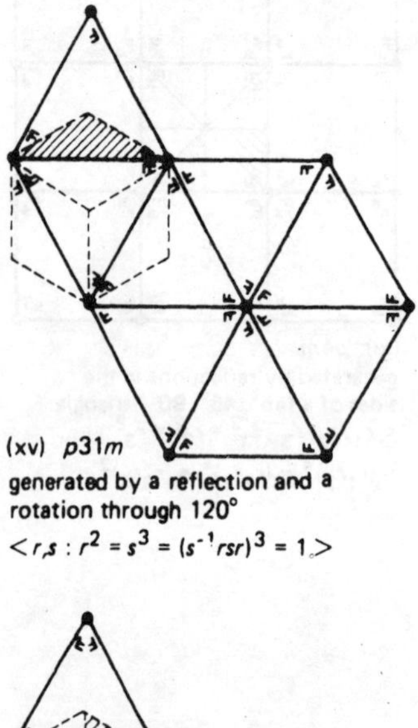

(xv) *p*31*m*
generated by a reflection and a rotation through 120°
$< r, s : r^2 = s^3 = (s^{-1} r s r)^3 = 1 >$

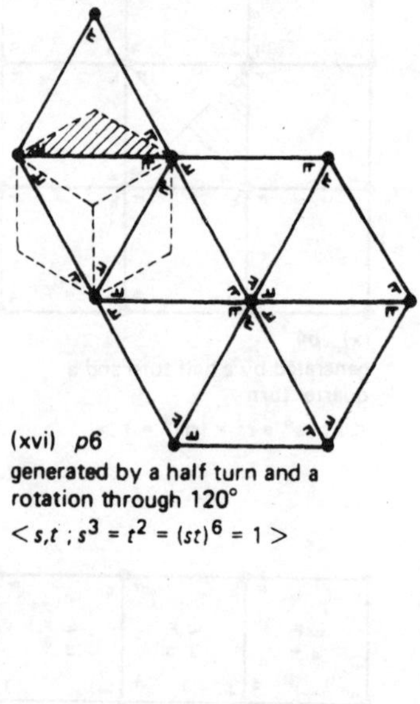

(xvi) *p*6
generated by a half turn and a rotation through 120°
$< s, t ; s^3 = t^2 = (st)^6 = 1 >$

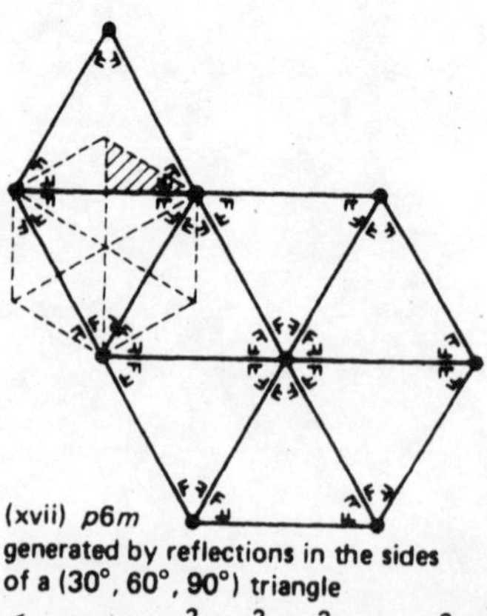

(xvii) *p*6*m*
generated by reflections in the sides of a (30°, 60°, 90°) triangle
$< r_1, r_2, r_3 : r_1^2 = r_2^2 = r_3^2 = (r_1 r_2)^3 = (r_2 r_3)^6 = (r_3 r_1)^2 = 1 >$

Fig. 5.13 (*cont.*)

In each part of Fig. 5.13 the shaded region is a fundamental region in the sense that when the elements of the group in question are applied to it the resulting regions completely fill the plane without overlap. No doubt you will be able to think of different systems of generators and relations to those given.

In view of the remarks at the end of Section 6.1 it is worthwhile reiterating the earlier remark that the enumeration of the 230 space groups was completed more than 20 years before the assumption of regularity of crystal structure on which it was based was confirmed ([64]).

Exercise Try to find all *frieze patterns.* These are the 2-dimensional repeating patterns whose symmetry groups are discrete and infinite but also leave a line in $\mathbb{R}^2$ fixed. In such groups the subgroup of translations must be isomorphic to the infinite cyclic group. There are seven distinct frieze patterns; their symmetry groups fall into four isomorphism classes. We give you two of the seven to get you started (see Fig. 5.14). The first has translations and the

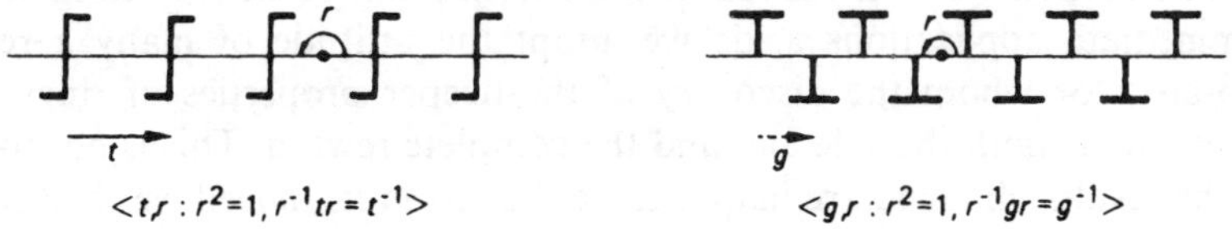

Fig. 5.14

second has glide reflections, and not *vice versa.* Although their symmetry groups are abstractly isomorphic there is (trivially) no isomorphism between the groups under which isometries in one group map to isometries *of the same kind* in the other group. This finer analysis distinguishes the symmetry groups of these patterns and hence the patterns themselves.

A derivation of these patterns can be found in [69].

6

Structure theorems of group theory

6.1 Introduction

In much of what we have done so far, applications of the work have been at the backs of our minds. In Chapters 2 and 3 we gave applications of algebra to number theory whilst in Chapter 4 we applied our theory to geometrical construction problems and to polynomial equations over $\mathbb{C}$. In Chapter 5 we saw the beginnings of a connection between solving equations by radicals and groups of permutations, a theme we shall greatly enlarge upon in Chapter 7.

In this chapter we offer the reader a different point of view. Putting thoughts of immediate applications aside we adopt the attitude of many a research algebraist, for whom the discovery of the deeper properties of rings, fields, groups, etc. is both the sole aim and the complete reward. This is not to imply that the results obtained are inapplicable. Far from it. Indeed we shall see how mere intellectual curiosity leads us, quite naturally, to the concept of soluble group (Section 6.5)—which, to aid his researches, Galois would surely have pounced on (see 7.9.3, 7.10.5) if only he hadn't been obliged to invent it for himself (see 6.5.7) some decades earlier.

We invite the reader to pause from time to time to experience a sense of wonderment that such deep and beautiful results as are described here are hidden in the very simple list, 5.3.1, of axioms.

6.2 Normaliser. Centraliser. Sylow's theorems

In this section we penetrate the subgroup structure of finite groups by proving the three famous theorems of Sylow* (1872). These theorems yield a tremendous amount of information about the subgroups of prime power order in any finite group. This information will be made great use of in this section and in Section 6.6.

We begin with

Definition 6.2.1 Let G be a group such that every element of G has p-power order for some fixed prime p. Then G is called a ***p*-group**.

Remarks

(i) 6.2.1 allows for infinite p-groups. Such objects do exist but, on the whole, they are not especially nice—see the remark following 6.2.9. We shall essentially deal only with finite p-groups.

* Peter Ludvig Mejdell Sylow (12 December 1832 – 7 September 1918).

(ii) By Lagrange's Theorem any finite group of order p^n is a (finite) p-group.

Example 6.2.2 S_3 is not a 2-group, nor a 3-group. For each prime p the multiplicative group of all complex pth power roots of 1 is an infinite p-group.

Problem 1 Can you show that a finite p-group must have order a power of p? (This can't be difficult—or can it?)

Definition 6.2.3
(i) Let H be a subset of the group G. The subset $g^{-1}Hg = \{g^{-1}hg: h \in H\}$ is called the **conjugate** of H by g in G. We denote $g^{-1}Hg$ briefly by H^g.
(ii) If H, K are subsets of G we say that K is **conjugate to** H in G if there exists in G an element g such that $H^g = K$. It then follows that $K^{g^{-1}} = H$. Hence we may fairly say that H and K are **conjugate** in G.
(iii) If $H^g = K$ where H (and hence K) has one element, $H = \{x\}$, $K = \{y\}$ say, we write $x^g = y$ rather than $\{x\}^g = \{y\}$ and say that y is **conjugate to** x in G. Copying (ii) we see that x is then conjugate to y in G and say briefly that x and y are **conjugate** in G.
(iv) The subset $N_G(H) = \{g: g \in G \text{ and } H^g = H\}$ is called the **normaliser** of H in G. If H comprises the single element x we have $N_G(\{x\}) = \{g: g \in G$ and $g^{-1}xg = x\} = \{g: g \in G \text{ and } xg = gx\}$. We call $N_G(\{x\})$ the **centraliser*** of x in G and denote it by $C_G(x)$.

Remarks
(i) Conjugacy, as defined in 6.2.3, determines an equivalence relation on the set of all subsets of G and also on the set of all subgroups of G. 6.2.3 (iii) determines an equivalence relation on G (exercise 3). In each case the corresponding equivalence classes are called **conjugacy classes**.
(ii) It follows easily that $N_G(H) \leqslant G$. If $H \leqslant G$ then $N_G(H)$ is the unique largest subgroup of G in which H is a normal subgroup (exercise 6). $C_G(x)$ is also a subgroup of G.

Examples 6.2.4
(i) In S_3 the subsets $\{(12)\}$, $\{(23)\}$, $\{(31)\}$ (equivalently, the elements (12), (23), (31)) are conjugate since $(13)^{-1}(12)(13) = (23)$ and $(12)^{-1}(23)(12) = (31)$. [What about the pair (12), (31)?]
(ii) In S_4 the normaliser of $V = \{I, (12), (34), (12)(34)\}$ is $\langle(12), (34), (1324)\rangle$, a subgroup of order 8.
(iii) In S_4 the centraliser of (123) is $\langle(123)\rangle$, a subgroup of order 3.
(iv) Any normal subgroup is conjugate only to itself.

One result we shall use several times below is

Theorem 6.2.5 The number of distinct conjugates of the subset H in the group G is equal to $|G : N_G(H)|$ and hence divides $|G|$.

* Cf. exercise 17.

Proof For any $x, y \in G$ we see that $x^{-1}Hx = y^{-1}Hy$ iff $yx^{-1}Hxy^{-1} = H$, that is, iff $xy^{-1} \in N_G(H)$, that is, iff $N_G(H)x = N_G(H)y$. This is enough.

We obtain immediately (cf. 5.6.10(iii))

Theorem 6.2.6 A group of order p^α (>1) has non-trivial centre.

Proof Let $C_1, C_2, \ldots, C_r$ be the equivalence classes of conjugate elements (i.e. conjugacy classes) of G where C_1 is the class containing just the identity elements of G. Let $C_2, \ldots, C_t (t \leq r)$ denote the remaining equivalence classes, if any, which contain just one element. By exercise 11, $\bigcup_{1\leq i\leq t} C_i = \zeta(G)$, the centre of G. Since $G = \bigcup_{1\leq i\leq r} C_i$ and the C_i are pairwise disjoint we see that*

$$|G| = \sum_{k=1}^{t} |C_k| + \sum_{l=t+1}^{r} |C_l|$$

(This equality is called the **class equation** of G.) But each of the $|C_i|$ divides $|G|$ (by 6.2.5) and hence is a power of p, possibly p^0. Now $p \| G|$ and $p \| C_l|$ for $l = t+1, \ldots, r$. Thus $p \Big| \sum_{k=1}^{t} |C_k| = t$. It follows that $t > 1$ and hence that G has non-trivial centre.

Remark A group of order $p^\alpha q^\beta$, p and q distinct primes, may easily have trivial centre—S_3 for example (see 5.6.10(iii)).

Corollary 6.2.7 A group of order p^2 (where p is a prime) is necessarily abelian.

Proof By the previous theorem $\zeta(G) \neq \langle e \rangle$. Thus $|\zeta(G)| = p$ or $|\zeta(G)| = p^2$. In the latter case $\zeta(G) = G$ and G is abelian immediately. We show that the other possibility cannot occur. Suppose then that $|\zeta(G)| = p$. Then $|G/\zeta(G)| = p$, which implies that $G/\zeta(G)$ is a cyclic group, by Lagrange's Theorem. Taking the elements of $G/\zeta(G)$ to be $\zeta(G)$, $g\zeta(G)$, $(g\zeta(G))^2 = g^2\zeta(G), \ldots, (g\zeta(G))^{p-1} = g^{p-1}\zeta(G)$, G is the disjoint union of these cosets. Suppose $x, y \in G$. Then for some i, j and for suitable $a, b \in \zeta(G)$ we have $x = g^i a$, $y = g^j b$. But then $xy = g^i a g^j b = g^i g^j ab = g^j g^i ba = g^j b g^i a = yx$. Consequently G is abelian, whence $G = \zeta(G)$. Thus $|\zeta(G)| = |G| > p$, a contradiction.

Remark Examples prove there is no hope of extending 6.2.7 to groups of order p^3. (See exercise 16.)

Theorems 6.2.8 and 6.2.12 are described in [31] as constituting the second most important result in finite group theory. The first of these extends 5.11.4

* $|C_l|$ denotes the number of elements in C_l.

as far as possible in the case of non-abelian groups (cf. the example of A_4 given after 5.11.5).

Theorem 6.2.8 (Sylow's First Theorem, 1872) Let G be a finite group* of order $p^\alpha s$ where p is a prime and $(p, s) = 1$. Then for each $\beta \in \mathbb{Z}$ such that $0 \leq \beta \leq \alpha$, G has a subgroup of order p^β.

Proof We prove the result by induction on $|G|$. Clearly the result is true if $|G| = 1$ so we assume that $|G| = p^\alpha s$ and that the result holds for all groups of smaller order. We use the class equation $|G| = \sum_{k=1}^{t} |C_k| + \sum_{l=t+1}^{r} |C_l|$ again.

If, for some l, we have $p \nmid |C_l|$, where C_l is the set of conjugates of g, say, then $p \nmid |G : C_G(g)|$ and so $p^\alpha \| C_G(g)|$. Further, $|C_G(g)| < |G|$ since $g \notin \zeta(G)$. By the induction hypothesis we can infer that $C_G(g)$, and hence G, contains a subgroup of order p^β.

If, for each l, we have $p \| C_l|$ then, as in the proof of 6.2.6, we find $p \| \zeta(G)|$. By 5.11.4 $\zeta(G)$ has a subgroup Y, say, of order p. Now $Y \lhd G$ (exercise 5.10.14). Look at the factor group G/Y. Clearly $|G/Y| = p^{\alpha-1}s$ and so, by the induction hypothesis, G/Y has a subgroup of every order p^γ for $0 \leq \gamma \leq \alpha - 1$. It then follows from exercise 5.11.8 that G has a subgroup of order $p^{\gamma+1}$ as required.

Remark G can have no subgroup of order p^β with $\alpha < \beta$ by Lagrange's Theorem.

We can now answer Problem 1. It is interesting to see how much machinery has been necessary in order to solve what might have seemed, at first glance, a rather simple problem.

Theorem 6.2.9 Let G be a finite group. Then $|G| = p^\alpha$ for some α if and only if the order of every element of G is a power of p.

Proof One way round is easy (given Lagrange's Theorem!): If G, of order p^α, contains an element of order n then $n | p^\alpha$ by Lagrange's Theorem.

The other way round is also easy (now!): If G contains only elements of p-power order then $|G| = p^\alpha$ for some α; otherwise there would be a prime $q \neq p$ such that $q \| G|$ and then by 6.2.8 a subgroup and hence an element of order q in G.

Remark From 6.2.6 and 6.2.9 every finite p-group (other than the trivial group) has non-trivial centre. There do exist infinite p-groups with trivial centre. The first example was constructed by A G Kurosh† in 1939.

* We may assume G non-abelian—if it is abelian use 5.11.5.

† Aleksandr Gennadievich Kurosh (19 January 1908 – 18 May 1971).

Sylow's investigations led to certain subgroups in any group G being named after him.

Definition 6.2.10 Let G be a finite group and let $|G| = p^\alpha s$ where p is a prime and $(p, s) = 1$. Each subgroup of order p^α in G is called a **Sylow p-subgroup** of G.

Examples 6.2.11
(i) The Sylow 2-subgroups of S_3 are $\{I, (12)\}$, $\{I, (23)\}$, and $\{I, (31)\}$. There is just one Sylow 3-subgroup, namely that generated by (123). For each other prime there is only one Sylow p-subgroup, namely the trivial subgroup.
(ii) The Sylow 2-subgroups of S_4 all have order 8. One such is the subgroup in 6.2.4(ii). Another is the conjugate of this by the element (13). In fact, as the following theorem shows, S_4 has precisely 3 Sylow 2-subgroups which are moreover conjugate to one another.

Remark The important part of Sylow's First Theorem says that, in every group G, Sylow p-subgroups do exist. Lagrange's Theorem, on the other hand, assures us that in a given group certain types of subgroup *don't* exist! Thus Lagrange's Theorem gives necessary conditions and Sylow's First Theorem sufficient conditions for the existence in a group of subgroups of certain types.

We put **Sylow's Second and Third Theorems** together in

Theorem 6.2.12 Let G be a finite group of order $p^\alpha s$ where $(p, s) = 1$ and let P be a Sylow p-subgroup. Then each Sylow p-subgroup of G is conjugate (and hence isomorphic) to P. Further, the number of such conjugates divides $|G|/p^\alpha$ and is congruent to 1 (mod p).

We shall need two preliminary lemmas.

Lemma 6.2.13 Let P be a Sylow p-subgroup of G and let $g \in G$ have p-power order. If $P^g = P$, then $g \in P$.

Proof Since $P \lhd N_G(P)$ and since $g \in N_G(P)$ we have $gP \in N_G(P)/P$. Further $\langle gP \rangle$ is a subgroup of some p-power order since g and hence gP each have p-power order. Let S be the inverse image of $\langle gP \rangle$ under the homomorphism from $N_G(P)$ onto $N_G(P)/P$. Then since $\langle gP \rangle$ and P have p-power order it follows from exercise 5.11.8 that S has p-power order. Since P is a Sylow p-subgroup of G we conclude that $S = P$. Thus $\langle gP \rangle$ is the trivial group; that is, $g \in P$.

Lemma 6.2.14 Let $T \leqslant G$ and $H \leqslant G$. Then the number of distinct conjugates of the form H^t where $t \in T$ is equal to $|T : T \cap N_G(H)|$.

The proof of this generalisation of 6.2.5 is left to exercise 7.
We can now give

Proof of Theorem 6.2.12 Let P be as given and let $K=\{P=P_0, P_1, \ldots, P_r\}$ denote the set of distinct conjugates of P in G. Clearly each is a Sylow p-subgroup of G. We place an equivalence relation on K by writing $P_i \sim P_j$ iff $P_i^a = P_j$ for some $a \in P$. Then $\{P\}$ is an equivalence class. However, no other $\{P_k\}$ is: otherwise we should have $P_k^b = P_k$ for each $b \in P$ and hence $P \leqslant P_k$ (a contradiction) by 6.2.13.

Next the number of conjugates of P_k $(k \neq 0)$ under P is, by 6.2.14, $|P: P \cap N_G(P_k)|$ and this is a power of p bigger than p^0 [why?]. Thus we see immediately that the conjugates $P_1, \ldots, P_r$ split into classes each with a multiple of p elements. Consequently the number of conjugates of P in G (including P itself) is of the form $1+mp$.

Suppose Q is any Sylow p-subgroup of G not listed in K. We apply the same technique as that just used except that we split K into classes by putting $P_i \sim P_j$ iff $P_i^c = P_j$ for some $c \in Q$. An identical argument to that in the second paragraph above shows that this time $r+1$ is a multiple of p which contradicts the formula above. This shows that all the Sylow p-subgroups are listed in K; that is each is conjugate, in G, to P.

Finally we observe that conjugate subgroups are clearly isomorphic (exercise 4) and that $1+mp=r+1=|G:N_G(P)|$. That $1+mp \mid |G:P| = |G|/p^\alpha$ follows immediately.

Remarks Sylow's Theorems can be generalised in several ways, on the one hand to infinite groups under certain extra hypotheses [36]. (One then defines a Sylow p-subgroup to be a p-subgroup* not contained in any bigger p-subgroup, rather than as one of 'maximum p-power order'.) On the other hand one can define the concept of Sylow π-subgroup where π is a set of primes (see exercise 6.5.20 and the concept of Hall π-subgroup in Section 6.5).

We now proceed to some applications. Others will be given later.

We begin with a definition whose importance will be revealed in Sections 6.5 and 6.6. See also 7.10.5 and exercise 7.9.2.

Definition 6.2.15 A group G is called **simple** iff it has no normal subgroups other than $\langle e \rangle$ and G.

One of the main uses of the Sylow Theorems is to prove, for certain special n, that no group of order n is simple. We give three examples.

Examples 6.2.16 (i) Let $|G|=42$. Then G is not simple.

* A *p-subgroup*: a subgroup which is itself a p-group.

Proof By 6.2.12, the number of Sylow 7-subgroups of G is of the form $1+7k$ and divides $42/7=6$. Thus $k=0$ and G has just one Sylow 7-subgroup P, say. But then all conjugates of P, being themselves Sylow 7-subgroups of G, must coincide with P. Thus P is normal in G.

(ii) Let $|G|=56$. Then G is not simple.

Proof The number of Sylow 7-subgroups of G is of the form $1+7k$ and divides 8. Thus there are 1 or 8 of them. The number of Sylow 2-subgroups of G is of the form $1+2k$ and divides 7. Thus there are 1 or 7 of them. Thus neither prime dividing 56 gives us a normal subgroup as immediately as in the last example. Let us do a finer analysis. Suppose G had 8 Sylow 7-subgroups. Since 7 is a prime each pair of these subgroups can only have $\langle e\rangle$ in common. [Are you convinced?] Thus these 8 subgroups yield $8(7-1)=48$ non-identity elements of G. Now G has a subgroup of order 8 (by 6.2.8), the order of each element in this subgroup being a power of 2. Thus this subgroup of 8 elements is totally disjoint from the subset of 48 non-identity elements of G already listed. It follows that if G has 8 Sylow 7-subgroups then G has room for only one Sylow 2-subgroup which would have to be normal.

Of course if G has only one Sylow 7-subgroup we proceed as in case (i).

(iii) If $|G|=48$ then G is not simple.

Proof Let H be a Sylow 2-subgroup. Then $|H|=16$. If H is the only Sylow 2-subgroup of G there is nothing more to prove. Otherwise there will be 3 such. Let K ($\neq H$) be one of these. Consider $H\cap K$. If $H\cap K$ has order 1 or 2 or 4 (and hence index 16 or 8 or 4) in H, it is not difficult to check that there are 16^2 or $8\cdot 16$ or $4\cdot 16$ distinct elements of the form hk (where $h\in H$ and $k\in K$) in G (exercise 25). None of these is a possibility since $|G|=48$. Thus $H\cap K$ has index 2 in H and in K. Consequently H and K are subgroups of $N_G(H\cap K)$. Since $H\neq K$ we see that $H<N_G(H\cap K)$. Thus $16\,|\,|N_G(H\cap K)|\,|\,|G|$ and $|N_G(H\cap K)|>16$. Thus $|N_G(H\cap K)|=48$, whence $H\cap K\lhd G$, as required.

Exercises

1 Show that a subgroup and a factor group of a p-group is again a p-group. Show that if $H\lhd G$ and if H and G/H are p-groups then so is G.

2 Give an example of a finite group G generated by two elements of order 2 but where G is not a 2-group.

3 Show that conjugacy determines an equivalence relation on (a) the set of all subsets of a group G; (b) the set of all subgroups of G; (c) the set of elements of G.

4 Show that conjugate subgroups are isomorphic as groups.

5 Let $H \leqslant G$. Show that $N_G(H) \leqslant G$ and that $H \leqslant N_G(H)$. Find examples, where G is a group, H a subgroup and K a subset but not a subgroup of G, such that (a) $N_G(H) = G$; (b) $N_G(H) = H$; (c) $N_G(K) = G$; (d) $N_G(K) = \langle e \rangle$.

6 Let $H \leqslant G$. Show that $H \lhd N_G(H)$; indeed $N_G(H)$ is the unique largest subgroup of G in which H is normal.

7 Prove Lemma 6.2.14.

8 Let G be a finite group and $H < G$. Show that $\bigcup_{g \in G} H^g \subsetneqq G$. [Hint: Use the fact that $H \leqslant N_G(H)$, 6.2.5 and $|H| = |H^g|$. Now count!]

9 Find the centraliser of (1234) in S_4. How many conjugates has (1234) in S_4?

10 Show that, in every group G, $\zeta(G) = \bigcap_{x \in G} C_G(x)$.

11 Show that $x \in \zeta(G)$ iff $|G : C_G(x)| = 1$.

12 Let G be a group of order p^n. Show that G has a series $G = G_0 > G_1 > \cdots > G_n = \langle e \rangle$ of subgroups such that, for each i $(1 \leqslant i \leqslant n)$ $G_i \lhd G$ and $|G_{i-1} : G_i| = p$. [Hint: Induction on n. Given G, find $G_{n-1} \lhd G$ such that $|G_{n-1}| = p$. Use the induction hypothesis to find in G/G_{n-1} a series $\frac{G}{G_{n-1}} > \frac{G_1}{G_{n-1}} > \cdots > \frac{G_{n-1}}{G_{n-1}}$ where each $\frac{G_i}{G_{n-1}} \lhd \frac{G}{G_{n-1}}$ and where $\left|\frac{G_{i-1}}{G_{n-1}} : \frac{G_i}{G_{n-1}}\right| = p$. Use exercise 5.11.9 and 5.10.10(iii) to show $G > G_1 > \cdots > G_{n-1} > G_n$ is a series as required.]

13 Verify the class equation for S_3, S_4 and D_4. (D_4 is the group of 8 symmetries of the square.)

14 Use the class equation to show that if $H \lhd G$ $(H \neq \langle e \rangle)$ and if G is a p-group then $H \cap \zeta(G) > \langle e \rangle$. [Hint: $H \cap C_i \neq \varnothing \Rightarrow H \supseteq C_i$.]

15 Let G be a p-group of order p^n. We know (from 6.2.8) that G has at least one subgroup of order p^{n-1}. Show that every such subgroup S is normal in G. [Hint: If $x \in \zeta(G) \backslash S$ then $\langle S, x \rangle = G$ and $S \lhd \langle S, x \rangle$. Otherwise $\zeta(G) \leqslant S$. Pass to $\frac{S}{\zeta} \leqslant \frac{G}{\zeta}$.] Doesn't exercise 12 already prove this?

16 Exhibit, for some prime p, a non-abelian group of order p^3. [Hint: try $p = 2$.]

17 Let G be a group and S a subgroup. Define $C_G(S)$, the *centraliser* of S in G by $C_G(S) = \{g : g \in G \text{ and } gx = xg \text{ for all } x \in S\}$. Show that $C_G(S)$ is a subgroup of G and that $C_G(S) \lhd N_G(S)$. Prove that $N_G(S)/C_G(S)$ is isomorphic to a subgroup of Aut (S). (See exercise 5.9.18.)

18 Show that any conjugate of a Sylow p-subgroup is a Sylow p-subgroup.

19 Let P be a Sylow p-subgroup and H a subgroup of the group G. Is it true that $P \cap H$ is a Sylow p-subgroup of H? [Hint: try S_3.]

20 Show that a Sylow p-subgroup P of a group G is a Sylow p-subgroup of its normaliser. Show further that P is the only Sylow p-subgroup of its normaliser.

21 Let S be a p-subgroup and P a Sylow p-subgroup of G. Show that S is contained in some conjugate of P. Deduce that one could equally well define a Sylow p-subgroup to be 'a p-subgroup of G not contained in any larger p-subgroup of G'.

22 Show that $N_G(N_G(P)) = N_G(P)$, P being a Sylow p-subgroup of G.

23 Show that no group with order 135, 30 or 96 can be simple.

24 Find two subgroups of order 4 which are not conjugate in S_4. (This contradicts 6.2.12, surely?)

25 Let H, K be subgroups of a finite group G. Show that the set $\{hk : h \in H, k \in K\}$ contains exactly $\dfrac{|H| \cdot |K|}{|H \cap K|}$ elements.

6.3 Direct products

In this section we provide a tool which can, on the one hand, be used to dissect given groups in to more easily managed pieces but which, used the other way round to build up 'big' groups out of little ones, can be useful for the purpose of constructing counterexamples. (See exercises 21 and 22 below, exercise 5.11.7 and exercise 6.5.11.) There are many such tools in existence in group theory, this one being just about the simplest of its kind to describe. Despite this simplicity, it is far from useless, being *exactly the right tool* to penetrate the structure of all finitely generated abelian groups (see 6.4.3 and 6.4.4 and the Remarks at the end of Section 6.4).

To keep things straightforward we shall deal only with the case of a finite number of groups. The infinite case presents two faces (which coincide in the finite case) but not, at this stage, any extra difficulty (exercise 20).

Definition 6.3.1 Let $G_1, G_2, \ldots, G_n$ be given groups and set $G = G_1 \times G_2 \times \cdots \times G_n$, the cartesian product of the sets G_i $(1 \leq i \leq n)$. We define on G a binary operation $\cdot$ by putting $(g_1, g_2, \ldots, g_n) \cdot (h_1, h_2, \ldots, h_n) = (g_1h_1, g_2h_2, \ldots, g_nh_n)$. $\langle G, \cdot \rangle$, which is clearly a group (see exercise 1), is called the **direct product** of the groups G_i $(1 \leq i \leq n)$.

We easily prove

Theorem 6.3.2 The group G contains subgroups H_i ($1 \leqslant i \leqslant n$) such that (i) for each i, $H_i \cong G_i$; (ii) for each i, $H_i \lhd G$; (iii) $G = \langle H_1, H_2, \ldots, H_n \rangle$ (see 5.6.11); (iv) for each i, $H_i \cap \langle H_1, \ldots, H_{i-1}, H_{i+1}, \ldots, H_n \rangle = \langle e \rangle$.

Proof We define H_i to be the set $\{(e, \ldots, e, h_i, e, \ldots, e) : h_i \in G_i\}$. It is easy to check that H_i is a subgroup of G. The map $(e, \ldots, e, h_i, e, \ldots, e) \mapsto h_i$ is clearly an isomorphism from H_i onto G_i. The equality $(g_1 \ldots, g_i, \ldots, g_n)^{-1} \cdot (e, \ldots, e, h_i, e, \ldots, e) \cdot (g_1, \ldots, g_i, \ldots, g_n) = (e, \ldots, e, g_i^{-1} h_i g_i, e, \ldots, e)$ shows that $H_i \lhd G$. The equality $(g_1, g_2, \ldots, g_i, \ldots, g_n) = (g_1, e, e, \ldots) \cdot (e, g_2, e, \ldots) \cdot \ldots \cdot (e, e, \ldots, e, g_n)$ shows that $G = \langle H_1, H_2, \ldots, H_n \rangle$. Finally (iv) is a simple set-theoretic matter (see exercise 2).

Remarks on notation and terminology
(i) When each G_i in 6.3.1 is abelian and when we are using + for the binary operation on each G_i we usually replace $\cdot$ by $\oplus$, so that $(g_1, g_2, \ldots, g_n) \oplus (h_1, h_2, \ldots, h_n) = (g_1 + h_1, g_2 + h_2, \ldots, g_n + h_n)$, and refer to G as the *direct sum* of the G_i: $G = G_1 \oplus \ldots \oplus G_n$.
(ii) In 6.3.1 we began with a set of groups G_i and formed a new group G. The G_i are themselves not subgroups of G; they are only isomorphic to the groups H_i which are *actually inside* G. For this reason the group G of 6.3.1 is sometimes called the **external direct product** of the G_i. By way of contrast, any group G satisfying the properties (ii), (iii), (iv) of 6.3.2 is called the **internal direct product** of the H_i. Clearly the word 'internal' is applicable; to justify the use of the expression 'direct product' requires the following

Theorem 6.3.3 Let G possess subgroups H_i with the properties (ii), (iii), (iv) as in 6.3.2. Then G is isomorphic to the direct product of the H_i, considered as groups in their own right.

Proof We must show that $G \cong H_1 \times H_2 \times \cdots \times H_n$. First, let $h_i \in H_i \leqslant G$ and $h_j \in H_j \leqslant G$ where $i \neq j$. Clearly $h_i^{-1} h_j^{-1} h_i h_j = h_i^{-1}(h_j^{-1} h_i h_j) \in H_i$ since $H_i \lhd G$. But also $h_i^{-1} h_j^{-1} h_i h_j = (h_i^{-1} h_j^{-1} h_i) h_j \in H_j$ [why?]. Thus $h_i^{-1} h_j^{-1} h_i h_j \in H_i \cap H_j = \langle e \rangle$ [why?]. It follows that, for all $h_i \in H_i$, $h_j \in H_j$ with $i \neq j$, $h_i h_j = h_j h_i$. Since $G = \langle H_1, \ldots, H_n \rangle$ we can clearly write every element g of G in the form of a product $g = x_{i_1} x_{i_2} \ldots x_{i_k}$, say, where each x_{i_j} is in one of the H_i. Using the fact, just proved, that each element of H_1 commutes with every element of $H_2, H_3, \ldots, H_n$, we can, in the expression for g, simply move all the x_{i_j} which belong to H_1 to the left and then take their product to reduce the number of elements from H_1 in the product for g to one, h_1 say. Next we similarly collect all elements of H_2 in this new product into one element, h_2, lying immediately to the right of h_1. Repeating this for the elements from $H_3, H_4, \ldots, H_n$ shows that g can be expressed in the form

$$g = h_1 h_2 \ldots h_n \qquad (*)$$

where each $h_i \in H_i$ (h_i may or may not be the identity of H_i). The final step in this initial stage of the proof is to show that the representation (*) of g as a product is unique. That is, if also $g = k_1k_2 \ldots k_n$ with $k_i \in H_i$ $(1 \leq i \leq n)$, we wish to show that $h_1 = k_1$, $h_2 = k_2, \ldots, h_n = k_n$. We leave this to exercise 9.

The alleged isomorphism is now easy to find. One defines $\theta: H_1 \times \cdots \times H_n \to G$ by $(h_1, h_2, \ldots, h_n)\theta = h_1h_2 \ldots h_n$.

The above remarks show in turn that θ is onto and 1–1. Thus only the homomorphism property remains. This again we leave to exercise 9.

Remark 6.3.2 and 6.3.3 demonstrate the intimate connection between the concept of internal and external direct product and the reader may ask why we bother with both. The answer, as the reader will see if he tries to do without one or the other concept in what follows, is simply 'for ease of exposition': we can (and will) be careless in talking of groups as 'being direct products' omitting the words 'isomorphic to' or 'of its subgroups', and we shall occasionally write = when ≅ might be more accurate.

Exercises

1 Show that the pair $\langle G, \cdot \rangle$ in 6.3.1 is a group.

2 Prove 6.3.2(iv).

3 Show that the direct product of any (finite) number of abelian groups is abelian. Does this result remain valid if the word 'abelian' is replaced by (a) 'finite'; (b) 'cyclic'; (c) 'matrix'?

4 (a) Is $G \cong C_6 \oplus C_{10}$ cyclic? If not, what is the largest order of any of the elements of G? (Clearly the answer is 60 if G *is* cyclic.)
(b) Is $A_5 \cong S_3 \times D_5$ (D_5 being the group of symmetries of a regular pentagon)?

5 A group G is a **proper direct product** if $G \cong A \times B$ with neither A nor B the trivial group. Are the following groups proper direct products (or sums)? (a) S_4; (b) D_6; (c) D_n (for each $n \geq 3$); (d) $\langle \mathbb{C}, + \rangle$; (e) C_{p^n}, p a prime.

6 Show that $C_r \times C_s = C_{rs}$ if $(r, s) = 1$.

7 Show that $G_1 \times G_2 \cong G_2 \times G_1$.

8 Show that
(a) If $G \cong A \times B$ and $B \cong C \times D$ then $G \cong A \times C \times D$;
(b) If $G \cong A \times B \times C$ and $D \cong B \times C$ then $G \cong A \times D$.

9 Complete the proof of 6.3.3 where indicated.

10 Write $\langle \mathbb{Z}_{24}, \oplus \rangle$ in as many different ways as you can as a direct sum of two or more groups.

11 How many subgroups has $\mathbf{Z}_2 \oplus \mathbf{Z}_2 \oplus \mathbf{Z}_2$? [Hint: try $\mathbf{Z}_2 \oplus \mathbf{Z}_2$ first where the answer is *not* 4.]

12 Is it true that every subgroup of $A \times B$ has the form $X \times Y$ where $X \leqslant A$ and $Y \leqslant B$?

13 Write $\dfrac{\mathbf{Z} \oplus \mathbf{Z}}{N}$ as a direct sum of two cyclic groups (if possible!) when N is (a) the subgroup of $\mathbf{Z} \oplus \mathbf{Z}$ generated by $(2, 2)$; (b) the subgroup of $\mathbf{Z} \oplus \mathbf{Z}$ generated by $(2, 3)$.

14 Show that $\langle \mathbf{C}^\times, \cdot \rangle \cong \langle \mathbf{R}^+, \cdot \rangle \times \mathbf{R}/\mathbf{Z}$ where $\mathbf{R}/\mathbf{Z}$ is defined in the same manner as the group of exercise 5.6.16.

15 Show that the map $\psi: G_1 \times G_2 \to G_1$ given by $(g_1, g_2)\psi = g_1$ is a homomorphism of $G_1 \times G_2$ onto G_1. What is $\ker \psi$?

16 Let $N_1 \lhd G_1, N_2 \lhd G_2$ and $N = \langle (n_1, e_{G_2}), \ (e_{G_1}, n_2) : n_1 \in N_1, n_2 \in N_2 \rangle \subseteq G_1 \times G_2$. Show that $N = N_1 \times N_2$, $N \lhd G_1 \times G_2$ and that $\dfrac{G_1 \times G_2}{N} \cong \dfrac{G_1}{N_1} \times \dfrac{G_2}{N_2}$.

17 Let $N_1, N_2 \lhd G$ and let $N = N_1 \cap N_2$. Show that $\dfrac{G}{N}$ is isomorphic to a subgroup of $\dfrac{G}{N_1} \times \dfrac{G}{N_2}$. [Hint: try mapping $G \to \dfrac{G}{N_1} \times \dfrac{G}{N_2}$ by $g \to (gN_1, gN_2)$. Show that this is a homomorphism with kernel N.] Deduce that if $N_1 \cap N_2 = \langle e \rangle$ then $\dfrac{G}{N_1} \times \dfrac{G}{N_2}$ contains a subgroup isomorphic to G.

18 Find the centre of $S_3 \times D_8$. More generally relate $\zeta(A)$ and $\zeta(B)$ to $\zeta(A \times B)$.

19 Let G be a group of order $p^2 q^2$ where p, q are distinct primes and where $q \nmid p^2 - 1$ and $p \nmid q^2 - 1$. Show that G is a direct product of a group of order p^2 and one of order q^2. Deduce that G is abelian.

20 Let $G_1, G_2, \ldots$ be an infinite set of groups. Their cartesian product set $G = G_1 \times G_2 \times \cdots$ comprises all *infinite vectors* $(g_1, g_2, \ldots)$ where $g_i \in G_i$. Show that with multiplication defined by $(g_1, g_2, \ldots) \cdot (h_1, h_2, \ldots) = (g_1 h_1, g_2 h_2, \ldots)$, G becomes a group. Show that the subset $\bar{G}$ of G comprising those infinite vectors which contain only finitely many non-identity elements is a subgroup of G. ($\bar{G}$ is called the *restricted direct product* and G the *unrestricted direct product* of the G_i.)

21 Show that there exist groups A, B such that each is isomorphic to a subgroup of the other and yet $A \not\cong B$. [Hint: Use infinite direct products involving C_2 and C_4.]

22 Show that there exist groups A, B such that each is a homomorphic image of the other and yet $A \not\cong B$. [Hint: Same as for exercise 21.]

6.4 Finite abelian groups

We now show how the relatively simple concepts of direct sum and finite cyclic group enable us to characterise and then classify finite abelian groups. Thus for these groups, as for finite fields, the algebraist's dream is realised. (See the end of Section 4.5 and Remarks in Section 3.10.) Actually it is possible similarly to characterise and classify all finitely generated abelian groups but we shall stick to the finite case mainly because it is particularly easy and the main aim of this section is to introduce the reader to the *spirit* of the development as much as to the theorems themselves.

We begin, using the additive notation, with

Theorem 6.4.1 Let A be a finite abelian group. If, for each prime p dividing $|A|$, we let S_p denote the set of all elements of p-power order in A (including 0) then each S_p is a subgroup of A and A is their direct sum.

Proof That each S_p *is* a subgroup (indeed a normal subgroup) of A we leave to exercise 1. Next let $x \in A$ $(x \neq 0)$. Then x has order n where $n \mid |A|$. Suppose that $n = p_1^{\alpha_1} p_2^{\alpha_2} \ldots p_r^{\alpha_r}$ where the p_i are distinct primes and each $\alpha_i \geq 1$. Write, for each i $(1 \leq i \leq r)$, $q_i = n/p_i^{\alpha_i}$. Then the positive gcd of the q_i is 1 and so there exist (exercise 1.4.6) integers $k_1, \ldots, k_r$ such that $k_1 q_1 + \cdots + k_r q_r = 1$. It follows that $x = (k_1 q_1 + \cdots + k_r q_r)x = k_1 q_1 x + \cdots + k_r q_r x$. Now $q_i x$ (and hence $k_i q_i x$) has order dividing [even equal to ?] $p_i^{\alpha_i}$. Thus x has been expressed as a sum of elements with orders dividing $p_1^{\alpha_1}, p_2^{\alpha_2}, \ldots, p_r^{\alpha_r}$ respectively. It follows that $x \in \langle S_{p_1}, S_{p_2}, \ldots, S_{p_r} \rangle \leq A$. But x was any element of A. Thus $\langle S_p : p \mid |A| \rangle = A$.

Finally A will be the direct sum of the S_p, if only we can show that, for each $q \mid |A|$, $S_q \cap \langle S_p : p \mid |A| \text{ and } p \neq q \rangle = \langle 0 \rangle$. But this is easy because any element of S_q has order a power of q whilst any element of $\langle S_p : p \mid |A| \text{ and } p \neq q \rangle$ has order coprime to q. Thus the only element common to both subgroups is 0.

We now show how to split each of these **p-primary components** S_p into cyclic components.

Theorem 6.4.2 Let S_p be a finite abelian p-group. Then S_p is a direct sum of cyclic groups (of p-power order).

Proof Let s be any of the elements of maximal order p^α, say, in S_p and let T be as large a subgroup of S_p as possible satisfying $\langle s \rangle \cap T = \langle 0 \rangle$. Then $\langle s, T \rangle = \langle s \rangle \oplus T$. For: (i) $\langle s \rangle$ and T generate $\langle s, T \rangle$!; (ii) $\langle s \rangle$ and T are normal in $\langle s, T \rangle$ and (iii) $\langle s \rangle \cap T = \langle 0 \rangle$ (by choice!). If $\langle s \rangle \oplus T < S_p$ we can find an $x \in S_p$ such that $x \notin \langle s \rangle \oplus T$. Since $p^\alpha x = 0$ [why?] $p^\alpha x \in \langle s \rangle \oplus T$. There then exists $\beta \in \mathbb{Z}^+$ so that $p^\beta x \in \langle s \rangle \oplus T$ but $p^{\beta-1} x \notin \langle s \rangle \oplus T$. Put $y = p^{\beta-1} x$.

Now $py \in \langle s \rangle \oplus T$. Hence $py = ls + t$ $(l \in \mathbb{Z}, t \in T)$. Then $0 = p^\alpha y = p^{\alpha-1} ls + p^{\alpha-1} t$. Thus $p^{\alpha-1} ls \in \langle s \rangle \cap T = \langle 0 \rangle$. It follows that $p \mid l$, $pk = l$, say, and hence that $p(y-ks)=py-ls=t\in T$. However, $y-ks\notin T$. [Why not?] Thus $\langle T, y-ks\rangle > T$. Consequently $\langle T, y-ks\rangle\cap\langle s\rangle>\langle 0\rangle$. [Why?] That is, for some $m, n\in\mathbb{Z}$ and $v\in T$ we have $0\neq ms=v+n(y-ks)$. Here $p\nmid n$ (or else $ms\in\langle s\rangle\cap T=\langle 0\rangle$, a contradiction). Thus $(n,p)=1$. Also $ny=ms-v+nks\in\langle s\rangle\oplus T$ and [by choice] $py\in\langle s\rangle\oplus T$. Thus $y\in\langle s\rangle\oplus T$ [why?], a contradiction. Thus $\langle s\rangle\oplus T=S_p$ after all. An argument using induction now completes the proof of the theorem. ($|T|<|S_p|$ so T is a direct sum of p-power cycles. Now use exercise 6.3.8(a).)

Combining 6.4.1 and 6.4.2 we find (exercise 6.3.8(a) again)

Theorem 6.4.3 Let A be a finite abelian group. Then A is the direct sum of cyclic subgroups of various prime power orders, the primes involved being those which divide $|A|$.

Remarks
(i) If A is cyclic of prime power order it is not decomposable into a direct sum of two or more non-trivial groups (exercise 6.3.5(e)).
(ii) Since every finite direct sum of (prime power) cycles is a finite abelian group we have characterised finite abelian groups as precisely those expressible as direct sums of (prime power) cyclic groups. In order to effect a *classification* of these groups we need to explain how to tell them apart. This follows from considering the question: Given an abelian group A, is its decomposition into a direct sum of prime power cyclic groups unique? The immediate answer is no; from $A=\langle s\rangle\oplus\langle t\rangle=\langle s\rangle\oplus\langle u\rangle$ where $\langle s\rangle$, $\langle t\rangle$, $\langle u\rangle$ are cycles one cannot infer $\langle t\rangle=\langle u\rangle$ (exercise 2). Recalling from Section 3.10 that an algebraist is a person who cannot see any difference between *isomorphic* systems, never mind *equal* ones, we see that the next theorem yields as much as the algebraist could ask for.

Theorem 6.4.4 (The Fundamental Theorem of Finite Abelian Groups*) Let A be a finite abelian group. Then any two decompositions of A into direct sums of cyclic groups of prime power order contain the same number of summands of each order.

Remark It is easy to check that if $A=S_{p_1}\oplus\cdots\oplus S_{p_r},=S'_{p_1}\oplus\cdots\oplus S'_{p_r}$ are decompositions of the type provided by 6.4.1 then $S_{p_1}=S'_{p_1},\ldots,S_{p_r}=S'_{p_r}$.

Thus we can concentrate on the case where A is an (abelian) p-group. Perhaps the pictorial representation in Fig. 6.1 of an abelian p-group which decomposes as a direct sum of 2 p-cycles, 3 p^2-cycles, 2 p^4-cycles and one p^5-cycle will help the reader through the following proof.

* First proved in 1878.

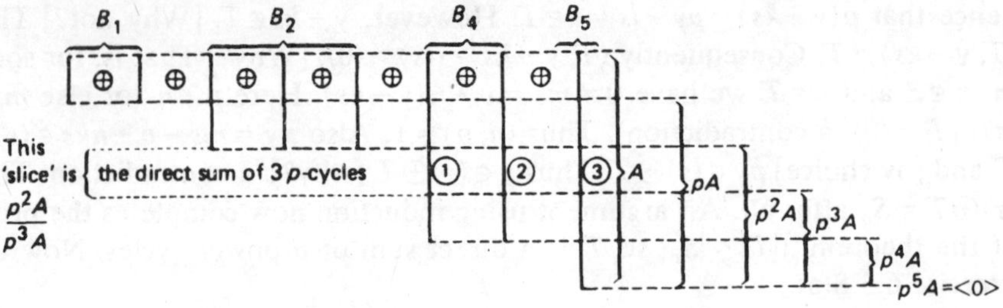

Fig. 6.1

Proof of Theorem 6.4.4 Let a decomposition of A into a direct sum of p-power cycles be given. For each i define B_i to be the direct sum of all cyclic summands of order p^i in this decomposition. Then $A = B_1 \oplus B_2 \oplus \cdots \oplus B_s$ (say) (exercise 6.3.8(b)) and, for each $j \in \mathbb{Z}^+$, $p^jA = p^jB_1 \oplus p^jB_2 \oplus \cdots \oplus p^jB_s = p^jB_{j+1} \oplus \cdots \oplus p^jB_s$. (Here $p^jA = \{p^ja : a \in A\}$. It is clear that p^jA is a subgroup of A.) Now consider the factor group* $\dfrac{p^jA}{p^{j+1}A}$. This group is isomorphic to $p^jB_{j+1} \oplus \dfrac{p^jB_{j+2}}{p^{j+1}B_{j+2}} \oplus \cdots \oplus \dfrac{p^jB_s}{p^{j+1}B_s}$ (exercise 6.3.16), a direct sum of (a direct sum of) p-cycles. Now the number of summands in this direct sum of cycles is $b_{j+1} + b_{j+2} + \cdots + b_s$ where b_t is the number of cyclic summands in B_t. Thus the total number of cyclic summands in $\dfrac{p^jA}{p^{j+1}A}$ is equal to the number of cyclic summands in A which have order at least p^{j+1}. Thus the number of cycles of each order p^α in a direct decomposition of A into p-power cycles depends only on A.

Remarks

(i) In view of our stressing, in Chapters 1 and 3, the concept of unique factorisation in $\mathbb{Z}$ and in various number rings, it is appropriate to observe that 6.4.3 and 6.4.4 together yield yet another result of this type. It follows that two finite abelian groups are equal in the algebraist's eyes (i.e. isomorphic) iff they decompose into the same number of indecomposable pieces (i.e. prime power cycles) of each kind. Another unique factorisation theorem for finite abelian groups is given in exercise 12.

(ii) The structure of an infinite abelian group A can be very involved. But if A is finitely generated we have: *A is the direct sum of a finite number of infinite cyclic groups and a finite number of cyclic p-groups for various primes p. The number of summands of each type completely characterises A.* ([27, p. 96]; [36, p. 106].)

* We keep to the fractional notation even in the case of abelian groups. The *difference notation*, $p^jB - p^{j+1}B$, risks confusion with $p^jB \backslash p^{j+1}B$.

Example 6.4.5 There are 10 different types (i.e. isomorphism classes) of abelian groups of order 1008 $(=2^4 \cdot 3^2 \cdot 7)$, namely

$C_2 \oplus C_2 \oplus C_2 \oplus C_2 \oplus C_3 \oplus C_3 \oplus C_7$		$C_2 \oplus C_2 \oplus C_2 \oplus C_2 \oplus C_9 \oplus C_7$	
$C_2 \oplus C_2 \oplus C_4$	$\oplus C_3 \oplus C_3 \oplus C_7$	$C_2 \oplus C_2 \oplus C_4$	$\oplus C_9 \oplus C_7$
$C_2 \oplus C_8$	$\oplus C_3 \oplus C_3 \oplus C_7$	$C_2 \oplus C_8$	$\oplus C_9 \oplus C_7$
$C_4 \oplus C_4$	$\oplus C_3 \oplus C_3 \oplus C_7$	$C_4 \oplus C_4$	$\oplus C_9 \oplus C_7$
C_{16}	$\oplus C_3 \oplus C_3 \oplus C_7$	C_{16}	$\oplus C_9 \oplus C_7$

The largest order of an element in each of these groups is respectively
42, 84, 168, 84, 336 and 126, 252, 504, 252, 1008.

Exercises

1 Prove that the subsets S_p of 6.4.1 are subgroups, as claimed.

2 Show by means of an example that if A is the internal direct sum of subgroups X and Y and also the internal direct sum of X and Z then it is possible that $Y \neq Z$. [Hint: $C_2 \oplus C_2$.]

3 Find the number of (isomorphism classes of) abelian groups of order (a) 360; (b) 218; (c) p^5 where p is a prime. In each case give the maximum of the orders of the elements in the respective groups.

4 Are there more (isomorphism classes of) abelian groups of order 11^3 than of order $2^2 3^2$?

5 How many different abelian groups of order 720 have (a) exactly 3, and (b) exactly 4, subgroups of order 2? (Careful with (b).)

6 Write $\langle M(16), \odot \rangle$ as a proper direct product of cycles—or prove such a decomposition impossible.

7 Let A be a finite abelian group. Show that A has a finite series of subgroups $A = A_0 > A_1 > \cdots > \langle 0 \rangle$ such that each A_i/A_{i+1} is a (cyclic) group of prime order.

8 Two finite abelian groups A, B have exactly the same number of elements of order m for each $m \in \mathbb{Z}^+$. Show that $A \cong B$.

9 Show that if a finite abelian group A has, for each $m \mid |A|$, exactly m elements such that $ma = 0$ then A is cyclic. [Hint: consider $C_2 \times C_2$.]

10 Show that, in an infinite abelian group, the elements of infinite order, together with 0, do not necessarily form a subgroup. [Hint: $C_2 \times C_\infty$.] Give an example other than $\langle \mathbb{Z}, + \rangle$ in which they *do*.

11 A finite abelian group A is generated by two elements a, b such that $4a = 4b = 0$ and $2a = 2b$. Does this mean that A cannot be expressed as a proper direct sum?

12 Let A be a finite abelian group. Show that A can be expressed as a direct sum of cyclic groups of orders $n_1, n_2, \ldots, n_r$ (say) where $n_1|n_2| \ldots |n_r$. [Hint: Try a specific example, say $C_2 \oplus C_2 \oplus C_3$, first and see if you can generalise the argument. Recall that $C_r \oplus C_s = C_{rs}$ if $(r, s) = 1$.] This is yet another kind of unique factorisation theorem since A determines the n_i uniquely. (Proof?)

13 (Cf. exercise 4.5.5 and exercise 9.) Let $\langle F, +, \cdot \rangle$ be a finite field. Prove that $\langle F^\times, \cdot \rangle$ is a cyclic group. [Hint: assume not and deduce that some equation of the form $x^p - 1 = 0$ has more than p solutions in F.]

14 Let G, H, K be finite abelian groups. Show that if $G \oplus G \cong H \oplus H$ then $G \cong H$ and that if $G \oplus H \cong G \oplus K$ then $H \cong K$.

6.5 Soluble* groups. Composition series

After Ruffini (1799, 1813) and Abel (1824) had shown the impossibility of finding a universal formula in terms of $+$, $\cdot$, $-$, $\div$ and $\sqrt[n]{}$ which would yield the roots of any given quintic equation (see Section 7.9) the question of exactly which particular equations were soluble* by radicals† was answered by Galois in 1832. The answer is: those equations whose associated group satisfies an easily stated condition (see 6.5.7). Groups having this property were christened 'soluble' by Jordan in 1867. However, even had it been of no significance for the theory of equations the solubility property would undoubtedly have attracted the attention of research workers in group theory because of the following philosophy. We have just seen what nice groups finite abelian groups are in that their structure can be explicitly written down. Given a quite arbitrary (finite) group G one can investigate it by looking for special (e.g. Sylow) subgroups or by mapping it homomorphically onto a member of some class of groups about which we are fairly knowledgeable. Into this latter class fall the finite abelian groups, so a good ploy might be to map G onto as 'large' a finite abelian group A, say, as possible. What A tells us about G may not be much but it may well be better than nothing. Now this homomorphism produces a kernel K, say, and we can apply the same procedure to K that we have just applied to G. That is, we find a subgroup N of K such that N is normal in K and K/N is as big an abelian homomorphic image of K as possible. Continuing in this way we may (or may not!) reach the identity subgroup of G. If we do, G is called a soluble group (Definition 6.5.5). It may be thought of as being made up of 'slices' (like a sliced loaf!),

* Solvable, in USA.
† Definition 7.7.2.

each of which is abelian. The prospect thus generated would surely be enough to set one about the study of such groups. (Presumably, if the theory had developed this way, the groups considered would not be called soluble but maybe multi- or poly-abelian?)

If G is a group, if $N \lhd G$ and if G/N is abelian then, for every pair of elements $a, b \in G$, we have $aNbN = bNaN$, that is $a^{-1}b^{-1}abN = N$, which means that $a^{-1}b^{-1}ab \in N$. This helps motivate the following

Definition 6.5.1 Let G be a group and $a, b \in G$. The element $a^{-1}b^{-1}ab$, denoted briefly by $[a, b]$, is called the **commutator** of a and b.

Remarks

(i) 'Commutator' is a good word to use: for $[a, b]$ is a kind of measure as to how near a and b come to commuting, $[a, b]$ being the identity element of G if and only if $ab = ba$.

(ii) It is perfectly natural to look at the set of all commutators in a group G. The sad fact is that this subset may not form a subgroup of G (see [26, p. 38]) so we move to the next best thing.

Definition 6.5.2 Let G be a group. The subgroup *generated* by the set $\{[a, b]: a, b \in G\}$ is called the **commutator subgroup** or **derived subgroup** of G. It is denoted by G' or $G^{(1)}$ or $[G, G]$.

More generally, for subgroups A, B of a group G we denote by $[A, B]$ the subgroup of G generated by the set $\{[a, b]: a \in A, b \in B\}$.

Examples 6.5.3

(i) Since $[a, a] = e$ the identity element of each group is a commutator. In an abelian group (and only in an abelian group) is e the only commutator. Hence G is abelian iff $G' = \langle e \rangle$.

(ii) $[a, b] = [b, a]^{-1}$; $[a, b]^g = [a^g, b^g]$ for all $a, b, g \in G$.

(iii) In A_5 $[(12)(35),(234)] = (12)(35)(432)(12)(35)(234) = (12345)$ is a commutator.

(iv) The derived group of the 8-gon group in 5.3.4(l) comprises I, a^2, a^4, a^6.

(v) $S'_n = A_n$ for all $n \geqslant 1$; $A'_1 = A'_2 = A'_3 = \langle e \rangle$, $A'_4 = \{I, (12)(34), (13)(24), (14)(23)\}$, $A'_n = A_n$ for all $n \geqslant 5$. (See exercise 4 and 6.6.1.)

To show that G' does what we are asking of it we prove

Theorem 6.5.4 For each group G, $G' \lhd G$. Further, if $N \lhd G$ then G/N is abelian iff $N \geqslant G'$. (Thus G/G' is the largest abelian homomorphic image of G.)

Proof G' is, by definition, a subgroup. Each element x of G' can be expressed as a product $x = [g_1, g_2][g_3, g_4] \ldots [g_{2r-1}, g_{2r}]$ of commutators of G. Now for each $g \in G$ we have (see 6.5.3(ii)) $x^g = [g_1^g, g_2^g][g_3^g, g_4^g] \ldots [g_{2r-1}^g, g_{2r}^g] \in G'$. Thus $G' \lhd G$.

We saw earlier that if G/N is abelian then $G' \leqslant N$. Conversely, if $G' \leqslant N$ then G/N is abelian: for, given a, b and G we have $a^{-1}b^{-1}ab \in G' \leqslant N$. Consequently $aNbN = bNaN$, as required.

Proceeding further, we make the formal

Definition 6.5.5 For each non-negative integer n define $G^{(n+1)}$ to be $[G^{(n)}, G^{(n)}]$, the derived subgroup of $G^{(n)}$, where $G^{(0)} = G$. The non-increasing sequence $G = G^{(0)} \geqslant G^{(1)} \geqslant \cdots \geqslant G^{(n)} \geqslant \cdots$ of subgroups of G is called the **derived series** of G. If, for some n, we have $G^{(n)} = \langle e \rangle$ then G is said to be a **soluble** group and if m is the least such integer G is said to be **soluble of length m**.

Examples 6.5.6

(i) $S_1 = \langle I \rangle$; S_2 is soluble of length 1 (that is, S_2 is abelian!); S_3 is soluble of length 2; S_4 is soluble of length 3. The remaining S_n $(n \geqslant 5)$ are not soluble at all (see example 6.5.3(v)).

(ii) $A_2 = \langle I \rangle$; A_3, A_4 are soluble of lengths 1 and 2 respectively. A_n is not soluble if $n \geqslant 5$.

(iii) The 8-gon group of example 5.3.4(l) is soluble of length 2.

(iv) Let $U_n(\mathbb{R})$ denote the set of all $n \times n$ matrices of the form

$$A = \begin{pmatrix} 1 & a_{12} & a_{13} & \cdots & a_{1n} \\ 0 & 1 & a_{23} & \cdots & a_{2n} \\ 0 & 0 & 1 & \cdots & a_{3n} \\ \vdots & \vdots & \vdots & \vdots & \vdots \\ 0 & 0 & 0 & \cdots & 1 \end{pmatrix}$$

where the a_{ij} are real numbers.

It is not difficult to check that A^{-1} exists and that if

$$A^{-1} = \begin{pmatrix} 1 & b_{12} & b_{13} & \cdots & b_{1n} \\ 0 & 1 & b_{23} & \cdots & b_{2n} \\ 0 & 0 & 1 & \cdots & b_{3n} \\ \vdots & \vdots & \vdots & \vdots & \vdots \\ 0 & 0 & 0 & \cdots & 1 \end{pmatrix}$$

then $b_{i,i+1} = -a_{i,i+1}$. It follows that the derived group $U_n(\mathbb{R})^{(1)}$ comprises all matrices whose form is that of A except that the elements $a_{12}, a_{23}, \ldots, a_{n-1,n}$ on the first 'superdiagonal' are all 0. In a similar manner one finds that $U_n(\mathbb{R})^{(k)}$ comprises matrices in which the first $2^k - 1$ superdiagonals are full of 0s.

It then follows easily that the group $U_n(\mathbb{R})$ is soluble of length $-[-\log_2 n]$.

Note 6.5.7 6.5.5 is a more recent definition of the term 'soluble'. 6.5.8 below gives an equivalent definition (for finite groups): *A (finite) group G is soluble iff it has a series*

$$G=G_0>G_1>\cdots>G_n=\langle e\rangle \qquad (*)$$

of subgroups such that each G_{i+1} is a proper normal subgroup of G_i with the property that each quotient $|G_i|/|G_{i+1}|$ is a prime. Since Galois had shown that an equation $f(x)=0$ is soluble by radicals iff its (Galois) group had a series as in $(*)$ it was a natural step for Jordan* to call such groups soluble.

For yet another variant lying midway between $(*)$ and 6.5.5, see exercise 9.

Remark We cannot infer from the facts that $G_1 \lhd G$ and $G_2 \lhd G_1$ that $G_2 \lhd G$. That is, normality is not transitive (exercise 5.10.17).

That the two definitions of soluble are equivalent for finite groups is given by

Theorem 6.5.8 Let G have a series of type $(*)$. Then $G^{(n)}=\langle e\rangle$. Conversely, if $G^{(k)}=\langle e\rangle$ for some k, then G has a series of type $(*)$.

Proof To prove the first part we use induction. Clearly $G^{(0)}\leq G_0$. Suppose $G^{(k)}\leq G_k$. Then $G^{(k+1)}=[G^{(k)},G^{(k)}]\leq[G_k,G_k]\leq G_{k+1}$ by 6.5.4, since G_k/G_{k+1} is abelian.

For the converse, let $G=G^{(0)}>G^{(1)}>\cdots>G^{(k)}=\langle e\rangle$ be the derived series of G. For each i, $G^{(i)}/G^{(i+1)}$ is a finite abelian group and hence, by exercise 6.4.7, has a series of (normal) subgroups $G^{(i)}/G^{(i+1)}=A_0>A_1>\cdots>\langle e\rangle$ such that each factor group A_j/A_{j+1} is prime cyclic.

Using methods like those called for in exercise 6.2.12 we see that the inverse images of these A_i under the natural homomorphism from $G^{(i)}$ onto $G^{(i)}/G^{(i+1)}$ form a series of normal subgroups of $G^{(i)}$ with corresponding factor groups again of prime order. Inserting these inverse images between $G^{(i)}$ and $G^{(i+1)}$ for each i refines the derived series of G into a series of type $(*)$ for G, as required.

Remarks

(i) From this characterisation and from exercise 6.2.12 we see immediately that every finite p-group is soluble.

(ii) Subgroups and homomorphic images of soluble groups are again soluble (exercise 7). Further, given a group G and a normal subgroup N such that N and G/N are soluble one can show that G is necessarily soluble (exercise 17).

* Camille Jordan (5 January 1838 – 22 January 1921).

In (∗) we see that the subgroups G_i form a sort of maximal series—the G_i are packed together as closely as possible; since the factor groups are of prime order there is, by Lagrange's Theorem, no room for any more insertions. Such a series is, according to the following definition, an example of a composition series.

Definition 6.5.9 Let G be a group and $G = G_0 > G_1 > \cdots > G_n = \langle e \rangle$ a decreasing series of subgroups such that, for each i $(0 \leq i \leq n-1)$ $G_{i+1} \lhd G_i$ and G_i/G_{i+1} is a simple group. Such a series is called a **composition series** for G and the groups G_i/G_{i+1} are called **composition factors** of G. n is called the **length** of the series.

Remarks

(i) By 5.10.10(iv) one can equally describe each subgroup G_{i+1} in 6.5.9 as being a maximal normal subgroup of G_i, maximal in the sense that *no bigger normal subgroup* of G_i can be placed strictly between G_i and G_{i+1}.
(ii) Clearly every finite group has a composition series; an infinite group may not have (exercise 12).
(iii) Series (∗) is a composition series for G; clearly finite groups of prime orders must be simple.

Examples 6.5.10

(i) $S_3 > A_3 > \langle e \rangle$ is a composition series for S_3. The composition factors here are isomorphic to C_2 and C_3.
(ii) $S_4 > A_4 > A_4' > C_2 > \langle e \rangle$ where $C_2 = \{I, (12)(34)\}$ is a composition series for S_4. Composition factors are C_2, C_3, C_2, C_2.
(iii) $G = \langle a, b \rangle > \langle a \rangle > \langle a^2 \rangle > \langle a^4 \rangle > \langle e \rangle$ is one composition series for the 8-gon group of 5.3.4(1). $\langle a, b \rangle > \langle a^2, b \rangle > \langle a^4, b \rangle > \langle a^4 \rangle > \langle e \rangle$ is another. In each series the composition factors are C_2, C_2, C_2, C_2.
(iv) If S is a simple group then $S > \langle e \rangle$ is the only composition series for S.

In the above terminology Galois' theory says that an equation is soluble by radicals if and only if the Galois group of the equation has a composition series with prime cyclic factors. Now there will, in general, be many ways of picking a composition series for a given finite group G. Beginning with $G_0 = G$ one simply takes, for each i, G_{i+1} to be any one of the (possibly many) maximal proper normal subgroups of G_i, continuing in this way until one reaches (as one must, since G is finite) the trivial subgroup. Suppose that one of the composition factors in a series so obtained is not prime cyclic. Does this mean that the corresponding equation is not soluble by radicals? Surely not. Maybe if we had chosen the subgroups G_i in a different manner we would have obtained a composition series in which each of the factors *is* prime cyclic.

Thus it looks as if we have to search through the various composition series of G until either we find a series with all factors cyclic, or else we prove that there is no such series—by examining each and every one!

Fortunately, Jordan proved in 1869 that the orders of the various composition factors of a given group G are invariant from one composition series to another. Thus G will have one composition series with all its factors prime cyclic iff this is the case with *all* composition series of G. That is, the first series one examines completely answers the problem regarding solubility.

Jordan's result was strengthened by Hölder* in 1889 to give, for *any* finite group, the

Theorem 6.5.11 (The Jordan–Hölder Theorem) Let G be a finite group with composition series

$$G = H_0 > H_1 > \cdots > H_r = \langle e \rangle$$

and

$$G = K_0 > K_1 > \cdots > K_s = \langle e \rangle$$

Then $r = s$ and the r factor groups H_i/H_{i+1} can be put in 1–1 correspondence with the r factor groups K_j/K_{j+1} in such a manner that corresponding groups are isomorphic.

In order to prove this we make first the

Definition 6.5.12 Let H, K be subsets of the group G. We denote by HK the set $\{hk: h \in H, k \in K\}$.

Next we note the

Lemma 6.5.13 If $H \leqslant G$ and $K \lhd G$ then HK is a subgroup of G.

Proof We leave this to exercise 13.

The tool required to prove 6.5.11 is

Theorem 6.5.14 (The Second Isomorphism Theorem) (cf. 4.3.2) Let $H \leqslant G$, $K \lhd G$. Then $K \lhd HK$, $H \cap K \lhd H$ and $HK/K \cong H/H \cap K$.

Proof Clearly $K \lhd HK$ since $K \lhd G$ and $K \leqslant HK \leqslant G$. We define $\theta: H \to HK/K$ by $h\theta = hK$. Clearly θ is a mapping. Rather easily we see that for $g, h \in H$ we have $(gh)\theta = ghK = gK \cdot hK = g\theta \cdot h\theta$. Also θ maps H onto HK/K since every element of HK/K is of the form hkK where $h \in H$ and $k \in K$. But $hkK = hK$ since $k \in K$.

* Otto Ludwig Hölder (22 December 1859 – 29 August 1937).

Thus θ is a homomorphism from H onto HK/K. According to 5.11.6 we see that $HK/K \cong H/\ker\theta$. Now if $h \in \ker\theta$ then $hK = eK = K$. Thus $h \in K$, that is $\ker\theta \subseteq H \cap K$.

The reverse inclusion is easy.

Now for the

Proof of Theorem 6.5.11 We proceed by induction on $|G|$. Thus we suppose

$$G = H_0 > H_1 > \cdots > H_r = \langle e\rangle \qquad \text{(i)}$$

and

$$G = K_0 > K_1 > \cdots > K_s = \langle e\rangle \qquad \text{(ii)}$$

are two composition series for G.

Clearly we may suppose $H_1 \neq K_1$. (If $H_1 = K_1$ an application of the induction hypothesis to $H_1 = K_1$ soon finishes the task.) From this we infer that $H_1K_1 = G$ [why?] and hence that $\dfrac{G}{K_1} = \dfrac{H_1K_1}{K_1} \cong \dfrac{H_1}{H_1 \cap K_1}$. The proof that $\dfrac{G}{H_1} \cong \dfrac{K_1}{K_1 \cap H_1}$ is similar.

Consider now the series

$$G = H_0 > H_1 > H_1 \cap K_1 > L_2 > \cdots > L_t = \langle e\rangle \qquad \text{(iii)}$$

and

$$G = K_0 > K_1 > H_1 \cap K_1 > L_2 > \cdots > L_t = \langle e\rangle \qquad \text{(iv)}$$

where

$$H_1 \cap K_1 = L_1 > L_2 > \cdots > L_t = \langle e\rangle$$

is a composition series for $H_1 \cap K_1$.

Those parts of series (i) and (iii) which begin at H_1 are composition series for H_1; similarly with those parts of (ii) and (iv) in relation to K_1. By induction we may infer that the factors

$$\left(\frac{G}{H_1}\right), \frac{H_1}{H_2}, \frac{H_2}{H_3}, \ldots, \frac{H_{r-1}}{H_r} \qquad \text{from (i)} \qquad \text{(A)}$$

and the factors

$$\left(\frac{G}{H_1}\right), \frac{H_1}{H_1 \cap K_1}, \frac{H_1 \cap K_1}{L_2}, \ldots, \frac{L_{t-1}}{L_t} \qquad \text{from (iii)} \qquad \text{(B)}$$

can be put into 1–1 correspondence in such a way that corresponding factors are isomorphic groups.

Similar remarks apply to the factors

$$\left(\frac{G}{K_1}\right), \frac{K_1}{K_2}, \frac{K_2}{K_3}, \ldots, \frac{K_{s-1}}{K_s} \qquad \text{from (ii)} \qquad \text{(C)}$$

and the factors

$$\left(\frac{G}{K_1}\right), \frac{K_1}{K_1 \cap H_1}, \frac{K_1 \cap H_1}{L_2}, \ldots, \frac{L_{t-1}}{L_t} \quad \text{from (iv)} \qquad \text{(D)}$$

Finally, since the groups in (B) and (D) can, by the above isomorphisms $\frac{G}{K_1} \cong \frac{H_1}{H_1 \cap K_1}$ and $\frac{G}{H_1} \cong \frac{K_1}{K_1 \cap H_1}$, clearly be put in 1–1 correspondence in the above manner, so can the factors of (A) and (C). This proves the theorem.

Remarks

(i) The Jordan–Hölder theorem is yet another sort of unique factorisation theorem (see Remarks at the end of Section 6.4). Notice that we do not claim $\frac{G}{H_1} \cong \frac{G}{K_1}, \frac{H_1}{H_2} \cong \frac{K_1}{K_2}, \ldots$, etc., only that the factors $\frac{H_i}{H_{i+1}}$ and $\frac{K_j}{K_{j+1}}$ are isomorphic in pairs, *in some order* (see exercise 18(b)). Comparing this situation with that in $\mathbb{Z}$ we see that we can regard G as a sort of (in general non-commutative) 'product of primes', namely the simple composition factors. If, in particular, the Jordan–Hölder theorem is applied to cyclic groups of finite order we obtain a splendidly long-winded proof of the uniqueness part of the fundamental theorem of arithmetic!

(ii) As two finite groups G_1 and G_2 may have isomorphic composition factors and yet not be isomorphic (see exercise 18) the Jordan–Hölder theorem is mainly of use in a negative sense: if H_1 and H_2 have distinct sets of composition factors then H_1 and H_2 cannot possibly be isomorphic.

We close this section by mentioning two of the more recent developments in the theory of finite soluble groups. First we indicate how Sylow's theorems can be generalised in the case of soluble groups.

Let π be a finite set of (positive) primes and π' denote the complementary set of (positive) primes not in π. A **π-number** (respectively **π'-number**) is any positive integer (including 1) whose prime factors belong to the set π (respectively, to π'). A finite group K is called a **π-group** iff $|K|$ is a π-number. A subgroup H of a finite group G is called a **Hall π-subgroup** of G iff (i) H is a π-group and (ii) $|G:H|$ is a π'-number. In particular if $\pi = \{p\}$ any finite group G possesses Hall π-subgroups, namely its Sylow subgroups. However in general a group need not have a Hall π-subgroup. For instance the group A_5 of order $60 = 2^2 \cdot 3 \cdot 5$ has no Hall $\{3, 5\}$-subgroup (exercise 20). However theorems like Sylow's can be proved for π-subgroups of soluble groups. In fact in 1928 Philip Hall* proved

Theorem H Let G be a finite soluble group and π be a finite set of primes. Let A be any π-subgroup (including $\langle e \rangle$) of G. Then

(i) G has Hall π-subgroups: there is even one containing A;

(ii) any two Hall π-subgroups of G are conjugate.

* Philip Hall (11 April 1904 – 30 December 1982).

The proof of this theorem appears in most modern introductory texts on group theory. (We refer the reader to [34] or [36] to name just two sources.) On the other hand the other result we wish to mention, proved as recently as 1963, possibly never will. We first reveal the intimate connection with finite simple groups.

By 1900, several infinite 'families', together with five 'sporadic' examples, of non-abelian finite simple groups were known (see next section). All were of even order. Burnside* (1897) suggested an investigation into the existence of non-abelian simple groups of odd order be carried out and in 1911 he conjectured that there is no such group. Assuming Burnside to be correct this implies immediately that every group of odd order is soluble. For if G is any group of odd order then G has a composition series $G = G_0 \rhd G_1 \rhd \cdots \rhd G_r = \langle e \rangle$ with simple, odd-order, composition factors. By Burnside's conjecture each factor must be simple abelian and hence prime cyclic (exercise 6.6.1). Thus G is soluble. Conversely if a non-abelian group G (whatever its order) is soluble, then it cannot be simple since G' is a normal proper and non-trivial subgroup of G. The theorem that all non-abelian simple groups must have even order was proved by Feit and Thompson, two algebraists from the USA (and then in their early thirties) in a mammoth 254-page paper, 'Solvability of groups of odd order' which occupied an entire edition of the *Pacific Journal of Mathematics* in 1963. (For his subsequent work in determining all minimal finite simple groups, that is all finite simple groups in which every proper subgroup is soluble, John Thompson was awarded the Fields Medal† at the 1970 International Congress of Mathematicians.)

Exercises

1 Let G be a group and let $x \in G$. Show that $x \in \zeta(G)$ iff $[x, g] = e$ for all $g \in G$.

2 Show that for all $a, b, c \in G$, $[a, bc] = [a, c][a, b]^c = [a, c][a, b][[a, b], c]$ and $[ab, c] = [a, c]^b[b, c] = [a, c][[a, c], b][b, c]$.

3 Let $a, b, c \in G$. Is it necessarily true that $[[a, b], c] = [a, [b, c]]$? [Hint: Look at S_3.]

4 Show that there are (even) permutations a, b, c, d in A_5 such that $[a, b]$ is (123) and $[c, d]$ is $(12)(34)$. Deduce that every element of A_5 is a commutator. Deduce that A_5 is not soluble.

5 Show that if $A \lhd G$ and $B \lhd G$ then $[A, B] \lhd G$ and $[A, B] \leqslant A \cap B$.

6 Let $\theta: G \to G\theta$ be a homomorphism. Show that $[a, b]\theta = [a\theta, b\theta]$ for all $a, b \in G$. Deduce that $(G')\theta = (G\theta)'$.

* William Burnside (2 July 1852 – 21 August 1927).

† For details of this 'Nobel Prize' of Mathematics see [99]. There is a photograph in *Bulletin of the American Math. Soc.*, Vol. 40, 1934, p. 189. See also [96].

7 Show that subgroups and homomorphic images of soluble groups are again soluble. [Hint: for homomorphisms extend exercise 6.]

8 (a) Show that the direct product of two soluble groups is again soluble. Find the solubility length of the direct product in terms of the lengths of its factors.

(b) Let $H \lhd G, K \lhd G$ be such that $\frac{G}{H}$ and $\frac{G}{K}$ are soluble. Show that $\frac{G}{H \cap K}$ is soluble. [Hint: use exercise 6.3.17.]

9 In 1864 Jordan, in effect, stated: The group G is soluble iff G has a series $G = G_0 \geqslant G_1 \geqslant \cdots \geqslant G_k = \langle e \rangle$ such that for each i $(1 \leqslant i \leqslant k)$ $G_i \lhd G_{i-1}$ and G_{i-1}/G_i is abelian. Prove this. [Hint: For the given series prove by induction that $G^{(i)} \leqslant G_i$. The other half of the proof is immediate.]

10 Show that the solubility length of S_4 is 3 by finding the derived series of S_4. Do the same for the group D_5 of symmetries of the regular pentagon.

11 Show that a direct product of infinitely many soluble groups need not be soluble. [Hint: find groups G_i with solubility length i, for each $i \in \mathbb{Z}^+$.]

12 Show that the infinite cyclic group has no (finite) composition series. Show that if one allows infinite series with simple factors then $\langle \mathbb{Z}, + \rangle$ has composition series but no analogue of the Jordan--Hölder theorem can hold. [Hint: Look at the series $\mathbb{Z} > 2\mathbb{Z} > 4\mathbb{Z} > \cdots$ and $\mathbb{Z} > 3\mathbb{Z} > 9\mathbb{Z} > \cdots$.]

13 Let $H \leqslant G$ and $K \leqslant G$. Show that HK need not be a subgroup of G. Show that HK is a subgroup of G if one of H and K is normal in G and that $HK \lhd G$ if both H, K are normal in G.

14 Write down as many distinct composition series as you can for the cyclic group of order 180. Find the composition factors in each case. Hence verify the Jordan–Hölder theorem directly for this group. Do the same for the group $S_3 \times D_5$.

15 Let G be a finite group and let $H \lhd G$. Show that G has a composition series in which H appears as a term.

16 Does S_4 have a sequence $S_4 = H_0 > H_1 > H_2 > H_3 > \langle e \rangle$ in which each H_i/H_{i+1} is prime cyclic and each H_i is normal in S_4?

17 Let $N \lhd G$. Show that $\left(\frac{G}{N}\right)^{(i)} = \frac{G^{(i)}N}{N}$. Deduce that if $\frac{G}{N}$ is soluble of length m and if N is soluble of length n then G is soluble of length $\leqslant m + n$.

18 (a) Give an example of two finite groups G, H which have the same composition factors and yet are not isomorphic. [Try order 4.]
(b) Show that C_6 has two composition series $C_6 = G_0 > G_1 > G_2 = \langle e \rangle$ and $C_6 = G_0 > H_1 > H_2 = \langle e \rangle$ such that $\frac{G_0}{G_1} \not\cong \frac{G_0}{H_1}$ and $\frac{G_1}{G_2} \not\cong \frac{H_1}{H_2}$.

19 Show that in a soluble group G the length of each composition series for each proper subgroup H of G is shorter than the length of each composition series of G (cf. exercise 6.6.10).

20 Show that if one defines a Sylow π-subgroup to be a maximal π-subgroup (see the remarks preceding 6.2.15) then every finite group has Sylow π-subgroups. By considering Sylow $\{3, 5\}$-subgroups of A_5 show that such Sylow π-subgroups may not be pairwise isomorphic. Show that A_5 has no Hall $\{3, 5\}$-subgroup.

6.6 Some simple groups

Because of the intimate connection between soluble groups and simple groups we complement the previous section with just a few remarks on simple groups. Clearly every (cyclic) group of prime order is simple since, by Lagrange's Theorem, it can have no subgroups other than itself and the trivial subgroup. Conversely, if an abelian group is simple then it must be (cyclic) of prime order (see exercise 1). We establish the existence of infinitely many non-abelian finite simple groups by proving

Theorem 6.6.1 For each integer $n \geqslant 5$ the alternating group A_n is simple.*

The strategy is to prove that
(i) If $H \neq \langle e \rangle$ and if $H \lhd A_n$ then H contains a 3-cycle;
(ii) H contains all 3-cycles;
(iii) A_n is generated by 3-cycles.
It follows immediately that $H = A_n$ whence A_n is simple.

Proof
(i) Let $H \lhd A_n$ and let h be an element of prime order. Write h as a product of disjoint cycles. For ease of reading we use letters rather than integers in expressing the elements of A_n.
(a) If $|h| = p \geqslant 5$ and if $h = (a_1a_2 \ldots a_p) \ldots (r_1r_2 \ldots r_p)$ we have $(a_1a_2a_3)h(a_3a_2a_1) \cdot h^{-1} = (a_2a_3a_p)$ so that H contains a 3-cycle.
(b) If $|h| = 3$ and if $h = (abc)(def) \ldots$ we have $(abcde)h(edcba) \cdot h^{-1} = (bdecf) \in H$ whence a 3-cycle can be found in H by using step (a).
(c) If $|h| = 2$ then either $h = (ab)(cd)$ or $h = (ab)(cd) \cdot (ef)(gh) \ldots$ using an even number of transpositions. In the first case $(bde)h(edb) \cdot h = (aebdc) \in H$ whence a 3-cycle can be found in H by using step (a).
In the second case $(bde)h(edb) \cdot h = (afc)(bde)$ whence a 3-cycle can be found in H by using step (b).
Thus whatever order h has, it leads us to a 3-cycle in H.

* Galois knew that A_5 is simple.

(ii) Let $\alpha = (xyz)$ be a 3-cycle in S_n. We know (exercise 5.5.6(ii)) that all 3-cycles in S_n are conjugate and, by counting, that there are $\frac{n(n-1)(n-2)}{3}$ of them. Thus* $|S_n : C_{S_n}(\alpha)| = \frac{n(n-1)(n-2)}{3}$ whence $|C_{S_n}(\alpha)| = 3(n-3)!$ Now $C_{S_n}(\alpha)$, being a group of permutations, comprises either all even permutations or half even and half odd permutations (see exercise 2). Since $n \geqslant 5$, there exist letters l, m distinct from x, y, z. Then $(xyz)(lm)$ is an odd permutation which clearly lies in $C_{S_n}(\alpha)$. Hence $C_{S_n}(\alpha)$ comprises permutations half of which are even and half of which are odd. Thus there are exactly $\frac{1}{2} \cdot 3(n-3)!$ permutations in A_n which commute with (xyz). That is, $|C_{A_n}(\alpha)| = \frac{1}{2} \cdot 3(n-3)!$ and so $|A_n : C_{A_n}(\alpha)| = \frac{n(n-1)(n-2)}{3}$. Thus conjugating (xyz) only by elements of A_n already yields all the $\frac{n(n-1)(n-2)}{3}$ 3-cycles which exist in S_n, and so starting with one 3-cycle in H we find, since $H \lhd A_n$, that all the 3-cycles of S_n are in H.

(iii) Every element of A_n is expressible as a product of an even number of transpositions. Now $(ab)(bc) = (acb)$ whereas $(ab)(cd) = (ab)(bc)(bc)(cd) = (acb)(bdc)$ so that every element of A_n is a product of 3-cycles. Thus A_n is generated by the set of all 3-cycles.

This completes the proof of 6.6.1.

Of the groups mentioned in 6.6.1 A_5, with $\frac{5!}{2} = 60$ elements, is the smallest. Are there any groups with order less than 60 which are also simple? Certainly all groups of prime order are. But what about the remainder? No group of order p^n with $n > 1$ can be simple since every such group is either abelian (when trivially it is not simple) or has a proper non-trivial centre (see Section 6.2). 17 of the remaining groups have order pq where p, q are distinct primes. We prove

Theorem 6.6.2 No group G of order pq, where p, q are distinct primes, can be simple.

Proof Suppose WLOG that $p > q$. The number of Sylow p-subgroups is of the form $1 + kp$ and divides q (see 6.2.12). This is impossible unless $k = 0$. Thus there is just one Sylow p-subgroup of G. Consequently it is normal in G and we are done.

* Theorem 6.2.5.

Remark One can take this analysis further. It follows easily that G is soluble. In the case that $q \nmid p-1$ we can even prove that G is a cyclic group. (See exercise 3.)

We leave it to the reader to check that all groups of order p^2q have a proper non-trivial normal subgroup and hence are soluble. We go one stage further.

Theorem 6.6.3 No group G of order p^2q^2 is simple.

Proof Suppose WLOG that $p > q$. The number of Sylow p-subgroups is of the form $1+kp$ and divides q^2 (by 6.2.12). Thus $1+kp = 1$ or q or q^2. In the first two cases the only possibility is $k = 0$ whence G has a unique, and hence normal, Sylow p-subgroup of order p^2. If $1+kp = q^2$ then $kp = q^2-1$ whence $p|q-1$ or $p|q+1$. Since $p > q$ the former is impossible whilst in the latter case $p = 3$ and $q = 2$ is the only possible solution. Thus $|G| = 36$, and, if G is simple, the number of Sylow 3-subgroups must be 4. These 4 Sylow 3-subgroups are all conjugate (by 6.2.12) and so, if P is one of them, $|G : N_G(P)| = 4$ (6.2.5). Let us denote $N_G(P)$ briefly by H, and write $G = H \cup Hg_1 \cup Hg_2 \cup Hg_3$ as a disjoint union of the four cosets of H in G. We now define a homomorphism θ (see 5.10.2(iii)) from G into the group of 24 permutations on the set $\{H, Hg_1, Hg_2, Hg_3\}$ by setting, for each $g \in G$,

$$g\theta = \begin{pmatrix} H & Hg_1 & Hg_2 & Hg_3 \\ Hg & Hg_1g & Hg_2g & Hg_3g \end{pmatrix}$$

That θ is indeed a homomorphism was noted in exercise 5.10.19. In particular θ is mapping the group G of order 36 into a group of order 24. Hence the kernel of θ is non-trivial. Also $\ker\theta \subseteq H$. Thus G has a non-trivial proper normal subgroup, as required.

We leave it to the reader to prove (exercise 5) that, apart from the prime cyclic groups and the trivial group (!) there are no simple groups of order less than 60 and to deduce that all groups of order less than 60 are soluble.

There also exist ([36, p. 292]) infinite classes of (finite) matrix groups over finite fields which are simple. Some of these are isomorphic to some of the alternating groups introduced above. On the other hand there do exist pairs of non-isomorphic simple groups which have the same order. In particular there is such a pair of order 20 160, one of the two groups being A_8. Let us now show that amongst all groups of order 60 there is, up to isomorphism, only one simple group.

Theorem 6.6.4 Let G be simple and of order 60. Then $G \cong A_5$.

Proof By 6.2.12 the number of Sylow 5-subgroups is 1 or 6. Since G is simple there must be 6. Hence we have already 24 non-identity elements of

G. The number of Sylow 2-subgroups is either 3, 5, or 15. There cannot be 3—or else G would have (cf. the proof of 6.6.3) a subgroup of index 3; and hence there would be a homomorphism of G into S_3 with kernel not equal to G. Suppose there are 15 Sylow 2-subgroups—each of order 4. If no two intersect in more than $\langle e \rangle$ we pick up 45 non-identity elements of G disjoint from the 24 we already have. This is impossible. Thus, if G has 15 Sylow 2-subgroups then at least one pair must intersect non-trivially—in a subgroup of order 2. Calling this pair of subgroups X, Y we see that if $G = \langle X, Y \rangle$ then $X \cap Y$ is central and hence normal in G, which is impossible. Thus $\langle X, Y \rangle$, which contains at least $\frac{4}{2} \cdot 4$ elements (exercise 6.2.25), must have index 3 or 5. The former is impossible, as shown above. Thus the latter holds. Note also that if G has 5 Sylow 2-subgroups we have immediately a subgroup of index 5.

Thus we have reduced the problem to showing that if G is a simple group of order 60 and with a subgroup of index 5 then $G \cong A_5$.

The by-now usual argument shows that G is (isomorphic to) a subgroup of S_5. Clearly $|S_5 : G| = 2$ so G is normal in S_5. It follows that $G \cap A_5 \lhd A_5$ and hence that $G \cap A_5 = A_5$ or $\langle e \rangle$. This latter is easily seen to be impossible (exercise 6) whence $G = A_5$ follows.

Having given in 6.6.1 one infinite family of non-abelian finite simple groups we close with a few words on some of the other non-abelian finite groups known to be simple. Matrix groups, with their entries coming from finite fields, yield several infinite families of finite simple groups. One such family is obtained as follows. Let n be any integer greater than 1 and let F be any field, finite or infinite. The set of all $n \times n$ matrices with entries in F and with determinant 1 forms a group denoted by $SL_n(F)$—the **special linear group** of degree n over F. This group is in general not simple since it might have non-trivial centre Z, say. (In fact Z comprises all matrices of the form αI where I is the $n \times n$ identity matrix and $\alpha \in F$ is such that $\alpha^n = 1$, because of the determinant having to be 1.) However, the factor group $\dfrac{SL_n(F)}{Z}$ can, except when $n=2$, $F = \mathbb{Z}_2$ and $n=2$, $F = \mathbb{Z}_3$, be shown to be simple ([36, p. 294]). In case F is an infinite field we thus have an example of an infinite simple group; if F is a finite field we obtain a finite simple group. (These groups were introduced in Jordan's *Traité* in 1870.) There are other families of simple matrix groups which you can find by looking up the terms **orthogonal, symplectic** and **unitary groups**. Besides these infinite families there were also known five groups, the smallest being of order 7920 and the largest of order 244 823 040, discovered by É Mathieu* (in 1861 and 1873) which did not appear to be members of any infinite family. Dickson† discovered some further families of finite simple groups around 1905 and then not a single one was

* Emile Léonard Mathieu (15 May 1835–19 October 1890).

† Leonard Eugene Dickson (22 January 1874–17 January 1954). Author of 18 books including the monumental *History of the Theory of Numbers*.

found until 1955 when Chevalley found yet more families. Variants of these were soon found and the list of finite simple groups was then presumed, by some, to be complete (all that was lacking was a proof!) when Janko found a group of 7×7 matrices with entries in the field $\mathbb{Z}_{11}$ and of order 175 560. Was this a member of another infinite family? At first it was thought that it might be. Now we know otherwise, (see for example [90])—there are no more finite simple groups to be found. The proof that besides the various infinite families of finite simple groups mentioned above there are just 26 sporadic ones, the largest being of order approximately $8 \cdot 10^{53}$(!), is due to the efforts of many mathematicians working in concert. Their combined contributions to the proof amount to several thousand pages. For an introduction to this proof see [16].

A very readable article, written just before the completion of the classification theorem, is in volume 84, number 9 of the *American Mathematical Monthly*.

As further study of simple groups is far from simple we conclude our chapter on 'raw' group theory here and look, instead, at its most famous application.

Exercises

1 Let G be a simple abelian group. Prove that G is finite of prime order (and hence cyclic).

2 Show that if H is a subgroup of S_n then either (i) all the elements of H are even permutations or (ii) exactly half are even and the other half are odd permutations. [Hint: recall the proof of 5.5.11.]

3 Show that if p, q are primes, if $p > q$ and if $q \nmid p-1$ then a group G of order pq has a normal subgroup of order p and one of order q. Show that G is the direct product of these two subgroups and is, in particular, cyclic. Show that whether $q | p-1$ or not, G is soluble of length at most 2.

4 Show that any group G of order p^2q has a proper and non-trivial normal subgroup. Deduce that G is soluble. [Note: you may not 'assume WLOG that $p > q$'. Why not?]

5 Show that the only finite simple groups amongst the non-trivial groups with orders less than 60 are the cyclic groups of prime order. Deduce that every group with order less than 60 is soluble.

6 Show that S_5 has one subgroup of index 2. Is this true of S_4; S_3; D_4?

7 Consider the set P of all even permutations on the set $\mathbb{Z}^+$ of positive integers. Thus P is by definition the set of all those permutations on $\mathbb{Z}^+$ each of which can be expressed as a product of a (finite) even number of transpositions. In particular, any element of P moves only finitely many elements of $\mathbb{Z}^+$. Show that P is a group and that it is an infinite simple group. Show that P is not a finitely generated group. [Hint: Think of P as the set-theoretic union of the increasing sequence $A_1 < A_2 < A_3 < \cdots$ of alternating groups. Note that if $N \lhd P$ then $N \cap A_n \lhd A_n$ for each n.]

There do exist infinite simple groups which are finitely generated. The first was discovered as recently as 1951 by Graham Higman.

8 Using 5.9.6 and exercise 5.9.20 prove that to every finite group G there exists a finite simple group S such that S contains a subgroup isomorphic to G. (We say that S contains an *isomorphic copy* of G or (cf. 3.10.2(v)) that S *embeds* G.)

Prove that there is an infinite simple group which contains an isomorphic copy of every finite group. [Exercise 7 helps.]

9 Let G be a finite group with all its Sylow subgroups abelian. Need G be soluble? [Hint: which is the smallest non-soluble* you know?]

10 Exhibit a group G with subgroup H such that all composition series for G have length 1 whereas all composition series for H have length 35. (The number 35 has no special significance!)

11 Show that if $n < 60$ then any two groups of order n have identical sets of composition factors. What about the case $n = 60$?

* Or, insoluble.

7
A brief excursion into Galois Theory

7.1 Introduction

At the start of Chapter 5 we indicated Lagrange's approach to the problem of finding 'algebraic' (or 'radical') formulae—that is, formulae involving (only) the operations $+, -, \times, \div$ and $\sqrt[n]{}$ for various n—which would yield the zeros* of any given polynomial. His investigations in the case of degree ≥ 5 proved inconclusive and it was left to Ruffini† (1799, 1813) to indicate and then Abel (1824) to demonstrate the non-existence of such a formula for the zeros of the general quintic $x^5+ax^4+bx^3+cx^2+dx+e$ in terms of the (literal) coefficients a, b, c, d, e. (See 7.9.5.)

Nevertheless, there certainly exist *specific* polynomials of degree 5 (and higher) which are irreducible over $\mathbb{Q}$ and for which all the zeros *can* be expressed in radical form (see exercises 4.6.6, 4.6.7 and Section 7.6). In particular Gauss showed, in the *Disquisitiones*, how, for each positive integer n, the zeros of the polynomial x^n-1 could be so expressed (see 7.8.2. and 7.10.1). But it fell to Evariste Galois to discover a criterion—in terms of certain groups of permutations—by which one can decide (in theory, if not always easily in practice) whether or not the zeros of a given polynomial with *numerical* coefficients are expressible in radical form.

We shall not try to follow Galois' presentation. For one thing the details (as distinct from the ideas, which are fairly easily described—see Section 7.12) of his original memoire were not readily comprehensible—a fact Galois partly ascribed to the novelty and nature of the material—even to mathematicians of the calibre of Poisson, Lacroix and others. (Those wishing to see a detailed discussion of Galois' method should consult [83].)

Accordingly, the approach taken here follows a different path from that of Galois, being essentially that inspired by Dedekind (1894) and Weber‡ (1893, 1895) and coming to full fruition under Emil Artin (1938 and 1942). (It is interesting to note that the 'old' methods took a long time to die out; an account

* In this chapter we shall talk of zeros (rather than roots) of polynomials to avoid possible confusion with the various $\sqrt[n]{}$ (including roots of unity) which occur.
† Paolo Ruffini (22 September 1765–10 May 1822).
‡ Heinrich Weber (5 May 1842 17 May 1913).

Evariste Galois *(25 October 1811–31 May 1832)*
Galois was born at Bourg-la-Reine where his father was at one time mayor. Whilst still at school he wrote his first paper (on continued fractions). In July 1829 Galois' father committed suicide and in August Galois failed in an attempt to enter the Ecole Polytechnique, because of his inability, apparent even in his early school life, to submit to the generally required methods of procedure. Later that year he entered the Ecole Normale Supérieur. In 1829/30 Galois had published several research papers including a note on the solubility by radicals of equations of prime degree and one on finite fields. But others, unfortunately, were 'lost' and Galois felt he was being persecuted. He then joined the revolution of 1830 and in December of that year was expelled from the ENS. In 1831, at Poisson's request, Galois submitted a third revised edition of his memoire on the solubility of equations by radicals to the Academy of Sciences. When he learnt of its rejection—since Poisson and others couldn't understand it as written—Galois was in prison. After his release he was persuaded into a duel, some say over a love affair. He spent the night before the fatal duel writing many letters, one containing his mathematical ideas. His work, described in modern terms in this chapter, was published by Liouville in 1846 but not really appreciated for another 25 years.

Galois also contributed to the theory of numbers and sketched out his ideas relating to elliptic functions and abelian integrals.

---

[11] of Lagrange's and Galois' work wholly in the 'old' style was published as late as 1930.)

However, rather than follow, unerringly, the route described by Artin in [3],we shall, in the hope of maintaining some sense of discovery and involvement on the part of the reader, adopt a slightly more 'experimental' approach in which the results we aim for will be determined more by perceived need—although, if we see a result or idea which looks interesting or of possible future use, we shall record it. One consequence of this is that not all results will be stated as sharply as possible.

7.2 Radical towers and splitting fields

We have already seen (4.2.20, 4.2.21), on being given the polynomial* $f(x)$ in the polynomial ring $F[x]$, F being a field, how we can construct a larger field§† S_f, the *splitting field*‡ of $f(x)$ over F, which contains F and in which $f(x)$ splits into a product of linear factors. To say that all the zeros of $f(x)$ are radically expressible *over* F is, then, simply to say that S_f can be chosen inside some suitable radical tower over F according to the following definition.

Definition 7.2.1 The field R is a **radical tower** over the field F iff there is a sequence $F=E_0 \subset E_1 \subset \cdots \subset E_s = R$ of subfields** of R where, for each i $(1 \leq i \leq s)$, $E_i = E_{i-1}(r_i)$, r_i being a zero of an equation of the form $x^{n_i} - a_i = 0$, where $a_i \in E_{i-1}$. (Clearly there is no loss of generality in assuming that each n_i is a prime. Prove this!) We shall call each E_{i+1} a **radical extension** of E_i.

Thus, formally, we arrive at

Definition 7.2.2 The polynomial $f(x) \in F(x)$ is **soluble by radicals** (over F) iff its splitting field S_f (over F) is contained in some radical tower R over F.

In this format, Galois' idea, following that of Lagrange (but from a different perspective), of (essentially) looking at certain permutations of the zeros of $f(x)$ in S_f (see exercise 11) is, for the most part, reinterpreted in terms of those automorphisms of S_f which fix F elementwise. (See Section 7.4.) That is, the new idea is to study the Galois group $\mathrm{Gal}(S_f/F)$—see exercise 5.3.7. (One advantage of this change is that one is then studying the *whole* of some group, namely $\mathrm{Gal}(S_f/F)$, rather than some obscure (?) subgroup of some group of permutations. See the Remark on p. 221.) Later we shall refer to $\mathrm{Gal}(S_f/F)$ as the *Galois group of* $f(x)$ *over* F.

Earlier (4.2.21) we asserted the uniqueness, up to isomorphism, of S_f. As we are now calling on this uniqueness to define $\mathrm{Gal}(S_f/F)$ unambiguously, we really ought to offer a proof of it.

How could one set about such a proof? Since, from 4.2.21, S_f can be expressed in the form $S_f = F(\alpha_1, \alpha_2, \ldots, \alpha_n)$, the α_i being the roots of $f(x)$ in S_f, an induction argument, taking the α_i one at a time, suggests itself. So surely, we must start with

Lemma 7.2.3 Let S_1, S_2 be splitting fields for $f(x)$ over F and let $\gamma_1 \in S_1, \gamma_2 \in S_2$ be zeros of the same irreducible§§ factor $g(x)$ (of $f(x)$) in $F[x]$. Then the identity

* It seems preferable to use the notation $f(x)$, rather than f, here.
§ A more accurate notation would be $S_f(F)$, or similar.
† There should be little chance of confusing this notation with that for symmetric groups.
‡ Also called a splitting extension of F (by $f(x)$).
** We shall use $\subset$, $\subseteq$ etc. to denote subfields and, later, $<$, $\leqslant$ etc. to denote subgroups.
§§ This condition is forced on us by the desired conclusion. See exercise 13.

map $\iota: F \to F$ can be extended to an isomorphism $\kappa: F(\gamma_1) \to F(\gamma_2)$ such that $\kappa(\gamma_1) = \gamma_2$ and $\kappa(t) = t$ for all $t \in F$.

At this point it is best to admit to a little hindsight. It turns out that 7.2.3 is not general enough for us successfully to apply an induction argument to extend the map ι to an isomorphism between S_1 and S_2. (Exercise 16 asks you to determine why not. When you see the reason, the appropriate modification, 7.2.3′, immediately suggests itself.)

Recalling that our approach is 'experimental', and that we are not, initially, attemping to make our lemmas and theorems as 'tight' as possible, it will be no surprise to find that 7.2.3′ contains a large slice of redundant hypothesis. Can you spot it? In addition yet *more* hindsight would allow a yet smoother and cleaner presentation. However, in the interests of not losing sight of our immediate target, we shall refrain from such generalisation.

Lemma 7.2.3′ Let F_1, F_2 be fields and let $\lambda: F_1 \to F_2$ be an isomorphism between them. Let $f_1(x) = a_0 + a_1x + \cdots + a_nx^n \in F_1[x]$ and let $f_2(x) = \lambda(a_0) + \lambda(a_1)x + \cdots + \lambda(a_n)x^n$ be the corresponding polynomial in $F_2[x]$. Let $g_1(x)$ be an irreducible factor of $f_1(x)$ in $F_1[x]$ and $g_2(x)$ the corresponding (irreducible—exercise 17(b)) factor of $f_2(x)$ in $F_2[x]$. Finally, suppose, for $i = 1, 2$, that γ_i is a zero of $g_i(x)$ (and hence of $f_i(x)$) in the splitting field S_i of $f_i(x)$ over F_i.

Then λ can be extended to an isomorphism μ between $F_1(\gamma_1)$ and $F_2(\gamma_2)$ such that $\mu(\gamma_1) = \gamma_2$ and $\mu(t) = \lambda(t)$ for all $t \in F_1$.

Proof We first note that the isomorphism λ easily extends*, as implied in the statement of the lemma, to an obvious isomorphism $\bar{\lambda}: F_1[x] \to F_2[x]$ which, in turn, gives rise to an isomorphism $\Lambda: \dfrac{F_1[x]}{[g_1(x)]} \to \dfrac{F_2[x]}{[g_2(x)]}$ (exercise 17(a)).

Since, for $i = 1, 2$, each $g_i(x)$ is irreducible in $F_i[x]$, $g_i(x)$ is the minimum polynomial $M_{\gamma_i}(x)$ of γ_i over F_i. (Prove this.) Hence, by 4.3.4, we have for $i = 1, 2$ an isomorphism $\sigma_i: \dfrac{F_i[x]}{[g_i(x)]} \to F_i(\gamma_i)$. We now define μ to be the composition

$$F_1(\gamma_1) \xrightarrow{\sigma_1^{-1}} \frac{F_1[x]}{[g_1(x)]} \xrightarrow{\Lambda} \frac{F_2[x]}{[g_2(x)]} \xrightarrow{\sigma_2} F_2(\gamma_2)$$

of the isomorphisms σ_1^{-1}, Λ and σ_2. Noting that $\sigma_i(x + [g_i(x)]) = \gamma_i$ $(i = 1, 2)$, we readily see that $\mu(\gamma_1) = \gamma_2$, as claimed.

* To prove this is a tedious chore which I happily delegate to *you*!

Did you spot the redundancy? It is, of course, that no use was made of the fact that S_1 and S_2 are splitting fields. (Can you write down the exact hypotheses used in the proof of 7.2.3′?)

As an example of 7.2.3′ in action, consider the isomorphism $\lambda: \mathbb{Q}(i) \to \mathbb{Q}(i)$ given by complex conjugation. Take $f_1(x) = g_1(x) = x^2 - i$, so that $f_2(x) = g_2(x) = x^2 + i$. Let $\gamma_1 = \varepsilon = e^{2\pi i/8}$ be a primitive 8th root of unity. The zeros of $x^2 + i$ are then ε^3 and ε^7. According to 7.2.3′ λ can be extended to an isomorphism μ of $\mathbb{Q}(\varepsilon)$ with itself in which $\mu(\varepsilon) = \varepsilon^3$ (or ε^7).

To extend the isomorphism μ above to one between S_1 and S_2 we proceed by induction—on the dimension $[S_1 : F_1]$ of S_1 over F_1 (see 4.5.6). We shall need to note that $[S_1 : F_1]$ is finite (exercise 12). Our aim is then to prove

Theorem 7.2.4 Continuing with the notation of 7.2.3′, there is an isomorphism $\nu: S_1 \to S_2$ such that $\nu(t) = \lambda(t)$ for all $t \in F_1$.

Proof If $[S_1 : F_1] = 1$ then $f_1(x)$ factorises completely into linear factors in $F_1[x]$. Hence, trivially, the same is true for $f_2(x)$ in $F_2[x]$. Consequently $S_2 = F_2$, so that we can take $\nu = \lambda$ in this case.

Now suppose $[S_1 : F_1] = m > 1$ and that the theorem is true for all splitting extensions $L \supseteq K$ for which $[L:K] < m$. Since $[S_1 : F_1] > 1$, $f_1(x)$ has at least one irreducible factor $h_1(x)$, say, of degree ≥ 2 in $F_1[x]$. Let $h_2(x)\{=\bar{\lambda}(h_1(x))\}$ be the corresponding (irreducible) polynomial in $F_2[x]$.

To make use of the induction hypothesis we note that, for $i = 1, 2$, S_i is a *splitting field for $f_i(x)$ over $F_i(\gamma_i)$* and that $[S_i : F_i(\gamma_i)] < [S_i : F_i]$. (See exercise 10.)

Thus by the induction assumption, taking $F_1(\gamma_1)$, $F_2(\gamma_2)$ rather than F_1, F_2 as the 'base' fields, we can extend the isomorphism $\mu: F_1(\gamma_1) \to F_2(\gamma_2)$ of 7.2.3′ to an isomorphism $\nu: S_1 \to S_2$ such that $\nu(u) = \mu(u)$ for each $u \in F_1(\gamma_1)$—hence, in particular, for each $t \in F$.

Taking $F_1 = F_2 = F$ and λ to be the identity map $\iota: F_1 \to F_2$ we achieve our aim, namely

Theorem 7.2.5 Let F be a field, $f(x) \in F[x]$ and S_1, S_2 be splitting fields for $f(x)$ over F. Then $S_1 \cong S_2$ under some isomorphism ν. Furthermore ν can be chosen so that it acts as the identity map when restricted to F.

Having tidied up that point, let us think how we might find something out about $\mathrm{Gal}(S_f/F)$, given that S_f is itself contained in some radical tower R over F. Although there are infinitely many different possible choices for $R(\supseteq S_f)$, each such R is constructed in the same general way from a succession of simple-looking radical extensions starting with F. We might hope that the existence *in each tower R* of these simple 'slices' is somehow reflected in the structure of $\mathrm{Gal}(S_f/F)$. All this suggests one line of attack on $\mathrm{Gal}(S_f/F)$: first find out how $\mathrm{Gal}(R/F)$ is related to its component parts—namely the various $\mathrm{Gal}(E_i/E_{i-1})$—and then how the structure of $\mathrm{Gal}(R/F)$ is inherited, if at all, by its smaller brother $\mathrm{Gal}(S_f/F)$. (An immediate, but rather trifling, relationship between certain Galois groups is given in exercise 21.)

Exercises

1 Use the method of exercise 4.6.6 to try to show that the zeros of $x^{11}-1$ are expressible in radical form over $\mathbb{Q}$. What is the main problem you come across?

2 Let $F=\mathbb{Q}$, $E_1=\mathbb{Q}(\sqrt{2})$, $E_2=E_1(\sqrt{(1+\sqrt{2})})$. Find the minimum polynomial over F of $\sqrt{(1+\sqrt{2})}$. Is E_2 its splitting field?

3 Find a polynomial $f(x)$ in $\mathbb{Q}(x)$ having $\sqrt{3}+\sqrt[4]{(1+\sqrt[3]{5/2})}$ as a zero. Exhibit a radical tower over $\mathbb{Q}$ which contains the splitting field of $f(x)$.

4 Find a splitting field S_f and a radical tower containing S_f for each of the following polynomials over $\mathbb{Q}$: (i) x^4+5x^2+6; (ii) x^4-10x^2+1; (iii) x^5-1; (iv) x^7-1; (v) x^9-1; (vi) x^6+1; (vii) x^3+x+1.

5 Confirm that $\mathbb{Q}(\sqrt{2},\sqrt{3})$ is a splitting field for x^4-10x^2+1 over $\mathbb{Q}$. Find (i) a polynomial of degree 6; (ii) a polynomial $a_0+a_1x+a_2x^2+a_3x^3+x^4$, with none of the $a_i=0$; for which $\mathbb{Q}(\sqrt{2},\sqrt{3})$ is also a splitting field.

6 Show that if $K\subseteq L\subseteq M$ with M a radical tower over K, then M is a radical tower over L. Do you think L is necessarily a radical tower over K? (See exercise 7.10.3.)

7 Show that each polynomial $f(x)=a_0+a_1x+\cdots+a_nx^n$ in $\mathbb{R}[x]$ is soluble by radicals over $\mathbb{R}$. Do you think it might also be soluble by radicals over $\mathbb{Q}(a_0,a_1,\ldots,a_n)$? Is $x^2+(\mathrm{e}+\sqrt{\pi})x+\pi\sqrt{\mathrm{e}}$ soluble by radicals over $\mathbb{Q}(\sqrt{\mathrm{e}},\sqrt{\pi})$?

8 Let F be a field of characteristic $\neq 2$. Show that each polynomial ax^4+bx^2+c $(a\neq 0)$ is soluble by radicals over F.

9 (a) Let K be a finite field. Show that K is a splitting extension and a radical tower over its prime subfield.
(b) Find splitting fields for $x^3-\hat{3}$ over $\mathbb{Z}_7$ and for $x^3-\hat{5}$ over $\mathbb{Z}_{13}$.

10 Let M be a field such that $F\subseteq M\subseteq S_f$. Show that S_f is a splitting field of some polynomial over M. [Hint: What polynomial springs to mind?!]

11 Let $S=F(\alpha_1,\alpha_2,\ldots,\alpha_n)$ be a field extension of F. Show that each automorphism σ of S which is the identity on F is completely determined once $\sigma(\alpha_i)$ is known for each $i(1\le i\le n)$. Show further that each element of $\mathrm{Gal}(S_f/F)$ gives rise to a permutation on the zeros of $f(x)$ in S_f. Finally show that distinct elements of $\mathrm{Gal}(S_f/F)$ give rise to distinct permutations.

12 (a) Prove, with $F[x]$, $f(x)$, S_f and R as in 7.2.2., that $[S_f:F]$, $[R:F]$, $|\mathrm{Gal}(S_f/F)|$ and $|\mathrm{Gal}(R/F)|$ are finite. (b) Show that, in fact, $[S_f:F]\le n!$ where $n=\deg(f(x))$.

13 Two of the (complex) zeros of $f(x)=x^4-2x^3+5x^2-4x+6$ are $\mathrm{i}\sqrt{2}$ and $1+\mathrm{i}\sqrt{2}$. Is there an element of $\mathrm{Gal}(S_f/\mathbb{Q})$ which maps one onto the other? Answer the same question for the zeros $\sqrt{2}$ and $\mathrm{i}\sqrt{2}$ of the polynomial x^4-4.

14 Let $f(x)=f_1(x)f_2(x)\ldots f_r(x)$ be a product of distinct irreducibles in $F[x]$. Show that under any element of $\mathrm{Gal}(S_f/F)$, each zero of $f_i(x)$ can only map to a zero of $f_i(x)$.

15 Show that the isomorphism κ of 7.2.3 is the only isomorphism between $F(\gamma_1)$ and $F(\gamma_2)$ which is the identity on F and for which $\kappa(\gamma_1)=\gamma_2$.

16 Explain why 7.2.3 has to be modified (to 7.2.3′).

17 (a) Establish the isomorphisms $\bar{\lambda}$, Λ of 7.2.3′. (b) Prove (see 7.2.3′) that if $g_1(x)$ is irreducible in $F_1[x]$, then $g_2(x)$ is irreducible in $F_2[x]$.

18 In 7.2.3′ put $F_1=F_2=\mathbb{Q}(\sqrt{2})$ and let $\lambda: F_1\to F_2$ be defined by $\lambda(\sqrt{2})=-\sqrt{2}$. Let S_1 be the splitting field of $g_1(x)=x^2-\sqrt{2}$ over F_1. Determine the splitting field S_2 of $g_2(x)$ (as in 7.2.3′) over F_2 and describe the isomorphism μ.

19 (a) Does the automorphism σ of $\mathbb{Q}(\sqrt{2})$ given by $\sigma(\sqrt{2})=-\sqrt{2}$ extend to an automorphism of $\mathbb{Q}(\sqrt{(1+\sqrt{2})})$? In how many ways?
(b) Does the automorphism τ of $\mathbb{Q}(\sqrt{6})$ given by $\tau(\sqrt{6})=-\sqrt{6}$ extend to an automorphism of $\mathbb{Q}(\sqrt{2},\sqrt{3})$? In how many ways?

20 Let $K\subseteq L$ be fields with $[L:K]<\infty$. Show that if $\sigma: L\to L$ is a *1–1 homomorphism which acts as the identity on K, then σ maps L *onto* L. [Hint: Let $\alpha\in L$ and let $g(x)$ be the minimum polynomial of α. Let Z be the set of zeros of $g(x)$ in L. Show that $\sigma(Z)\subseteq Z$. Since σ is 1–1, σ maps Z *onto* Z. Or: show that a basis for L over K maps to a set of elements (of $\sigma(L)$) which is linearly independent over K.]

21 Show that if $K\subseteq L\subseteq M$ are fields, then $\mathrm{Gal}(M/L)\leq\mathrm{Gal}(M/K)$.

7.3 Examples

As it is always a good idea when beginning a mathematical investigation, let us start by looking at some particular examples. All are radical extensions or towers—some, but not all, are splitting extensions.

Example 7.3.1 (i) Let $F=\mathbb{Q}$ and $r=\sqrt{2}$, the positive square root of 2. Set $R=\mathbb{Q}(r)=\{a_0+a_1\sqrt{2}: a_0, a_1\in\mathbb{Q}\}\subset\mathbb{R}$. Each automorphism σ of R automatically 'fixes' F elementwise (i.e. $\sigma(t)=t$ for all $t\in F$—exercise 3.10.8). On the other hand $\{\sigma(r)\}^2=\sigma(r^2)=\sigma(2)=2$, so that $\sigma(r)=\pm\sqrt{2}$. It follows that

$$\sigma(a_0+a_1\sqrt{2})=\sigma(a_0)+\sigma(a_1)\sigma(\sqrt{2})=a_0+a_1(\pm\sqrt{2}).$$

So, evidently, there are just two automorphisms of $\mathbb{Q}(\sqrt{2})$ (each leaving $\mathbb{Q}$ fixed). [See exercise 3.10.1(i).]
(ii) Let $F=\mathbb{Q}$ and $r=\sqrt[3]{2}$, the real cube root of 2. Set $R=\mathbb{Q}(r)=\{a_0+a_1r+a_2r^2: a_0, a_1, a_2\in\mathbb{Q}\}\subset\mathbb{R}$—see 4.3.5.

* Usually called a *monomorphism*. See 7.4.3.

Again each automorphism σ of R automatically fixes $\mathbb{Q}$ elementwise. But that isn't *all* that σ fixes. Indeed, since $r^3=2$, we find $\{\sigma(r)\}^3=\sigma(r^3)=\sigma(2)=2$. Since R contains only real numbers, $\sigma(r)$ must be the (unique) real cube root of 2. That is, $\sigma(r)=r$. But then σ fixes *all* elements in R. In other words, σ is the identity automorphism of R.

On the other hand, roots of unity yield nice results. As an example:
(iii) Let $F=\mathbb{Q}$ and and $R=\mathbb{Q}(r)$ where $r(\neq 1)$ is a (primitive) complex pth root of unity, p being a prime. Then $r=e^{2k\pi i/p}$ (for some k, $1\leq k<p$) and r has minimum polynomial $x^{p-1}+x^{p-2}+\cdots+x+1$ (cf.: 1.9.17(ii)). Yet again, if σ is an automorphism of R, then σ fixes F elementwise, and so $\sigma(r)$ also satisfies $x^{p-1}+x^{p-2}+\cdots+x+1=0$. (Cf. exercise 7.2.14.) Thus $\sigma(r)=e^{2j\pi i/p}$ for some $j(1\leq j<p)$, that is, $\sigma(r)=r^u$ for some $u(1\leq u<p)$. Furthermore *each such u produces a different automorphism of R* (exercise 2).

As 'two-step' examples we offer:
(iv) Let $F=\mathbb{Q}$, $E_1=F(\sqrt{2})$, $E_2=E_1(\sqrt{3})=R$. Then

$$R=\{a_0+a_1\sqrt{2}+a_2\sqrt{3}+a_3\sqrt{6}:a_0,\,a_1,\,a_2,\,a_3\in\mathbb{Q}\}.$$

Here there are at most 4 automorphisms of R. For: each such automorphism σ fixes F; $\sigma(\sqrt{2})=\sqrt{2}$ or $-\sqrt{2}$ and $\sigma(\sqrt{3})=\sqrt{3}$ or $-\sqrt{3}$, and this information suffices to determine the action of σ on each element of R. We leave you (exercise 6(a)) to check that each of the potential automorphisms is indeed one. (In due course see 7.4.7.)
(v) Let $F=\mathbb{Q}$, $E_1=F(\omega)$ and $E_2=E_1(r)=R$, where $\omega=(-1+i\sqrt{3})/2$ is a complex cube root of 1 and r is the real cube root of 2. Here

$$R=\mathbb{Q}(\omega,r)=\{a_0+a_1r+a_2r^2+a_3\omega+a_4\omega r+a_5\omega r^2:a_0,\,a_1,\,a_2,\,a_3,\,a_4,\,a_5\in\mathbb{Q}\}.$$

Once again each automorphism σ of R fixes F. Also, as in (ii), $(\sigma(r))^3=2$, but, this time, $\sigma(r)=r$ or ωr or $\omega^2 r$. Similarly $\sigma(\omega)=\omega$ or ω^2. As in (iv) these observations determine the action of σ on every element of R. It follows that there are at most 6 automorphisms of R (all fixing $\mathbb{Q}$). Again we leave the checking (fairly lengthy if done naïvely!) to you that there are exactly 6. (7.4.7, together with 7.5.1, gives this result quickly.)

Other interesting examples include:
(vi) Let $F=\mathbb{Q}(\sqrt{2})$, $E_1=F(\sqrt[4]{2})$, $E_2=E_1(\sqrt[8]{2})=R$. Put $\gamma=\sqrt[8]{2}$, the positive (real) eighth root of 2. Then γ has minimum polynomial $x^4-\sqrt{2}$ over F and so the only automorphisms of R fixing F are those determined by $\gamma\to\gamma$ and $\gamma\to-\gamma$.
(vii) Let $F=\mathbb{Z}_2$ and let R be the field of 4 elements. Since each non-zero element of R satisfies the equation $x^3-\hat{1}=0$ we see that R is both a radical extension and a splitting extension of F. Denoting the elements of R by 0, 1, α and $1+\alpha$ allows us to check easily that the map given by $\alpha\to 1+\alpha$ determines an automorphism of R. Thus there are exactly two automorphisms of R (each 'fixing' $\mathbb{Z}_2$).
(viii) Let $F=\mathbb{Q}$, $E_1=F(\sqrt{2})$, $E_2=E_1(\sqrt[3]{(3+\sqrt{2})})$. Then $\gamma(=\sqrt[3]{(3+\sqrt{2})})$ has minimum polynomial $x^3-(3+\sqrt{2})$ over E_1 and x^6-6x^3+7 (why?) over F. The zeros of this sextic are $\sqrt[3]{(3+\sqrt{2})}$, $\sqrt[3]{(3-\sqrt{2})}$ and four complex (i.e. non-real)

zeros. One can show (exercise 16) that $\sqrt[3]{(3-\sqrt{2})} \notin E_2$. As a consequence (as in (ii)) the only automorphism of E_2 is the identity map.

Exercises

1 Let $n \in \mathbb{Z}^+$. Find the number of automorphisms of $\mathbb{Q}(\sqrt[n]{2})$ if (i) n is odd; (ii) n is even.

2 Show, using 7.2.3′, that, in 7.3.1(iii), for each u such that $1 \leq u \leq p-1$, there *is* an automorphism σ of $\mathbb{Q}(r)$ for which $\sigma(r)=r^u$.

3 Let $\omega \neq 1$ be a cube root of unity. Does $\mathbb{Q}(\omega)$ have an automorphism σ for which $\sigma(\omega)=-\omega$?

4 Let α be a non-real cube root of 2. Is the subfield $\mathbb{Q}(\alpha)$ of $\mathbb{C}$: (i) a radical tower over $\mathbb{Q}$; (ii) a splitting extension of x^3-2 over $\mathbb{Q}$?

5 Describe informally as in 7.3.1: the automorphisms (all fixing $\mathbb{Q}$) of (i) $\mathbb{Q}(\mathrm{i}\sqrt{3})$; (ii) $\mathbb{Q}(\mathrm{i}, \sqrt{3})$; (iii) $\mathbb{Q}(\mathrm{i}, \sqrt[4]{2})$; (iv) $\mathbb{Q}(\sqrt{2}, \sqrt{-2})$ and of (v) $\mathbb{Q}(\mathrm{i}\sqrt[4]{3})$ fixing $\mathbb{Q}(\sqrt{3})$; (vi) $\mathbb{Q}(\omega, \sqrt{-3})$ fixing $\mathbb{Q}(\sqrt{-3})$; (vii) $\mathbb{Q}(\sqrt[3]{5}, \mathrm{i}\sqrt{3})$ fixing $\mathbb{Q}(\mathrm{i}\sqrt{3})$.

6 (a) Show that there are exactly 4 automorphisms of $\mathbb{Q}(\sqrt{2}, \sqrt{3})$ and exactly 6 automorphisms of $\mathbb{Q}(\omega, \sqrt[3]{2})$. [Hint: Use the methods of 7.2.3′ and 7.2.4.] (b) Informally, how many automorphisms are there of $\mathbb{Q}(\sqrt{2}, \sqrt{3}, \sqrt{5})$?

7 How many automorphisms are there of $\mathbb{Q}(\sqrt[3]{2}, \sqrt[3]{3}, \omega)$ (i) over $\mathbb{Q}$; (ii) over $\mathbb{Q}(\sqrt[3]{2})$; (iii) over $\mathbb{Q}(\omega)$?

8 Write down in full the permutations of the zeros corresponding to the automorphisms of R in 7.3.1(v).

9 Indicate informally how many automorphisms there are of the splitting field over $\mathbb{Q}$ of (i) x^2-3; (ii) x^3-3; (iii) x^4-3; (iv) x^5-3?

10 How many elements are there in $\mathrm{Gal}(S_f/\mathbb{Q})$ if $f(x)=$: (i) x^4+1; (ii) x^4+x^2+1; (iii) x^3+2x-1; (iv) $(x^2+2x-1)^2(x^3-2)$?

11 If $F \subseteq E$ are fields with $[E:F]=2$ show that, if the characteristic of $F \neq 2$, there exists a non-trivial automorphism of E fixing F.

12 Let p be an odd prime. Write the splitting field S_f of x^p-2 over $\mathbb{Q}$ as a sequence of radical extensions. What is $|\mathrm{Gal}(S_f/\mathbb{Q})|$?

13 Find a radical tower R of least dimension over $\mathbb{Q}$ containing α where $\alpha=\sqrt{2}+\sqrt[3]{5}$. How many automorphisms does R have? Let $f(x)$ be the minimum polynomial of α over $\mathbb{Q}$. How many elements are there in: (i) $\mathrm{Gal}(S_f/\mathbb{Q}(\sqrt{2}))$; (ii) $\mathrm{Gal}(S_f/\mathbb{Q}(\sqrt[3]{5}))$?

14 Let $f(x)$ be the minimum polynomial of $\alpha=\sqrt{(1+\sqrt{2})}$ in $\mathbb{Q}[x]$. Find the orders of $\mathrm{Gal}(S_f/\mathbb{Q})$ and $\mathrm{Gal}(S_f/\mathbb{Q}(\sqrt{2}))$.

15 What is the order of Gal($\mathbb{Q}(\cos 72°)/\mathbb{Q}$)?

16 Show that the six zeros, in $\mathbb{C}$, of x^6-6x^3+7 are as stated in 7.3.1(viii) and that $\sqrt[3]{(3-\sqrt{2})} \notin \mathbb{Q}(\sqrt[3]{(3+\sqrt{2})})$. [Hint: Assume $z=\sqrt[3]{(3-\sqrt{2})} \in \mathbb{Q}(\sqrt[3]{(3+\sqrt{2})}) = \mathbb{Q}(t)=K$, say. Then $7^{1/3}=a1+bt+ct^2$ where $a,b,c \in \mathbb{Q}(\sqrt{2})$. Cube up to get $7=A1+Bt+Ct^2$ where $A,B,C \in \mathbb{Q}(\sqrt{2})$. But $7 \in \mathbb{Q} \subseteq \mathbb{Q}(\sqrt{2})$. Hence $A=7$, $B=C=0$. Deduce that if $a \neq 0$ then $b=c=0$, whilst if $a=0$ then $b=0$ or $c=0$. Hence $7^{1/3}=a$ (if $a \neq 0$) or bz or cz^2 (if $a=0$). If $7^{1/3}=cz^2$ then $c^3=(3+\sqrt{2})/(3-\sqrt{2})$. Setting $c=(a+b\sqrt{2})/d$ (where $a,b,d \in \mathbb{Z}$ and $(a,b,d)=1$), try to show that this equality is impossible in $\mathbb{Q}(\sqrt{2})$.]

7.4 Some Galois groups: their orders and fixed fields

We can make a number of interesting observations about the above examples. For one of these it is helpful if we first make (cf. exercise 5.3.7).

Definition 7.4.1 Let L be a field and let Σ be a (non-empty) set of automorphisms of L. The **fixed field**, *Fix(Σ), of Σ is the set $\{t \in L : \sigma(t)=t \text{ for all } \sigma \in \Sigma\}$.

One would *expect* Fix(Σ) to be a subfield of L. Exercise 2 invites you to confirm this.

Let us reconsider 7.3.1.

Example 7.4.2 In (i), $R=\mathbb{Q}(r)$ *is* a splitting field of $f(x)=x^2-2$ over $F=\mathbb{Q}$, $|\text{Gal}(R/F)|=2=[R:F]$ and $\text{Fix}(\text{Gal}(R/F))=F$;
In (ii), $R=\mathbb{Q}(r)$ is *not* a splitting field of $f(x)=x^3-2$ over $F=\mathbb{Q}$, $|\text{Gal}(R/F)|=1<[R:F]=3$ and $\text{Fix}(\text{Gal}(R/F))=R$;
In (iii), $R=\mathbb{Q}(r)$ *is* a splitting field of $f(x)=x^p-1$ over $F=\mathbb{Q}$, $|\text{Gal}(R/F)| = p-1=[R:F]$ and $\text{Fix}(\text{Gal}(R/F))=F$;
In (iv), $R=\mathbb{Q}(\sqrt{2},\sqrt{3})$ *is* a splitting field of $f(x)=x^4-10x^2+1$ over $F=\mathbb{Q}$, $|\text{Gal}(R/F)|=4=[R:F]$ and $\text{Fix}(\text{Gal}(R/F))=F$;
In (v), $R=\mathbb{Q}(\omega,r)$ *is* a splitting field of $f(x)=x^3-2$ over $F=\mathbb{Q}$, $|\text{Gal}(R/F)]=6=[R:F]$ and $\text{Fix}(\text{Gal}(R/F))=F$;
In (vi), $R=\mathbb{Q}(\sqrt[8]{2})$ is *not* a splitting field of $f(x)=x^8-2$ over $F=\mathbb{Q}(\sqrt{2})$, $|\text{Gal}(R/F)|=2<[R:F]=4$ and $\text{Fix}(\text{Gal}(R/F))=E_1$.
In (vii), R *is* a splitting field of $f(x)=x^2+x+1$ over $F=\mathbb{Z}_2$, $|\text{Gal}(R/F)|=2=[R:F]$ and $\text{Fix}(\text{Gal}(R/F))=F$.
In (viii), $R=\mathbb{Q}(\sqrt[3]{(3+\sqrt{2})})$ is *not* a splitting field of $f(x)=x^6-6x^3+7$ over $F=\mathbb{Q}$, $|\text{Gal}(R/F)|=1<[R:F]=6$ and $\text{Fix}(\text{Gal}(R/F))=R$.

The assertions in (i), (iii), (iv), (v) and (vii) regarding fixed fields and the orders of the Galois groups look attractive. Have we happened to choose particularly nice examples or is there something 'deeper' at work here? No prizes for

* Again, a better notation would be Fix(Σ, L), or similar.

guessing the answer! Note, too, that we have given no example of a splitting extension which is not itself also a radical tower. Is *this* another universal truth? We leave the reader to ponder this for a while. (A plausible (?) proof there there is *no such example* is: Surely each splitting field $S_f \supseteq F$ which is contained in some radical tower $F \subset E_1 \subset \cdots \subset E_s$ is itself radical over F via the sequence $F \subseteq E_1 \cap S_f \subseteq \cdots \subseteq E_s \cap S_f = S_f$? In due course see exercise 7.10.3.)

Let us begin by looking at the apparent bounds placed on the orders of the Galois group. Since not all the parts of 7.4.2 concern splitting extensions we shall learn from our experience with 7.2.3′ and 7.2.4 and keep our options open by working as generally as seems prudent. Although we are going to *count* automorphisms rather than merely show their existence, it is not too surprising to find that the same line of argument used earlier will prove useful. Accordingly we shall only sketch the proof of 7.4.3. We hope the brevity here will help you more readily see what makes the earlier proof tick.

Lemma 7.4.3 Suppose $K_1 \subseteq L_1$ and $K_2 \subseteq L_2$ are fields with $[L_1:K_1] < \infty$. Further suppose that θ is an isomorphism from K_1 onto K_2. Then there are at most $[L_1:K_1]$ ways of extending θ to a monomorphism* of L_1 into L_2.

Proof We first check that the induction step will work.

Choose γ_1 in $L_1 \backslash K_1$ and let $g_1(x) \in K_1[x]$ be the minimum polynomial of γ_1. Suppose that in L_2 there exists an element β whose minimum polynomial $g_2(x)$ (over K_2) is $\theta(g_1(x))$ (cf. the map λ in 7.2.3′ – and note that, this time, there is no guarantee that such a β exists – exercise 7.2.19). Then, as in 7.2.3′ we can find an isomorphism $\mu: K_1(\gamma_1) \rightarrowtail K_2(\beta)$ extending θ and such that $\mu(\gamma_1) = \beta$. Since each root of $g_2(x)$ *which lies in* L_2 can give rise to such an extension of θ and since, for each such extension, $\mu(\gamma_1)$ is necessarily a zero of $g_2(x)$, we see that no more than $\deg g_2(x)$ $(= \deg g_1(x) = [K_1(\gamma):K_1])$ such extensions are possible. [As an obvious, but important, aside let us note here that there will be *exactly* $[K_1(\gamma_1):K_1]$ such extensions if $g_2(x)$ has $\deg g_2(x)$ *distinct zeros in* L_2.]

We now begin the induction—cf. 7.2.4.

If $[L_1:K_1] = 1$ there is clearly only one 'extension' of θ—itself! If $[L_1:K_1] = m > 1$ choose $\gamma_1 \in L_1 \backslash K_1$ as above. Then $\theta: K_1 \rightarrow K_2$ can be extended to a map $\mu: K_1(\gamma_1) \rightarrowtail K_2(\beta)$ in at most $[K_1(\gamma_1):K_1]$ ways. By induction, replacing $K_1 \subseteq L_1$ and $K_2 \subseteq L_2$ by $K_1(\gamma_1) \subseteq L_1$ and $K_2(\beta) \subseteq L_2$, each such μ can be extended to a map $\nu: L_1 \rightarrow L_2$ in at most $[L_1:K_1(\gamma_1)]$ ways. It follows immediately that there are *at most* $[K_1(\gamma_1):K_1][L_1:K_1(\gamma_1)]$ $(= [L_1:K_1])$ ways of extending the isomorphism $\theta: K_1 \rightarrow K_2$ to a monomorphism $\nu: L_1 \rightarrow L_2$. (Incidentally *all* monomorphisms from L_1 to L_2 must be obtainable by this 'climbing up' method. Can you see why?)

An immediate and interesting corollary is:

Corollary 7.4.4 Let $K \subseteq L$ be fields with $[L:K] < \infty$. Then $|\text{Gal}(L/K)| \leq [L:K]$.

* We prefer to use this (standard) terminology for a homomorphism which is 1–1 but not necessarily onto, to the word 'embedding' introduced in 3.10.2.

Proof Exercise 15.

This result helps to explain the relationships between $|\text{Gal}(R/F)|$ and $[R:F]$ in all sections (splitting field or not) of 7.4.2. Can we go further and show that equality always holds whenever R is a splitting field over F?

From the aside in the proof of 7.4.3 it is clear that to turn the *inequalities* in the statements of 7.4.3 (resp. 7.4.4) into *equalities* we only need to assume that the polynomial $g_2(x)$—and its equivalents which are implicitly present at the induction step—has as many *distinct* zeros in L_2 (resp. L) as its degree. This requirement of distinctness strongly suggests that we should introduce

Definition 7.4.5 Let F be a field, $f(x) \in F[x]$ and S_f be its splitting field. We say that $f(x)$ is **separable** over F iff *each irreducible factor* of $f(x)$ in $F[x]$ has no repeated zeros in S_f.

Simply declaring that each of the $g(x)$ appearing in the proof of 7.4.3 be separable looks like a distinct 'fiddle', especially if it is a bit difficult to identify exactly which $g(x)$ are involved. (Perhaps we could just assume that 'all polynomials arising' are separable.) However, in the case in which we are presently interested, namely the splitting field of a given polynomial, the $g(x)$ which arise can be easily identified—as being factors of $f(x)$ which are irreducible over various fields containing F. And, of course, *their* separability is then subsumed by that of $f(x)$ itself.

The following result is an immediate consequence of these remarks.

Theorem 7.4.6 Let $\lambda: F_1 \to F_2$ be an isomorphism, let $f_1(x)$ and $f_2(x)$ correspond under the natural extension $\bar{\lambda}: F_1[x] \mapsto F_2[x]$ of λ and let S_1 and S_2 denote, as usual, splitting fields for $f_1(x)$ and $f_2(x)$ over F_1 and F_2, respectively. If $f_1(x)$ is separable over F_1 then $f_2(x)$ is separable over F_2 (exercise 20) and λ can be extended to an isomorphism of S_1 onto S_2 in exactly $[S_1:F_1]$ distinct ways.

The result we are looking for then follows if we specialise 7.4.6 to the case where $F_1 = F_2 = F$ (say), $f_1(x) = f_2(x) = f(x)$ (say), $S_1 = S_2 = S_f$ and take λ to be the identity map $\iota: F \to F$. For then we are counting the number of automorphisms of S_f which leave F fixed. That is, we have

Theorem 7.4.7 Let S_f be the splitting field of the separable polynomial $f(x)$ over the field F. Then $|\text{Gal}(S_f/F)| = [S_f:F]$.

7.4.4 and 7.4.7 explain fully the relationships between the $|\text{Gal}(R/F)|$ and the $[R:F]$ in 7.4.2. In addition, 7.4.7 confirms, without the need for extensive calculation, that the $p-1$ (respectively 4, 6) 'maps' σ of 7.3.1 (iii) (respectively (iv), (v)) are indeed automorphisms of R.

What about fixed fields? The various parts of 7.4.2 suggest that, if $|\text{Gal}(S_f/F)| = [S_f:F]$, then $\text{Fix}(\text{Gal}(S_f/F)) = F$. That this is true generally is the content of

Theorem 7.4.8 Let F, $f(x)$ and S_f be as in 7.4.7. Then $\text{Fix}(\text{Gal}(S_f/F)) = F$.

Proof Suppose $\mathrm{Fix}(\mathrm{Gal}(S_f/F)) = H \supsetneqq F$. Then $\mathrm{Gal}(S_f/F) = \mathrm{Gal}(S_f/H)$ – exercise 21. Consequently, by 7.4.7, $[S_f:F] = [S_f:H]$. But $[S_f:F] = [S_f:H][H:F] > [S_f:H]$ (see 4.6.1) – a contradiction, since $[H:F] > 1$, by assumption.

The information given in 7.4.2 (ii), (vi), (viii) regarding the fixed fields of these non-splitting (over F) extensions is at least partly explained by 7.4.4.

Exercises

1 Let S_f be the splitting field over $\mathbb{R}$ of the minimum polynomial over $\mathbb{R}$ of $e + i\pi$. Write down this polynomial and find $\mathrm{Gal}(S_f/\mathbb{R})$.

2 (i) Prove that $\mathrm{Fix}\,\Sigma$ (see 7.4.1) *is* a subfield of L. (ii) Show that $\mathrm{Fix}\,\Sigma = \mathrm{Fix}\,G$, where G is the subgroup of $\mathrm{Aut}\,L$ generated by Σ.

3 Let G be a group of automorphisms of the field L. Prove that the prime subfield P of L is contained in $\mathrm{Fix}\,G$ and that $\mathrm{Aut}\,L = \mathrm{Gal}(L/P)$.

4 Find the fixed fields over $\mathbb{Q}$ of: (i) $\mathrm{Aut}(\mathbb{Q}(\sqrt[4]{2}))$; (ii) $\mathrm{Aut}(\mathbb{Q}(\sqrt[3]{2}, \sqrt[4]{3}, \omega))$; (iii) $\mathrm{Aut}(\mathbb{Q}(\sqrt[3]{2}, \sqrt[4]{3}, \mathrm{i}))$.

5 Find the fixed fields of $\mathrm{Aut}(\mathbb{Q}(\sqrt[3]{2}, \sqrt[4]{3}, \mathrm{i}))$ over (i) $\mathbb{Q}(\mathrm{i})$; (ii) $\mathbb{Q}(\sqrt[3]{2})$.

6 How many subfields has $F = \mathbb{Q}(\sqrt{2}, \sqrt{3})$? How many can act as $\mathrm{Fix}\,\sigma$ for some automorphism σ of F?

7 What are the fixed fields of the automorphisms $\rho, \sigma, \tau : \mathbb{Q}(x) \mapsto \mathbb{Q}(x)$ given by: (i) $\rho(x) = -x$; (ii) $\sigma(x) = x/2$; (iii) $\tau(x) = 1 + x$? [Hint: for (ii) consider $f(x)/g(x) = h(x)$. If $\deg f(x) = \deg g(x)$, look at $h(x) - r$ for suitable $r \in \mathbb{Q}$.]

8 Let $G_i\,(1 \le i \le n)$ be subgroups of $\mathrm{Gal}(L/K)$. Prove* $\mathrm{Fix}\langle \cup G_i \rangle = \cap(\mathrm{Fix}\,G_i)$. Is it necessarily true that $\mathrm{Fix}(\cap G_i) = \langle \cup \mathrm{Fix}\,G_i \rangle$?

9 Let S_f be the splitting field of $x^8 - 1$ over $\mathbb{Q}$. What are the fixed fields of: (i) σ; (ii) σ^2; where σ is the automorphism determined by $\sigma(\xi) = \xi^5$ where $\xi = e^{2\pi i/8}$?

10 Let S_n, regarded as the group of all permutations on $x_1, x_2, \ldots, x_n$, give rise to automorphisms of $\mathbb{Q}(x_1, x_2, \ldots, x_n)$ in the obvious way (c.f. 5.5.9). Take a guess at what $\mathrm{Fix}(S_n)$ is.

11 Prove the assertions in 7.4.2 (i), (iii), (iv) regarding their fixed fields.

12 Prove the assertions in 7.4.2 (vi), (viii) that R is not a splitting field and that the fixed field is larger than F.

* In what follows $\langle X \cup Y \rangle (\langle \cup X_i \rangle)$ etc. will denote the smallest subgroup or subfield (as appropriate) which contains X and Y (all the X_i).

13 Prove the assertions in 7.4.2 (ii), (iii), (v), (viii) concerning $|\mathrm{Gal}(R/F)|$ and $[R:F]$.

14 Determine whether or not $[R:\mathbb{Q}]=|\mathrm{Gal}(R/\mathbb{Q})|$ if: (i) $R=\mathbb{Q}(\sqrt{(1+\sqrt{2})})$; (ii) $R=\mathbb{Q}(\sqrt{(2+\sqrt{3})})$; (iii) $R=\mathbb{Q}(\sqrt{(1+\sqrt{3})}, \mathrm{i})$.

15 Prove 7.4.4.

16 Confirm, using 7.4.7, the orders of the Galois groups given in 7.4.2 (iii), (iv), (v).

17 Given fields $K\subseteq L\subseteq M$ do we always have $|\mathrm{Gal}(M/K)|=|\mathrm{Gal}(M/L)|\cdot|\mathrm{Gal}(L/K)|$?

18 Show that for each $k\in\mathbb{Z}^+$ the polynomial $x^k+x^{k-1}+\cdots+x+1$ is separable over $\mathbb{Q}$.

19 Prove that if $f(x)\in F[x]$ is separable over F and if $F\subset E$, then $f(x)$ is separable over E.

20 Let $\lambda: F_1\to F_2$ be an isomorphism of fields. Show that if $f_1(x)\in F_1[x]$ and $f_2(x)=\bar{\lambda}(f_1(x))\in F_2[x]$ (see 7.2.3′) then $f_1(x)$ is separable over F_1 iff $f_2(x)$ is separable over F_2.

21 Show, as asserted in 7.4.8 that, if $\mathrm{Fix}(\mathrm{Gal}(S_f/F))=H$, then $\mathrm{Gal}(S_f/F)=\mathrm{Gal}(S_f/H)$.

22 Let $F=\mathbb{Z}_p(t)$ (cf. exercise 4.5.13) where t is an 'indeterminate'. Let $f(x)=x^p-t\in F[x]$ and let S_f be a splitting field for $f(x)$ over F. Show that if α is a zero of $f(x)$ in S_f, then $(x-\alpha)^p=x^p-t$ in $S_f[x]$.

7.5 Separability and normality

In Section 7.4 we were virtually *forced* to introduce the idea of separability for polynomials. Before continuing we ought to determine if this imposes any serious restrictions upon us, there being little point in developing a theory which is applicable to only a small number of examples. Fortunately separability is *not* a very strong restriction. There is, as they say, 'a lot of it about' (cf. exercises 2–4 in Section 1.11).

We have

Theorem 7.5.1 Let F be any field of characteristic 0. Then each polynomial $f(x)$ in $F[x]$ is separable over F.

Proof Let $g(x)$ be an irreducible factor in $F[x]$ of $f(x)$ (and so of degree ≥ 1) and let $g'(x)$ be its (usual) derivative. (See exercise 1.11.2.) Since $\deg g'(x)<\deg g(x)$, the irreducibility of $g(x)$ forces the gcd $(g(x), g'(x))$ to be 1. Consequently (cf. exercises 1.9.17, 1.10.3) there exist, in $F[x]$, polynomials $s(x)$, $t(x)$ such that $1=s(x)g(x)+t(x)g'(x)$ in $F[x]$. Regarding this as an equality in

$S_f[x]$ we obtain $1 = s(\alpha)g(\alpha) + t(\alpha)g'(\alpha)$ for each α in S_f. If, in particular, $x - \alpha$ is a repeated factor of $g(x)$ in $S_f[x]$, so that $x - \alpha$ is a factor of $g'(x)$ in $S_f[x]$ (exercise 1.11.2), we obtain $1 = s(\alpha) \cdot 0 + t(\alpha) \cdot 0$ in S_f, a manifest contradiction.

A similar result can be obtained if F is a finite field (see exercise 13(c)). Note, however, that there do exist examples of infinite fields F of prime characteristic such that *not all* polynomials in $F[x]$ are separable over F—see exercises 4.5.13. and 7.4.22.

Although Galois, in a different piece of work, was responsible for the introduction of fields of finite characteristic he appears to have ignored any discussion of separability in his work on (zeros of) polynomial equations, by assuming that the zeros of his polynomials, which he supposed to exist 'somewhere', were always pairwise distinct.

7.5.1 (and exercise 13(c)) tells us that, more often than not, an irreducible polynomial $f(x) \in F[x]$ will factor into a product of *distinct* linear factors *in* $S_f[x]$. Of course, $S_f[x]$ depends heavily upon F and $f(x)$: it is specially constructed to meet the needs of $f(x)$. It is, therefore, mildly surprising that, given half a chance, an irreducible polynomial in $F[x]$ will also try to split in $S_f[x]$. Formally this becomes

Theorem 7.5.2 Let S_f be the splitting field of the separable polynomial $f(x)$ over F and let $g(x) \in F[x]$. If $g(x)$ is irreducible in $F[x]$ and has *at least one zero* in $S_f \backslash F$, then $g(x)$ splits (completely) into a product of distinct linear factors in $S_f[x]$.

Proof Let $\beta_1 \in S_f \backslash F$ be a zero of $g(x)$. Denote by $\beta_1, \beta_2, \ldots, \beta_l$ the distinct values of $v(\beta_1)$ as v ranges over the (finitely many—why?) elements of $\mathrm{Gal}(S_f/F)$. Setting $h(x) = \prod_{j=1}^{l} (x - \beta_j) = h_0 + h_1 x + \cdots + h_l x^l$, say, *each* v permutes these factors and so maps each h_i to itself. Consequently, by 7.4.8, each coefficient lies in F; i.e. $h(x) \in F[x]$. If $g(x) \nmid h(x)$ in $F[x]$ the irreducibility of $g(x)$ leads (as in the proof of 7.5.1) to $1 = s(x)g(x) + t(x)h(x)$ for suitable $s(x), t(x) \in F[x]$ and hence, as in 7.5.1, again to $1 = s(\beta_1)g(\beta_1) + t(\beta_1)h(\beta_1) = 0$. Thus $g(x) \mid h(x)$ in $F[x]$ and hence in $S_f[x]$. Since $h(x)$ splits into distinct linear factors in $S_f[x]$—by definition!—so too must $g(x)$. (In fact $g(x) = h(x)$. Can you see why?)

This property surely deserves special attention.* So we make

Definition 7.5.3 Let $K \subseteq L$ be fields with $[L:K] < \infty$. Suppose that each irreducible polynomial $g(x) \in K[x]$ either has no zeros in L or factorises into a product of linear factors in $L[x]$. Then L is said to be a **normal extension** of K (or to be **normal over** K).

* In fact we've already given it very special attention. See exercise 7.5.9 for a pleasant surprise!

(We shall see in exercise 7.10.1 that this name is not idly chosen!)

We can succinctly summarise where we've got to so far if we extend 7.4.5 to

Definition 7.5.4 Let $K \subseteq L$ be fields with $[L:K] < \infty$. If, for $\alpha \in L$, the minimum polynomial $M_\alpha(x)$ of α over K is separable over K, then we say that α is **separable** over K. If, for all $\alpha \in L$, α is separable over K, then we say that L is a **separable extension** of K (or is **separable over** K.)

Our summary is then given by

Theorem 7.5.5 Let $F, f(x)$ and S_f be as in 7.5.2. Then S_f is a (finite*) normal and separable extension of F.

Proof Bring together 7.5.2 and exercise 7.2.12.

In fact, after you have established the result in exercise 9, you will be able to accept

Theorem 7.5.6 Let $K \subseteq L$ be fields. Then (i) L is the splitting field of a separable polynomial over K iff (ii) L is a (finite) normal and separable extension of K.

Comments 7.5.7 (a) Further conditions, equivalent to 7.5.6(i), (ii) are noted in exercise 8.

(b) Although, in 7.5.6, (i) is equivalent to (ii), it is interesting to see (in (ii)) how we can get away from a lopsided dependence on a particular polynomial $f(x)$ and its splitting field S_f to a more symmetrical setting in which no polynomial dominates.

(c) Noticing the word 'separable' in each part of 7.5.6, one is naturally led to ask if (i) and (ii) remain equivalent if the word 'separable' is removed. Exercise 9 gives the answer.

It turns out that extensions which are finite, normal and separable will play a major role in what is still to come. Accordingly it is worthwhile making

Definition 7.5.8 If $K \subseteq L$ are fields satisfying (either of) the conditions of 7.5.6, we shall describe L as a **Galois extension** of K. Using this terminology we can rewrite 7.5.5 as

Theorem 7.5.5′ Let S_f be a splitting field of the separable polynomial $f(x)$ over the field F. Then S_f is a Galois extension of F.

Exercises

1 Let $f(x) \in F[x]$ where the characteristic of F is 0 and $\deg f(x) \geq 1$. Let $d(x)$ denote the gcd of $f(x)$ and $f'(x)$ in $F[x]$ (see exercises 1.10.3 and 1.11.2). Show

* i.e. finite-dimensional (see 4.5.6).

that $f(x)/d(x)$ (trivially separable over F) has the same splitting field over F as $f(x)$. Test this on exercise 7.3.10 (iv).

2 Let $K \subseteq L$ be fields with $[L:K] < \infty$. Show that L is normal over K iff L contains (an isomorphic copy of) S_f for each irreducible $f(x) \in K[x]$ which has a zero in L.

3 Let $K \subseteq L \subseteq M$ be fields. Show that if M is normal over K then M is normal over L. Give an example to show that L need not be normal over K. Show that L *is* normal over K if, in addition, $\sigma(L) \subseteq L$ for every automorphism σ of M which fixes K.

4 Use the radical tower $\mathbb{Q} \subset \mathbb{Q}(\sqrt{2}) \subset \mathbb{Q}(\sqrt[4]{2})$ to show that if $K \subset L \subset M$ and if L is normal over K and M is normal over L, then M need not be normal over K.

5 Let $K \subseteq L$ with $[L:K] < \infty$. Let $F = \text{Fix}(\text{Gal}(L/K))$. Show that L is a normal extension of F.

6 Let $K \subseteq L \subseteq M$ be fields with M separable over K. Show that M is separable over L and L is separable over K.

7 Artin [3] defines L to be a normal extension of K if $[L:K] < \infty$ and if $\text{Fix}(\text{Gal}(L/K)) = K$. Show that if L is normal in Artin's sense then it is in our sense—but not necessarily conversely—unless the separability of L over K is also assumed.

8 Show that if $[L:K] < \infty$ then conditions (i) and (ii) of 7.5.6 are also equivalent to: (iii) $|\text{Gal}(L/K)| = [L:K]$; (iv) the fixed field of $\text{Gal}(L/K) = K$.

9 Let $K \subseteq L$. Show that L is a normal extension of K iff L is a splitting field over K of some $g(x) \in K[x]$. [Hint: 'Only if': Since, by definition, we have $[L:K] < \infty$, we also have $L = K(\alpha_1, \ldots, \alpha_n)$ for suitable α_i in L. L is the splitting field of $M_{\alpha_1}(x) \ldots M_{\alpha_n}(x)$ where $M_{\alpha_i}(x)$ is the minimum polynomial over K of α_i. 'If': Let $h(x)$ be irreducible in $K[x]$ with a zero α in L. Let $M(\supseteq L)$ be a splitting field for $h(x)$ regarded as an element of $L[x]$. Let β be any zero of $h(x)$ in $M \setminus L$. Use 7.2.3′ to show $K(\alpha) \cong K(\beta)$ by an isomorphism which acts identically on K and sends α to β. L is a splitting field for $g(x)$ over $K(\alpha)$ and $L(\beta)$ is a splitting field of $g(x)$ over $K(\beta)$. Show $L \cong L(\beta)$. Deduce that $\beta \in L$.]

10 Let L_1, L_2 be normal extensions of K with $K \subseteq L_1, L_2 \subseteq M$. Show that $L_1 \cap L_2$ and $\langle L_1 \cup L_2 \rangle$ (see footnote on page 286) are both normal over K. Assuming only that L_1 is normal over K prove that $\langle L_1 \cup L_2 \rangle$ is normal over L_2.

11 Let $L = K(\alpha_1, \alpha_2, \ldots, \alpha_n)$ where $[L:K] < \infty$. Show that if the minimum polynomial of each α_i splits in $L[x]$ then L is a normal extension of K.

12 Give an example of a separable extension which is not normal and an example of a normal extension which is not separable.

In the following assume that each field has characteristic $p \neq 0$.

13 (a) Let $f(x) \in K$ be irreducible. Show that $f(x)$ is *not* separable in $K[x]$ iff $f(x)$ is of the form $a_0 + a_1x^p + \cdots + a_nx^{np}$, where the $a_i \in K$.
(b) Assume that K is a finite field. Prove that the derivative of $f(x) \in K[x]$ is 0 iff $f(x) = (g(x))^p$ for some $g(x) \in K[x]$.
(c) Prove that 7.5.1 holds if F is a finite field (of characteristic p).

14 Define the map $\phi: K \rightarrow K$ by $\phi(a) = a^p$ for each $a \in K$. Show that ϕ is a 1–1 homomorphism of K into K—called the *Frobenius monomorphism* of K. Show that the fixed field of ϕ is the prime subfield P of K and that if K is a finite field then ϕ is an automorphism of K. [Hint: To show that $\text{Fix}\,\phi = P$ note that if $c^p = c$ then c is a zero of $x^p - x$, which has at most p zeros in a field.]

7.6 Subfields and subgroups

7.4.7 and 7.4.8 told us something about the order and fixed field of $\text{Gal}(S_f/F)$. To find out more about the *structure* of this group we are led, naturally, to investigate its subgroups (and their corresponding fixed fields in S_f).

Having just indicated (7.5.7) how we can free ourselves from direct consideration of a specific polynomial and its splitting field let us reinterpret 7.4.7 and 7.4.8—and even extend them somewhat—in these terms.

With very little effort we obtain

Theorem 7.6.1 Let $K \subseteq L$ be a Galois extension and suppose that $K \subseteq C \subseteq L$, C being a field. Then
(i) $|\text{Gal}(L/C)| = [L:C]$; (ii) $\text{Fix}(\text{Gal}(L/C)) = C$;
(iii) The map $\Phi: \mathscr{F} \rightarrow \mathscr{G}$ given by $\Phi(C) = \text{Gal}(L/C)$ is a 1–1 *inclusion reversing* map from the set $\mathscr{F} = \{C: K \subseteq C \subseteq L\}$ of all fields between K and L into the set $\mathscr{G}$ of all subgroups of Gal (L/K).

For a pictorial interpretation of Φ see 7.6.5 below.

Proof (i) and (ii). These are easy: We know, by 7.5.6, that L is a splitting field over K of some separable polynomial $f(x) \in K[x]$. It follows trivially that L is a splitting extension of $f(x)$ over C and that $f(x)$ is separable over C (exercise 7.4.19). Hence (i) and (ii) follow from 7.4.7 and 7.4.8.
(iii) Given $K \subseteq C_1$, $C_2 \subseteq L$ and $\text{Gal}(L/C_1) = \text{Gal}(L/C_2)$, 7.6.1(ii) tells us at once that $C_1 = \text{Fix}(\text{Gal}(L/C_1)) = \text{Fix}(\text{Gal}(L/C_2)) = C_2$.

Maintaining our 'experimental' approach *may* lead us to some dead ends, but surely we must follow up the obvious question suggested by 7.6.1(iii): Is there a 'corresponding' 1–1 mapping from $\mathscr{G}$ to $\mathscr{F}$? Put another way:

Given $H \leq \text{Gal}(L/K)$ does there exist a field D such that $K \subseteq D \subseteq L$ and $\text{Gal}(L/D) = H$?

In view of 7.6.1(iii) and the finiteness (why?) of $\mathscr{G}$ we need only check whether or not Φ maps $\mathscr{F}$ *onto* $\mathscr{G}$.

Now *if* there were such a D *then*, by 7.6.1(ii), we'd have $\text{Fix}\, H = \text{Fix}(\text{Gal}(L/D)) = D$. So let's look at $\text{Gal}(L/\text{Fix}\, H))$. Surely this *must* be equal to H?

Now, by 7.6.1(iii), Φ sends $\text{Fix}\, H$ to $\text{Gal}(L/\text{Fix}\, H))$, a group all elements of which surely fix $\text{Fix}\, H$ elementwise! Because H is *a* group of automorphisms of L fixing $\text{Fix}\, H$ and $\text{Gal}(L/\text{Fix}\, H))$ is the set of *all* automorphisms of L fixing $\text{Fix}\, H$, there seems no obvious reason why $H \subsetneqq \text{Gal}(L/\text{Fix}\, H))$ should not be possible. Fortunately the next result, called *Artin's Lemma* but known, in essence, to Dedekind, does just what we want. [*Warning!!* The proof is so beautiful you may feel impelled to tell it to your friends!]

Lemma 7.6.2 Let L be a field and let H be a finite group of automorphisms of L. Then $[L:\text{Fix}\, H] \leq |H|$.

Proof Let $H = \{\theta_1 = e, \theta_2, \ldots, \theta_n\}$ so that $|H| = n$, and suppose that $[L:\text{Fix}\, H] > |H|$. Then there will exist a set $\{\alpha_1, \alpha_2, \ldots, \alpha_{n+1}\}$ of $n+1$ elements of L which are linearly independent (see 4.5.2 and 4.5.6) over $\text{Fix}\, H$. Consider the following system of n homogeneous linear equations in $n+1$ unknowns $x_1, x_2, \ldots, x_{n+1}$ with coefficients in L:

$$\left.\begin{array}{l} \theta_1(\alpha_1)x_1 + \theta_1(\alpha_2)x_2 + \cdots + \theta_1(\alpha_{n+1})x_{n+1} = 0 \\ \theta_2(\alpha_1)x_1 + \theta_2(\alpha_2)x_2 + \cdots + \theta_2(\alpha_{n+1})x_{n+1} = 0 \\ \quad\vdots \qquad\qquad \vdots \qquad\qquad\qquad \vdots \\ \theta_n(\alpha_1)x_1 + \theta_n(\alpha_2)x_2 + \cdots + \theta_n(\alpha_{n+1})x_{n+1} = 0 \end{array}\right\} \qquad (*)$$

By a well-known theorem of linear algebra, (see, for example, [18]) the system (*) has a solution $(a_1, a_2, \ldots, a_{n+1})$, say, where $a_i \in L (i = 1, 2, \ldots, n+1)$ and not *all* the a_i are equal to zero.

From amongst all these non-trivial solutions we choose one having the least number r, say, of non-zero 'components' a_i. By reordering the α_i, if necessary, we may assume that this non-trivial solution is $(a_1, a_2, \ldots, a_r, 0, \ldots, 0)$ where $r \geq 1$. First note that $r > 1$ (since $\alpha_1 a_1 = 0$ is impossible—why?). Next observe that $(1, b_2, \ldots, b_r, 0, \ldots, 0)$, where $b_i = a_i/a_1$, is also a solution. (Why?) That is, for each $i (1 \leq i \leq n)$, we have

$$\theta_i(\alpha_1)1 + \theta_i(\alpha_2)b_2 + \cdots + \theta_i(\alpha_r)b_r = 0 \qquad (**)$$

Now not *all* the b_k belong to $\text{Fix}\, H$. (Why not? Concentrate on the first equality in (**)). Suppose that b_j is one such element. It follows (why?) that there exists $\theta \in H$ such that $\theta(b_j) \neq b_j$. Applying θ to the equalities (**) we obtain the new system of equalities in L:

$$\theta(\theta_i(\alpha_1))\theta(1) + \theta(\theta_i(\alpha_2))\theta(b_2) + \cdots + \theta(\theta_i(\alpha_r))\theta(b_r) = \theta(0) = 0 \quad (1 \leq i \leq n) \qquad (***)$$

Now, $H = \{\theta_1, \theta_2, \ldots, \theta_n\} = \{\theta\theta_1, \theta\theta_2, \ldots, \theta\theta_n\}$—see exercise 5.4.5. So, to each $i (1 \leq i \leq n)$, there exists $j(i) (1 \leq j(i) \leq n)$ such that $\theta_i = \theta\theta_{j(i)}$. Consequently, for each $i (1 \leq i \leq n)$ the ith equality of (**) and the $j(i)$th equality of (***) are:

$$\theta_i(\alpha_1)1 \quad + \theta_i(\alpha_2)b_2 \quad + \cdots + \theta_i(\alpha_r)b_r \quad = 0$$

and

$$\theta_i(\alpha_1)\theta(1) + \theta_i(\alpha_2)\theta(b_2) + \cdots + \theta_i(\alpha_r)\theta(b_r) = 0,$$

respectively. Subtracting gives:

$$\theta_i(\alpha_2)[b_2 - \theta(b_2)] + \cdots + \theta_i(\alpha_r)[b_r - \theta(b_r)] = 0 \qquad (1 \le i \le n).$$

That is $(0, b_2 - \theta(b_2), \ldots, b_r - \theta(b_r), 0, \ldots 0)$ is a non-zero (why?) solution of the system of equations (*) and yet has fewer non-zero components.

This contradiction to our choice of r shows that the assumption $[L:\text{Fix}\,H] > |H|$ is untenable. Consequently, $[L:\text{Fix}\,H] \le |H|$.

As an immediate corollary we get

Corollary 7.6.3 Let L and H be as in 7.6.2. Then $[L:\text{Fix}\,H] = |H|$ and $H = \text{Gal}(L/\text{Fix}\,H)$. In particular L is a (finite) normal and separable (hence Galois) extension of Fix H.

Proof Trivially $H \le \text{Gal}(L/\text{Fix}\,H)$. But, by 7.4.4, $|\text{Gal}(L/\text{Fix}\,H)| \le [L:\text{Fix}\,H]$. (Why is 7.4.4 applicable?) Consequently $|H| \le |\text{Gal}(L/\text{Fix}H)| \le [L:\text{Fix}\,H] \le |H|$, the last inequality coming from 7.6.2. The final remark follows from exercise 7.5.8.

Specialising to Galois extensions we get

Theorem 7.6.4 Let $K \subseteq L$ be a Galois extension. Then there exists a 1–1 and onto mapping $\Phi: \mathcal{F} \to \mathcal{G}$ given by: $\Phi(C) = \text{Gal}(L/C)$, with $\Phi^{-1}(H) = \text{Fix}\,H$.

Proof If $H \le \text{Gal}(L/K)$ then $H = \text{Gal}(L/\text{Fix}\,H)$ (by $7.6.3) = \Phi(\text{Fix}\,H)$ (see 7.6.1(iii)).

This result is a part of the so-called Fundamental Theorem of Galois Theory as normally presented—see 7.10.6.

At this point we offer just two specific examples of the above correspondence in action. Wanting one which is neither too trivial nor too involved we arrive at what is probably the most used illustration.

Example 7.6.5 The field $S = \mathbb{Q}(r, i)$, where $r = \sqrt[4]{2}$ is the real positive 4th root of 2 and $i^2 = -1$, is a splitting field over $\mathbb{Q}$ of the polynomial $x^4 - 2$ (cf. 7.4.2(v)) and so is a Galois extension of $\mathbb{Q}$ (7.5.5′).

To show you that converting problems concerning subfields of fields into problems about subgroups of groups is a worthwhile activity, let us ask you *now* to describe all the subfields of $\mathbb{Q}(r, i)$ which contain $\mathbb{Q}$. When you have found a few how confident will you be that you then have them all?

To determine $G = \mathrm{Gal}(S/\mathbb{Q})$, note that each automorphism σ of S will automatically fix $\mathbb{Q}$ and will map i to i or $-i$ and r to r or ir or $-r$ or $-ir$ (cf. 7.3.1(v)). Thus $|\mathrm{Gal}(S/\mathbb{Q})| \leqslant 8$. But, clearly, $[S:\mathbb{Q}] = 8$ and so, by 7.4.7, we must have $|\mathrm{Gal}(S/\mathbb{Q})| = 8$. We can describe these automorphisms succinctly by their actions on i and r as follows:

	σ_1	σ_2	σ_3	σ_4	σ_5	σ_6	σ_7	σ_8
r	r	ir	$-r$	$-ir$	r	ir	$-r$	$-ir$
i	i	i	i	i	$-i$	$-i$	$-i$	$-i$

Setting $b = \sigma_2$ we find $b^2 = \sigma_3$, $b^3 = \sigma_4$, $b^4 = \sigma_1 = e$, the identity element. Setting $a = \sigma_5$ we find $a^2 = e$. Further $(ab)(r) = a(b(r)) = a(ir) = -ir = (b^3a)(r)$ whilst $(ab)(i) = a(b(i)) = a(i) = -i = (b^3a)(i)$. Consequently $ab = b^3a$. Thus G is seen to be the dihedral group of order 8, its three subgroups of order 4 and five subgroups of order 2 corresponding to subfields of $\mathbb{Q}(r, i)$ which have dimensions 2 and 4 respectively over $\mathbb{Q}$. This correspondence, determined by the mapping Φ, is best understood via the following diagrams in which fields and groups in the same 'place' in each diagram correspond under the map Φ of 7.6.4. (Notice that we place 'larger' fields above 'smaller' but reverse this principle for $\mathcal{G}$.)

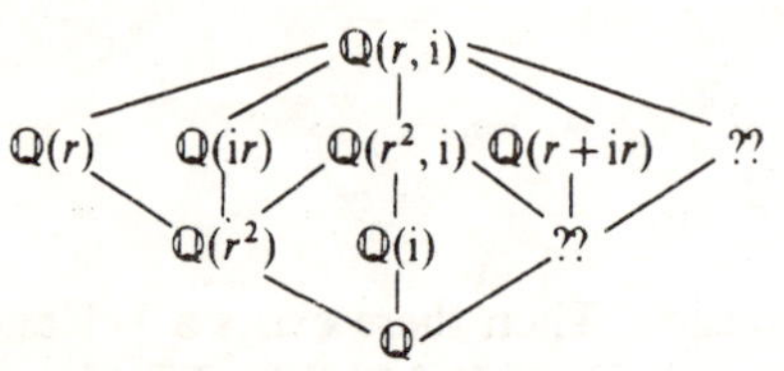

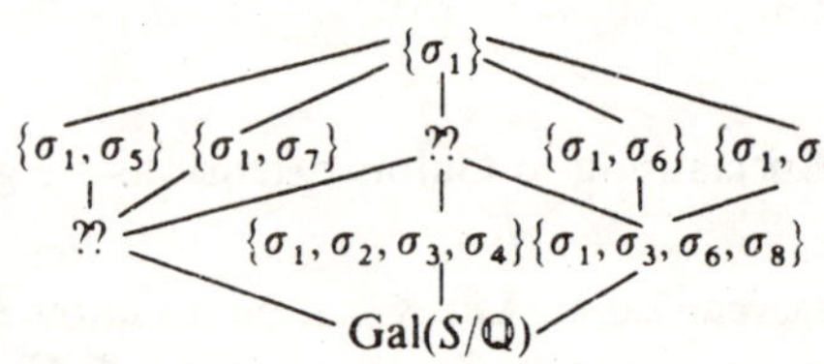

You are asked, in the exercises, to fill in the ??

How can we confirm this correspondence? Naïvely we take a typical element

$$a_0 1 + a_1 r + a_2 r^2 + a_3 r^3 + a_4 i + a_5 ir + a_6 ir^2 + a_7 ir^3 \qquad (a_i \in \mathbb{Q}) \qquad (*)$$

using the obvious basis $\{1, r, \ldots, ir^3\}$ of $\mathbb{Q}(r, i)$ over $\mathbb{Q}$. Applying σ_6 and equating the result to (*), we find that if (*) is not changed by σ_6 then (*) takes the special form $a_0 1 + a_1(r + ir) + a_3(r^3 - ir^3) + \alpha_6(ir^2)$.

Now you might just recognise that $ir^2 \in \mathbb{Q}(r + ir)$ – since $(r + ir)^2 = r^2(1 + i)^2 = 2ir^2$. Further $(1 + i)^3 = 2(i - 1) = -2(1 - i)$. Hence $r^3 - ir^3 \in \mathbb{Q}(r + ir)$, too. So, the fixed field of $\{\sigma_1, \sigma_6\}$ is $\mathbb{Q}(r + ir)$.

Alternatively: If, in general, $\sigma \in \mathrm{Gal}(L/K)$ has order n and if $t \in L$ then σ maps $t + \sigma(t) + \cdots + \sigma^{n-1}(t)$ to itself. (In the above example, σ_6 has order 2 and so maps $r + ir$ to itself.) Consequently $\mathbb{Q}(r + ir) \subseteq \mathrm{Fix}\{\sigma_1, \sigma_6\}$. Hence, if $\mathbb{Q}(r + ir) \subsetneqq \mathrm{Fix}\{\sigma_1, \sigma_6\}$, then the subgroup of $\mathrm{Gal}(S/\mathbb{Q})$ fixing $\mathbb{Q}(r + ir)$ is strictly bigger than $\mathrm{Fix}\{\sigma_1, \sigma_6\}$. But you can easily check that no other σ_i fixes $r + ir$. Hence $\mathbb{Q}(r + ir) = \mathrm{Fix}\{\sigma_1, \sigma_6\}$.

The next example is even more impressive—at least in what it tells us.

Example 7.6.6 Let S be the splitting field over $\mathbb{Q}$ of the polynomial $x^p - 1$, where p is a prime. We know (7.3.1(iii) and exercise 7.3.2) that $|\text{Gal}(S/\mathbb{Q})| = [S:\mathbb{Q}] = p-1$. Put $\zeta = \cos(2\pi/p) + i\sin(2\pi/p)$ and set $a = \zeta + 1/\zeta = 2\cos(2\pi/p)$. Then $a \in \mathbb{R}$ and $\mathbb{Q} \subset \mathbb{Q}(a) \subset \mathbb{Q}(\zeta) = S$. Suppose we now choose p to be of the form $2^k + 1$. (Then, necessarily, $k = 2^n$ for some integer n—exercise 4.6.5.) Consequently $[S:\mathbb{Q}] = 2^k$ and so $[\mathbb{Q}(a):\mathbb{Q}] = 2^\iota$ for some integer ι (why?). Since Gal$(S/\mathbb{Q})$ is abelian [in fact it's cyclic! Try to prove it abelian but, if stuck look at exercise 7.8.6] we deduce from exercise 6.4.7 that Gal$(S/\mathbb{Q})$ has a decreasing sequence of subgroups

$$S = S_0 > S_1 > \cdots > S_\iota = \langle e \rangle,$$

each of index 2 in its predecessor. Hence, by 7.6.4, we see that there is a corresponding sequence

$$\mathbb{Q} = \mathbb{Q}_0 \subset \mathbb{Q}_1 \subset \cdots \subset \mathbb{Q}_\iota = \mathbb{Q}(a) \text{ with } [\mathbb{Q}_i:\mathbb{Q}_{i-1}] = 2 \qquad \text{for all } i.$$

Hence, by exercise 4.6.3, we can deduce that the point $(a, 0)$ is obtainable by straightedge and compass construction. Thus we have proved the 'if' part of Gauss's assertion concerning constructible polygons. (See Section 4.6.)

Exercises

1 Suppose that $L \supseteq M \supseteq K$ and $G \geq H \geq \langle e \rangle$ where $G = \text{Gal}(L/K)$. Write $G(M)$ for Gal(L/M) and $F(H)$ for the subfield Fix(H) of L. Prove that $G(F(H)) \geq H$, $F(G(M)) \supseteq M$ and (omitting brackets) that $FGF(H) = F(H)$, $GFG(M) = M$.

2 Suppose $K \subseteq L$ with $[L:K] < \infty$ and let Gal(L/K) have fixed field $F(\supseteq K)$. Show that there is a 1–1 correspondence between the set $\mathscr{F}$ of subfields of L containing F and the set $\mathscr{G}$ of subgroups of Gal(L/K). Deduce that the mapping Φ of 7.6.1 is always *onto* $\mathscr{G}$.

3 (Dedekind's Independence Theorem: DIT) Let H and K be fields and let $\phi_1, \ldots, \phi_r$ be distinct monomorphisms* of H into K. Prove that $\phi_1, \ldots, \phi_r$ are linearly independent over K—i.e. (cf. 4.5.2(ii)): If $a_1, \ldots, a_r \in K$ are such that $a_1\phi_1(h) + a_2\phi_2(h) + \cdots + a_r\phi_r(h) = 0_K$ for all $h \in H$, then all the $a_i = 0_K$.
[Hint: Suppose false and let $\phi_1, \ldots, \phi_n$ $(n \leq r)$ be a smallest non-empty set such that $a_1\phi_1 + \cdots + a_n\phi_n$ is the zero map $H \to K$ with no $a_i = 0$. Since $\phi_i(1_H) = 1_K$, $n \geq 2$. Since $\phi_1 \neq \phi_n$ there is some $h \in H$ such that $\phi_1(h) \neq \phi_n(h)$. For all $g \in H$,

$$a_1\phi_1(hg) + \cdots + a_n\phi_n(hg) = a_1\phi_1(h)\phi_1(g) + \cdots + a_n\phi_n(h)\phi_n(g) = 0.$$

But also

$$a_1\phi_n(h)\phi_1(g) + a_2\phi_n(h)\phi_2(g) + \cdots + a_n\phi_n(h)\phi_n(g) = 0$$

(why?). Subtract this from the previous equality to get $a_1(\phi_1(h) - \phi_n(h))\phi_1(g) + \cdots + ?? = 0$. But $\phi_1(h) \neq \phi_n(h)$.]

* See 7.4.3.

4 (1st application of DIT) Let $k_1, \ldots, k_n$ be non-zero elements of a field K and let $\tau_1, \ldots, \tau_n$ be distinct automorphisms of K. Then there exists $k \in K$ such that $k_1\tau_1(k) + \cdots + k_n\tau_n(k) \neq 0_K$.

Check this theorem with the four automorphisms: Identity, $\sigma(\sqrt{2} \to -\sqrt{2}, \sqrt{3} \to \sqrt{3})$, $\tau(\sqrt{2} \to \sqrt{2}, \sqrt{3} \to -\sqrt{3})$ and $\sigma\tau$ of $\mathbb{Q}(\sqrt{2}, \sqrt{3})$ given that $k_1 = 1 + 0\sqrt{2} - 2\sqrt{3} + 1\sqrt{6}$, $k_2 = -3 - 1\sqrt{2} + 2\sqrt{3} + 0\sqrt{6}$, $k_3 = -1 + 0\sqrt{2} - 2\sqrt{3} - 1\sqrt{6}$ and $k_4 = 3 + 1\sqrt{2} + 2\sqrt{3} + 0\sqrt{6}$. [This result is helpful in giving an alternative proof of part of 7.10.4.]

5 (2nd application of DIT) Prove directly, without using 7.6.3, that if G is a finite group of automorphisms of the field K and if F_0 denotes Fix(G), then $|G| \leq [K:F_0]$. [Hint: Suppose $m = [K:F_0] < |G| = n$ and that $x_1, \ldots, x_m$ is a basis for K over F_0. Let $G = \{\tau_1, \ldots, \tau_n\}$. Note that there are $k_1, \ldots, k_n \in K$, not all equal to 0_K, such that

$$\tau_1(x_j)k_1 + \cdots + \tau_n(x_j)k_n = 0 \qquad (j = 1, \ldots, m).$$

Let $a \in K$. Then $a = \alpha_1 x_1 + \cdots + \alpha_m x_m$ for suitable $\alpha_i \in F_0$. Multiply the above equations by $\alpha_1, \ldots, \alpha_m$ respectively in order. Using $\alpha_j\tau_i(x_j) = \tau_i(\alpha_j x_j)$ (why equality?) add the m resulting equalities to get $\tau_1(a)k_1 + \cdots + \tau_n(a)k_n = 0$, contradicting exercise 3.]

6 Let K be a Galois extension of F with Galois group G. Given fields E_1, E_2 such that $F \subseteq E_1, E_2 \subseteq K$ with $\mathrm{Gal}(K/E_i) = H_i \leq G (i = 1, 2)$, show that $\mathrm{Gal}(K/\langle E_1 \cup E_2 \rangle) = H_1 \cap H_2$, $\mathrm{Gal}(K/E_1 \cap E_2) = \langle H_1 \cup H_2 \rangle$.

7 Suppose $G = \{\phi_1, \ldots, \phi_n\}$ is a (finite) group of automorphisms of the field L with fixed field K. Suppose that $\psi \in \mathrm{Aut}(L)$ and that ψ fixes K. Show that $\psi \in G$.

8 Suppose $K \subseteq L \subseteq M$ are fields with $[L:K] < \infty$. Show that $[\mathrm{Gal}(M/K):\mathrm{Gal}(M/L)] \leq [L:K]$ and that equality holds if M is a Galois extension of K. [Hint: Show that elements in the same coset of $\mathrm{Gal}(M/L)$ in $\mathrm{Gal}(M/K)$ give rise to the same mapping of L into M. Now use 7.4.3 with $K_1 = K_2 = K$ etc.]

9 Determine the subgroups and subfields omitted from the diagram of 7.6.5.

10 Determine $\mathrm{Gal}(S_f/\mathbb{Q})$ where: (i) $f(x) = x^3 - 2$; (ii) $f(x) = x^5 - 2$; (iii) $f(x) = x^6 - 2$. Draw the subgroup and subfield 'lattices' as in 7.6.5.

11 Determine $\mathrm{Gal}(S_f/\mathbb{Q})$ where: (i) $f(x) = x^4 - 3$; (ii) $f(x) = x^4 - 4$. Draw the subgroup and subfield lattices.

12 Find $\mathrm{Gal}(S_f/\mathbb{Q})$ if $f(x) = (x^2 + 2x - 1)^4(x^3 - 2)$. (Cf. exercise 7.3.10)

13 Determine $\mathrm{Gal}(S_f/F)$ where $f(x) = x^4 + \hat{2}$ and F is $\mathbb{Z}_3$.

14 Draw the subfield lattice for the splitting field of $x^3 + x + 1$ of exercise 7.2.4(vii). (Its Galois group is S_3.)

15 How many subfields has $\mathbb{Q}(\sqrt{2}, \sqrt{3}, \sqrt{5})$ got?

16 Let $f(x)=x^8-1\in\mathbb{Q}[x]$. Check that $\mathrm{Gal}(S_f/\mathbb{Q})$ is $C_2\times C_2$. Show $\mathbb{Q}(i)$, $\mathbb{Q}(\sqrt{2})$ and $\mathbb{Q}(i,\sqrt{2})$ are subfields of S_f. What are the others?

17 Let L be a Galois extension of K and suppose that $\mathrm{Gal}(L/K)$ is cyclic. Show that for each $d|[L:K]$ there exists just one field T such that $K\subseteq T\subseteq L$ and $[L:T]=d$.

7.7 The groups Gal(*R*/*F*) and Gal(S_f/*F*)

Whilst our investigations so far have followed a fairly natural path and produced at least one very powerful looking by-product, (7.6.4), it is surely time to return to our motivating problem, the solubility of equations by radicals.

To keep our main aim firmly in view, we work, once again, in terms of splitting fields of separable polynomials rather than talk of Galois extensions.

We consider, first, the groups of the title where R, S_f are as in 7.2.2. The following result is both readily spotted—and proved!

Lemma 7.7.1 Suppose that $F\subseteq S_f\subseteq R$ are as in 7.2.2 and suppose that $\theta\in\mathrm{Gal}(R/F)$. Then $\theta(S_f)=S_f$.

Proof Let $s\in S_f=F(\alpha_1,\alpha_2,\dots,\alpha_n)$ where $\alpha_i(1\le i\le n)$ are the zeros of $f(x)$ in S_f. Since θ maps F identically we see that, for each α_i, $\theta(\alpha_i)$ is a zero of $f(x)$, i.e. $\theta(\alpha_i)=\alpha_{j(i)}$ for some $j(i)$. $\theta(S_f)\subseteq S_f$ is then immediate. [What about equality?]

Thus, in order to relate $\mathrm{Gal}(R/F)$ to $\mathrm{Gal}(S_f/F)$ it seems reasonable to associate, with $\theta\in\mathrm{Gal}(R/F)$, the restriction $\tilde{\theta}$ (say) of θ to $\mathrm{Gal}(S_f/F)$. ($\tilde{\theta}$ is certainly meaningful—by 7.7.1.) This suggests that there should be a (natural) homomorphism τ, say, from $\mathrm{Gal}(R/F)$ to $\mathrm{Gal}(S_f/F)$. Let's confirm this.

Lemma 7.7.2 Let $F\subseteq S_f\subseteq R$ be as in 7.2.2. The map $\Omega:\mathrm{Gal}(R/F)\rightarrowtail\mathrm{Gal}(S_f/F)$ given by $\Omega(\theta)=\tilde{\theta}$ is a homomorphism.

Proof Let $\phi,\psi\in\mathrm{Gal}(R/F)$ so that $\tilde{\phi},\tilde{\psi}\in\mathrm{Gal}(S_f/F)$. We find, for all $s\in S_f$: $\widetilde{\phi\psi}(s)=(\phi\psi)(s)=\phi(\psi(s))=\phi(\tilde{\psi}(s))=[\text{why?}]\tilde{\phi}(\tilde{\psi}(s))=(\tilde{\phi}\tilde{\psi})(s)$. [You should *try* to answer the question 'why?' at *each* equality—especially the one indicated!] Thus the map Ω defined by $\Omega(\theta)=\tilde{\theta}$ is a homomorphism.

7.7.2 may look nice but, as it stands, it is not all that much use, for there seems little likelihood of being able to determine much about the structure of $\mathrm{Gal}(S_f/F)$ from that of $\mathrm{Gal}(R/F)$ unless Ω above maps $\mathrm{Gal}(R/F)$ *onto* $\mathrm{Gal}(S_f/F)$. (In fact $\mathrm{Gal}(R/F)=\langle e\rangle\not\cong\mathrm{Gal}(S_f\;/F)$ is possible—exercise 4.)

Let us see what it might cost us (in terms of hypotheses) to ensure that Ω maps $\mathrm{Gal}(R/F)$ *onto* $\mathrm{Gal}(S_f/F)$: One criterion is immediate: If $V(\supseteq F)$ is the fixed field of $\mathrm{Gal}(R/F)$, then each θ is required to fix all of V. But, by 7.4.8, $\mathrm{Fix}(\mathrm{Gal}(S_f/F))=F$. Consequently we are *forced* to assume that $V\cap S_f=F$.

Is $V \supsetneqq F$? More than likely! (See exercises 3, 4.) But let us press on and see what turns up. Certainly Gal(R/F) = Gal(R/V). Also R is a normal extension of V (by 7.6.3). If we denote by T the smallest subfield of R containing V and S_f, we see that T is a splitting extension of $f(x)$ over V—so that V is also the fixed field of Gal(T/V). As we shall see below (cf. 7.7.3) this is enough to ensure that the natural map (cf. 7.7.2) from Gal(R/V) (= Gal(R/F)) to Gal(T/V) is onto. This looks promising! Can we now map from Gal(T/V) onto Gal(S_f/F)?

Since $T \supseteq S_f$ there is a homomorphism, as in 7.7.2, from Gal(T/V) into Gal(S_f/F). Is it onto? Sadly ... in general, no! [Since $T = V(\alpha_1, \alpha_2, \ldots, \alpha_n)$, each element of Gal$(T/V)$ determines a permutation on the set of αs which, itself, determines a corresponding element of Gal(S_f/F). Unfortunately there is no guarantee that these permutations on the α_i yield *every* element of Gal(S_f/F).]

So, what can we do? Since our choice of R is unlimited (whereas we are stuck with $F, f(x)$ and S_f) perhaps we can choose a new R for which $V = F$? It may be a long shot, but it's worth a try! First, to save time if we are wrong, let us check if indeed Ω (of 7.7.2) *will* be onto if Fix(Gal(R/F)) is assumed to be no bigger than F.

In fact we obtain

Theorem 7.7.3 Assume* that Fix(Gal(R/F)) = F (so that R is a Galois extension of F, by 7.6.3). Then the map Ω is a homomorphism from Gal(R/F) *onto* Gal(S_f/F). Further $\ker \Omega$ = Gal(R/S_f). Consequently Gal$(R/S_f) \lhd$ Gal(R/F) and Gal$(S_f/F) \cong$ Gal(R/F)/Gal(R/S_f).

Proof Let $\psi \in$ Gal(S_f/F). Since R is a Galois extension of F (by 7.6.3 with $L = R$ and H = Gal(R/F)), R is, by 7.5.6, a splitting field of a suitable separable polynomial $h(x)$, say, over F and, hence, of $h(x)$ over S_f (trivially).

On taking $F_1 = F_2 = S_f$ and $\lambda = \psi$, 7.2.4 tells us that we may extend the automorphism ψ of S_f to an automorphism ν of R. Since ψ acts identically on F so does ν. That is, $\nu \in$ Gal(R/F)—and, by construction, $\tilde{\nu} = \psi$. Next: $\ker \Omega$ = Gal(R/S_f). For, if $\nu \in$ Gal(R/F), then $\nu \in \ker \Omega$ iff $\tilde{\nu}$ is the identity map on S_f; that is iff $\nu \in$ Gal(R/S_f).

Finally: the isomorphism then follows from 5.11.6.

So, our immediate task is to show that the *given* radical tower R can be replaced, when necessary, by a radical tower *which is also normal over F*. Can we take a hint from 7.3.1(ii) and (v) where adjoining a cube root of unity changed the rather sterile $\mathbb{Q}(\sqrt[3]{2})$ into the splitting field (and hence normal extension) $\mathbb{Q}(\omega, \sqrt[3]{2})$ of $x^3 - 2$ over $\mathbb{Q}$?

Perhaps it is sufficient to adjoin to R, a primitive nth root of unity, where $n = \prod_{i=1}^{s} n_i$? (See 7.2.1) Unfortunately this won't do. As exercise 8 shows, there is

* In fact R can be *any* Galois extension of F which contains S_f.

still no guarantee that the new field will be normal (i.e. a splitting field) over F. Of course normality (perhaps without radicality) is easily obtained—given exercise 7.5.9. (See also 7.5.7(c).)

To get the desired type of extension note first that (in 7.2.1) $R = F(r_1, r_2, \ldots, r_s)$. For each i let $M_{r_i}(x)$ denote the minimum polynomial of r_i *over* F (why is there such a polynomial?) and put $m(x) = \prod_{i=1}^{s} M_{r_i}(x)$. Let $S_m = R(u_1, u_2, \ldots, u_t)$ be the splitting field of $m(x)$ *over* R. It follows (exercise 10) that S_m is a splitting field—hence a normal extension (see 7.5.7(c))—for $m(x)$ *over* F. But there's an obvious question. Have we just destroyed the required radical property? In fact we have

Theorem 7.7.4 *S_m is a radical tower over F.*

Proof We have $F \subseteq S_f \subseteq F(r_1, r_2, \ldots, r_s) \subseteq S_m$. 7.2.3′ and 7.2.4 show that if u is any zero (in S_m) of the minimum polynomial $M_{r_i}(x)$ of r_i over F then an element of Gal(S_m/F) exists which maps r_i to u.

Consequently, $S_m = F(\eta_1(r_1), \ldots \eta_1(r_s), \eta_2(r_1), \ldots, \eta_2(r_s), \ldots, \eta_v(r_1), \ldots, \eta_v(r_s))$ where $\eta_1 = e, \ldots \eta_v$ are the elements of Gal(S_m/F). But this implies that S_m is a radical tower over F. For, S_m contains the increasing sequence

$$F \subseteq F_{1,1} \subseteq F_{1,2} \subseteq \cdots \subseteq F_{1,s} \subseteq F_{2,1} \subseteq \cdots \subseteq F_{2,s} \subseteq \cdots \subseteq F_{v,s} = S_m \qquad (*)$$

of fields where, $F_{i,j} = F(\eta_1(r_1), \ldots, \eta_1(r_s), \ldots, \eta_i(r_1), \ldots, \eta_i(r_j))$ so that* $F_{k,l} = F_{k,l-1}(\eta_k(r_l))$ where $\eta_k(r_l)^{n_l} = \eta_k(a_l) \in F_{k,l-1}$.

(Note that the n_l arising here are precisely those appearing in the originally given tower in 7.2.1.)

Exercises

1 Establish the equality $\theta(S_f) = S_f$ in 7.7.1.

2 Show that there are fields $\mathbb{Q} \subset K \subset L \subset \mathbb{C}$ with Gal($L/\mathbb{Q}$) = Gal($K/\mathbb{Q}$) = $\langle e \rangle$. [Hint: $K = \mathbb{Q}(\sqrt[3]{2})$, $L = \mathbb{Q}(\sqrt[n]{2})$, $n = ?$]

3 Let $K = \mathbb{Q}(\sqrt[3]{2}, \omega)$, $L = \mathbb{Q}(\sqrt[3]{2}, \sqrt[5]{5}, \omega)$. Find Gal($K/\mathbb{Q}$) and Gal($L/\mathbb{Q}$) and show that Fix(Gal($L/\mathbb{Q}$)) $\supset$ Fix(Gal($K/\mathbb{Q}$)).

4 Noting that $\mathbb{Q} \subset \mathbb{Q}(\sqrt{2}) \subset \mathbb{Q}(\sqrt[3]{(3+\sqrt{2})})$ is a radical tower containing the splitting field over $\mathbb{Q}$ of $x^2 - 2$, compare Gal($\mathbb{Q}(\sqrt{2})/\mathbb{Q}$) with Gal($\mathbb{Q}(\sqrt[3]{(3+\sqrt{2})})/\mathbb{Q}$). (See exercise 7.3.16.)

5 Check, in the discussion following 7.7.2, that T is indeed a splitting extension of $f(x)$ over V and that Fix(Gal(T/V)) = V.

* Read $F_{i,0}$ as $F_{i-1,s}$.

6 Find three radical towers over $\mathbb{Q}$, with distinct Galois groups, all containing the splitting field $\mathbb{Q}(\sqrt{2})$.

7 Prove that if $K \subseteq L \subseteq M$, if M is normal over K and if σ is an isomorphism of L into M which fixes K, then σ is the restriction ψ of some automorphism of M which also fixes K.

8 Let $\mathbb{Q} \subset \mathbb{Q}(\sqrt{d}) \subset \mathbb{Q}(\sqrt{(1+\sqrt{d})})$, where $d \in \mathbb{Z}$, be given. Show that if $d=2$ then the extension $\mathbb{Q}(\sqrt{(1+\sqrt{d})}, \mathrm{i})$ is a normal extension of $\mathbb{Q}$, whereas if $d=3$ it is not. [Note: we are adjoining i, a primitive 4th root of unity, since the given radical tower comprises two quadratic extensions.]

9 Given that $K \subseteq L \subseteq M$, where L is normal over K, show that $\mathrm{Gal}(M/L)$ is a normal subgroup of $\mathrm{Gal}(M/K)$.

10 Confirm the assertion in the paragraph preceeding 7.7.4 that the splitting field of $m(x)$ *over* R is actually a splitting field for $m(x)$ *over* F.

7.8 The groups Gal($F_{i,j}/F_{i,j-1}$)

Recall that our main aim is to express the 'big' group (which is now $\mathrm{Gal}(S_m/F)$)—and hence the group of *real* interest, $\mathrm{Gal}(S_f/F)$—in terms of the (we hope) 'simpler' groups* $\mathrm{Gal}\,(F_{i,j}/F_{i,j-1})$. Unfortunately, as 7.3.1 (ii) shows, without roots of unity present, the groups $\mathrm{Gal}(F_{i,j}/F_{i,j-1})$ may be *too* simple, namely they may be of order 1!

There seems nothing for it†, therefore, but to adjoin to S_m the n_jth roots of 1 for all n_j, after all. Does this destroy the radicality and normality we've worked hard to get?

Fortunately not! Indeed if $S=S_m(\zeta)$ is obtained from S_m by adjoining a primitive nth root ζ of unity (where n is the lcm of the n_i occurring in 7.2.1—see the remark at the end of Section 7.7) we see that S is certainly a radical tower over S_m *and* a splitting field over F of the polynomial $m(x)(x^n-1)$. (Prove this!) There's one last rearrangement required which it is easy to spot as being helpful. In order that, when considering $\mathrm{Gal}(F_{i,j}/F_{i,j-1})$, we already have an n_jth root of 1 available in $F_{i,j-1}$, we place ζ at the 'beginning' rather than at the 'end' of the radical chain. That is, we consider the sequence—it is (still) a radical tower over F—

$$F \subseteq F(\zeta) \subseteq F_{1,1}(\zeta) \subseteq F_{1,2}(\zeta) \subseteq \cdots \subseteq F_{v,s}(\zeta)=S_m(\zeta)=S. \tag{7.8.1}$$

Note that, in 7.8.1, $F(\zeta)$ is a splitting (and radical) extension of F. Further each $F_{i,j}(\zeta)$ is a splitting (and radical) extension of $F_{i,j-1}(\zeta)$ since $F_{i,j} = F_{i,j-1}(\eta_i(r_j))$, where $\eta_i(r_j)^{n_j} \in F_{i,j-1}$ and $F_{i,j-1}$ contains a primitive n_jth root ζ of unity.

What, then, are the Galois groups of these extensions? To deal with the extension $F(\zeta)$ of F we prove (cf. 7.3.1(iii)):

* See footnote on p 299.
† But see exercise 1!

Theorem 7.8.2 Let‡ L be a splitting field of x^n-1 over a subfield K. Then Gal(L/K) is abelian.

Proof Let $1, \varepsilon, \ldots, \varepsilon^{n-1}$ be the zeros of x^n-1 in L, so that ε is a primitive nth root of 1. Then $L=K(\varepsilon)$. Let $M_\varepsilon(x)$ be the minimum polynomial of ε over K. Then each zero of $M_\varepsilon(x)$ is one of the ε^i. Now, for general n, not all the ε^i will be zeros of $M_\varepsilon(x)$—exercise 2. Nevertheless, if $\phi \in \text{Gal}(L/K)$, then $\phi(\varepsilon)=\varepsilon^j$ for some j (coprime to n—why?). [Note that $\phi(\varepsilon^k)$ is then determined for all k—viz. $\phi(\varepsilon^k)=\varepsilon^{kj}$.] Suppose that $\phi_k, \phi_l \in \text{Gal}(L/K)$ are given by $\phi_k(\varepsilon)=\varepsilon^k$ and $\phi_l(\varepsilon)=\varepsilon^l$. Then $\phi_k\phi_l(\varepsilon)=\phi_k(\varepsilon^l)=\varepsilon^{lk}=\varepsilon^{kl}=\phi_l(\varepsilon^k)=\phi_l\phi_k(\varepsilon)$. This is enough to prove that Gal(L/K) is abelian, as claimed. (Why?)

For dealing with all the other extensions, we obviously need (c.f. 7.3.1(v)):

Theorem 7.8.3 Let L be a splitting field of x^n-a over a subfield K where $a \in K$ and where we assume that K *contains n distinct nth roots of unity*. Then Gal(L/K) is abelian.

Proof (Briefly) If ε is a primitive nth root of 1 and if α is *one* of the zeros of x^n-a in L, then the zeros of x^n-a in L are precisely the elements $\alpha, \varepsilon\alpha, \ldots, \varepsilon^{n-1}\alpha$. In particular, $L=K(\alpha)$. (Why?) The minimum polynomial of α has some of these $\varepsilon^i\alpha$ as *its* zeros. Thus, again, if $\phi_j \in \text{Gal}(L/K)$ is such that $\phi_j(\alpha)=\varepsilon^j\alpha$, then $\phi_j\phi_k(\alpha)=\phi_j(\varepsilon^k\alpha)=\varepsilon^{j+k}\alpha=\phi_k\phi_j(\alpha)$. Hence Gal($L/K$) is abelian.

Exercises

1 Let $F=E_0 \subset E_1 \subset \cdots \subset E_s$ where $E_i=E_{i-1}(r_i)$ and where $r_i^{p_i}=a_i \in E_{i-1}$, p_i being some prime in $\mathbb{Z}$. Let $M(x)=\prod_{i=1}^{s} M_{r_i}(x)$, where $M_{r_i}(x)$ is the minimum polynomial of r_i *over* F. Let S_M be the splitting field of $M(x)$ over E_s. Prove that, for each p_i, S_M contains all complex p_ith roots of unity. [Hint: What are the zeros in S_M of the minimum polynomial $N_{r_i}(x)$ of r_i *over* E_{i-1}?]

2 Find the minimum polynomial (notation $\Phi_n(x)$) for the primitive nth roots of unity for each n such that $1 \le n \le 12$. (The polynomials you are dealing with are called *cyclotomic polynomials*. C.f. [15], Theorem 15.3.) [Hint: $\Phi_4(x) = (x^4-1)/(x^2-1)=x^2+1$. In fact $x^n-1=\prod_{d|n}\Phi_d(x)$.]

3 Find Gal($S_f/\mathbb{Q}$) where $f(x)=x^{12}-1$. Determine all the subgroups of Gal($S_f/\mathbb{Q}$) and all fields between S_f and $\mathbb{Q}$.

4 Let S_f be the splitting field of x^n-1 over $\mathbb{Q}$. Prove that Gal($S_f/\mathbb{Q}$) is isomorphic to the multiplicative group $\mathbb{Z}_n$, say, of invertible elements of $\mathbb{Z}_n$.

5 Find the Galois group over $\mathbb{Q}$ of: (i) x^4+1; (ii) x^5+1.

‡ To avoid problems concerning separability we need only assume characteristic $K \nmid n$. (See [15].) For simplicity we here assume characteristic $K=0$.

6 Prove that the groups in 7.8.3 are cyclic for all n and that those in 7.8.2 are cyclic if n is prime. [Cf. exercise 4.]

7 Let $f(x)=x^p-a$ be irreducible in $\mathbb{Q}[x]$, p being a prime. Show that the Galois group of x^p-a over $\mathbb{Q}$, that is, $\mathrm{Gal}(S_f/\mathbb{Q})$, is isomorphic to the group of transformations on $\mathbb{Z}_p$ given by $y \to iy+j (i,j \in \mathbb{Z}_p, i \neq \hat{0})$.

8 Show that the Galois group of x^n-2 over $\mathbb{Q}$ is not abelian if $n>2$.

9 Let F be any field of characteristic 0. Show that the Galois group of $f(x)=x^n-a$ over F has an abelian normal subgroup with abelian factor group.

10 Let $m,n \in \mathbb{Z}^+$ with $(m,n)=1$. Put $f(x)=x^m-1$, $g(x)=x^n-1$ and $h(x)=x^{mn}-1$. Show that $\mathrm{Gal}(S_h/\mathbb{Q}) = \mathrm{Gal}(S_f/\mathbb{Q}) \times \mathrm{Gal}(S_g/\mathbb{Q})$.

7.9 A necessary condition for the solubility of a polynomial equation by radicals

We now come to the first half of Galois' result concerning the existence of radically expressible zeros of polynomials. We shall suppose, as Galois did implicitly, that all fields are of characteristic 0 so that all polynomials considered are automatically separable (by 7.5.1).

Suppose, then, that F, $f(x)$ and R are as given in 7.2.2 and that the normal radical tower S over F of 7.8.1 has been built around R. For convenience let us relabel the terms of 7.8.1 as

$$F = E_0 \subseteq E_1 \subseteq E_2 \subseteq \cdots \subseteq E_w \subseteq S. \tag{7.9.1}$$

(From now on these *new* E_i replace those given in 7.2.1.) Since S is a Galois extension of F we know (by 7.6.4) that there exists a 1–1 correspondence between the subgroups of $\mathrm{Gal}(S/F)$ and the subfields of S which contain F. In particular the (increasing) sequence of fields in (7.9.1) corresponds to a (decreasing) sequence of subgroups

$$\mathrm{Gal}(S/F) \geq \mathrm{Gal}(S/E_1) \geq \mathrm{Gal}(S/E_2) \geq \cdots \geq \mathrm{Gal}(S/S) = \langle e \rangle. \tag{7.9.2}$$

Since S is a splitting extension of F, S is also a splitting extension of each field V for which $F \leq V \leq S$. In particular, it is a splitting extension of each $E_{i-1} (1 \leq i \leq w)$. Furthermore, as already observed, each E_i is a splitting extension of its predecessor E_{i-1}. Consequently, as in the proof of 7.7.3 (see also exercise 7.10.1!) each subgroup in 7.9.2 is normal in its predecessor. Since each of the corresponding factor groups is, by 7.8.2 and 7.8.3, abelian, we deduce that $\mathrm{Gal}(S/F)$ is itself soluble. (See exercise 6.5.9.)

Since we also have $F \subseteq S_f \subseteq S$, where S_f is a splitting extension of F, the same argument (which we have already given in 7.7.3!) shows that $\mathrm{Gal}(S_f/F)$ is a homomorphic image of $\mathrm{Gal}(S/F)$. Thus, from exercise 6.5.7, we obtain our criterion.

Theorem 7.9.3 Let F, $f(x)$ and S_f be as usual with characteristic $F=0$. If S_f is contained in some radical tower over F, then $\mathrm{Gal}(S_f/F)$ is a soluble group.

Comment 7.9.4 (See exercise 3) This part of Galois' result (7.10.5) is already sufficient to show the existence of polynomials of degree 5 whose zeros are not (all*) expressible in terms of radicals. We shall leave examples of specific polynomials until Section 7.11, and content ourselves with merely stating here (but see exercise 2) the famous result of Abel and Ruffini, namely:

Niels Henrik Abel *(5 August 1802 – 6 April 1829)*
Niels Abel, born in Findö, Norway, was the second child of a poor but gifted churchman. Abel's talents, at first dormant, blossomed under the care of a schoolteacher, Berndt Michael Holmböe, seven years Abel's senior. Together they studied the works of Euler, Lagrange and Laplace. Whilst still at school Abel believed he had discovered how to solve the general quintic equation by radicals. He himself later found the error and in 1824 he published, at his own expense, his proof of impossibility.

After studying at the University of Christiania (Oslo) Abel travelled to Paris where he was ignored by the leading mathematicians, and then to Berlin where he was befriended by Crelle who published Abel's papers in his newly founded journal. Crelle tried to get Abel a professorship at the University of Berlin but because of illness and poverty Abel had to return to Christiania. A few days after he died of tuberculosis a letter from Crelle arrived telling Abel of his appointment to a position in Berlin.

Besides his work on equations Abel did fundamental work on the theory of convergence of power series. He is also regarded as a cofounder, along with Jacobi, of the theory of elliptic functions (although Gauss, in work unpublished in his lifetime, anticipated many of their results).

Theorem 7.9.5 Let $f(x)=a_nx^n+a_{n-1}x^{n-1}+\cdots+a_1x+a_0\in F[x]$, where F is a field of characteristic 0 and the a_i are 'general' coefficients. [In formal terms they

* See exercise 7.10.8.

are elements of the field $F(a_0, a_1, \ldots, a_n)$ of fractions of the polynomial ring $F[a_0, a_1, \ldots, a_n]$ in the distinct letters $a_0, a_1, \ldots, a_n$.]

For each integer $n \geq 5$ the equation $f(x)=0$ is not soluble by radicals over $F(a_0, \ldots, a_n)$. That is, there exists no universal 'formula' in terms of $+, \cdot, -, \div$, the various $\sqrt[n]{\ }$ and the coefficients $a_n, a_{n-1}, \ldots, a_0$ which will always yield the solutions of $f(x)=0$ on substituting specific numbers for the a_i.

Note 7.9.6 Despite the content of 7.9.5 and its interesting corollary, exercise 7.9.5, it is still possible that each *individual* equation $f(x)=0$ over $\mathbb{Q}$ might succumb to a solution by radicals via a 'trick' peculiar to itself. However, in Section 7.11, we'll see how to construct any number of specific quintic polynomials in $\mathbb{Q}[x]$ which are not soluble by radicals over $\mathbb{Q}$. Of course the existence of even one such is sufficient to prove 7.9.5 (at least for $n=5$), but you are invited to prove 7.9.5 directly in exercise 2.

Exercises

1 Let F be any field and let $L=F(x_1, \ldots, x_n)$, where the x_i are 'indeterminates'. Let S_n, regarded as the group of all permutations on $x_1, \ldots, x_n$, give rise to automorphisms of L in the obvious way (cf. 5.5.9). Let $s_j (1 \leq j \leq n)$ be the jth elementary symmetrical polynomial on the x_i (see Section 4.7). Show that $K=F(s_1, \ldots, s_n)$ is the fixed field of S_n as follows:

(i) Show that L is a splitting field for the polynomial

$$y^n - s_1 y^{n-1} + \cdots + (-1)^{n-1} s_{n-1} y + (-1)^n s_n \in K[y]$$

and deduce that L is a Galois extension of K.
(ii) Use exercise 7.2.12(b) to deduce that $[L:K] \leq n!$
(iii) $n! = |S_n| = [L:\text{Fix } S_n] \leq [L:K] \leq n!$

Deduce the following extension of 4.7.1: Every symmetric *rational* function in n indeterminates with coefficients in a field F is expressible as a rational function, with coefficients in F, of the $s_j (1 \leq j \leq n)$.

2 Let F be any field. Use exercise 1 to prove the insolubility by radicals of the 'general' quintic equation

$$f(x) = a_0 + a_1 x + a_2 x^2 + a_3 x^3 + a_4 x^4 + x^5 \in F(a_0, a_1, \ldots a_4)[x],$$

the a_i being indeterminates over F.

3 The first sentence of 7.9.4 is so vague as to be almost worthless. Why?

4 Deduce from exercise 1 that, given any finite group G, there is a pair $F \subseteq E$ of fields such that $\text{Gal}(E/F) \cong G$.

5 Find specific real numbers $a_0, a_1, \ldots, a_4$ such that the polynomial

$$x^5 + a_4 x^4 + a_3 x^3 + a_2 x^2 + a_1 x + a_0$$

is not soluble by radicals over the subfield $\mathbb{Q}(a_0, \ldots, a_4)$ of $\mathbb{R}$.

[This is probably rather hard. Let's just say that the uncountability of $\mathbb{R}$ (exercise 2.6.15) might help! A starting question might be: Can you find a real number which is not a zero of any polynomial in $\mathbb{Q}(\pi)[x]$? Does this generalise?]

7.10 A sufficient condition for the solubility of a polynomial equation by radicals

One naturally asks if the necessary condition stated in 7.9.3 is also sufficient. It would certainly be very satisfying to be able to prove:

Theorem 7.10.1 Let S_f be the splitting field of $f(x) \in F[x]$, where F has characteristic 0. If $\text{Gal}(S_f/F)$ is a soluble group, then there exists a radical tower over F containing S_f.

Taking a hint from Section 7.9, we first adjoin to S_f some roots of 1 by taking the splitting field over S_f of the polynomial $x^n - 1$ for some n. But which n shall we choose? Let us postpone a decision for the moment. In fact, in our first result, the value of n seems to be immaterial!

Lemma 7.10.2 Let $F, f(x), S_f$ be as indicated and ζ be a primitive nth root of unity. If we let $F(\zeta)$ be a splitting field of $x^n - 1$ over F and T be a splitting field of $f(x)$ over $F(\zeta)$, then $T \cong S_f(\zeta)$ and $\text{Gal}(S_f(\zeta)/F(\zeta))$ is (isomorphic to) a subgroup of $\text{Gal}(S_f/F)$.

Proof By definition $T = F(\zeta)(\alpha_1, \alpha_2, \ldots, \alpha_n) = F(\zeta, \alpha_1, \ldots, \alpha_n)$ where $\alpha_1, \alpha_2, \ldots, \alpha_n$ are the zeros of $f(x)$ in T. Hence $T = F(\alpha_1, \alpha_2, \ldots, \alpha_n, \zeta) \cong S_f(\zeta)$, S_f being the splitting field of $f(x)$ over F. If $\theta \in \text{Gal}(S_f(\zeta)/F(\zeta))$, then θ fixes F and permutes the α_i amongst themselves (exercise 7.2.11). Hence each θ maps S_f to itself and so gives rise to an element $\tilde{\theta}$ of $\text{Gal}(S_f/F)$. The map $\theta \to \tilde{\theta}$ is then a homomorphism of $\text{Gal}(S_f(\zeta)/F(\zeta))$ into $\text{Gal}(S_f/F)$ (as in 7.7.2). Finally, if $\tilde{\theta}$ fixes (F and) all the α_i, then θ fixes F, the α_i and ζ. Hence $\theta = e$. That is, the map $\theta \to \tilde{\theta}$ is 1–1 (and into).

We deduce immediately that: *If* $\text{Gal}(S_f/F)$ *is a soluble group then so is* $\text{Gal}(S_f(\zeta)/F(\zeta))$. Consequently there is a chain of subgroups

$$\text{Gal}(S_f(\zeta)/F(\zeta)) = G_0 > G_1 > \cdots > G_n = \langle e \rangle, \qquad (*)$$

each G_i being normal in its predecessor G_{i-1} with G_{i-1}/G_i being cyclic of prime order. (See 6.5.8.)

We know from 7.6.4 that (*) gives rise to a corresponding (increasing) sequence of subfields

$$F(\zeta) = F_0 \subset F_1 \subset F_2 \subset \cdots \subset F_n = S_f(\zeta)$$

so that $G_i = \mathrm{Gal}(S_f(\zeta)/F_i)$, $\mathrm{Fix}(G_i) = F_i$ and with each $[F_i:F_{i-1}]$ prime. Does the normality of each G_i in G_{i-1} and the primality of each $|G_{i-1}/G_i|$ imply anything special about the field extension $F_i \supset F_{i-1}$? We shall see that it does!

Let us first see what the relation of normality implies. Since $S_f(\zeta)$ is a splitting field (for $f(x)$) over $F(\zeta)$, it is also a splitting field over (F_i and) F_{i-1} for each $i(1 \le i \le n)$. There then follows

Lemma 7.10.3 For each $i(1 \le i \le n)$ and for each $\phi \in \mathrm{Gal}(S_f(\zeta)/F_{i-1})$ we have: (i) $\phi(F_i) = F_i$; (ii) F_i is a splitting extension of F_{i-1}.

Proof (i) Let $\theta \in G_i = \mathrm{Gal}(S_f(\zeta)/F_i)$ and $\phi \in G_{i-1} = \mathrm{Gal}(S_f(\zeta)/F_{i-1})$. Then, since $G_i \lhd G_{i-1}$, we have $\phi^{-1}\theta\phi \in G_i$. Hence, for $v_i \in F_i$, $(\phi^{-1}\theta\phi)(v_i) = v_i$. Consequently $\theta\phi(v_i) = \phi(v_i)$—for all $\theta \in G_i$. Hence $\phi(v_i) \in \mathrm{Fix}(G_i) = F_i$ by 7.6.4. Thus $\phi(F_i) \subseteq F_i$. To prove equality, note that $[F_i:F_{i-1}] = [\phi(F_i):\phi(F_{i-1})] = [\phi(F_i):F_{i-1}]$ (why?) $\subseteq [F_i:F_{i-1}]$. That $\phi(F_i) = F_i$ follows immediately.
(ii) $|\mathrm{Gal}(F_i/F_{i-1})| = |\mathrm{Gal}(S_f(\zeta)/F_{i-1})|/|\mathrm{Gal}(S_f(\zeta)/F_i)|$ (exercise 7.6.8) $= [S_f(\zeta):F_{i-1}]/[S_f(\zeta):F_i] = [F_i:F_{i-1}]$ by 4.6.1. The result claimed then follows immediately from exercise 7.5.8.

Thus 'all' we need to do is to show, for each $i(1 \le i \le n)$, that $F_i = F_{i-1}(\delta)$ where $\delta^p \in F_{i-1}$ for some prime p. Surely we won't be so lucky? We *are*!!

Theorem 7.10.4 Let p be a prime and let K be any field containing the p distinct pth roots of 1. Let L be a splitting field over K of a separable polynomial and suppose that $[L:K] = p$. Then $L = K(\delta)$ where $\delta^p \in K$ (so that L is the splitting field of $x^p - \delta^p$ over K).

Proof Since $[L:K] = p$ we have $|\mathrm{Gal}(L/K)| = p$ and so $\mathrm{Gal}(L/K) = \{e, \theta, \theta^2, \ldots, \theta^{p-1}\}$ is cyclic of prime order. Let $c \in L \backslash K$. Define $c_0 = c$, $c_1 = \theta(c)$, $c_2 = \theta^2(c)$, etc. so that $\theta^p(c) = c$. Let $\varepsilon(\neq 1)$ be one of the pth roots of unity in K and put* $\delta = c_0 + \varepsilon c_1 + \varepsilon^2 c_2 + \cdots + \varepsilon^{p-1} c_{p-1}$. Then $\theta(\delta) = \varepsilon^{-1}\delta$. Consequently $\theta(\delta^p) = \delta^p$ so that $\delta^p \in \mathrm{Fix}(\mathrm{Gal}(L/K)) = K$ (since $\mathrm{Fix}(\mathrm{Gal}(L/K)) \ge K$ whilst $[L:K] = p$). But $\delta \notin K$ since $\theta(\delta) \neq \delta$. Hence $L = K(\delta)$ since $[L:K] = p$, as required. (Isn't that nice? Well it *would* be, except for one thing. Why is $\delta \neq 0$?)

So the big question is: can we assume that $\delta \neq 0$? Exercise 10 tells you 'essentially yes'.

The main content of sections 7.9 and 7.10 is thus

Theorem 7.10.5 (Galois) Let $f(x) \in F[x]$, where char $F = 0$ and let S_f be the splitting extension of $f(x)$ over F. Then $f(x)$ is soluble by radicals over F *if and only if* $\mathrm{Gal}(S_f/F)$ is a soluble group.

* Elements of this type were used by Lagrange and Galois and go under the name 'resolvent'. See §§ 5.2 and 7.12.

This result provides the unification which Lagrange was seeking. It tells us that, because the symmetric groups S_2, S_3, S_4 are all soluble, there *are* general formulae (that is, involving literal coefficients) for obtaining the zeros of polynomial equations of degrees 2, 3 and 4.

The above is just one consequence of what is now called the *Fundamental Theorem of Galois Theory*, namely:

Theorem 7.10.6 Let L be a finite normal and separable (i.e. Galois) extension of a subfield K. The map $\Phi:\mathcal{F}\to\mathcal{G}$ from the set $\mathcal{F}$ of fields lying between L and K to the set $\mathcal{G}$ of subgroups of $\mathrm{Gal}(L/K)$ given by $\Phi(M)=\mathrm{Gal}(L/M)$ is a 1–1 correspondence. In addition M is a normal extension of K iff $\mathrm{Gal}(L/M)$ is a normal subgroup of $\mathrm{Gal}(L/K)$. In this case $\mathrm{Gal}(M/K)\cong\mathrm{Gal}(L/K)/\mathrm{Gal}(L/M)$.

Proof Put together 7.6.4 and exercise 7.10.1.

Exercises

1 Let $F\subseteq K\subseteq E$ with E a Galois extension of F. Prove, in full, that K is a normal extension of F iff $\mathrm{Gal}(E/K)$ is a normal subgroup of $\mathrm{Gal}(E/F)$.

2 Find a normal extension N of $\mathbb{Q}$ such that $N\subset\mathbb{Q}(\omega,\sqrt[3]{2})$. Is N a radical tower over $\mathbb{Q}$?

3 By considering the subgroups of the Galois group of $f(x)=x^7-1$ over $\mathbb{Q}$, show that there exist normal extensions which are not radical towers over $\mathbb{Q}$. (Hence, even if $f(x)$ is soluble by radicals over $\mathbb{Q}$, S_f need not itself be a radical tower.)

4 Show that if K is a radical extension of prime degree (>2) over $\mathbb{Q}$, then K is not a normal extension of $\mathbb{Q}$.

5 Let M be a Galois extension of F and let $\alpha\in\mathrm{Gal}(M/F)$. Suppose that $F\subseteq K, L\subseteq M$. Show that

$$\alpha(K)=L \quad\text{iff}\quad \alpha\,\mathrm{Gal}(M/K)\alpha^{-1}=\mathrm{Gal}(M/L)(\subseteq\mathrm{Gal}(M/F)).$$

6 Let K be a field. Define automorphisms $s,t\in\mathrm{Aut}(K(x))$ by: $s(x)=1/x$ and $t(x)=1-x\in K(x)$. Show that the subgroup $S=\langle s,t\rangle$ of $\mathrm{Aut}(K(x))\cong S_3$. Show that $k=(x^2-x+1)^3/x^2(x-1)^2$ is in $\mathrm{Fix}\,S$. Deduce that $[K(x):\mathrm{Fix}\,S]\leq[K(x):K(k)]$. But $kx^2(x-1)^2-(x^2-x+1)^3=0$. Hence x satisfies a polynomial of degree 6 over $K(k)$. Deduce that $\mathrm{Fix}\,S=K(k)$, precisely.

7 Show that if $f(x)$ is irreducible in $F[x]$, then $\mathrm{Gal}(S_f/F)$ is transitive on the zeros of $f(x)$ in S_f. [A group G of permutations on a set X is *transitive* on X iff, for each pair y, z of elements in X, there is a permutation in G under which y is sent to z.]

8 Let $f(x) \in F[x]$ with $f(x)$ irreducible and characteristic $F=0$. Show that if an extension field E of F containing *one* zero of $f(x)$ is a subfield of some radical tower R, say, then S_f is a subfield of some (other) radical tower. Loosely: If one zero of $f(x)$ is radically expressible, then they all are.

9 Let $f(x)$ be irreducible of prime degree p over F (of characteristic 0) with splitting field S_f over F. It can be shown, [15], that $G=\mathrm{Gal}(S_f/F)$ is soluble iff G is isomorphic to some subgroup of the group of exercise 7.8.7 which includes all transformations with $i=1$. Use this result to prove the 'only if' part of what Galois saw ([83]) as his most important result, namely: *Let $f(x)$ be as above and let a, b be zeros of $f(x)$ in S_f. Then $f(x)$ is soluble by radicals iff $S_f=F(a, b)$.*

10 In 7.10.4 replace δ by $\delta_i=c_0+\varepsilon^i c_1+(\varepsilon^2)^i c_2+\cdots+(\varepsilon^{p-1})^i c_{p-1}$ $(0 \leq i \leq p-1)$. Prove that $\theta(\delta_i)=\varepsilon^{-i}\delta_i$ so that $\delta_i^p \in K$. Adding the p equalities obtain $\delta_0+\cdots+\delta_{p-1}=pc_0$. Deduce that not *all* δ_j can be in K—and so $L=K(\delta_i)$, for some i.

11 Noting that $\mathbb{C}$ is a splitting field over $\mathbb{R}$ for each polynomial in $\mathbb{R}[x]$, and that $\mathrm{Gal}(\mathbb{C}/\mathbb{R})=C_2$, how can you reconcile 7.10.5 with exercise 7.9.5?

12 (a) Give two examples of fields $K \subseteq L$ with $[L:K]=10$ and such that: in (i) *all* intermediate fields are normal over K; in (ii) *not* all intermediate fields are normal over K.
(b) Find the Galois groups over $\mathbb{Q}$ of: (i) x^3-3; (ii) x^4+4; (iii) $x^{60}-1$; (iv) $\mathbb{Q}(\sqrt{2}+\sqrt[3]{5})$. In (i), (ii), (iii) identify the normal subfields of the corresponding splitting field.

13 Why are the following soluble by radicals over $\mathbb{Q}$? (i) $x^6+x^4-2x^2-2$; (ii) $x^8-4x^5-2x^3+8$; (iii) $x^6-2x^5+4x^4-8x^3+16x^2-32x+64$?

7.11 Non-soluble polynomials: grow your own!

To show the existence of particular polynomials $g(x)$ not all* of whose zeros are expressible in radical form only requires us to find a polynomial whose Galois group is not soluble. A good way to do this is to return to Galois' idea (see exercise 7.2.11) of regarding $\mathrm{Gal}(S_f/F)$ as a group of permutations on the zeros of $f(x)$ in S_f. This suggests, in the first instance, that we look for a polynomial with Galois group $\cong A_5$ (since A_5 is not soluble—see 6.6.1). However, it seems easier (see exercise 6) to construct, instead, a polynomial whose Galois group is (isomorphic to) S_5.

To be certain what is required, let us suppose that $f(x)$ is irreducible in $F[x]$ with $S_f=F(\alpha_1, \ldots, \alpha_n)$ being, as usual, the splitting field of $f(x)$ over F, the α_i being the *distinct* zeros of $f(x)$ in S_f. Since each element θ of $\mathrm{Gal}(S_f/F)$ effects a

* Hence 'none of ...'. See exercise 7.10.8.

permutation θ_p say, on the set $\{\alpha_1, \ldots, \alpha_n\}$ (exercise 7.2.11), the obvious map $\Gamma:\text{Gal}(S_f/F) \rightarrow S_n$ generates the 'obvious' theorem, namely

Theorem 7.11.1 The map Γ is a 1–1 homomorphism of $\text{Gal}(S_f/F)$ into S_n. That is, $\text{Gal}(S_f/F)$ is *isomorphic* to a subgroup of S_n.

Proof Exercise 7.2.11.

Since S_n is generated (see exercise 5.6.25) by the two cycles $(12 \ldots n)$ and (12), the next two lemmas almost suggest themselves.

Lemma 7.11.2 Let $F, f(x), S_f$ be as usual with $f(x)$ irreducible over F and of degree p, p being a prime. Then $\text{Gal}(S_f/F)$ contains a p-cycle.

Proof Let α be a zero of $f(x)$ in S_f. Then $[S_f:F]=[S_f:F(\alpha)][F(\alpha):F]$. Since $[F(\alpha):F]=p$ (see 4.3.4) we see that $p \mid [S_f:F]=|\text{Gal}(S_f/F)|$. Hence $\text{Gal}(S_f/F)$ has an element of order p. But $\text{Gal}(S_f/F)$ is a subgroup of S_p (why?) and the only elements of order p in S_p are p-cycles. (Exercises 1 (below) and 5.6.8(b).)

Perhaps the easiest way to ensure that $\text{Gal}(S_f/F)$ also contains a transposition is to specialise $F[x]$ to $\mathbb{Q}[x]$ and prove

Lemma 7.11.3 If $f(x) \in \mathbb{Q}[x]$ is irreducible and has exactly 2 non-real zeros, then $\text{Gal}(S_f/F)$ contains a 2-cycle.

Proof Here we may, of course, assume that $S_f \subseteq \mathbb{C}$. Then complex conjugation fixes all the real zeros in S_f and permutes the complex ones (by exercise 2). This suffices.

Combining 7.11.1, 7.11.2 and 7.11.3 we have:

Theorem 7.11.4 If $f(x) \in \mathbb{Q}[x]$ is irreducible of degree p (p being a prime ≥ 5) and $f(x)$ has exactly 2 non-real zeros, then $\text{Gal}(S_f/\mathbb{Q}) = S_p$. In particular S_p is not a soluble group and $f(x)$ is not soluble by radicals.

Using 7.11.4 you can construct your own non-soluble equations *ad nauseam*!

To do so first note that a real polynomial $f(x)$ of degree 5 has three real zeros if $f'(x)$ has just two (distinct) zeros a, b, (say) such that (say) $f(a) > 0$ and $f(b) < 0$ as in Figure. 7.1

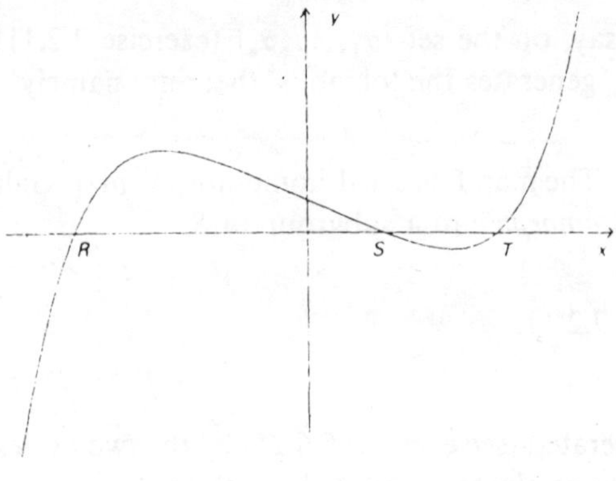

Fig. 7.1 Part of the graph of $y=x^5-4x+2$, with zeros R, S, T near to -1.52, 0.51, and 1.24

To keep the arithmetic easy let's assume that $f'(x)=x^4-c$ where c is some positive real number to be determined. Then $f(x)=x^5/5-cx+d$ for some constant d. Set $\gamma=c^{1/4}$, the real positive 4th root of c. Now $f'(x)=0$ iff $x=\pm\gamma$. Also $f(-\gamma)=-\gamma^5/5+\gamma^4\gamma+d=\frac{4}{5}\gamma^5+d$ whilst $f(\gamma)=-\frac{4}{5}\gamma^5+d$. Thus if $d\geq 0$ then $f(-\gamma)>0$ whilst $f(\gamma)<0$ if $d<\frac{4}{5}\gamma^5$. Since we want $f(x)$ to be irreducible in $\mathbb{Q}(x)$ we choose c, d so that Eisenstein's criterion will check this requirement quickly. Now $5f(x)=x^5-5cx+5d$, so to use Eisenstein let's choose $d=1$ and then $c=16$, say. Consequently $x^5-80x+5$ is a polynomial of the required kind! We leave other choices, for example $c=\frac{4}{5}$, $d=\frac{2}{5}$, to you.

Exercises

1 Show that the only elements of order p in S_p are p-cycles.

2 Let M be a subfield of $\mathbb{C}$ which contains, to each α, its complex conjugate $\bar{\alpha}$. Show that complex conjugation on $\mathbb{C}$ induces an automorphism of order 2 on M.

3 Find two polynomials $f(x)$, $g(x)$ in $\mathbb{Q}[x]$ such that $\mathrm{Gal}(S_f/\mathbb{Q})$ and $\mathrm{Gal}(S_g/\mathbb{Q})$ are isomorphic (as abstract groups) but are not isomorphic as permutation groups. [$\{I, (ab), (cd), (ab)(cd)\}$ and $\{I, (ab)(cd), (ac)(bd), (ad)(bc)\}$ are two such groups. Can you see why they can't be regarded as being isomorphic as permutation groups?]

4 Can you prove the following? (Cf. 7.11.4.) Let $f(x)\in\mathbb{Q}[x]$ be irreducible of degree 5 and have five real zeros including the pair $a+b\sqrt{2}$ and $a-b\sqrt{2} (b\neq 0)$, where $a, b\in\mathbb{Q}$. Then $f(x)$ isn't soluble by radicals over $\mathbb{Q}$.

5 (Cf. 7.11.4.) Find an irreducible polynomial of degree 6 in $\mathbb{Q}[x]$ which has just two non-real zeros and yet *is* soluble by radicals over $\mathbb{Q}$.

6 Let F be a field (characteristic $\neq 2$). Let $f(x) \in F[x]$ be separable and let $S = F(\alpha_1, \ldots, \alpha_n)$ be a splitting field. Put $\delta = \prod_{1 \leq i < j \leq n} (\alpha_i - \alpha_j)$. Show that (i) $\delta^2 \in F$ and (ii) $\mathrm{Gal}(S/F)$ is isomorphic to a subgroup of A_n iff $\delta \in F$.

7 Exhibit a field F, containing $\mathbb{Q}$, and a polynomial $f(x) \in F[x]$ such that $\mathrm{Gal}(S_f/F) = A_n$.

8 Find $f(x)$, in $\mathbb{Q}[x]$, of degree 5 and with all six coefficients non-zero such that $f(x)$ is not soluble by radicals over $\mathbb{Q}$. [Hint: $f(x+1)$.]

9 $x^5 - 4x + 2$ and $x^5 - 6x + 3$ are frequently chosen as examples of polynomials of degree 5 which are not soluble by radicals over $\mathbb{Q}$. Confirm they are indeed both insoluble over $\mathbb{Q}$.

10 True or false? A soluble irreducible equation of prime degree over $F(\subseteq \mathbb{R})$ has exactly one real zero—or *all* zeros real. [Hint: Use the result of exercise 7.10.9.]

11 Test the following for solubility over $\mathbb{Q}$.
(i) $x^5 - 8x + 3$; (ii) $x^5 - 7x + 3$; (iii) $x^6 - x^5 + x^4 - x^3 + x^2 - x + 1$; (iv) $x^5 - 3x^3 - 3x + 3$; (v) $x^5 - 4x^2 + 2$; (vi) $x^5 - 4x^3 + 2$; (vii) $x^5 - 4x^4 + 2$; (viii) $x^p - 4x + 2$ ($p \geq 5$ prime); (ix) $\frac{6}{5}x^5 - \frac{5}{4}x^4 + \frac{7}{3}x^3 - \frac{5}{2}x^2 + x + \frac{13}{100}$. [Hint: for (ix) Eisenstein shows irreducibility. Use the rational root test on the derivative to help locate local max and min.].

12 (a) Find $\mathrm{Gal}(GF(p^n)/\mathbb{Z}_p)$. [Hint: $L = GF(p^n)$ is a Galois extension of $\mathbb{Z}_p$. Hence $|\mathrm{Aut}(L)| = [L:\mathbb{Z}_p] = n$. Show that, for L, the Frobenius automorphism ϕ (exercise 7.5.14) has order n.]
(b) Draw the subgroups/subfields diagram for the finite field of order p^{12}.

7.12 Galois theory—old and new

Having spent some time seeing where our noses led *us*, let us quickly review where Galois' nose led *him*.

Galois was interested in equations of the form $f(x) = a_0 + a_1 x + \cdots + a_n x^n = 0$ where the a_i are given numbers (as distinct from the 'unknowns' used in 7.9.5) whose n zeros are pairwise distinct.* [For ease of communication we shall suppose—as in the presentations [12] and [11] of 1926 and 1930 respectively—that we are working inside $\mathbb{C}$ and we shall let F denote the least subfield of $\mathbb{C}$ (namely $\mathbb{Q}(a_0, a_1, \ldots, a_n)$) which contains $\mathbb{Q}$ and all the a_i. As already mentioned in Section 7.5, Galois seems not to have worried too much about where or what the zeros of $f(x)$ were.]

Let us designate the zeros of $f(x)$ by $\alpha_1, \alpha_2, \ldots, \alpha_n$. Whereas, for Lagrange, the α_i were *formally* distinct symbols, so that $\alpha_1 + \alpha_2\alpha_3$ and $\alpha_2 + \alpha_3\alpha_1$, for example, were likewise formally distinct, Galois had to accept that $\alpha_1 + \alpha_2\alpha_3$ and $\alpha_2 + \alpha_3\alpha_1$

* There is no loss in generality in this latter assumption; see exercise 7.5.1.

could be numerically equal. Nevertheless, under the (obviously necessary) assumption that the α_i were *numerically* distinct, Galois observed (without proof) that integers $z_1, z_2, \ldots, z_{n!}$ could be found so that the quantity

$$t = z_1\alpha_1 + z_2\alpha_2 + \cdots + z_n\alpha_n \tag{7.12.1}$$

would take $n!$ distinct *numerical* values on formally permuting the α_i (equivalently the z_i) in all $n!$ possible ways.

With t so chosen, Galois showed that all the α_i $(1 \leq i \leq n)$ could be expressed as rational functions of (i.e. a ratio of polynomials in) t. [In our notation, this means $\mathbb{Q}(\alpha_1, \alpha_2, \ldots, \alpha_n) = \mathbb{Q}(t)$. (Why not just '$\subseteq$'?)] Galois now forms 'the equation for t'—that is, the equation

$$R(x) = (x - t_1)(x - t_2)\ldots(x - t_{n!}) \tag{7.12.2}$$

where the t_i are the $n!$ distinct values of t $(= t_1)$ obtained from the $n!$ permutations of the α_i. Each such permutation permutes the t_i and hence fixes the coefficients of $R(x)$: that is, they are symmetric functions of the α_i. [Hence, in our language, they belong to F. Why?]

Now $R(x)$ may factorise (into a product of irreducibles) in $F[x]$, $R(x) = R_1(x)R_2(x)\ldots R_m(x)$, say. Suppose (without loss of generality) that $R_1(x) = (x - t_1)(x - t_2)\ldots(x - t_k)$ is the factor with t $(= t_1)$ as a zero. Then

Definition *The Galois group of* $f(x)$ over $F[x]$ *is the set of permutations on the* $\alpha_1, \ldots, \alpha_n$ *which arise when* t *is mapped into* $t_1, t_2, \ldots, t_k$ *in turn.*

One can show that this finite set has the properties required for it to be a group and that the group is independent of the particular t chosen at the outset.

Galois now says that if he 'adjoins to the given equation' the zero of some auxiliary equation (that is, if we replace the 'base' field F by a larger field F_1, say, and think of $f(x)$ as being in $F_1[x]$), then either the Galois group of $f(x)$ (over $F_1[x]$) will be the same as that of $f(x)$ (over $F[x]$) or will be a subgroup of the latter—cf. the proof of 7.10.2. [The reason is easily seen: In $F_1[x]$ the previously irreducible $R_1(x)$ may, itself, now factorise into a product of factors *irreducible in* $F_1[x]$. If $R_1(x) = R_{11}(x)R_{12}(x)\ldots R_{1m}(x)$ with $R_{11}(x) = (x - t_1)(x - t_2)\ldots(x - t_l)$, say, where $l \leq k$, the corresponding Galois group comprises only those permutations of the α_i which arise on mapping t to $t_1, \ldots, t_l$.

If, by a succession of such adjunctions—i.e. by repeatedly enlarging the (current) base field F—we arrive at a field F_2, say, in which we can factor $R(x)$ into a product of linear factors, then the Galois group of $f(x)$ (over F_2) will be trivial and, since t $(= t_1)$ belongs to F_2, so too will all the zeros $\alpha_1, \alpha_2, \ldots, \alpha_n$ of $f(x)$.

So the question is: What must the above sequence of subgroups (from that of the Galois group of $f(x)$ over F down to $\langle e \rangle$) look like if we insist that each 'adjunction' is of a p_ith root (p_i some prime) of a quantity previously constructed (or given)?

Galois' answer, is that the (Galois) group must have a series of subgroups as in (*) in 6.5.7: that is G must be soluble.

Although, as we noted earlier, Galois' ideas were still being followed as late as 1930 [11], the 'modern' viewpoint sees the subject of Galois Theory as that of investigating the connection between the subfields of certain types of extensions (in the first instance Galois extensions) K of a given field F and those subgroups of the automorphism group of K which 'fix' F. The conditions relating to the solubility of equations by radicals then result from an *application* of the general theory. [Indeed Artin, [3], will have nothing to do with equation solving, leaving applications to an appendix *written by a coauthor*!] Of course Artin's approach didn't suddenly emerge from nowhere; one can trace various ideas via, for example, Dedekind (who obtained the result of Artin's lemma working throughout in $\mathbb{C}$), Weber (who discussed groups and fields of 'things') and Steinitz (who noticed that a wholesale generalisation of Galois' ideas to these general fields wouldn't necessarily go through and introduced the concepts of normality and separability to clarify when it would). The account of this thread of history given in [108] is very interesting but do follow this up by reading the account [126] of later development by one who was involved.

Naturally enough, attempts have been made to extend Galois Theory to other settings, for example to commutative rings, to division rings and to differential equations (see, for instance, [4]). Even in its most 'primitive' setting there are still huge gaps in our knowledge. Perhaps the most difficult problem is the determination of precisely which (finite) groups can be Galois groups of the form $\mathrm{Gal}(S_f/\mathbb{Q})$. It is believed that every finite group is of this kind. Shafarevich showed (1954) that every finite *soluble* group is a Galois group over $\mathbb{Q}$ so, naturally, attention turned to the finite simple groups. In 1984 Thompson showed that the 'monster', the largest of the 26 sporadic simple groups, is a Galois group over $\mathbb{Q}$. Since then the majority of the sporadic groups have been proved to be Galois over $\mathbb{Q}$ by one or more of Hoyden-Siedersleben, Hunt, Matzat and Pahlings.

Exercises

1 (The Primitive Element Theorem) *Let* $F \subseteq E$ *be fields of characteristic* 0. *Suppose* $E = F(a, b)$, *where* a, b *are algebraic over* F (see p. 153). *Then there exists* $d \in E$ *such that* $E = F(d)$. (That is, E can be generated over F by a single element. We say E is a *simple* extension of F.) Proceed as follows. Let $f(x)$, $g(x)$ be the minimum polynomials over F of a, b respectively and let $K = F(a_1, \dots, a_m, b_1, \dots, b_n)$, where $a = a_1, b = b_1$, be a splitting field for $f(x)g(x)$ over F, the a_i, b_j being its distinct zeros. Choose $c \in F$ such that $c \neq 0$ and $c \neq (a_i - a_k)/(b_j - b_l)$ for all $i \neq k$ and $j \neq l$. [Why can we do this?] Claim: $a + bc$ ($= d$, say) works as required. For, clearly, $F(d) \subseteq F(a, b)$ ($= E$). Conversely, let $h(x) = f(d - cx) \in F(d)[x]$. Prove that $h(b) = g(b) = 0$ in E. Setting $t(x) = \gcd(h(x), g(x)) \in F(d)[x]$ we have $\deg t(x) \geq 1$. [Why? What if $t = 1$?] In $K[x]$, $t(x) | g(x) = (x - b_1) \dots (x - b_n)$. Hence any other zero of $t(x)$ in K is a $b_j (j > 1)$. But then $h(b_j) = 0$, so that $f(d - cb_j) = 0$, i.e. $d - cb_j = a_k$ for some k. This is impossible (why?). Hence $t(x) = x - b$, whence $b \in F(d)$.

(This result was for years one of the cornerstones in the proofs of the main results of Galois Theory. In particular the vital concept of normal extension was made to depend on it. Artin showed how to reshape the definition of normal extension without having to introduce the somewhat artificially chosen primitive element.)

2 Deduce from exercise 1 that, if $K \subseteq L$ where $[L:K] < \infty$ and characteristic $K = 0$, then $L = K(\alpha)$ for suitable $\alpha \in K$.

3 (a) Find a suitable primitive element for (i) $\mathbb{Q}(\sqrt{2}, \sqrt{3})$; (ii) $\mathbb{Q}(\sqrt{2}, \sqrt{3}, \sqrt{5})$; (iii) the splitting field over $\mathbb{Q}$ of $x^3 - 2$.
(b) In all honesty (!) which 'presentation' of the splitting field of $(x^2 - 2)(x^2 + 1)$ is the more natural (*i*) $\mathbb{Q}(\sqrt{2}, \mathrm{i})$ or (ii) $\mathbb{Q}(\sqrt{2} + \mathrm{i})$?

4 Which result in Section 7.10 corresponds to Galois' idea of 'reducing the group of the equation' by extending the 'base' field?

5 Use exercise 2 to show, under those assumptions, that $|\mathrm{Gal}(L/K)| \leqslant [L:K]$.

Partial solutions to the exercises

The following 'answers' are only rarely complete. They are in general meant to give a good clue without doing all the work for the reader—and they should certainly not be taken as the epitome of good mathematical style. Where a counterexample is given, you should aim to produce another.

Exercises in Chapter 0

1 (a) False; (c) true; (e) false ($e^{\pi} > \pi^{e}$: *Math. Gazette*, p. 220, Oct. 1987).

2 (i) True; (iii) false; (v) false; (viii) true; (x) true.

3 $B \cap C = \{3\}$; $A \cup B = \{x : x \in \mathbb{R}^{+}$ and $x = 1$ or $x = 2$ or $x > \sqrt{7}\}$.

4 $A \cap B = \{(u, v), (u, -v)\}$ where $u = -2 + \sqrt{5}$ and $v = 2\sqrt{(-2 + \sqrt{5})}$.

5 (b) Let $x \in (A \cup B) \cap C$. Then ($x \in A$ or $x \in B$) and $x \in C$. Hence ($x \in A$ and $x \in C$) or ($x \in B$ and $x \in C$). Consequently $x \in (A \cap C) \cup (B \cap C)$.

6 $\bigcap_{r \in \mathbb{R}^{+}} T_r = \{0\}$; $\bigcup_{n=1}^{\infty} R_n = \{x : x \in \mathbb{R}$ and $-1 < x\}$.

7 (i) If $x \in A^c \cap B^c$ then $x \notin A$ and $x \notin B$. Hence $x \notin A \cup B$ and so $x \in (A \cup B)^c$. Deduce (iii) from (i) and (ii). Do (iv) similarly.

8 Suppose $\{\{a\}, \{a, b\}\} = \{\{c\}, \{c, d\}\}$ and that $a \neq c$. Then $\{a\} \neq \{c\}$. Therefore $\{a\} = \{c, d\}$ and so $a = c (= d)$, contradiction! Consequently $a = c$ and so $\{a\} = \{c\}$. If now $b \neq d$ then $\{a, b\} \neq \{c, d\}$. Therefore $\{a, b\} = \{c\}$. Hence ....

9 (ii) No! Take, for example, $A = \{1, 2\}$, $B = \{a, b\}$ and $S = \{(1, a), (2, b)\}$.

10 (i) 1024 ($= 2^{10}$).

11 Assume m is odd. Then $m = 2t + 1$ for some integer t. Hence $m^2 = 4t^2 + 4t + 1$, a contradiction! Consequently m must be even.

15 Note that each integer is of the form $4k$ or $4k + 1$ or $4k + 2$ or $4k + 3$. Then use $(4k)^2 = 8(2k^2)$, $(4k + 1)^2 = 8(2k^2 + k) + 1$, etc.

16 Informally: For each integer x and each integer y, the product xy is the square of an integer.

17 (i) is true; (ii) is false.

Exercises 1.2

1 Writing x for $b+c$ we deduce $(a+(b+c))+d=(a+x)+d=a+(x+d)$ $=a+((b+c)+d)$. The others are similar.

3 $(a-b)\cdot c$ means $(a+(-b))\cdot c$. So $(a-b)\cdot c=a\cdot c+(-b)\cdot c=a\cdot c$ $+(-(b\cdot c))=a\cdot c-b\cdot c$. (The reasons for each step are left to you.)

5 From $c\cdot a=c\cdot b$ we obtain $c\cdot a+(-(c\cdot b))=0$ (using Axiom A4) $=c\cdot a$ $+c\cdot(-b)$ (by 1.2.1(ii)) $=c\cdot(a+(-b))$ Axiom D). Since $c\neq 0$ (given) we deduce $a+(-b)=0$ (by property Z). Consequently $a=b$ (using Axioms A2 (where?) A4 and A3).

7 Let z and $-f$ be the functions defined for all $x\in C[0,1]$ by: $z(x)=0$; $(-f)(x)=-(f(x))$. Then $z, -f\in C[0,1]$. z is the function satisfying (the analogue of) the A3 axiom; given f, then $-f$ satisfies A4. To show axiom Z *isn't* satisfied:

$$\text{let } F(x)=\begin{cases}\frac{1}{2}-x & \text{if } 0\le x\le\frac{1}{2}\\ 0 & \text{if } \frac{1}{2}\le x\le 1\end{cases}\qquad G(x)=\begin{cases}0 & \text{if } 0\le x\le\frac{1}{2}\\ x-\frac{1}{2} & \text{if } \frac{1}{2}\le x\le 1\end{cases}$$

Then $F\cdot G=z$ but $F\neq z$ and $G\neq z$. (Draw yourself a picture of $F(x)$ and $G(x)$!)

9 $a\neq 0\Rightarrow a<0$ or $a>0$ (i.e. $-a\in N$ or $a\in N$: see 1.2.4.). Then $(-a)\cdot(-a)$ $=a\cdot a\in N$. Hence $0<a^2$.

10 If, for example, $-a\in N$ and $b\in N$ then $-(a\cdot b)=(-a)\cdot b\in N$. Hence $-(a\cdot b)\neq 0$, and so neither is $a\cdot b$. (Now tidy up this argument!)

11 (i) Use $(a_1-b_1)\cdot(a_2-b_2)>0$.

13 $\mathbb{Q}$. (Here take N as the positive rationals.)

14 If there is no '0' then A3, A4 will not be satisfied. Note also that Axiom I *can't* fail if the set under consideration has no '1'. So you might look for a subset of $\mathbb{Z}$, say, containing neither 0 nor 1.

15 All ten are satisfied.

16 (ii) Let U be the set of all positive integers n such that $1+3+\cdots$ $+(2n-1)=n^2$. For $n=1$ the left-hand side (lhs) is the sum of all odd integers from 1 to $2n-1$ (i.e. from 1 to 1 in this case). Hence the sum is 1—and this is the value of the rhs when $n=1$. Hence $1\in U$. Now suppose $k\in U\subseteq N$; that is, suppose lhs = rhs if $n=k$. For $n=k+1$ lhs is $1+3+\cdots+(2k-1)$ $+(2(k+1)-1)$. By the induction hypothesis this sum is $k^2+(2(k+1)-1)$, that is, $(k+1)^2$. Thus $k+1\in$ U. Hence $U=N$ by property I.

17 Fact: $1+\frac{1}{2}+\frac{1}{3}+\cdots$ can be made as big as you wish by taking sufficiently many terms. (See, for example, G. H. Hardy's famous *Pure Mathematics*)

18 Put $S_n=1+\frac{1}{2}+\cdots+(\frac{1}{2})^n$. Then $S_{n+1}=(\frac{1}{2})S_n+1$.

19 Let U be the set of positive integers n for which $1 \leq n$. Then $1 \in U$. Now let $k \in U \subseteq N$. Then $(k+1)-1=k \in N$ so that $1<k+1$. By property I, $U=N$. Hence if c is an integer with $0<c$ then $1 \leq c$. Hence $c<1$ is impossible. Finally, prove $0<t-k<1$ using exercise 8.

20 If $b=0$ then $ab=0$ $(\neq 1)$. If $b<0$ then $b=-c$ for some $c>0$. This implies $1=ab=a(-c)=-(ac) \in N$, a contradiction. Hence $0<b$. By exercise 19, $1 \leq a, 1 \leq b$. If $1<b$ then, by 1.2.6, $(1 \leq)a1<ab=1$ – a contradiction.

22 Look to see if the argument purporting to show that $S(k) \Rightarrow S(k+1)$ *always* holds is valid in the case $k=1$.

23 (i) Show that $u_{5k}=5u_{5k-4}+3u_{5k-5}$.

Exercises 1.3

1 If $b=ua$ and $a=vb$ then $b=u(vb)=(uv)b$. Hence, if* $b \neq 0$ then $1=uv$ [why?]. Thus u, v are units. [*What if $b=0$?]

3 Use induction on the number r of factors in the product $b_1 b_2 \dots b_r = b_1(b_2 \dots b_r)=x$, say, the case of $r=2$ being given. Begin by saying: 'If $a|x$ then $a|b_1$ or $a|b_2 \dots b_r$'.

5 Suppose $x=5\alpha+u$, $y=5\beta+v$, $z=5\gamma+w$, where $u, v, w \in \{-2, -1, 1, 2\}$. (Why these remainders?) Show that $5 \nmid u^2+v^2+w^2$.

6 The first *six* irreducibles are 5, 9, 13, 17, 21, 29. Note that 9 is irreducible (essentially because $3 \notin H$) but is not prime in H (since $693=9 \cdot 77=21 \cdot 33$ and yet $9 \nmid 21$ and $9 \nmid 33$ *in* H). On the other hand each H-prime is an H-irreducible. In fact, if $h \in H$, then h is prime in H iff h is prime in $\mathbb{Z}$.

7 Prove that if $N_n=q_1 q_2 \dots q_t$, where each q_i is of the form $4k+1$, then so is N_n—a contradiction. Thus at least *one* of the q_i must be of the form $4k+3$ and hence must be one of the p_j. Finish the proof as in 1.3.10.

8 [Cf. exercise 7] For the $6k+5$ case consider $N_n=6p_1p_2 \dots p_n-1=q_1 \dots q_t$. Show that each q_i is of the form $6k+1$ or $6k+5$. A similar proof for primes of the form $8k+7$ fails because numbers of the form $8k+7$ need not have any prime factors of that form. (What is the smallest example of this?) [In fact there *do* exist infinitely many primes of the form $8k+7$—see [42], p. 53.]

Exercises 1.4

2 Since c is a common divisor of a and b and since d is a *greatest* common divisor, we have $c|d$. Likewise $d|c$. Now use exercise 1.3.1.

4 There exist $\alpha, \beta, \gamma, \delta \in \mathbb{Z}$ such that $\alpha a+\gamma c=1$, $\beta b+\delta c=1$. Then $1=(\alpha a+\gamma c)(\beta b+\delta c)=\alpha\beta ab+c(\alpha\delta a+\gamma\beta b+\gamma\delta c)$. Hence, if $d|ab$ and $d|c$, then ...?

5 Show that if $d|(a,c)$ then $d|a$ and $d|a+b$. Hence $d|\ldots$? But $(a,b)=1$.

6 Let $u=(a,(b,c))$. Then $u|a$ and $u|(b,c)$. Thus u divides a and b and c. Hence u divides (a,b) and c. Consequently $u|((a,b),c)$. Conversely ...?

8 Assume without loss of generality that $0\le r_1\le r_2<|b|$. From $b(m_1-m_2)=r_2-r_1$ we deduce $b|r_2-r_1$. But this forces $r_2-r_1=0$—since $0\le r_2-r_1<|b|$.

10 If S doesn't comprise 0 alone then there exists in S a non-zero integer a, say. Then $0(=a-a)\in S$ and so $0-a\in S$. Hence S contains at least one *positive* integer. Let c be the smallest positive integer in S. [How do we know such a c exists?] Now $c-(-c)=2c\in S$ and likewise (or, preferably, by induction!) $mc\in S$ for all $m\in\mathbb{Z}$. Conversely, suppose $w\in S$. Write $w=mc+r$ where $0\le r<c$. Then $r\in S$ (since w and mc are both in S). Hence, by choice of c, $r=0$, whence $w=mc$. Thus each multiple of c is in S and, conversely, each member of S is a multiple of c, as claimed.

11 If $d=(s,t)$ then $dc|sa$ and $dc|tb$ so that $dc|sa+tb=c$. Hence $d(>0)=1$.

12 (i) If $d|sa$ and $d|tb$ then $d|sa+tb$. (ii) $3\cdot 3+(-4)\cdot 2=1$ but $(3\cdot 2,(-4)\cdot 3)\ne 1$—for example.

13 Put $c=(a,b)$ and $d=(b,r)$. Then $c|a$, $c|b$, so that $c|a-mb=r$. Hence $c|b$ and $c|r$ and so $c|d$. Likewise $d|b$ and $d|r$ imply $d|a$. Consequently $d|c$. Thus $d=c$.

14 Having found $(901,527)=17=12\cdot 527-7\cdot 901$, suppose $17=527u+901v$. Then $527\cdot(12-u)=901\cdot(v+7)$. Hence $31\cdot(12-u)=53\cdot(v+7)$. Therefore $53|u-12$ (why?); that is $u=12+53m$, $m\in\mathbb{Z}$. So u can only be equal to ...?

15 The common divisors of 441 and 693 *in H* are 1, 9, 21. So 21 is numerically the greatest but is not a multiple (*in H*) of 9.

16 (i) If $k=rs$ then $2^r-1|2^k-1$. (Cf. $x-1|x^s-1$ for polynomials. Now put $x=2^r$.) (ii) List the divisors of $2^{k-1}p$ assuming p is a prime.

17 Put

$$\binom{p}{i}=\frac{p(p-1)\ldots(p+1-i)}{1\cdot 2\ldots i}=u.$$

Then $u\cdot i!=p(p-1)\ldots(p+1-i)$. Now clearly u is an integer. (One can prove this by induction or observe that it is the number of ways of choosing i things from a set of p objects.) Also $p|u\cdot i!$ But $p\nmid i!$ [Why not?] Hence $p|u$ as required.

Exercises 1.5

1 No—since each of the factors is prime. See 1.5.1.

4 If $a=md_1=nd_2$ and if $\alpha d_1+\beta d_2=1$ then $a=\alpha d_1(nd_2)+\beta d_2(md_1)$.

5 $sa+tb=1\Rightarrow sac+tbc=c\Rightarrow a(sc+tm)=c$—assuming $bc=ma$. Therefore $a|c$.

6 Let $a=p_1^{\alpha_1}p_2^{\alpha_2}\dots p_r^{\alpha_r}$ and $b=q_1^{\beta_1}q_2^{\beta_2}\dots q_s^{\beta_s}$ where the p_i, q_j are pairwise distinct primes. Suppose, further, that $c=u_1^{\gamma_1}u_2^{\gamma_2}\dots u_t^{\gamma_t}$, where the u_k are pairwise distinct primes, and that all the $\alpha_i, \beta_j, \gamma_k$ are positive integers. Then $c^2=p_1^{\alpha_1}p_2^{\alpha_2}\dots p_r^{\alpha_r}q_1^{\beta_1}q^{\beta_2}\dots q_s^{\beta_s}=u_1^{2\gamma_1}u_2^{2\gamma_2}\dots u_t^{2\gamma_t}$. By 1.5.1 the u_k must be the p_i and q_j in some order and each α_i, β_j is some $2\gamma_k$ and hence even. Thus both a and b are squares.

Exercises 1.6

2 (b) $u\odot v=(0, 2, 6, -1, -13, 7, 14, -2, -15, 6, 0, 0, \dots)$

5 $1+3x+x^2-x^3=(1+x)(1+2x-x^2)$. So take $u=(1, 1, 0, 0, \dots)$, $v=\dots$?

6 Assuming $u=(u_0, u_1, \dots)$, $v=(v_0, v_1, \dots)$ have zero entries from u_m and v_n onwards respectively, then $u\oplus v$ has only zero entries from the tth place onwards, where t is $\max\{m, n\}$, the greater of m and n. (Note that it's easiest to regard u_0, for example, as being in the 0th place.) For the product, the infinite succession of zeros starts at the $((m-1)(n-1)+1)$th place. (Correct?)

Exercises 1.7

2 $u\odot(v\oplus w)$ has as its kth term $\sum_{i=0}^{k} u_i(v_{k-i}+w_{k-i})=\sum_{i=0}^{k} u_iv_{k-i}+\sum_{i=0}^{k} u_iw_{k-i}$.

4 Take $N_1(N_2)$ as the set of all polynomials of the form $a_rx^r+a_{r+1}x^{r+1}+\dots+a_sx^s$ (where $0\le r\le s$) such that $a_r>0(a_s>0)$.

5 If $(a_0, a_1, \dots)\odot(b_0, b_1, \dots)=(1, 0, 0, \dots)$, then $a_0b_0=1$. Hence $a_0\neq 0$. Conversely, given $a_0\neq 0$, use this to define $b_0, b_1, b_2, \dots$ inductively by $1=a_0b_0$; $0=a_0b_t+a_1b_{t-1}+\dots+a_{t-1}b_1+a_tb_0$ (for $t>0$). Thus, at each stage, b_t is defined in terms of the given a_i and the already evaluated b_k $(0\le k\le t-1)$.

Exercises 1.8

2 Formally x is $((0,1,0,\dots),(0,0,\dots),(0,0,\dots),\dots)$ whilst y is represented formally by $((0,0,\dots),(1,0,0,\dots),(0,0,\dots),\dots)$.

Exercises 1.9

1 $(x^2-2)(x^2+2)$, $(x-\sqrt{2})(x+\sqrt{2})(x^2+2)$ in $\mathbb{Z}[x]$ and $\mathbb{R}[x]$ respectively.

3 (i) reducible: 2 and $(1+x+3x^2)$ are non-units in $\mathbb{Z}[x]$; (ii) irreducible—note that 2 is a unit in $\mathbb{Q}$; (iii) irreducible; (iv) reducible (why?).

5 (i) Yes. First show that p is neither the zero element nor a unit in $\mathbb{Z}[x]$. Finally note that if $p|f(x)g(x)$ in $\mathbb{Z}[x]$, then p divides $f_0g_0, f_0g_1+f_1g_0$, etc. in $\mathbb{Z}$. If

$p \nmid f(x)$ and $p \nmid g(x)$ in $\mathbb{Z}[x]$ then there exist a least i and j such that f_i and g_j are not divisible by p. Then (cf. the proof of 1.9.10) the coefficient of x^{i+j} in $f(x)g(x)$ is not divisible by p—contradiction.

8 (b) $\frac{1}{6}(3x^2+2x+42)$.

9 Assuming that $f \neq 0$ and $g \neq 0$ (so that degree is defined) we have $\deg(f \cdot g) = \deg f + \deg g$; $\deg(f+g) \leq \max\{\deg f, \deg g\}$—if, also, $f \neq -g$.

10 (a) Irreducible (Eisenstein with $p=2$); (c) Eisenstein is no use with f as given. But f_{-1} succumbs to Eistenstein and proves f_{-1} and hence f irreducible.

11 $f(x)=x^4+1 \Rightarrow f(x+1)=x^4+4x^3+6x^2+4x+2$—irreducible by Eisenstein ($p=2$ again!) Hence $f(x)=g(x)h(x)$ is impossible—by the usual argument.

12 Let $g(x)=1+x^p+\cdots+x^{(p-1)p}$. Then $g(x)(x^p-1)=x^{p^2}-1$. Suppose $g(x+1)h(x+1)=k(x+1)$, say, where $k(x+1)=[(x+1)^{p^2}-1]=x^{p^2}+\cdots+p^2x$ has coefficients $\binom{p^2}{i}$ all of which (except $\binom{p^2}{0}$) are multiples of p. Likewise all coefficients of $h(x+1)=(x+1)^p-1$ are divisible by p—except for that of x^p—see exercise 1.4.17. Comparing coefficients in $g(x+1)h(x+1)$ and $k(x+1)$ show that all the coefficients in $g(x+1)$—except that of $x^{p(p-1)}$—are multiples of p. Now use Eisenstein.

14 Take $F=1+x$. Clearly F is irreducible in $\mathbb{Z}[x]$—but no prime divides 1 and 1. (Note that $(A \Rightarrow B)$ *doesn't* imply $(\sim A \Rightarrow \sim B)$—as some students occasionally believe!)

15 $6+6x+3x^2$ is reducible in $\mathbb{Z}[x]$—but irreducible in $\mathbb{Q}[x]$ (Eisenstein, $p=2$ yet again!!)—essentially because 3 is a unit in $\mathbb{Q}$ but not in $\mathbb{Z}$. Proof fails within first three lines. No G, H with $\deg G$, $\deg H < n$ may be available.

17 (Cf. the proof of 1.4.4.) Since $f, g \in S$ we see that S contains a non-zero polynomial. Let h be any polynomial in S from amongst those of smallest degree in S. (Such an h obviously exists once we know S contains *some* non-zero polynomial.) Let $h_0 = \alpha h (\alpha \in \mathbb{Q})$ be monic. Now let $w \in S$. Write $w = mh_0 + r$ where $m, r \in \mathbb{Q}[x]$ and either (i) $r=0$ or (ii) $r \neq 0$ and $\deg r < \deg h_0$. Then $r = w - mh_0 \in S$ (since

$$w = s_1 f + t_1 g \text{ and } h_0 = s_2 f + t_2 g \Rightarrow r = (s_1 - ms_2)f + (t_1 - mt_2)g$$

clearly lies in S). But, by choice of h_0, we are forced to deduce $r=0$ whence $w = mh_0$.

Exercises 1.10

1 (a) $m=4x^3-\frac{1}{2}x^2-\frac{1}{2}x+\frac{5}{4}$; $r=-\frac{9}{40}$. (b) $m=\frac{3}{2}x^3+\frac{5}{4}x^2+x$; $r=\frac{1}{2}x^2+\frac{1}{2}x+3$.

2 $f(x)=x$, $g(x)=2x$. Second part: There exist $m, r \in \mathbb{Q}[x]$ such that $f=mg+r$. Writing $m=M/z$, $r=R/z$ for suitable $z \in \mathbb{Z}$ we get the required result—with $\deg R = \deg r < \deg g$ (if $r \neq 0$).

3 (a) x^2+2x+1; (b) Repeated division algorithm seems to yield 392/1849 as a gcd of the given polynomials. Hence the unique *monic* gcd is ...?

Exercises 1.11

2 Note that f' is just the derivative of f in the usual calculus sense. Thus, if $f=(x-a)^2g$, then $f'=2(x-a)g+(x-a)^2g'$. Hence $(x-a)|f'$.

3 Let $f=x^4-4x^3+4x^2+17$. Then $f'=4x(x-1)(x-2)$. But none of 0, 1, 2 is a root of f.

4 With $f=x^4+4x^2-4x-3$, we have $(f,f')=1$ in $\mathbb{Q}[x]$. Hence $1=sf+tf'$ with $s,t\in\mathbb{Q}[x]$. If α were a repeated complex root of f we would have $f(\alpha)=f'(\alpha)=0$ in $\mathbb{C}$, whence $1=s(\alpha)f(\alpha)+t(\alpha)f'(\alpha)=0$. Contradiction.

5 $f-g$ is of degree at most n (if it's not zero), but has $n+1$ roots. Now use 1.11.4.

7 (a) By the rational root test the only possibilities are ± 1, $\pm\frac{1}{2}$, $\pm\frac{1}{4}$. (b) $\frac{r}{s}$ is a root here iff $\frac{s}{r}$ is a root in part (a). (d) No roots here. (This polynomial plays a prominent role in Section 4.6.)

9 $x-(2-i)$ is a factor and so is $x-(2+i)$—by exercise 8. Hence $(x-2+i)(x-2-i)=(x-2)^2-i^2=x^2-4x+5$ is a factor of $x^4-3x^3+2x^2+x+5$. The other quadratic factor is therefore x^2+x+1 with roots $x=(-1\pm\sqrt{-3})/2$.

10 The only potential rational roots are ± 1—but neither *is*. So try $x^4-x^2+1=(x^2+ax\pm 1)(x^2+bx\pm 1)$. This shows $a+b=0$, $ab\pm 2=-1$. Hence no rational solution. (Or: Solve the quadratic in x^2!!)

Exercises 2.2

1 A binary relation on A is just a subset of $A\times A$. If $|A|=n$ then $|A\times A|=n^2$. Now use exercise 10 of Chapter 0.

2 (ii) Note that $n|(a-c)b+c(b-d)$; (iii) From $n|ma-mc$ we deduce $n|a-c$ provided $(n,m)=1$ (exercise 1.5.5). As an example $4\equiv 2 \pmod 2$ but $2\not\equiv 1 \pmod 2$.

4 (i) $3x\equiv 7 \pmod 8 \Rightarrow 9x\equiv 21\equiv 5 \pmod 8$. Hence $x=5, 13$ will do. (ii) $4x=7+8m$ is impossible—since $4\nmid 7$.

5 (i) t, a. (a? Yes—for, if not, there must exist $x,y\in\mathbb{Z}$ such that $x<y$ and $y<x$ and $x\neq y$. But there is no such pair satisfying even the first two of the three conditions; (iv) r, s.

8 E is all of $A\times A$ except (1,4), (2,4), (3,4), (4,1), (4,2), (4,3).

9 The singleton $S=\{(1,2)\}$ won't do—since condition t is satisfied, as we say, vacuously. (For every *two* distinct or identical pairs to be found in S (!!) the t condition holds.)

11 I claim the relation $\subseteq$ comprises 81 ordered pairs. Am I right?

12 Let $S=\{a,b\}$. Define $\leq$ by: $a\leq b$, $a\leq a$, $b\leq b$ and $\subseteq$ by: $b\subseteq a$, $b\subseteq b$, $a\subseteq a$. Then $a\leq b\subseteq b$ whilst $b\leq b\subseteq a$. Thus $a\leq\subseteq b$ and $b\leq\subseteq a$ but $a\neq b$. Hence $\leq\subseteq$ is not an order relation.

Exercises 2.3

1 Take $\{1\}$, $\{2\}$, $\{3\}$, $\{4\}$ and the rest of $\mathbb{Z}$ as equivalence classes.

2 Try $\{1\}$, $\{2,3\}$, $\{4,5,6\}$, ... etc.

4 (i) $(x_1,y_1)R(x_2,y_2)$ iff $(x_1-1)^2+(y_1-1)^2=(x_2-1)^2+(y_2-1)^2$.

5 52 (Am I correct?)

6 (i) $z_1|z_2|=z_2|z_1|$ iff $z_1/z_2=|z_1/z_2|$, i.e. iff z_1/z_2 is real and positive. Thus the equivalence classes are all half lines sprouting out of the origin.

Exercises 2.4

2 $17|x^2-1\Rightarrow 17|x-1$ or $17|x+1$. (Why?) Hence $x\equiv\pm 1 \pmod{17}$. That is, there are essentially two solutions in $\mathbb{Z}$—*exactly* two in $\mathbb{Z}_{17}$. $x^2\equiv 1$ has 4 solutions, namely $\hat{1}$, $\hat{3}$, $\hat{5}$, $\hat{7}$ in $\mathbb{Z}_8$. Put another way: all odd integers are solutions of the congruence $x^2\equiv 1 \pmod 8$.

4 (a) The argument goes through since $\mathbb{Z}_p$ is a field (as are $\mathbb{Q}$, $\mathbb{R}$, $\mathbb{C}$). In fact 1.11.5 is valid too. (Do 1.11.4 and 1.11.5 hold for integral domains—e.g. $\mathbb{Z}$?)

8 By inspection: $\hat{3}$ and $\hat{4}$ will each do. (Any more?)

10 (ii) $5\cdot 8\equiv 1 \pmod{13}$. So $\hat{8}^{-1}=\hat{5}$ in $\mathbb{Z}_{13}$; (iv) $8x=1+34k$ is impossible in $\mathbb{Z}$; (v) By the Euclidean Algorithm $1=128\cdot 8-3\cdot 341$.

11 (ii) True.

12 Choosing elements almost at random note that $\widehat{\tfrac{1}{2}}=\widehat{\tfrac{3}{2}}$ and $\widehat{\tfrac{1}{3}}=\widehat{\tfrac{4}{3}}$. But, by (proposed) definition $\widehat{\tfrac{1}{2}}\odot\widehat{\tfrac{1}{3}}=\widehat{\tfrac{1}{6}}$ whereas $\widehat{\tfrac{3}{2}}\odot\widehat{\tfrac{4}{3}}=\widehat{\tfrac{12}{6}}=\hat{0}\neq\widehat{\tfrac{1}{6}}$.

Exercises 2.5

1 (i) $3^{28}\equiv 1 \pmod{29}$. Hence $3^{56}\equiv 1$. Consequently $3^{60}\equiv 3^4=81\equiv 23$. Thus $x=23$.

3 $2^{10}-1=1023=3\cdot 11\cdot 31$. Hence $2^{10}\equiv 1(\text{mod } 341)$. It follows that $2^{340}\equiv(2^{10})^{34}\equiv 1(\text{mod } 341)$.

6 Suppose $n=n_1n_2$ is composite with $1<n_1,n_2<n$. Show $n_1|(n-1)!+1$. But $n_1|(n-1)!$ Hence $n_1|1$, contradiction.

8 In the range $1,2,\dots,p,\dots,2p,\dots,p^r$ those numbers *divisible* by p are $p,2p,\dots,p^{r-1}p$. Thus there are p^{r-1} of them. Hence there are p^r-p^{r-1} which are *not* divisible by (that is, coprime to) p.

9 Assuming $a^{(p-1)p^{k-1}}\equiv 1(\text{mod } p^k)$ look at $(a^{(p-1)p^{k-1}})^p=(1+h_kp^k)^p$ and then reduce mod p^{k+1} after expanding by the binomial theorem.

Exercises 2.6

1 (ii) Yes; (iii) no.

2 (a) (ii) No; (iii) yes: (b) 2^3 (from A to B).

3 f is not onto; the range of g is $\mathbb{R}^+$; h is 1–1.

4 g is not onto; h is both 1–1 and onto.

6 If $f(x)=g(x)$ for all x in $\mathbb{Q}$ then $f(i)=g(i)$ for $i=1,2,\dots,n+1$, where $n=\max\{\deg f,\deg g\}$. Hence $f=g$ by exercise 1.11.5 (which uses 1.11.4).

7 $\{0,7\}f^{-1}=\{\sqrt{5},-\sqrt{5}\}$. (Note that 7 is not a value of $5-x^2$ for $x\in\mathbb{R}$.)

8 (ii) $x\in f(S\cap T)\Rightarrow x=f(u)$ for $u\in S\cap T$. Therefore $x\in f(S)\cap f(T)$. For an example to show that equality does *not* hold in general, try to make $|S\cap T|$ 'small' whilst making $|f(S)|=|f(T)|$ 'large'.
(iv) $x\in f^{-1}(U\cap V)\Rightarrow f(x)\in U\cap V\Rightarrow x\in f^{-1}(U)\cap f^{-1}(V)$. Conversely, $x\in f^{-1}(U)\cap f^{-1}(V)\Rightarrow f(x)\in U$ and $f(x)\in V\Rightarrow f(x)\in U\cap V\Rightarrow x\in f^{-1}(U\cap V)$.

9 (ii) '1–1 ness': If $a(f\circ g)=b(f\circ g)$ then $(af)g=(bf)g$. Hence $af=bf$ (since g is 1–1). Consequently $a=b$ (since ?). Thus $f\circ g$ is 1–1.

10 $f\circ g\neq g\circ f$—since $1(f\circ g)\neq 1(g\circ f)$, for example.

11 Let $b\in B$. Then $b=af$ for some $a\in A$—since f is onto. If $(b,a_1)=(b,a_2)\in f^{-1}$ then $b=a_1f=a_2f$ by definition. But f is 1–1. Therefore $a_1=a_2$. Hence f^{-1} is a function from B to A. Since each a yields an $af\in B$, f^{-1} is onto.

12 If $a_1f=a_2f$ then $(a_1f)g=(a_2f)g$, that is $a_11_A=a_21_A$, In other words $a_1=a_2$. Therefore f is 1–1. Now suppose $a\in A$. Then $(af)g=a1_A=a$. Hence g sends af to a and so g is onto.

14 The answer to the second part is—No! (For if $f\circ g=1_A$ (A finite) then f is 1–1 (exercise 12) and hence onto (exercise 13). Likewise g is onto and, consequently, 1–1. Now if $b(g\circ f)\neq b$ and if $af=b$ (since f is onto) then $a(f\circ g\circ f)=((af)g)f=b(g\circ f)\neq b$, whereas $((af)g)f=(a1_A)f=af=b$.

Exercises 2.7

1 The negative integers provide an example. (Can you find others?)

2 In (iv) and (v) the identity functions are, respectively; (iv) I given by $I(x)=0$ for all $x \in \mathbb{R}$; (v) I given by $I(x)=x$ for all $x \in X$.

4 *If* $|X| \geq 2$, choose $a, b \in X (a \neq b)$. Now define $f, g \in E$ by $af = bf = a$; $ag = bg = b$. Then show that $a(f \circ g) \neq a(g \circ f)$.

7

	(i)	(iii)	(v)
A?	×	✓	✓
C?	×	✓	×
I?	×	×	×

9 In an $n \times n$ multiplication table there are $n(n+1)/2$ entries on or above the 'main diagonal'. Once these entries are inserted the rest are automatically determined—if the operation is given to be commutative. The number of commutative operations is therefore $n^{n(n+1)/2}$.

11 We need a non-associative (but commutative) example. Try (vi) of exercise 7.

12 The operation is associative and has an identity element but is not commutative. (Try $a=2, b=c=d=1$: now find a 'nicer' counterexample.)

Exercises 3.2

1 (b) is a cheat (sorry!) since + isn't a binary operation on the given set; (c) all axioms hold—it's a field; (e) M4 fails.

2 Note that A1, A2, A3, A4 must hold in each case. For the second example look at axiom D.

5 All elements of $\mathscr{P}(S)$, except S itself, are zero divisors (look at $X \cap X^c$). For $\langle \mathscr{P}(S), (p \boxplus, \odot \rangle$ we need (now) only check the axioms involving $\boxplus$. Clearly $\emptyset$ would have to be the element satisfying A3. Now try A4.

6 The first part is essentially exercise 1.2.7. For $\langle F, \oplus, \circ \rangle$ there is no need to check the $\oplus$ axioms again so just check M2 and D. For D try functions f, g, h where $f(t) = 1$ for each $t \in \mathbb{R}$. (In fact $\langle F, \oplus, \circ \rangle$ would be a ring except it just fails one 'half' of D.)

8 Write d and n in place of 0 and 1. We then have (with slight rearrangement)

$\oplus$	n	d
n	n	d
d	d	n

$\odot$	n	d
n	n	n
d	n	d

(Recall that 0 and 1 are only symbols; the tables define their relationship.)

This now looks like the set $\{E, O\}$ of exercise 1.2.15 (In fact n and d were chosen being the *last* letters of 'even' and 'odd'.) Thus, by exercise 1.2.15, all axioms are satisfied and, as defined, $\langle\{0,1\}, \oplus, \odot\rangle$ *is* a field.

9 This construction is important but proofs are quite straightforward. For example (z_R, z_S) is the additive identity element; $(-r, \boxminus s)$ is the additive inverse of (r, s), etc.

10 $R=\mathbb{Z}_4$, $a=\hat{2}$, $b=\hat{3}$, $c=\hat{1}$. Now find another example.

11 $\widehat{18}\odot\widehat{2n}=\widehat{36n}=\widehat{34n}\oplus\widehat{2n}=\widehat{2n}$. It is the only unity—see exercise 3.3.1(iii).

13 Try $\begin{pmatrix}0 & 1\\ 0 & 0\end{pmatrix}$ and $\begin{pmatrix}0 & 0\\ 1 & 0\end{pmatrix}$ in the first case—and with 'hats' on in the second case.

14 Both $\mathbb{Q}(\sqrt{d})$ and $\mathbb{Q}[\sqrt{d}]$ ($d\equiv 1 \pmod 4$ or not) clearly satisfy A1, A2, A3, A4, M2, D, M1, M3, Z since they are sets of complex numbers [Note that each is closed with respect to $+$ and $\cdot$] So to check $\mathbb{Q}(\sqrt{d})$ is a field we only need to check that the multiplicative inverse of $a+b\sqrt{d}$ is again in $\mathbb{Q}(\sqrt{d})$.

15 Here we are assuming that $f/g=fh/gh$ if $h\neq 0$ (the zero polynomial), etc. Note that f and g are to be manipulated as *formal symbols*—no discussion of the roots of g is required. In particular f/g is undefined when and only when g is the zero polynomial.

Exercises 3.3

1 (a) (i) $a+c=b+c$. By A4 there exists c^* such that $c+c^*=z_R$. Then $(a+c)+c^*=(b+c)+c^*$. By A2, $a+(c+c^*)=b+(c+c^*)$. By A4, $a+z_R=b+z_R$. Hence $a=b$, by A3. (a) (iii) If $e(f)$ satisfies M3 then $e\cdot f=f(e\cdot f=e)$. Therefore $e=f$. Suppose a', a'' satisfy M4. Now $a'\cdot(a\cdot a'')=(a'\cdot a)\cdot a''$ (by A2). Hence $a'\cdot e=e\cdot a''$ and so $a'=a''$. (b) If $c\cdot a=c\cdot b$ we have $c\cdot(a+b^*)=c\cdot a+(c\cdot b^*)$—by the first part—$=c\cdot b+(c\cdot b)^*$ [by 3.3.2(iii)] $=z_J$. Since $c\neq z_J$ we deduce $a+b^*=z_J$ (axiom Z). Therefore $b=z_J+b=(a+b^*)+b$ $=a+(b^*+b)=a+z_J=a$.

3 For A3 use $(0_R, 0)$; for A4 use $(-r_1, -z_1)$. M2 and D2 are messy but straightforward. They make heavy use of the associative and distributive laws in R and $\mathbb{Z}$. (Look out for their uses. Are any other properties of $\mathbb{R}$ or $\mathbb{Z}$ used?)

5 Suppose $b, c\in R$ are such that $b\cdot c=z$. Then either $b=z$ (and so we're finished) or $b\neq z$. In the latter case there exists, since M4 holds, an element $b'\in R$ such that $b'\cdot b=e$. But then $c=e\cdot c=(b'\cdot b)\cdot c=b'\cdot(b\cdot c)=b'\cdot z=z$.

7 $a+a=(a+a)^2=a^2+2aa+a^2=a+2a+a$. Therefore $2a=0_R$.

8 For x, y choose (almost!) any pair of matrices from $M_2(\mathbb{Z})$.

9 (a) $\frac{1}{ab}$ isn't equal to $\frac{1}{a} \cdot \frac{1}{b}$ (even though the notation may tempt you to believe it is). (b) The moral is: Use suggestive symbolism—but with care!

10 (cf. the proof of 3.3.4) If the elements of R are denoted by $0_R = f_0, f_1, \ldots, f_t$, then $f_0 f_1, f_1 f_1, \ldots, f_t f_1$ are $t+1$ distinct elements of R. ($f_i f_1 = f_j f_1 \Rightarrow (f_i - f_j) f_1 = 0_R \Rightarrow f_i - f_j = 0_R$—since $f_1 \neq 0_R$ and R has no zero divisors.) Thus $f_1 = f_i f_1$ for some i. Likewise, given $f_j \in R$, we have, using f_1 on the left, $f_j = f_1 f_k$. Then $f_1 f_k = f_i f_1 f_k$, and so $f_j = f_i f_j$—for any $f_j \in R$. Hence f_i is a *left* identity. In a similar way there exists a right identity, f_l, say, for R. Then $f_i f_l = f_i = f_l$. Hence f_i is a multiplicative identity for R. Finally, given $f_m (\neq 0) \in$ R, on multiplying all elements of R by f_m, there exists an f_n such that $f_m f_n = f_i$ so that f_n is a right inverse for f_m.

Exercises 3.4

1 (a) No; (c) Yes—each is!; (e) no—sum may fail; (g) yes.

2 (ii) $\{a/b;\ a,b \in \mathbb{Z}$ and $b = 2^\alpha 3^\beta$ for some $\alpha, \beta \in \mathbb{Z}^+ \cup \{0\}\} = \{c/6^n : c \in \mathbb{Z}, n \in \mathbb{Z}^+ \cup \{0\}\}$.

4 (Use 3.4.2 rather than 3.4.1) $\Rightarrow$: From $a, b \in S$ we find $-b \in S$. Consequently $a + (-b) \in S$, that is $a - b \in S$ (and $a \cdot b \in S$, given). $\Leftarrow$: From $a, b \in S$ we find $0_R = a - a \in S$, hence $-b = 0_R - b \in S$ and so $a - (-b) \in S$.

7 Even if R, S both have identity elements $1_R, 1_S$ so that $s1_R = s1_S = s$ for all $s \in S$ we cannot infer that $1_R = 1_S$ by cancellation. (Additive cancellation is always allowed, multiplicative cancellation only rarely.)

9 For all $t \in T, t \cdot 1_T = t \cdot 1_F$. But (since T contains *at least one non-zero element*—see 3.4.1(F) and 3.2.3(iv)) given $t \neq 0_F$, there exists $t_F^{-1} \in F$. But then $t_F^{-1} t 1_T = t_F^{-1} t 1_F$. Therefore $1_F 1_T = 1_F 1_F$ from which $1_T = 1_F$.

12 (i) Since $15s + 21t = 3(5s + 7t)$ we have $[15, 21] \subseteq [3]$. Since $3s = 15(-4s) + 21(3s)$ we have $[3] \subseteq [15, 21]$.

13 Only (b), (c), (f), (g) and (h)(ii) are subrings (S, say). (b) is not an ideal since $1 \in S, \frac{1}{2} \in \mathbb{Q}, \frac{1}{2} \notin S$. Only the first set in (c) is an ideal. (g) is the principal ideal [15]. (h)(ii) is an ideal.

14 (2nd part) 'only if' done in first part! 'If': Let $x \in R (x \neq 0)$. Then $xR (= [x])$ is an ideal containing $x (= x1_R)$ Hence $xR = R$ (and not $\{0\}$). In particular there exists $r \in R$ such that $xr = 1_R$. Consequently M4 holds and R is a field.

16 Any ideal containing the a_i *must* contain all the $z_i a_i$, $s_i a_i$, $a_i t_i$ and $u_i a_i v_i$ and all sums of such where $z_i \in \mathbb{Z}$, $s_i, t_i, u_i, v_i \in R$. Checking that the given set is an ideal—and hence the unique smallest ideal containing $a_1, \ldots, a_m$—is left to you. If R has unity 1_R, then, for example, $3a = a + a + a = a1_R + a1_R + a1_R = ar$ (where

$r=1_R+1_R+1_R\in R$). Since R is commutative, sa and uav can be written as as and $a(uv)$.

17 When R is commutative with a unity, $I_1=\{a_1r_1+\cdots+a_mr_m:r_i\in R\}$, $I_2=\{b_1s_1+\cdots+b_ns_n:s_j\in R\}$ as in 3.4.7(iv). Hence each element of I_1I_2 is a sum of elements of the form $(a_1r_1+\cdots+a_mr_m)(b_1s_1+\cdots+b_ns_n)$ and hence a sum of elements of the form $a_ir_ib_js_j=a_ib_jr_is_j$—since R is commutative. Thus $I_1I_2\subseteq[a_1b_1,\ldots a_1b_n,a_2b_1,\ldots,a_mb_n]$. The reverse inclusion is clear.

19 Given $a,b\in U$, suppose (without loss of generality) that $a\in I_i$, $b\in I_j$ with $i\leq j$. Then $a,b\in I_j$. Therefore $a+b,\ -a,ar,ra\,(r\in R)\in I_j\subseteq U$.

20 Let I be an ideal of $M_2(\mathbb{R})$. If $I\neq\begin{pmatrix}0&0\\0&0\end{pmatrix}$ then there exists $\begin{pmatrix}a&b\\c&d\end{pmatrix}\in I$ with not all of a, b, c, d equal to zero. By the given identity (or similar ones) we find $\begin{pmatrix}u&0\\0&0\end{pmatrix}\in I$ ($u=a$, b, c or d). We may, therefore, assume that $u\neq0$. Then, for $\alpha,\beta,\gamma,\delta\in\mathbb{R}$ we easily check that $\begin{pmatrix}\frac{1}{u}&0\\0&0\end{pmatrix}\odot\begin{pmatrix}u&0\\0&0\end{pmatrix}\odot\begin{pmatrix}\alpha&0\\0&0\end{pmatrix}=\begin{pmatrix}\alpha&0\\0&0\end{pmatrix}$ and $\begin{pmatrix}\beta&0\\0&0\end{pmatrix}\odot\begin{pmatrix}0&1\\0&0\end{pmatrix}=\begin{pmatrix}0&\beta\\0&0\end{pmatrix}\in I$. Likewise $\begin{pmatrix}0&0\\\gamma&0\end{pmatrix}$ and $\begin{pmatrix}0&0\\0&\delta\end{pmatrix}\in I$. Hence $I=M_2(\mathbb{R})$.

22 (b) If $[x,2]=\{pr:r\in\mathbb{Z}[x]\}$ then $x=pr_1$, $2=pr_2$ for suitable $r_1,r_2\in\mathbb{Z}[x]$. The latter implies that $p=1$ or -1 or 2 or -2 whilst the former rules out 2 and -2. Hence $p=1$ or -1. This implies $[x,2]=\mathbb{Z}[x]$ which is impossible since $1\notin[x,2]$. (Why not?)

Exercises 3.5

1 (Cf. exercise 1.5.6.) Care is needed if n is even. For example, $36=(-4)(-9)$ with coprime factors, neither of which is a square!

3 A solution for $n=uv$ yields a solution for v. Thus if FC is true for 4 and each odd prime p then it is also true for $n=4v$ and $n=pv$. But every integer $n\geq3$ either has an odd prime factor or is a power of 2.

Exercises 3.6

1 (b) From $b=ma$ and $c=na$ we obtain $bx+cy=a(mx+ny)$. Hence $a|bx+cy$. (cf. 1.3.4.)

2 If c is a unit in R then $cd=1_R=1_{R[x]}$ where $d\in R\subseteq R[x]$. Hence c is a unit in $\mathbb{R}[x]$. Conversely if $cp(x)=1_{R[x]}=1_R$ then $p(x)\in R$ (by a degree argument) and so c is a unit *in* R.

Suppose c is irreducible in R. Then $c \neq 0$, $\neq$ unit *in* R and, hence, not in $R[x]$. If $c = p(x)q(x) \in R[x]$, then $p(x)$, $q(x) \in R$ and one is a unit in R—hence in $R[x]$.

3 If $v = 2+\sqrt{3}$ then $uv = 1$ in $\mathbb{Z}[\sqrt{3}]$. But $u^k v^k = (uv)^k = 1$ so u^k is a unit for each $k \in \mathbb{Z}^+$. Since $0 < u < 1$ we have $u^i = u^j$ iff $i = j$.

4 In $M_2(\mathbb{R})$ the units are those matrices with non-zero determinant. In $(\mathbb{Z}[x])[y]$ the units are the units of $\mathbb{Z}[x]$ (by exercise 2) and these are just 1 and -1.

5 If $3 = (a+b\sqrt{-5})(c+d\sqrt{-5})$ then (taking norms) $9 = (a^2+5b^2)(c^2+5d^2)$. Now $a^2+5b^2 \neq 3$ (if $a, b \in \mathbb{Z}$). Hence $a^2+5b^2 = 1$ or 9. Consequently $a+b\sqrt{-5}$ (or $c+d\sqrt{-5}$)) $= \pm 1 + 0\sqrt{-5}$, i.e. a unit. However 3 is not prime in $\mathbb{Z}[\sqrt{-5}]$ since $3 \mid (1+2\sqrt{-5})(1-2\sqrt{-5}) = 21$ but $3 \nmid 1 \pm 2\sqrt{-5}$ in $\mathbb{Z}[\sqrt{-5}]$.

6 If $\dfrac{a+b\sqrt{-3}}{2} \cdot \dfrac{c+d\sqrt{-3}}{2} = 1+0\sqrt{-3}$ then $(a+b\sqrt{-3})(c+d\sqrt{-3}) = 4$. Therefore $(a^2+3b^2)(c^2+3d^2) = 16$. Consequently $a^2+3b^2 = 4$ (why not 1, 2, 8 nor 16?). It follows that $a = \pm 2$, $b = 0$, or $a = \pm 1$, $b = \pm 1$. These six possibilities *do* all yield units.

7 $\hat{4} \mid \hat{8}$ and $\hat{8} \mid \hat{4}$ in $2\mathbb{Z}_{12}$ (the even integers mod 12) but $\hat{4}, \hat{8}$ are not associates since $2\mathbb{Z}_{12}$ has no multiplicative identity! (Cf. exercise 3.2.11.)

Even better is the following, due to Al Hales. On $R = \mathbb{Z}_5 \times \mathbb{Z}$ define $\oplus$ and $\odot$ by: $(\hat{a}, m) \oplus (\hat{b}, n) = (\hat{a}+\hat{b}, m+n)$, $(\hat{a}, m) \odot (\hat{b}, n) = (\widehat{an+bm}, mn)$. (Cf. exercise 3.3.3.) You can check that R is a commutative ring with unity (... what?) and that $(\hat{1}, 0)$, $(\hat{2}, 0)$ divide each other in R but are not associates.

8 (i) in $\mathbb{Z}_{12}$ the units are $\hat{1}, \hat{5}, \hat{7}, \widehat{11}$. Irreducibles are $\hat{2}, \widehat{10}$. Primes are $\hat{2}, \hat{3}, \hat{9}, \widehat{10}$. (iv) This example is fascinating! First note that f is a unit iff $f(x) \neq 0$ for all x. There are no irreducibles: for any such f must have $f(a) = 0$ for some a. Defining I by: $I(a) = 0$; $I(x) = 1$ if $x \neq a$, we find $f(x)I(x) = f(x)$ for all x, i.e. $f = fI$. Now show that f is prime iff f has exactly one zero. (Show that if f has more than one zero, then you can find functions g and h such that $f = gh$ but $f \nmid g$ and $f \nmid h$.)

10 Yes! (2i) Let π be a prime and u and unit in $\mathbb{Z}[i]$. Then $u\pi \neq 0$, $\neq$ unit. Finally—if $u\pi t = \alpha\beta$ then $\pi \mid \alpha$ or $\pi \mid \beta$ so $u\pi \mid uv\alpha$ or $u\pi \mid uv\beta$, where $uv = 1$.

12 (a) Follow 1.3.7. End more formally by: 'It follows that $asc = bc$. But $bc = a$. Hence $asc = a = ae$. Therefore $a(sc-e) = z$ and so $sc - e = z$ (why?). Consequently $sc(= cs$, why?$) = e$; that is, c is a unit and a is irreducible.' (b) 2 is irreducible (see text) but 2 isn't prime since $2 \mid (1+\sqrt{-3})(1-\sqrt{-3})$ yet $2 \nmid 1 \pm \sqrt{-3}$ in $\mathbb{Z}[\sqrt{-3}]$. (c) Since α is irreducible the only divisors of α are u and αu where u $(= \pm 1)$ is a unit. Likewise for β. Hence the only common divisors are 1, -1. (Note that α, β are not squares since they are irreducibles.)

13 $8 = 2 \cdot 2 \cdot 2 = (1+\sqrt{-7})(1-\sqrt{-7})$.

Exercises 3.7

1 (a), (b), (c). They are all domains. That 3.7.1(I) holds for $\mathbb{Z}[\sqrt{-1}]$, $\mathbb{Z}[\sqrt{-2}]$ has already been noted (Theorem 3.7.4). The proof for $\mathbb{Z}[\sqrt{2}]$ is similar via 3.6.3. To check (II) look at $\mathbb{Z}[\sqrt{d}]$ $(d=-1,-2,2)$. Let $a=s+t\sqrt{d}$, $b=u+v\sqrt{d}$ $\in \mathbb{Z}[\sqrt{d}]$. Then $\frac{a}{b}=\frac{s+t\sqrt{d}}{u+v\sqrt{d}}\frac{u-v\sqrt{d}}{u-v\sqrt{d}}=x+y\sqrt{d}$, where $x, y \in \mathbb{Q}$. Now choose X, $Y \in \mathbb{Z}$ such that $0 \leq |x-X|, |y-Y| \leq \frac{1}{2}$. We obtain $a=(X+Y\sqrt{d})b+r$ where

$$r=\{(x-X)+(y-Y)\sqrt{d}\}b=a-(X+Y\sqrt{d})b \in \mathbb{Z}[\sqrt{d}].$$

But $N(r)=|(x-X)^2-d(y-Y)^2|N(b)=MN(b)$, say. For $d=2$, $M=|g^2-2h^2|$, where $0 \leq g^2, h^2 \leq \frac{1}{4}$. Thus, $0 \leq M \leq \frac{1}{2}$. (Can you see why?) (A similar proof copes with $d=-1,-2$ and 3 but not with $d=-3$. Again, can you see why?). (d) (II) For $a, b \neq 0$ write $a=mb+r$ where $-|b|/2 \leq r < |b|/2$. Then $|r| \leq \frac{1}{2}|b|$ so that $\delta(r) < \cdots$ what? (f) For (II) note that if $a=x^i(a_i+a_{i+1}x+\cdots)=x^iu_i$, $b=x^j(b_j+b_{j+1}x+\cdots)=x^ju_j$, where u_i, u_j are units then $a=x^{i-j}bu_j^{-1}u_i+0$ (if $j \leq i$). (What if $i<j$?)

2 Given η, define $\delta(a)=2^{\eta(a)}$. Then checking (I) and (II) of 3.7.1 is straightforward. Does δ (in 3.7.1) imply the existence of η as defined here? Try $R=\mathbb{Z}$. Now prove that $\eta(9)=2\eta(3)$ and $\eta(9) \leq \eta(3)$.

3 (ii) $m=3+\sqrt{-2}$, $r=16+5\sqrt{-2}$. The gcd is $4-3\sqrt{-2}$.

5 Write $Rf+Sg=1$ and then write $R=mg+r(m, r \in \mathbb{Q}[x]$ and $\deg r < \deg g$. $r=0$ is impossible. Why?) Then $1=rf+(S+mf)g$.

6 (i) Using $2 \cdot 3=\sqrt{-6} \cdot \sqrt{-6}$ show that 2 is irreducible but not prime in $\mathbb{Z}[\sqrt{-6}]$. (ii) If $r=a+b\sqrt{10}$ then $N(r)=|a^2-10b^2|$. Putting $a \equiv 0, \pm 1, \ldots, \pm 4, 5 \pmod{10}$ show that $a^2-10b^2=\pm 2$ has no solutions. Thus $N(r) \neq 2$. Show similarly that $N(r) \neq 5$. Hence show that $2, 5, \sqrt{10}$ are irreducible in $\mathbb{Z}[\sqrt{10}]$.

7 There is no contradiction. (Perhaps the factors with equal norms are associates?) The equality $a^2-2b^2=(a+b\sqrt{2})(a-b\sqrt{2})$ shows that elements with norm 1 are units. (Cf. exercise 9(ii).)

8 (i) Since $(77, 91, 143)=1$, $z=1$ will do. Which other z will do?

9 (i) From 3.7.1(I), $\delta(1) \leq \delta(1a)$ if $a \neq 0$. (ii) If *also* $ab=1$ then $\delta(a) \leq \delta(ab)$ $=\delta(1)$.

11 If $d|2$ and $d|1+\sqrt{-3}$ in $\mathbb{Z}[\sqrt{-3}]$ then $N(d)|4$. Show that $N(d)=1$ so that $d=\pm 1$. Any lcm, k say, would have to divide the common multiples 4 and $2(1+\sqrt{-3})$ and, hence, $N(k)=4$, 8 or 16. There is no contradiction; the pair $2+2\sqrt{-3}, 4$ doesn't have a gcd in $\mathbb{Z}[\sqrt{-3}]$ (exercise 3.6.12(d)). Now note the word 'each' on line 1 of the question!

12 If each irreducible is prime in D copy the proof of 3.7.11 to find D is a UFD. If D is a UFD and if $\pi c=ab$, with π irreducible, write a, b, c as products of irreducibles and use uniqueness of factorisation to show that π is an associate of an irreducible factor of a or b.

13 This is rather long. You can find the proof in [9], Vol. I.

Exercises 3.8

1 One of u, v must be odd, the other even: $u=2r+1, v=2s$, say.

2 Since $4k \equiv -1$, $4k-1 \equiv -2, \ldots, 2k+1 \equiv -2k \pmod p$ we see that, mod p, $-1 \equiv (1 \cdot 2 \cdot \cdots \cdot 2k)(2k+1 \cdot \cdots \cdot 4k) \equiv 2k! \cdot (2k!)(-1)^{2k} = (2k!)^2 = \{(\frac{p-1}{2})!\}^2$.

3 If $p=4k+3$ is *not* a prime in $\mathbb{Z}[\mathrm{i}]$ then it is not irreducible either. Hence $p=(a+ib)(c+id)$ with neither factor being a unit. Then from $p^2=(a^2+b^2)(c^2+d^2)$ would follow $p=a^2+b^2=4k+3$, contradicting exercise 1. If $p=4k+1$ is a prime in $\mathbb{Z}$, then $p=a^2+b^2=(a+ib)(a-ib)$ where $a \neq 0 \neq b$.

4 $(a^2+b^2)(c^2+d^2)=N(\alpha\beta)=(ac \mp bd)^2+(ad \pm bc)^2$.
$11\,009=(10^2+1^2)(10^2+3^2)$.

7 Note that at step (ix) we would need an equality of the form $r^4+s^4=t^4$.

9 Note that $(x^2+4, 16)=1$ in $\mathbb{Z}$ since x is odd. Deduce that $x+2\mathrm{i}$, $x-2\mathrm{i}$ are coprime in $\mathbb{Z}[\mathrm{i}]$. Note also that if $y^3=\alpha\beta$ with α, β coprime in $\mathbb{Z}[\mathrm{i}]$, then $\alpha=us^3$, $\beta=vt^3$, where $u, v \in \{1, -1, \mathrm{i}, -\mathrm{i}\}$. But u, v are each cubes of units, hence α, β are cubes.

Exercises 3.9

1 'Only if': Supposing that n is prime in $\mathbb{Z}$ and that $ab \in [n]$, then $ab=tn$ for some $t \in \mathbb{Z}$ and so $n|a$ or $n|b$. Therefore a (or b) $\in [n]$.

2 $[x]$ is prime since if $f_1 f_2 \in [x]$ then $f_1 f_2 = xg$ where we think of f_1, f_2, g as polynomials in x with coefficients in $\mathbb{Z}[y, z]$. Now f_1, f_2 can't *both* have non-zero 'constant' terms in $\mathbb{Z}[y, z]$. (Why not?) Thus $x|f_1$ or $x|f_2$.

3 (Last part) If $[a]$ is prime, if $a \neq 0$ and if $a \cdot 1 = b \cdot c$ in R, then $b \in [a]$ or $c \in [a]$; that is, $b=ar$ or $c=as$. Therefore $a \cdot 1 = arc$ or $a \cdot 1 = bas$. Thus c (or b) is a unit, so that a is irreducible.

4 There is no contradiction since R is not a ... what? (Note that $(1, 0) \cdot (0, 1) = (0, 0)$.)

5 Each ideal is principal (3.7.16). If $I=[f]$ with $f \neq 0$ then f is irreducible, by exercise 3. Hence $[f]$ is maximal—for if $[f] \subsetneq [g] \subsetneq \mathbb{Q}[x]$ then $f=gh$, where h is not a unit in $\mathbb{Q}[x]$ since $[g] \neq [f]$. (Couldn't g be a unit?)

6 (ii) If $\{3(a+b\sqrt{-5})+(-1+\sqrt{-5})(c+d\sqrt{-5})\}=[\alpha]$ then $1 \notin [\alpha]$—since $(3a-5d-c=1$ and $3b-d+c=0) \Rightarrow 3|1$ in $\mathbb{Z}$. Hence $N(\alpha) \neq 1$. But $(3=\alpha\gamma$ and

$-1+\sqrt{-5}=\alpha\delta) \Rightarrow (N(\alpha)|9$ and $N(\alpha)|6)$ whence $N(\alpha)=3$. Impossible! (iii) $AB=[3\cdot 3, 3(\sqrt{-5}+1), 3(\sqrt{-5}-1), (\sqrt{-5}+1)(\sqrt{-5}-1)]$. Thus $9, 6 \in AB$; hence $[3] \subseteq AB$. But $AB \subseteq [3] \ldots$.

8 (i) $ab \in [x]$ then $xr=ab=a_1 \ldots a_m b_1 \ldots b_n$, say, in terms of irreducibles. By uniqueness of factorisation x is an associate of an a_i or a b_j. This implies that $x|a$ or $x|b$; that is $a \in [x]$ or $b \in [x]$. (ii) Cf. exercises 3 and 5. (iii) Each PID is an UFD. Now use (i) and (ii).

10 Let $a \in I_1 \backslash I_2, b \in I_2 \backslash I_1$. Then $ab \in I_1 \cap I_2$ but ...

11 If $A \nsubseteq P$ and $B \nsubseteq P$ select $a \in A \backslash P, b \in B \backslash P$ and proceed as in exercise 10.

12 No contradiction—the hypotheses of 3.9.6 don't hold!

14 Prove first that each non-zero ideal I of $\mathbb{Z}_m$ is of the form $I=\{\hat{d} \odot \hat{z} : \hat{z} \in \mathbb{Z}_m\}$, where $d|m$ and d is the least positive integer such that $\hat{d} \in I$. Next show that if I is a prime ideal then d is prime in $\mathbb{Z}$. (Then I must be maximal—for if $\hat{r} \in \mathbb{Z}_m \backslash I$ then $d \nmid r$ in $\mathbb{Z}$. Therefore $\hat{1}=\hat{\alpha} \odot \hat{d} \oplus \hat{\beta} \odot \hat{r}$ for suitable $\hat{\alpha}, \hat{\beta} \in \mathbb{Z}_m$. (Cf. exercise 3.6.8(i).)

15 Let $f=x^k(a_k+a_{k+1}x+\cdots)=x^k u$, say, where $a_k \neq 0$. If the non-zero ideal I of $\mathbb{Q}[[x]]$ has the element f as one of its elements for which k is minimal then, certainly, $I=[x^k]$.

Exercises 3.10

1 (i) Yes!; (iii) no (check multiplication); (v) not 1–1. ($\hat{2}$ and $\hat{?}$ have the same image.); (vi) yes. (Perhaps you should try to find proofs!)

2 Checking M2: $(x \odot y) \odot z=(xy+x+y) \odot z=(xyz+xz+yz)+(xy+x+y)+z$ whereas $x \odot (y \odot z)=x \odot (yz+y+z)=xyz+xy+xz+x+yz+y+z$. $\langle \mathbb{Z}, \oplus, \odot \rangle$ is indeed a ring with 'z' $=-1$ and 'e' $=0$. Thus for an isomorphism θ we would need $z\theta=0$, $e\theta=1$. (Can you 'extend' this θ to the whole of $\mathbb{Z}$?)

3 Yes. This is part of the (well-known) 'block multiplication' of matrices.

4 Intuition: $\mathbb{Z}[\sqrt{2}]$ has an element whose square is $1+1$ (i.e. 2); $\mathbb{Z}[\sqrt{3}]$ surely hasn't? Proof: For any isomorphism θ we'd have $\theta(1)=1$ hence $\theta(2)=2$. Suppose $\theta(\sqrt{2})=a+b\sqrt{3}$. Then $2=(a+b\sqrt{3})^2$.

5 Define $\theta: H \to Q$ by $H(a,b,c,d)=\begin{pmatrix} a+ib & c+id \\ -c+id & a-ib \end{pmatrix}$.

6 Since $\begin{pmatrix} x & y \\ -y & x \end{pmatrix}$ 'corresponds' to $x+iy$ in 3.10.2 we should choose $x=\cos(2\pi/17)$, $y=\sin(2\pi/17)$—using De Moivre's theorem.

7 (i) $0_R\phi=(0_R+0_R)\phi=0_R\phi \oplus 0_R\phi$. Now add $\ominus(0_R\phi)$ to each side and use the associative law. (ii) Let $s \in S$. Since ϕ is *onto* S, there exists $r \in R$ such that $r\phi=s$.

Then $(1_R\phi)\odot s = 1_R\phi \odot r\phi = (1_R \cdot r)\phi = r\phi = s$. (iii) $0_S = 0_R\phi = (a+(-a))\phi = a\phi \oplus (-a)\phi$. Hence $(-a)\phi$ is an (hence *the unique*) additive inverse for $a\phi$. (vi) $\phi: R \to S$ is 1–1 and onto. Hence $\phi^{-1}: S \to R$ exists, is 1–1 and onto (2.6.5). Given $s_1, s_2 \in S$ there exist $r_1, r_2 \in R$ such that $s_i = r_i\phi$ $(i=1,2)$. If $(s_1 \odot s_2)\phi^{-1} = r$ then $r\phi = s_1 \odot s_2 = (r_1 \cdot r_2)\phi$. Therefore $(s_1 \odot s_2)\phi^{-1} = r = r_1 \cdot r_2$ (why?) $= s_1\phi^{-1} \cdot s_2\phi^{-1}$

8 For $\phi: \mathbb{Q} \to \mathbb{Q}$ show that $1\phi = 1$ and hence that $n\phi = n$ and $(n/m)\phi = n/m$ for all $m(\neq 0), n \in \mathbb{Z}$. For $\phi: \mathbb{R} \to \mathbb{R}$ note that if $0 < r - a$ then $0 < (r-a)\phi$; i.e. $a\phi < r\phi$.

10 (i) $\{a + b\sqrt{-1} : a, b \in \mathbb{Q}\}$; (ii) $\{x^t(b_0 + b_1 x + \cdots) : t \in \mathbb{Z},\ b_i \in \mathbb{Q}\}$.

12 For each prime p take $\mathbb{Q}_p = \{\frac{m}{n}; m, n \in \mathbb{Z}, p \nmid n\}$. (Why are no two such isomorphic?)

13 No. For example $\mathbb{Z}[x]$ has only two units whereas $\mathbb{Q}[x]$ has infinitely many.

14 (Last part) $\{2r, 2s\}$ determines the element r/s of $\mathbb{Q}$.

17 Let T be *any* subfield of F. By considering $S \cap T$ show that $S \subseteq T$. Now show that $S \cong T$ is impossible if $S \subset T$.

18 If $na = 0_R$ for all a, then $n \cdot 1 = 0_R$. Conversely, if $1 + 1 + \cdots + 1 = 0_R$ then $a(1 + 1 + \cdots + 1) = a0_R = 0_R$. If mn is the characteristic of R then $(m1)(n1) = (mn)1 = 0_R$. But if R is a domain then $m1 = 0_R$ or .... For the next part use the binomial expansion and exercise 1.4.17. For the last part try $\mathbb{Z}_p[x]$.

Exercises 3.11

2 (i) Certainly U must be an integral domain. Let u be a non-zero non-unit in U (if there are any such). Then $u = u_1 u_2 \ldots u_r$ is a product of irreducibles in $U[x]$ (why?), hence in U (see exercise 3.6.2). If $u = u_1 \ldots u_r = v_1 \ldots v_s$ as products of irreducibles of U (hence of $U[x]$; why?) then the u_i, v_j pair off as associates in $U[x]$, hence in U (why? why?). (ii) If $u \in U (u \neq 0)$ is an irreducible element with no multiplicative inverse in U form the ideal $I = \{xf + ug : f, g \in U[x]\}$. Continue as in exercise 3.4.22(b).

3 For the first part follow the text for $\mathbb{Z}[x]$. For the first part of the second paragraph regard $\mathbb{Q}[x, y]$ as $\mathbb{Q}[x][y]$, i.e. as polynomials in y with coefficients in the UFD $\mathbb{Q}[x]$. Note that x is prime and that $x | 3x^2$, $x | 4x^7$, $x | 2x$ but $x^2 \nmid 2x$. The final polynomial factorises.

4 Try $1 + i$.

5 Any 'rational' root is of the form r/s where $(r, s) = 1$ and $r | y^2$ and $s | 1$ in $\mathbb{Z}[y]$. The second polynomial has content $y^2 - 1$.

6 If $F_1=u_1P_1$, $F_2=u_2P_2$ with P_1, P_2 primitive and if f is a common divisor of degree ≥ 1 of F_1 and F_2, in $F_U[x]$, then there exists $g \in F_U[x]$ such that $g|P_1$ and $g|P_2$ in $F_U[x]$. Now (cf. 1.9.14), if $P_1=gh_1$ and $P_2=gh_2$ and if $g=(a/b)G$ then $G|P_1$ and $G|P_2$ in $U[x]$.

8 Is $\mathbb{Q}[[x]]$ a UFD? See 3.7.14 and exercise 3.9.15. In fact the two rings *aren't* isomorphic: $P(x,y)=1+xy+x^2y^2+x^3y^3+\cdots$ lies in one ring but not the other. (So what? They might be isomorphic in some unnatural way. Note that $P(x,y)$ has a multiplicative inverse in $\mathbb{Q}[y][[x]]$. $\mathbb{Q}[y][[x]]$ *is* a UFD. See [22].)

Exercises 3.12

1 $\mathrm{i} \in R^+ \Rightarrow \mathrm{i}^3=-\mathrm{i} \in R^+$.

2 Let α be *any* real number which is not a root of any polynomial in $\mathbb{Z}[x]$. (π and e are such numbers.) Set $f(x) \in R^+$ iff $f(\alpha)>0$ in $\mathbb{R}$. (Cf. exercise 1.7.4.) In particular look at, say, $x-3$ for $\alpha=e$ and $\alpha=\pi$. Now have fun!

6 $1 \in R^+$ implies that $1+1$, $1+1+1$, etc. are all in R^+. Hence $0 \in R^+$ iff $0=1+1+\cdots+1$ (n summands, for some n). Contradiction.

8 Clearly (?) F_D^+ is not empty. If $0<_F a/b$ and $0<_F c/d$ then $0<_D ab$ and $0<_D cd$. Therefore $0<_D abd^2+cdb^2=(ad+bc)(bd)$. Hence the sum of .... The answer to the question is 'no'. (Find a counterexample in $\mathbb{Q}$ with $a=b=-1$.)

10 Using P_2, $\mathbb{Z}[x]$ is not Archimedean since there is no $n \in \mathbb{Z}$ such that $x<n \cdot 1_{\mathbb{Z}[x]}$. Note that, with P_1, $\mathbb{Z}[x]$ *is* Archimedean.

11 Try $\mathbb{Z}[x]$ under P_2. (How do you know that there is no 'funny' isomorphism which shows that $\mathbb{Z}[x] \cong \mathbb{Z}$?)

Exercises 4.2

1 (a) Each is. Intuitively, for the first one: 'replace x by $x+2$ and add' is the same as 'add first and *then* replace x by $x+2$'. Same for multiplication. Formally, ...? (b) Look at $1\theta \cdot 1\theta$ and $\mathrm{i}\phi \cdot \mathrm{i}\phi$.

2 For (i), (ii), (iii) the image $\hat{1}\theta$ of $\hat{1}$ determines $\hat{n}\theta$ for each $\hat{n}$. Also $\hat{1}\theta=\hat{a} \Rightarrow (\hat{a})^2=\hat{1}\theta \cdot \hat{1}\theta=\hat{1}\theta=\hat{a}$. (In $\mathbb{Z}_8$ this implies $\hat{a}=\hat{0}$ or $\hat{1}$, for example.) For the final part think in terms of $n|m$.

3 Question asks: Is $\det(AB)=\det A \cdot \det B$; is $\det(A+B)=\det A+\det B$?

4 (i) $R\theta$ isn't empty (why not?). Given $s_1, s_2 \in R\theta$ let $r_1, r_2 \in R$ be such that $s_1=r_1\theta, s_2=r_2\theta$. Then $s_1 \oplus s_2=r_1\theta \oplus r_2\theta=(r_1+r_2)\theta \in R\theta$. Likewise for $\odot$. That $\ominus s_1 \in R\theta$ follows from 4.2.6(ii). Now use 3.4.2. For an example try $R=\mathbb{Z}$, $S=\mathbb{Q}$ or $\mathbb{Z}[x]$. (ii) Use $r_1\theta \ominus r_2\theta=(r_1-r_2)\theta$.

5 For $a, b \in R$, $(a+b)(\theta \circ \psi) = ((a+b)\theta)\psi = (a\theta \oplus b\theta)\psi = (a\theta)\psi \boxplus (b\theta)\psi = a(\theta\psi) \boxplus b(\theta\psi)$. (Why is each equality valid?) Multiplication is similar.

6 Let $1_R\theta = e$ and let $s \in S$. Since θ is *onto*, $s = r\theta$ for some $r \in R$. Now check that $e \odot s = s \odot e = s$. Thus e is a (hence *the*) multiplicative identity for S. 'Ontoness' is essential: try $z\theta = \begin{pmatrix} z & 0 \\ 0 & 0 \end{pmatrix} \in M_2(\mathbb{Z})$.

7 Given just ζ you can say almost nothing. (That's why arbitrary maps are of little use in comparison to homomorphisms.) If ζ *is* a homomorphism use exercise 6.

9 $x - (\frac{3}{7} + \frac{7}{3}i)$ would be in $\ker\theta$ if $\frac{3}{7} + \frac{7}{3}i$ were in $\mathbb{Q}$. Clearly $k(x) = (x - (\frac{3}{7} + \frac{7}{3}i))(x - (\frac{3}{7} - \frac{7}{3}i)) = x^2 - \frac{6}{7}x + \frac{2482}{441} \in \ker\theta$. Then $\{k(x)h(x) : h(x) \in \mathbb{Q}[x]\} \subseteq \ker\theta$. If $f(x) \in \ker\theta$ write $f(x) = m(x)k(x) + r(x)$ with $\deg r(x) \le 1$. Prove $r(x)\theta = 0$. In the $\mathbb{Z}[x]$ case $k(x)$ would not be in $\ker\theta$ but $441k(x)$ *would* be!

10 (One half) If $\ker\theta = \{0_R\}$ and if $r_1\theta = r_2\theta$ then $r_1 - r_2 \in \ker\theta$ by exercise 4(ii). Hence $r_1 - r_2 = 0_R$ and so θ is 1–1.

11 Since F has only two ideals $\{0\}$ and F, the images are: (i) isomorphic to F (see exercise 10); (ii) the ring with one element (exercise 3.2.4.)

12 $-1 + 67x + I$. $4 + 67x + x^2 + I$ is another.

13 $r_1 = r_1 + 0_R \in r_1 + I = r_2 + I$. Therefore $r_1 = r_2 + i$ for some $i \in I$. For the converse consult the proof of 4.2.14.

14 $0_R + I$ satisfies A3 and, given $a + I$, $-a + I$ satisfies A4.

15 For properties (r), (s), (t) use $r_1 - r_1 \in I$; $r_1 - r_2 \in I \Rightarrow r_2 - r_1 \in I$; $r_1 - r_2, r_2 - r_3 \in I \Rightarrow r_1 - r_3 \in I$, respectively.

16 $x + S = 1 + x + S$ (since $1 \in S$) but their squares $x^2 + S$ and ??? aren't equal.

17 (i) $\mathbb{Q}$—since x has been 'reduced to zero' in $\mathbb{Q}[x]$. [Define $\theta : \mathbb{Q}[x] \to \mathbb{Q}$ by $(f + [x])\theta = f(0)$]; (iii) $\mathbb{Z}_4$—since both x and 4 have been 'reduced to zero' in $\mathbb{Z}[x]$. [Define $\theta : \mathbb{Z}[x]/\mathbb{Z}[x, 4] \to \mathbb{Z}_4$ by $(f + [x, 4])\theta = \widehat{f(0)}$]. (i) gives a field, (v) doesn't.

18 'Only if': If $g = hk$ where $\deg h$ and $\deg k \ge 1$, then $h + [g]$, $k + [g]$ would be zero divisors. For the converse see the proof of 4.2.1. (What if $\deg g \le 1$ or $g = 0$?) $\mathbb{Z}_5[x]/[g]$ is the field $\{a + bx + [g];\ a, b \in \mathbb{Z}_5\}$, so it has 25 elements.

19 If $f_1, f_2 \in F \subseteq F[x]$ and $f_1 \ne f_2$ then $f_1 + [g] \ne f_2 + [g]$ since $f_1 - f_2 \notin [g]$.

20 Use the map $\theta : \mathbb{Q}[\sqrt{2}] \to R/I$ given by $(a + b\sqrt{2})\theta = a + bx + I$. Check θ is well defined, onto and 1–1. Finally

$$\begin{aligned}((a + b\sqrt{2})(c + d\sqrt{2}))\theta &= ((ac + 2bd) + (bc + ad\sqrt{2}))\theta \\ &= (ac + 2bd) + (bc + ad)x + I = (a + bx + I)(c + dx + I)\end{aligned}$$

– since $bdx^2 + I = bd(x^2 - 2) + 2bd + I = 2bd + I$.

21 After finding r, s using the Euclidean Algorithm you should find the required inverse to be $\frac{1}{13}(3x^2-5x+7)+I$.

22 $\mathbb{R}$ will do—but what is the *smallest* such extension?

Exercises 4.3

1 θ maps R *onto* the subring $R\theta \subseteq S$. Now use 4.3.1.

2 'Deduce': $r \in \ker\theta$ iff $r+A_1=0+A_1$ and $r+A_2=0+A_2$, i.e. iff $r \in A_1$ and $r \in A_2$. Note that $\dfrac{\mathbb{Z}}{[m]\cap[n]}$, which has mn elements since $(m,n)=1$, is isomorphic to a subring of $\dfrac{\mathbb{Z}}{[m]}\oplus\dfrac{\mathbb{Z}}{[n]}$.

3 Define $\eta: R/K \to S$ by $(r+K)\eta = r\theta$. Don't forget to check well-definedness, i.e. that $r_1+K=r_2+K \Rightarrow r_1\theta=r_2\theta$.

4 Θ is 1–1: if $A_1\theta=A_2\theta$ and if $a_1 \in A_1$, then $a_1\theta=a_2\theta$ for some $a_2 \in A_2$. Consequently $a_1-a_2 \in I \subseteq A_2$, so $a_1 \in A_2$. Therefore $A_1 \subseteq A_2$. By symmetry $A_2 \subseteq A_1$. Note that $I \subseteq B\theta^{-1}$ since $I\theta=\{0_S\}\subseteq B$.

Let $I \subseteq A \subseteq R$, where A is an ideal of R. Let $a \in A$ and $s \in S$. Since θ is onto, there exists $r \in R$ such that $r\theta = s$. Let $t=a\theta \in A\theta$. Then $ts=a\theta r\theta=(ar)\theta \in A\theta$. (Why?). Likewise $st \in A\theta$.

Deduction: If R/M had any ideals strictly between R/M and $\{0_{R/M}\}$ it would correspond to an ideal of R strictly between R and M (why?).

5 Since R is commutative and has a 1 so does R/P. Then R/P is a domain iff

$$\{x(=r_1+P), y(=r_2+P) \in R/P \text{ and } x\cdot y=0_{R/P}\} \Rightarrow \{x=0_{R/P} \text{ or } y=0_{R/P}\}$$

i.e. iff

$$r_1r_2+P=0_{R/P} \Rightarrow \{r_1+P=0_{R/P} \text{ or } r_2+P=0_{R/P}\}.$$

7 (i) Since P is a prime ideal, R/P is a domain (exercise 15) and hence a field (3.3.4). Now use the last part of exercise 4.

9 Use $F(\alpha)\cong\dfrac{F[x]}{[M_\alpha]}$

Exercises 4.4

1 If $y \in B$ put $D=y^2-2>0$. If $y \geq \frac{3}{2}$ put $v=\frac{13}{9}$. If $y<\frac{3}{2}$ put $v=y-D/3$ (so $0<v<y$). Then $v^2-2=D-\frac{2}{3}Dy+D^2/9>0$ since $9-6y+D>0$.

3 (i) Let $s_1=(a_1,a_2,\ldots)$, $s_2=(b_1,b_2,\ldots)$, $s_3=(c_1,c_2,\ldots)$ be Cauchy convergent sequences. For transitivity note that, given $k>0$, there exist M_1, M_2 such that $|a_i - b_i| < k/2$ for all $i > M_1$ and $|b_j - c_j| < k/2$ for all $j > M_2$. Then $|a_t - c_t| \leq |a_t - b_t| + |b_t - c_t| < k$ for all $t > \max\{M_1, M_2\}$. (iii) To show s is Cauchy first prove, by induction, that $1 \leq x_t \leq 2$ for each x_t. Then show that, for $n > m$, $|x_{n+1} - x_{m+1}| = |(x_n - x_m)(\frac{1}{2} - 1/x_n x_m)| < \frac{1}{2}|x_n - x_m|$. Hence $|x_{n+1} - x_{m+1}| < \frac{1}{2}|x_n - x_m| < \frac{1}{2}^m |x_{n-m+1} - x_1| < \frac{1}{2}^m$. If $\lim_{n \to \infty} x_n = l$ then $l = l/2 + 1/l$ so that $l = ?$

4 This is quite involved. See [50].

5 (i), (ii), (iii) of 4.4.1 are easy to check. The zero element is the upper section of *positive* rationals; $\ominus V$ is the set of all rationals t such that $-t \in \mathbb{Q} \setminus V$ except that if $\mathbb{Q} \setminus V$ contains a rational a such that $x \leq a$ for all $x \in \mathbb{Q} \setminus V$, then $-a$ is omitted from $\ominus V$.

7 $1 < s \Rightarrow 1 < s < s^2 < s^3 < \cdots$. Since $f(s) \geq s^n - |a_{n-1}| s^{n-1} - \cdots - |a_0|$, we have $f(s) \geq s^{n-1}(s - (|a_{n-1}| + \cdots + |a_0|)) > 0$—if, in addition, $s > |a_{n-1}| + \cdots + |a_0|$.

8 $(0,0)$, $(1,0)$ satisfy A3 and M3: For M4 (a,b) has multiplicative inverse $\left(\frac{a}{a^2+b^2}, \frac{-b}{a^2+b^2}\right)$ which is meaningless only when $a^2 + b^2 = 0$, i.e. when (a,b) is the element $(0,0)$.

9 The map $\phi: \mathbb{C} \to \mathbb{R}[x]/[x^2+1]$ given by $(a,b)\phi = a + bx + [x^2+1]$ shows this (cf. exercise 4.2.20). ϕ is (fairly clearly) well defined, onto, 1–1 (why?) and is easily checked to be a homomorphism.

11 No! Use $(1,0)$ to show that axiom Z fails.

12 Check the multiplication axioms.

13 (ii) Field! (Find, by experimenting, specific zero divisors in the other two cases.)

14 Is $1\theta = 1\theta \cdot 1\theta$?

17 What we're doing is constructing $\mathbb{Z}$ formally from $\mathbb{Z}^+ \cup \{0\}$, just as we built $\mathbb{Q}$ from $\mathbb{Z}$, $\mathbb{R}$ from $\mathbb{Q}$ and $\mathbb{C}$ from $\mathbb{R}$. $\langle m, n \rangle$ represents $m - n$ (even if $m < n$).

Exercises 4.5

1 $a1 + b\sqrt{2} + c\sqrt{3} + d\sqrt{5} = 0 \Rightarrow (a + b\sqrt{2})^2 = (c\sqrt{3} + d\sqrt{5})^2$. Deduce that $u = 2ab\sqrt{2} - 2cd\sqrt{15}$, u^2 and $\sqrt{30}$ are rational.

2 $(f_1 - g_1)w_1 + \cdots + (f_t - g_t)w_t = 0_E$ iff each $f_i - g_i = 0_F$.

3 Since $x^3 - 2$ is M_θ (the minimum polynomial for $\theta = \sqrt[3]{2}$) the result is immediate from 4.3.5 (i) and the uniqueness part of 4.3.4.

4 Let $u = p^n - 1$ and let $f_1, f_2, \ldots, f_u$ be the non-zero elements of the field (F, say). Let b be any one of the f_i. Then $f_1 b, \ldots, f_u b$ is the same set of elements (cf.

3.3.4). Thus $f_1 f_2 \ldots f_u = f_1 b f_2 b \ldots f_u b$. Cancelling $f_1 f_2 \ldots f_u$ we get $1 = b^u$. Thus $x^{u+1} - x$ has $u+1$ (distinct) roots in F and, clearly, no subfield of F has this property.

5 Writing $q-1 = mk + n (0 \le n < k)$ we find $\alpha^n = 1$, so that $n = 0$ and, hence, $k | q-1$. Clearly $(\alpha\beta)^{kl} = 1$. If $(\alpha\beta)^s = 1$ then $s | kl$. So $s = uv$, where $ux = k$ and $vy = l$. But $(\alpha\beta)^{uv} = 1 \Rightarrow (\alpha\beta)^{kv} \Rightarrow \beta^{kv} = 1$. But $\beta^{vy} = 1$. Hence $\beta^v = 1$ (why?). Thus $v = l$. Likewise $u = k$. The rest is reasonably straightforward.

6 Each cubic is irreducible. (Show neither has a root in $\mathbb{Z}_3$.) Thus the factor rings are fields by exercise 4.2.18. They are isomorphic by 4.5.8(iii).)

7 Clearly $x + [P]$ isn't of order 3, nor of order 5 since

$$(x + [P])^5 = x(x^4 + x + 1) + x^2 + x + [P] \ne 0 + [P].$$

You can check the order is 15.

8 Thinking of the 9 elements of H as the 9 polynomials $a + bx$ with $a, b \in \mathbb{Z}_3$ and x^2 '=' $-\hat{1}$ we get $(\hat{1} + x)^2 = \hat{1} + \hat{2}x + x^2$ '=' $-x$. Thus $(\hat{1} + x)^4$ '=' $-\hat{1}$. This looks hopeful!

9 (i) Since $m | p^n - 1 = u$, say, we have $x^u - 1 = (x^m - 1)g(x)$ where $g(x)$ is of degree $u - m$. Now $x^u - 1$ has u distinct roots in $GF(p^n)$ whilst $x^m - 1$ and $g(x)$ have at most m and $u - m$ roots respectively.

11 Recall: the non-zero elements of $GF(8)$ satisfy the equation $x^7 = 1$.

12 Note that, if $g(x) = r(x)f(x) + s(x)(x^{p^t} - x)$ for suitable $r(x), s(x) \in \mathbb{Z}_p[x]$, then $g(u) = r(u) \cdot 0 + s(u) \cdot 0 = 0$—so $g(x) \ne 1$.

13 Cf. exercise 7.4.22.

Exercises 4.6

1 (a) In Fig. 4.8 take $a = 1$. Then $c = 1/b$. (b) Use Fig. 4.8 to get $\sqrt{u}$ from u ($u = 3$, $u = 2 + \sqrt{3}$, etc.). (c) (a, b) is constructible from the pair $(0,0), (1,0)$ iff each of a, b lies in some field in some sequence $\mathbb{Q} = \mathbb{Q}_0 \subset \mathbb{Q}_1 \subset \cdots \subset \mathbb{R}$, where each $\mathbb{Q}_{n+1} = \mathbb{Q}_n(\sqrt{r})$ for some $r \in \mathbb{Q}_n$. The problem is to show (see (3.4.2(F)) that each of $a + b$, ab, etc. lies in some such sequence. (Cf. exercise 3.4.19.)

2 Suppose

$$(\alpha_{11}v_1 + \cdots + \alpha_{1s}v_s)u_1 + (\alpha_{21}v_1 + \cdots + \alpha_{2s}v_2)u_2 + \cdots + (\alpha_{r1}v_1 + \cdots + \alpha_{rs}v_s)u_s$$
$$= 0_K = 0_E = 0_F.$$

Then each bracket is equal to 0_E since the u_j form a basis for K over E. But then each $\alpha_{ij} = 0_F$ (why?)

3 First part: If (u, v) is the point of intersection of $ax + by + c = 0$ and $kx + ly + m = 0$ where $a, b, c, k, l, m \in \mathbb{Q}_n$ then (in general) $u = (bm - lc)/(al - bk)$

$\in \mathbb{Q}_n$. Likewise for v. Conversely: If $\mathbb{Q}_r \subseteq \mathbb{Q}_{r+1} \subseteq \cdots \subseteq \mathbb{Q}_t$ are subfields of $\mathbb{R}$ with $[\mathbb{Q}_{i+1}:\mathbb{Q}_i] \leq 2$ for each i and if $a_t, b_t \in \mathbb{Q}$, then the point $p_t = (a_t, b_t)$ is constructible from $P(=\{p_1, \ldots, p_r\})$. [*Proof*: If $[\mathbb{Q}_{i+1}:\mathbb{Q}_i]=2$ then there exists a (real) $r \in \mathbb{Q}_{i+1}$ such that $r^2 = a \in \mathbb{Q}_i$. Clearly $(r, 0)$ is constructible if $(a, 0)$ is.)

4 Note that $\mathbb{Q}\left(\frac{\sqrt{3}}{2}\right)$ isn't a subfield of $\mathbb{Q}(\cos 20°)$ (why not?). In any case $\left[\mathbb{Q}\left(\frac{\sqrt{3}}{2}\right):\mathbb{Q}\right]=2$. (So what?)

6 Put $u = 2\cos(2\pi/5) = z + 1/z$ where $z = e^{2\pi i/5}$. Then

$$u^2 + u - 1 = \left(z^2 + \frac{1}{z^2} + 2\right) + (z + 1/z) - 1 = 0,$$

since $z^4 + z^3 + z^2 + z + 1 = 0$.

7 Prove $f(x)$ is irreducible over $\mathbb{Q}$ directly—or note that, if $f(x) = g(x)h(x)$, then $y^i g(y) \cdot y^j h(y)$ where $i = \deg g(x)$ and $j = \deg h(x)$, yields a factorisation of $y^6 + y^5 + \cdots + 1$.

8 (a) Use the fact that $am - bn = 1$ for suitable $a, b \in \mathbb{Z}$. (b) For (ii): notice how the $\dfrac{\text{`}p-1\text{'}}{2}$ arises in exercises 6 and 7. Apply the same technique to (i).

9 (a) For (i) use angle or arc bisection. For (ii) use exercise 8. (b) 2° is not constructible; 3° is! (Reason?)

10 (a) Use Pythagoras' Theorem.

11 Given the angle θ, split θ into 2^n equal parts for any n. Then look at $\theta - \theta/2 + \theta/4 - \cdots \pm \theta/2^n$.

Exercises 4.7

1 (a) $s_1 s_2 - 3s_3$; (d) $5s_5 - 3s_4 s_1 + s_3 s_2$. (The method of exercise 4 is quite quick here.)

2 (i) Coefficient of x^2 in the new equation is $-((-3)^2 - 2\cdot 7)$; (iii) Use

$$\frac{1}{a} + \frac{1}{b} + \frac{1}{c} = \frac{s_2}{s_3}, \qquad \frac{1}{ab} + \frac{1}{bc} + \frac{1}{ca} = \frac{s_1}{s_3}$$

3 (i) follows since m is chosen as small as possible. Then $h(s_1, \ldots, s_k) = 0$ yields $h_0(\bar{s}_1, \ldots, \bar{s}_{k-1}) = 0$ on putting $x_k = 0$.

5 Put $U_r(x_1, \ldots, x_{r+t+1}, 0, \ldots, 0) = \sum_{j=0}^{r} u_j(x_1, \ldots, x_{r+t}) x_{r+t+1}^j$. Then

$$U_r(x_1, \ldots, x_r, 0, \ldots, 0) = 0 \Rightarrow u_0(x_1, \ldots, x_{r+t}) = 0.$$

Hence

$$U_r(x_1,\ldots,x_{r+t+1},0,\ldots,0)=x_{r+t+1}V(x_1,\ldots,x_{r+t+1}) \text{ [for some } V]=$$

$$x_{r+t+1}\sum_{k=0}^{r} v_k(x_1,\ldots,x_{r+t-1},x_{r+t+1})x_{r+t}^k.$$

Therefore

$$0=U_r(x_1,\ldots,x_{r+t-1},0,x_{r+t+1},0,\ldots,0) \text{ (why?)}$$
$$=x_{r+t+1}x_{r+t}v_0(x_1,\ldots,x_{r+t-1},x_{r+t+1}).$$

Consequently $U_r(x_1,\ldots,x_{r+t+1},0,\ldots,0)=x_{r+t+1}x_{r+t}W(x_1,\ldots,x_{r+t+1})$ for some W. Continuing in this way yields

$$U_r(x_1,\ldots,x_{r+t+1},0,\ldots,0)=x_{r+t+1}x_{r+t}\ldots x_1Z(x_1,\ldots,x_{r+t+1}),$$

for some Z. But this is impossible! (Why? Look at degrees.)

Exercises 5.2

1 The coefficient of x in the new equation is $s_1s_3-4s_4$.

2 The constant term in the equation for the first function is $s_4s_1^2+s_3^2-s_3s_2s_1$.

Exercises 5.3

1 (a) No: $\begin{pmatrix}0 & 0\\ 0 & 0\end{pmatrix}$ is a problem; (c) Yes: identity is 37; (d) No. 0 is the identity but a has no multiplicative inverse if $a=\ldots$ what?; (e) Yes: $\hat{6}$ is the identity; (f) Yes; (h) No; (j) Yes; (l) Yes: $f_1(x)$ is the identity.

2 (b) Not abelian; finite order (in fact how many elements?); (e) abelian—order 4; (g) not abelian—order 48 (why?); (k) abelian—infinite order; (l) not abelian (look at f_2f_4, f_4f_2)—order 6.

3 Clearly (C) is satisfied. Also

$$\begin{aligned}(g_1,h_1)\cdot\{(g_2,h_2)\cdot(g_3,h_3)\}&=(g_1,h_1)\cdot(g_2\circ g_3,h_2*h_3)\\&=(g_1\circ(g_2\circ g_3),h_1*(h_2*h_3))\\&=((g_1\circ g_2)\circ g_3,(h_1*h_2)*h_3)=\cdots\\&=\{(g_1,h_1)\cdot(g_2,h_2)\}\cdot(g_3,h_3)\end{aligned}$$

—hence associative. (e_G,e_H) is the identity and $(g,h)^{-1}=(g^{-1},h^{-1})$. Hence (N) and (I) hold.

5 Yes. Prove it! (e_G is still the identity; likewise for inverses.)

6 Order is surely 12 (or is it 24???).

7 Let Aut(F) denote the *set* of all automorphisms of F. Then $\langle \mathrm{Aut}(F), \circ \rangle$ is a group. (C and I follow from exercise 3.10.7 (vii) and (vi).)

8 Closure holds by exercise 3.8.4. Axioms A, N, I are fairly trivially satisfied.

10 For all $x \in \mathbb{R}$, $T_{a,b}(T_{c,d}(x)) = a(cx+d)+b = T_{ac,ad+b}(x)$. $T_{1,0}$ is the identity.

12 There are 3, two being the identity permutation and (123).

13 If $|S|>1$ then $\langle S, \circ \rangle$ is not a group. (Look at $e \circ x$, e being the supposed identity.)

14 No! Think along the lines of exercise 13.

Exercises 5.4

1 Since $x \cdot x = x$, we have $(x^{-1} \cdot x) \cdot x = x^{-1} \cdot (x \cdot x) = x^{-1} \cdot x$.

3 Let $a, b \in G$. Then $a = a^{-1}$, $b = b^{-1}$, and so $ab = (ab)^{-1} = b^{-1}a^{-1} = ba$.

5 $g_i g = g_j g \Rightarrow g_i g g^{-1} = g_j g g^{-1} \Rightarrow g_i = g_j$. Hence the $g_i g$ $(1 \le i \le n)$ comprise n *distinct* elements of G. But $|G| = n \ldots$.

9 (i) Look at a specific $a \in G$. Now, $aa^{-1} = e \Rightarrow (a^{-1}a)a^{-1} = a^{-1}e = a^{-1}$ (by $\mathrm{N_R}$). Also $a^{-1}(a^{-1})^{-1} = e$ by $\mathrm{I_R}$. Therefore

$$\begin{aligned} a^{-1}a &= (a^{-1}a)e = (a^{-1}a)(a^{-1}(a^{-1})^{-1}) = ((a^{-1}a)a^{-1})(a^{-1})^{-1} \\ &= a^{-1}(a^{-1})^{-1} = e. \end{aligned}$$

Now show $ea = a$. Therefore the *right* neutral and inverse elements are double sided.

Exercises 5.5

1 Try $1f=2$, $2f=3$, $3f=?$ (Is there a 'smaller' example?)

2 (α) (12463)(58)

3 (a) (1 5 4 7 3)(2 6); (b) (1 5 4 7 3)(2 6); (c) is?

4 $Y^{-1}XY = \begin{pmatrix} 1 & 2 & 3 & 4 & 5 & 6 & 7 \\ 1 & 3 & 4 & 7 & 5 & 2 & 6 \end{pmatrix}$

5 X and $Y^{-1}XY$ have the same 'cycle shape'. Indeed see exercise 6.

6 (i) $f^{-1}(x_1, \ldots, x_r)f$ sends each $x_i f$ successively to x_i, to x_{i+1}, to $x_{i+1}f$. i.e. $x_i f$ to $x_{i+1} f$ overall. (ii) To get the correct 'general' f, try experimenting with specific examples.

7 (i) If f^t is the identity and if $\beta = \alpha f^r$ $(0 \le r < t)$ then $\alpha = \beta f^{t-r}$.

8 If γ is common to the orbit of α and β then α, β are in the same orbit—namely γ's—by exercise 7(i).

9 For the first part see 5.5.8. The even permutations include (123) and (12)(34). By 5.5.11 you should find 12 altogether.

11 Note that the second (fourth) equality follows from the first (third).

12 To see if (i) can be changed into (ii) first change (i) (or (ii)), legally, by moving the blanks to the same 'place'. Then (ii) will be obtainable from (i)—legally—iff the corresponding permutation

$$\begin{pmatrix} 14 & 11 & 7 & 1 & 12 & 2 & B & 9 & 6 & \dots \\ 15 & 12 & 7 & 13 & 4 & 11 & B & 14 & 6 & \dots \end{pmatrix}$$

is even. (See Thomas Fournelle, The permutation game. *Pi Mu Epsilon Journal*, 5, 1973, 425–9).

Exercises 5.6

1 If S is a subgroup and if $a, b \in S$ then $a^{-1} \in S$, hence $a^{-1}b \in S$. (5.6.5 (ii) and (i).) For the converse, first show $e \in S$, then $b^{-1} \in S$, finally $ab \in S$.

2 Assume that $a, b \in S \Rightarrow ab \in S$. Deduce that $ab^d \in S$ for each $d \in \mathbb{Z}^+$ and, since $b^n = e$ for some $n \in \mathbb{Z}^+$, that $ab^{-1} \in S$.

3 If $A \nsubseteq B$ and $B \nsubseteq A$ look at ab with $a \in A \backslash B$ and $b \in B \backslash A$.

4 If $a^m = e$ and $b^n = e$ look at $(ab^{-1})^{mn}$. (Is the set in question non-empty?)

5 Use exercise 2. $|T| = 1.1.3!.10!$ (correct?)

7 Clearly $T \subseteq G$ and $\bar{\circ}$ coincides with the restriction $\bar{\circ}$ of $\circ$ to $T \times T$. Also $\langle T, \bar{\circ} \rangle = \langle T, \bar{\circ} \rangle$ *is* a group—it is given as a subgroup of $\langle S, \bar{\circ} \rangle$.

8 (a) Lagrange's Theorem will tell us we need only look for subgroups of orders 1, 2, 4, 8, 16. $\langle a^4 \rangle$ and $\langle ba^i \rangle$ $(i = 0, 1, \dots, 7)$ are those of order 2 (since $ba^iba^i = a^{-i}a^i = e$). $\langle a^2 \rangle$, $\langle a^4, ba^i \rangle$ $(i = 0, 1, \dots, 7)$ are those of order 4. (b) In S_4: The different cyclic ones are: $\langle(1234)\rangle$, $\langle(1243)\rangle$, $\langle(1324)\rangle$. (Why no more? What about non-cyclic ones?) In S_5 each $(1abcd)$ generates a 5-cycle where $\{a, b, c, d\} = \{2, 3, 4, 5\}$—surely not all different?

9 For S_6: Note that each element is a product of disjoint cycles of lengths $x, y, \dots$ involving ≤ 6 letters. Now show that one cannot write $6 = x + y + \cdots$ where $\operatorname{lcm}\{x, y, \dots\} = 12$.

11 The order is mn if $(m, n) = 1$. Otherwise be careful. For example, it's possible for a, b each to have order 6 and for ab to have order 1 or 2 or 3 or 6.

12 $(AB)^{2n} = \begin{pmatrix} 1 & -2n \\ 0 & 1 \end{pmatrix} \neq \begin{pmatrix} 1 & 0 \\ 0 & 1 \end{pmatrix}$.

14 (ii) $ab=b^{-1}(ba)b$. Now use part (i).

16 $\mathbb{Q}/\mathbb{Z}$ is infinite since, for each $n\in\mathbb{Z}^+$, $1/n$ has order n exactly.

17 If $x=(abc\ldots)(\ldots)(\ldots)\ldots\in S_n$ is a product of disjoint cycles then $x(ab)\neq(ab)x$.

18 $\zeta=\left\{\begin{pmatrix} a & 0\\ 0 & a\end{pmatrix}: a\in\mathbb{R}\backslash\{0\}\right\}.$

19 (i) (12), (13), (123){=(12)(13)}, (132){$=(123)^2$}, (23){=(123)(12)} all lie in the subgroup. Any more? (ii) The order is 12.

20 $\{a, ba\}$, $\{a^3, b\}$ are two such.

21 For $n=3$ choose 6, 10, 15. (Why? Factorise these integers into primes!)

22 If $\{a_i/b_i: 1\leq i\leq t\}$ generated $\mathbb{Q}^+$ and if $p\nmid b_i$, $1\leq i\leq t$, p being prime, how could you show that $1/p\in\mathbb{Q}^+$?

24 $U\subseteq S_\lambda\Rightarrow\langle U\rangle\leq S_\lambda$ so $\langle U\rangle\leq\bigcap_{\lambda\in\Lambda}S_\lambda$. But $\langle U\rangle$ is one of the $S_\lambda\ldots$.

25 $(12\ldots n)^{-1}(12)(12\ldots n)=(23)$ and $(23)(12)(23)=(13)$. Obtain (34) and (14); (45) and (15) etc. similarly. Then use products of transpositions.

Exercises 5.7

1 $S=SI$, $S(12)(34)=S(243)=S(143)$, $S(13)(24)$, $S(14)(23)$ are the distinct right cosets.

3 $g(h^*h)\in gH$. Therefore $gh^*H\subseteq gH$—and conversely since $gh=gh^*(h^{*-1}h)\in gh^*H$.

4 $c\in cH$. So, if $cH=dH$ then $c=dk$ for suitable $k\in H$. Conversely $d^{-1}c=k\in H\Rightarrow c=dk\in dH$. But $c\in cH$. Therefore $cH\cap dH\neq\emptyset$. Hence $cH=dH$ (by 5.7.4).

5 $(a\sim b$ and $b\sim c)\Rightarrow a^{-1}b, b^{-1}c\in H$. Therefore $a^{-1}bb^{-1}c\in H$ (*why*?). That is, $a\sim c$. (i) $a^{-1}b\in H\Rightarrow ab^{-1}\in H$? Try $a=(12)$, $b=(123)$, $H=$??. For (iii) think of S_3—again!

6 If $t\in xA\cap yB$ then $xA=tA$, $yB=tB$ (why?). Hence $xA\cap yB=tA\cap tB=t(A\cap B)$ (Why?) Then $|G:H\cap K|\leq|G:H||G:K|$ follows, by counting! If $|G:H\cap K|=|G:H||G:K|$ then each $xH\cap yK$ must be a coset of $H\cap K$, hence non-empty. Thus if $h\in H$, $k\in K$ we have $kH\cap hK\neq\emptyset$. Hence $h^{-1}k=k_1h_1$, where $h_1\in H$ and $k_1\in K$. From this $HK=KH$, which is therefore (proof?) a subgroup of G. Now prove, by similar means that HK is actually equal to G. (The converse, namely that if $G=HK$ then $|G:H\cap K|=|G:H||G:K|$ follows using exercise 7 and the equality $|H:H\cap K|=|HK:K|$.)

7 Given $G=g_1K\cup g_2K\cup\cdots\cup g_mK$ and $K=k_1H\cup k_2H\cup\cdots\cup k_nH$ as unions of distinct cosets, prove that $G=\bigcup_{i=1,j=1}^{i=m,j=n} g_ik_jH$ is a union of distinct cosets (cf. 4.6.1).

8 The a^iH $(i=0,1,\ldots,n)$ can't *all* be pairwise distinct. Hence $a^rH=a^sH$ for some r,s. Note that $t\nmid n$ *is* possible.

9 If $G=\{g_ih:1\le i\le r, h\in H\}$ then $G=\{(g_ih)^{-1}:1\le i\le r, h\in H\}$.

10 What is it that '... *divides* the order of a group.'?

11 The order is the lcm of $\{1, 2, 3, \ldots, 12\}$ $(=27\,720?)$

Exercises 5.8

1 If $a,b\in G=\langle x\rangle$, cyclic, then $a=x^r$, $b=x^s$ for some r,s. So $ab=x^rx^s=x^{r+s}$ $=?$

3 If $(x^3)^k=e$ then $10|3k$. The smallest such k is 10.

4 $y=x^t$ for any t such that $(t,15)=1$.

5 $\phi(n)$—since $C_n=\langle y\rangle$ iff $y=x^t$ with $(t,n)=1$ and $1\le t\le n$. For (ii), note that $\phi(n)=3$ is impossible (use exercise 2.5.8).

6 $(88,28)=4$. Hence $n=4$ will do. (So will $n=2$. Why?)

7 Let a $(\ne e)\in G$. Then $\langle a\rangle\le G$. Hence G is cyclic. But, also, $\langle a^2\rangle=e$ or $\langle a^2\rangle=G$. Hence G isn't *infinite* cyclic.

9 Try our old friend (the smallest non-abelian group) again!

10 If $\langle\mathbb{Q},+\rangle$ *were* cyclic with generator $x(\ne 0)$, then $x/2\notin\mathbb{Q}$. Now let $a=x/z$, $b=y/z$ where $x,y,z\in\mathbb{Z}$ and $z>0$. Let $(x,y)=u=rx+sy$ for suitable $r,s\in\mathbb{Z}$ and set $c=u/z$. Then $c\in\langle a,b\rangle$. Conversely, writing $x=\alpha u$, $y=\beta u(\in\mathbb{Z})$ we have $a=\alpha c$, $b=\beta c$. Hence $a,b\in\langle c\rangle$.

12 No! If $a+b\sqrt{2}$ were the proposed generator what power of it would equal, say, $\sqrt{2}$ or $1+\sqrt{2}$?

14 Clearly $\langle P,\cdot\rangle$ is infinite and a group. If $P=\langle t\rangle$ is cyclic, then $t^{p^k}=1$ for some k – impossible! Let S be a subgroup of P. Then either S contains p^nth roots of unity for arbitrarily large n or there is a k such that S contains no p^sth roots of unity for any $s>k$.

15 The given number is prime!

Exercises 5.9

1 Let $\langle G, \cdot\rangle = \langle\langle x\rangle, \cdot\rangle$ be the (multiplicatively written) infinite cyclic group generated by its element x. Define $\theta: \langle \mathbb{Z}, +\rangle \to \langle G, \cdot\rangle$ by $\theta(n) = x^n$. Then θ is an isomorphism. If $\langle G, \cdot\rangle$ and $\langle H, \odot\rangle$ are each infinite cyclic, use 5.9.4.

2 See 3.10.2 (iii) and exercises 3.10.7 (vi), (vii).

3 $|A_6| \neq |S_5|$.

4 The subgroup $\langle(1\ 2\ 3\ 4\ 5\ 6)\rangle$ is certainly abelian. Is either $\langle(123), (45)\rangle$ or $\langle(123), (34)\rangle$ non-abelian of order 6?

5 One is—the other isn't. (I leave you to find out which is which!)

6 One is isomorphic to C_4—the other is not.

7 $\langle M(14), \odot\rangle$ is abelian and has order $\phi(14)$. $\langle \mathbb{Z}_6, \oplus\rangle$ is cyclic of order 6.

8 $\phi: \langle \mathbb{Z}, \circ\rangle \to \langle \mathbb{Z}, +\rangle$ given by $\phi(z) = z + 7$ (or is it $z - 7$??) looks a likely candidate.

9 Look at the map given in 3.10.2(i).

10 Cf. exercise 3.10.6.

11 (a) Yes. Try the obvious(?) map. (b) Suppose $x = p_1^{\alpha_1} p_2^{\alpha_2} \ldots p_r^{\alpha_r}$ ($\alpha_i \in \mathbb{Z}$), the p_i being distinct (positive) primes. Show that there is an isomorphism between $\langle \mathbb{Q}^+, \cdot\rangle$ and $\langle \mathbb{Z}[x], +\rangle$ via $x \Leftrightarrow \alpha_1 + \alpha_2 x + \cdots + \alpha_r x^r$. Now try to use the fact that each of $\mathbb{Q}[\sqrt{2}]$, $\mathbb{Q}[\sqrt{3}]$ is, like $\mathbb{Z}$, a UFD.

12 No! (Use the fact that $\mathbb{Q}[x]$ is countable whilst $\mathbb{R}$ is not—or determine, if ϕ *were* an isomorphism with $\phi(r) = x$, what $\phi(\sqrt{r})$ would be.)

13 The *fields* $\langle \mathbb{Q}(\sqrt[3]{2}), +, \cdot\rangle$, $\langle \mathbb{Q}(\omega\sqrt[3]{2}), +, \cdot\rangle$ are isomorphic since each is isomorphic to $\mathbb{Q}[x]/[x^3 - 2]$. Hence their multiplicative subgroups of non-zero elements are isomorphic.

14 Let $|A| = 6$. If A has an element of order 6 then $A \cong C_6$. Otherwise A can only have elements of orders 1, 2 and 3. If it has *only* elements of orders 1 and 2 then A is abelian (exercise 5.4.3) and has a subgroup of order 4 ($\nmid 6$). Neither can all elements ($\neq e$) have order 3 (since $\{a^i b^j (0 \leq i, j \leq 2)\}$ produces 9 (distinct?) elements). So suppose $a, b \in A$ with $|a| = 2$, $|b| = 3$. Prove that $\{a^i b^j : 0 \leq i \leq 1, 0 \leq j \leq 2\}$ is a set of 6 distinct elements of A. Hence show that $a^{-1}ba = b$ or $b^2 (= b^{-1})$ [why?] In the former case A is abelian. In the latter $A \cong S_3$.

15 (i) No. (Look at the order of each element and then at ac and ca). (ii) No. (Assume so. Find the orders of a and T. Deduce the 'group' would be C_6.) (iii) If so then it must be cyclic of order 5 with O as the identity. Is it?

16 Define $\Phi: G \to H$ by $g\Phi = (g\psi)^{-1}$.

17 Taking the origin at the centre of the cube, note that the larger group contains the rotations $\{r_i\}$ of exercise 5.3.1 together with 24 isometries r_iR, where R denotes reflection in the origin. (What about the rR if R is a reflection in the y–z plane? Doesn't that yield more isometries of the cube?)

18 For the first part recall exercise 5.3.7. For $x, y \in G$, $(xy)\psi_g = g^{-1}xyg = g^{-1}xg \cdot g^{-1}yg = x\psi_g y\psi_g$. Now prove that ψ_g is 1-1 and onto. For all $x \in G$, $x(\psi_g \circ \psi_h) = (x\psi_g)\psi_h = h^{-1}g^{-1}xgh = x\psi_{gh}$. Now show $(\psi_g)^{-1} = \psi_{g^{-1}}$. Hence $\mathrm{Inn}(G)$ is a subgroup.

19 See, for example, [8].

20 For the deduction use the first part and 5.9.6.

21 There are only finitely many essentially different $n \times n$ multiplication tables. Alternatively ask yourself: How many subgroups (roughly) has S_n?

Exercises 5.10

1 (ii) Use $\det(AB) = \det A \cdot \det B$.

2 θ_m *can* be onto—for example if $\langle A, +\rangle = \langle \mathbb{Q}, +\rangle$ or if $\langle A, +\rangle = \langle C_n, \cdot\rangle$ and $(m, n) = \ldots$ guess what?

3 Let $u = g\psi$, $v = h\psi \in G\psi$ – a homomorphic image of G (abelian). Then $u \cdot v = g\psi \cdot h\psi = (gh)\psi = (hg)\psi = h\psi \cdot g\psi = v \cdot u$. Hence G is abelian. If $g^n = e$ then $(g\psi)^n = g^n\psi (= e\psi) = e$. Hence the order of $g\psi$ divides n. One specific instance is trivial (and *that's* a hint!)

4 Let $S_3\phi$ be a homomorphic image of S_3. By 5.10.6, $\ker \phi = \langle e\rangle$ or A_3 or S_3. Thus $S_3\phi = S_3$ or some group (!) of order 2 or the trivial group.

5 (ii) For any such ψ, $\hat{1}\psi$ must have order dividing both 5 and 12. Hence $\hat{1}\psi = \hat{0}$. (iii) $\langle \mathbb{Z}_6, \oplus\rangle \cong \langle M(14), \odot\rangle$ under ϕ, say. (See exercise 5.9.7.) Hence $\psi = \lambda \circ \phi : \mathbb{Z} \to M(14)$ will do—where $\lambda : \mathbb{Z} \to \mathbb{Z}_6$.

7 Note that ψ can't be onto (by 5.10.10 (iv)). Try $\psi : C_2 \to S_3$.

8 Map odd permutations to -1, etc.

9 If $x = u_1^{\pm 1} u_2^{\pm 1} \ldots u_r^{\pm 1}$, then $x\psi = (u_1\psi)^{\pm 1}(u_2\psi)^{\pm 1} \ldots (u_r\psi)^{\pm 1}$ is determined since the $u_i\psi$ are known.

10 If $x, y \in \ker \psi$ then $(x^{-1})\psi = (x\psi)^{-1} = e^{-1} = e$ whilst $(xy)\psi = x\psi \cdot y\psi = e \cdot e = e$. (Question: Is $\ker \psi$ non-empty?)

11 (i) $\Leftarrow$: Let $N = g^{-1}Ng$. Then, given $gn \in gN$, we find $gn = g(g^{-1}n_1g) = n_1g$ for suitable $n_1 \in N$. Thus $gN \subseteq Ng$. (ii) $\Rightarrow$: Let $m \in M$. Then $g^{-1}mg = n \in N$. Therefore $m = gng^{-1} \in gNg^{-1}$. Hence $M \subseteq gNg^{-1}$.

12 If $S \leq A$ (abelian) and if $s \in S$, $a \in A$, then $a^{-1}sa = sa^{-1}a \in S$. 5.9.2(iv) provides the smallest example of the desired sort.
13 (a) Each element of 5.3.4(1) is of the form $b^i a^j$ where $i=0,1$; $0 \leqslant j \leqslant 7$. Hence $a^{-j}b^{-i}a^t b^i a^j = a^{-j}(b^{-i}ab^i)^t a^j = a^{-j}a^{-t}a^j$ (since $b^{-1}ab = a^{-1}$).

14 For $g \in G$ and $c \in \zeta(G)$, $g^{-1}cg = cg^{-1}g = c \in \zeta(G)$.

15 For $g \in G$, $g^{-1}Sg$ is a subgroup of G of order 20—hence $g^{-1}Sg = S$. The number 20 has no significance.

16 If $a \in \langle H \cup K \rangle$, then $a = h_1k_1 \ldots h_rk_r$ $(h_i \in H, k_j \in K)$. Hence $g^{-1}ag = g^{-1}h_1g \cdot g^{-1}k_1g \ldots g^{-1}k_rg$. But $g^{-1}h_ig \in H$ and $g^{-1}k_jg \in K$ since $H, K \lhd G$.

17 $V \lhd A_4$ (example 5.3.4(m)). Now use 5.10.9.

18 Let $\theta \in \mathrm{Aut}(G)$ and $\psi_g \in \mathrm{Inn}(G)$. $\mathrm{Inn}(G) \leq \mathrm{Aut}(G)$ (exercise 5.9.18). For all $x \in G$, $x(\theta^{-1} \circ \psi_g \circ \theta) = (g^{-1}(x\theta^{-1})g)\theta = (g\theta)^{-1} \cdot x \cdot g\theta = x\psi_{g\theta}$. Therefore $\theta^{-1} \circ \psi_g \circ \theta = \psi_{g\theta} \in \mathrm{Inn}(G)$.

19 Clearly each Hg_ig is one of the Hg_k. Prove that $Hg_ig = Hg_jg$ iff $Hg_i = Hg_j$ (use exercise 5.7.4). Thus ρ_g is a permutation on X. Clearly $\rho_{uv} = \rho_u \circ \rho_v$. Hence $(uv)\psi = \rho_{uv} = \rho_u \circ \rho_v = u\psi \circ v\psi$. Now $k \in \ker \psi$ iff $Hg_ik = Hg_i$—i.e. iff $(h)g_ikg_i^{-1}(h^{-1}) \in H$—for all g_i and all h, that is, iff $k \in g^{-1}Hg$ for all $g \in G$. Finally, if $N \leq H$ with $N \lhd G$, then $N = \bigcap_{g \in G} g^{-1}Ng \subseteq \bigcap_{g \in G} g^{-1}Hg$.

20 The subgroup of index 4 leads to a homomorphism ψ of G into the group S_X of order 4! Consequently $\ker \psi$ is not G—nor $\langle e \rangle$ (why not?).

21 For $n \times n$ matrices A, B it is easy to check that $\mathrm{tr}(AB) = \mathrm{tr}(BA)$. It follows that $\mathrm{tr}(C^{-1} \cdot DC) = \mathrm{tr}(DC \cdot C^{-1}) = \mathrm{tr}D$.

Exercises 5.11

2 Put $x = \begin{pmatrix} 1234 \\ abcd \end{pmatrix} \in S_4$. Then $x^{-1}(12)(34)x = (ab)(cd)$ [cf. exercise 5.5.6] $\in V$. $S_4/V \cong C_6$ or S_3. But $(12)(123)V \neq (123)(12)V$.

3 $g^m \in N$ iff $(gN)^m = e_{G/N}$ iff $n \mid m$.

4 If $(aT)^k = T$ then $a^k \in T$. Therefore a^k (and hence a) has finite order. Consequently $a \in T$, i.e. $aT = T$.

5 $H \lhd G \Rightarrow H/N \lhd G/N$: If $H \lhd G$ then $(gN)^{-1}hNgN = (g^{-1}hg)N \in H/N$. $S \leq G/N \Rightarrow S = K/N$: Let $K = \{k : kN \in S\}$. Trivially $N \subseteq K$. Now use $(k_1k_2^{-1})N = k_1N(k_2N)^{-1}$. Hence $K \leq G$ (and $K/N = S$, by definition).

6 Let $g\lambda = \psi_g$. Then λ is a homomorphism, since $\psi_{gh} = \psi_g \circ \psi_h$ (exercise 5.9.18). $g\lambda$ is the identity map iff $x\psi_g = x$ (i.e. iff $g^{-1}xg = x$) for all $x \in G$.

7 (a) Take $N_1 < N_2 < G$, G being infinite cyclic: (b) Try $G = D_4$ with $|N_1| = |N_2| = 4$.

8 Let $t \in T$, so that $t = a\psi$ for some $a \in G$. Then $g\psi = t$ iff $g = ak$ where $k \in K$. Thus $|K|$ elements map to each element of T.

9 Given $gN = hN$ show that $gK = hK$. This shows that θ is well-defined. $\ker\theta = \{tN: tK = e_{G/K}\} = \{tN: t \in K\} = K/N$. By 5.11.6, $G/K \cong (G/N)/(K/N)$.

Exercises 6.2

1 Let $g \in G$. Then $(gH)^{p^\alpha} = e_{G/H} \Rightarrow g^{p^\alpha} \in H \Rightarrow (g^{p^\alpha})^{p^\beta} = e_H = e_G$ for some α, β.

2 No prizes for guessing which group!

3 (a) (Transitivity) For subsets U, V, W of G: $(U^g = V$ and $V^h = W) \Rightarrow U^{gh} = W$.

4 Let $S \leq G$. The map λ given by $u\lambda = g^{-1}ug$ $(u \in S)$ establishes the isomorphism between S and $g^{-1}Sg$.

5 If $x, y \in N_G(H)$ then $H^x = H$, $H^y = H$ and $H^{y^{-1}} = H$. Consequently $H^{xy^{-1}} = H$ so that $xy^{-1} \in N_G(H)$. (Have we yet used that H is a subgroup?); (b) Take $H = G$ (any G!)—or, for example, $\{I, (12)\}$ in S_3; (d) Look for a suitable subset of S_3 to act as K.

6 For $h \in H$ and $v \in N_G(H)$, $v^{-1}hv \in H^v = H$. Since $N_G(H)$ is the set of *all* g such that $H^g = H$, 'unique largest'-ness follows.

7 For $x, y \in T$, $H^x = H^y$ iff $H^{xy^{-1}} = H$, i.e. iff $xy^{-1} \in N_G(H) \cap T$.

8 $\bigcup_{g \in G} H^g$ is a union of $|G:N_G(H)| (\leqslant |G:H|)$ subsets, each of size $|H|$ and each containing e_G. So the union contains at most $|H| + (|G:H| - 1)(|H| - 1)$ elements.

9 Look at $g^{-1}(1234)g = (1234)$ where $g = \begin{pmatrix} 1234 \\ abcd \end{pmatrix}$. There are 4 such g.

10 $g \in \bigcap_{x \in G} C_G(x)$ means that g centralises every element x of G.

12 $\zeta(G) \neq \langle e \rangle$ (6.2.6). Hence there is $y \in \zeta(G)$ such that $|y| = p$. But $\langle y \rangle \lhd G$ (exercise 5.10.14). Set $G_{n-1} = \langle y \rangle$.

13 For S_4, if not S_3, exercise 5.5.6 might help. D_4 has classes $\{I\}, \{a^2\}$—and three others, each with two elements.

14 Let $x \in H \cap C_i$ $(x \neq e)$. Then $g^{-1}xg \in H \cap C_i$—since $H \lhd G$ and C_i is a complete conjugacy class. It follows that $H \supseteq C_i$. But $|C_i| = p^\alpha$ for some integer $\alpha \geq 0$ and H contains at least one class with just one element.

15 Let $x \in \zeta(G)\backslash S$. Then $\langle S, x\rangle = G$ (since $|\langle S, x\rangle| > |S|$) and so each element of G is of the form $x^i s$, where $s \in S$ (why?). Since $x^i s \cdot s_1 \cdot s^{-1} x^{-i} = s s_1 s^{-1} \in S$ we deduce that $S \lhd \langle S, x\rangle$. Note that exercise 12 says 'there exists'; exercise 15 asks you to show 'for all'.

16 Take $n=4$ in 5.3.4(l). (There is another example in Section 5.9.)

17 Let $t \in N_G(S)$, $x \in C_G(S)$ and $s \in S$. Prove that $s(t^{-1}xt) = t^{-1}(tst^{-1})xt = t^{-1}x(tst^{-1})t$ {why?} $= (t^{-1}xt)s$. It follows that $t^{-1}xt \in C_G(S)$. Let $t \in N_G(S)$. Then λ_t defined by $\lambda_t(s) = t^{-1}st$ gives an automorphism of S (cf. exercise 5.9.18). The map $\theta: N_G(S) \to \mathrm{Aut}(S)$ given by $t\theta = \lambda_t$ is a homomorphism with $\ker\theta = \{t : t^{-1}st = s \text{ for all } s \in S\}$.

18 Note that $|P| = |P^g|$ for each $g \in G$.

19 No!

20 If P is a Sylow p-subgroup in G then surely also in each T for which $P \le T \le G$. If P, Q are Sylow p-subgroups of $N_G(P)$ then $Q = P^t$ for some $t \in N_G(P)$ (by 6.2.12). But $P^t = P$.

21 Let $K = \{P = P_0, P_1, \ldots P_{kp}\}$ be the set of all Sylow p-subgroups. Put $P_i \sim P_j$ iff $P_i^s = P_j$ for some $s \in S$. As in 6.2.12 the conjugacy classes have $|S : S \cap N_G(P_i)|$ (as power of p) elements in them. Thus at least one P_i is in a class of its own. Consequently $S \le N_G(P_t)$ for some P_t. Now look at $P_t S$ to deduce $S \le P_t$. Hence $S \le P^g$ for some $g \in G$, by 6.2.12.

22 Suppose $x^{-1}N_G(P)x = N_G(P)$. Then $x^{-1}Px \le x^{-1}N_G(P)x = N_G(P)$. But $P = x^{-1}Px$ (by exercise 20). Therefore $x \in N_G(P)$.

23 $135 = 5 \cdot 3^3$. Look at the number of Sylow 3-subgroups. $30 = 2 \cdot 3 \cdot 5$. The number of Sylow 5-subgroups is 1 or 6. If 6, count the number of Sylow 3-subgroups. In the last case follow the proof of 6.2.16(iii).

24 One of the subgroups is example 5.3.4(m). (Recall that conjugate subgroups are isomorphic.) There is no contradiction (since $8|24$). [Do you see why I'm saying this?]

25 Write $hk = h \cdot h_1 k_1 = h^* k_1$ where $h_1 \in H \cap K$ and k_1 belongs to a set of coset representatives of K modulo $H \cap K$. Thus $\{hk : h \in H, k \in K\}$ contains at most $|H||K : H \cap K|$ distinct elements. (*Are* they pairwise distinct?)

Exercises 6.3

1 Generalize the solution to exercise 5.3.3.

3 (a) Yes. (The proof is easy.); (c) strictly, 'no' (n-tuples of matrices aren't matrices!), but if we allow the words 'isomorphic to' then the answer is 'yes'—for example, with $A \in GL_m(\mathbb{R})$, $B \in GL_n(\mathbb{R})$ associate $\begin{pmatrix} A & 0 \\ 0 & B \end{pmatrix}$ in $GL_{n+m}(\mathbb{R})$.

4 (a) No—the largest order is 30; (b) A_5 has no element of order 15.

5 (a) If $S_4 = A \times B$ with $|A| \le |B|$, then $|A| = 2$, 3 or 4. But then S_4 would have a non-trivial centre—contradicting exercise 5.6.17. (b) Does $C_2 \times D_3$ help? (c) Try to generalise (b). (e) No! Count the subgroups of order p.

6 Show that if C_r, C_s are generated by x, y respectively, then xy generates C_{rs}.

8 (a) Map $(a,(c,d))$ to (a,c,d).

9 *If* $g = h_1 \ldots h_i \ldots h_n = k_1 \ldots k_i \ldots k_n$ as given then, using the commutativity of H_i, H_j $(i \ne j)$, we get $k_i^{-1}h_i = k_1 \ldots \hat{k}_i \ldots k_n h_n^{-1} \ldots \widehat{h_i^{-1}} \ldots h_1^{-1} \in H_i \cap \langle H_1, \ldots, \hat{H}_i, \ldots, H_n \rangle = \langle e \rangle$. [Here $\hat{}$ denotes that the element below it is 'missing'.]

10 $\mathbb{Z}_{24} = \mathbb{Z}_3 \oplus \mathbb{Z}_8$ is one way. Any more?

11 I claim 16. Am I right?

12 No! Find an example inside $C_2 \times C_2$.

13 (a) Prove $\mathbb{Z}\oplus\mathbb{Z}$ is the internal direct sum $A \oplus B$ where $A = \langle(1,0)\rangle$, $B = \langle(1,1)\rangle$ are infinite cycles. Now $N \le B$. Hence $(\mathbb{Z}\oplus\mathbb{Z})/N \cong A \oplus (B/N) \cong \mathbb{Z}\oplus\mathbb{Z}_2$. (b) $\mathbb{Z}\oplus\mathbb{Z} = A \oplus B$ where $B = \langle(2,3)\rangle$ and $A = ??$

14 Map $a + ib (= re^{i\theta})$ to (r, θ). [This won't *quite* work—can you see how to amend the θ bit?]

15 $\ker \psi = \{(e, g_2) : g_2 \in G_2\} (\cong G_2)$.

16 $N = N_1 \times N_2$ is very easy. For the rest, take the map $G_1 \times G_2$ to $G_1/N_1 \times G_2/N_2$ given by $(g_1, g_2) \to (g_1N_1, g_2N_2)$. The kernel is $\{(n_1, n_2) : n_1 \in N_1, n_2 \in N_2\} = N$. In particular $N \lhd G_1 \times G_2$.

17 The kernel is $\{g \in G\colon gN_1 = N_1 \text{ and } gN_2 = N_2\} = \{g \in G\colon g \in N_1 \cap N_2\}$. If $N_1 \cap N_2 = \langle e \rangle$ the map is a 1–1 homomorphism into $G/N_1 \times G/N_2$.

18 $\zeta(A \times B) = \zeta(A) \times \zeta(B)$. The proof is straightforward.

19 G has $1 + kp$ Sylow p-subgroups, where $1 + kp \mid q^2$ (so that $1 + kp = 1$, q or q^2). Now q, q^2 are impossible since $p \nmid q^2 - 1$. Likewise for q. Thus G has unique and hence normal) Sylow r-subgroups for $r = p, q$ and so their direct product G is then abelian since the subgroups are (see 6.2.7).

20 Left to you. Note that, even if all the G_i are finite, G can still have elements of infinite order whereas $\bar{G}$ can't have.

21 and 22 $A = C_2 \times C_4 \times C_4 \times \ldots$; $B = C_4 \times C_4 \times C_4 \times \ldots$. These are not isomorphic. (Look for an element in A which has a property not possessed by any element of B. Try to phrase this formally.) Now let $\theta\colon A \to B$ map C_2 to $\langle e \rangle$ and the other C_4s in the obvious way; let $\psi\colon B \to A$ map the first C_4 to C_2 etc.

Exercises 6.4

1 If $|a|=p^\alpha$, $|b|=p^\beta$, then $p^{\alpha+\beta}(a-b)=0$.

2 Try $X=\langle(12)\rangle$, $Y=\langle(34)\rangle$, $Z=\langle(12)(34)\rangle$.

3 b) $218=2\times 109$. Hence $C_2\oplus C_{109}$ is the only possibility – and it's cyclic (why?). (c) The seven 'partitions' $1+1+1+1+1$, $1+1+1+2$, $1+2+2$ etc. of 5 lead to seven different isomorphism classes.

5 (a) To have just 3 elements of order 2 the 2-subgroup must be $C_2\oplus C_8$ or $C_4\oplus C_4$.

6 Since, for each odd integer a, $a^4\equiv 1 \pmod{16}$, $\langle M(16), \odot\rangle$ is certainly not cyclic. Hence $M(16)=C_2\times C_4$ or $C_2\times C_2\times C_2$. Which?

7 Example (which much better describes what happens than a formal explanation can!): If $A=C_9\oplus C_8\oplus C_5$ with generators x, y, z respectively, then one such series, in additive notation, is (with corresponding factors underneath):

$$A=\langle x, y, z\rangle > \langle 3x, y, z\rangle > \langle y, z\rangle > \langle 2y, z\rangle > \langle 4y, z\rangle > \langle z\rangle > \langle e\rangle.$$
$$C_3 \qquad C_3 \qquad C_2 \qquad C_2 \qquad C_2 \quad C_5$$

8 Clearly we can concentrate on the Sylow p-subgroups of A and B. Suppose $S_p(A)=C_1\oplus C_2\oplus\cdots\oplus C_n$. The elements with orders $\le p\ (\le p^2, \le p^3, \le\cdots)$ are precisely those in the subgroups $D_1\oplus D_2\oplus\cdots\oplus D_n$ $(E_1\oplus E_2\oplus\cdots\oplus E_n,\ldots)$ where each $D_i(E_i,\ldots)\le C_i$ is cyclic of order $p(p^2, p^3,\ldots)$. Thus if $S_p(B)=C'_1\oplus C'_2\oplus\ldots\oplus C'_n$ etc. one sees that $m=n$ and, more generally, that $S_p(A)$ has neither more nor less cyclic factors of order p^α than does $S_p(B)$.

9 Write A as a direct sum of its Sylow subgroups, S_p. Then each S_p is cyclic—or else S_p has more than p elements of order p.

11 A can be generated by a and $a-b$. Try $A=\langle a\rangle\oplus\langle a-b\rangle$ with a, a-b of orders 4 and 2 respectively. Does this work?

12 Again, an example to show the way. If A is the direct sum of cycles of orders $2, 2, 2, 2^2, 2^6$; $3^2, 3^3$; 5^2; $7^4, 7^5, 7^7$ choose the n_i 'backwards' as $n_5=2^6\cdot 3^3\cdot 5^2\cdot 7^7$, $n_4=2^2\cdot 3^2\cdot 7^5$, $n_3=2\cdot 7^4$, etc. For the last part see, for example, [13].)

13 Suppose $\langle F^*, \cdot\rangle$ is $S_p\times\cdots\times S_r$—the direct product of its Sylow subgroups—and that S_p, say, is not cyclic. Then S_p has more than p elements of order p, all satisfying $x^p=e$. This is impossible by 1.11.4.

14 Let $G=C_1\oplus\cdots\oplus C_r$, $H=D_1\oplus\cdots\oplus D_s$ where the C_i, D_j are prime power cycles. The pairing off of the direct summands *of $G\oplus G$ with those of $H\oplus H$* given by the uniqueness of the factors is easily seen to lead to a similar pairing of the factors of G and H.

Exercises 6.5

2 $[a, bc]=a^{-1}c^{-1}b^{-1}abc$: $[a, c][a, b]^c=a^{-1}c^{-1}acc^{-1}a^{-1}b^{-1}abc$.

4 $[(142), (235)]=?$

5 Use $g^{-1}[a, b]g=[g^{-1}ag, g^{-1}bg]$. Note that $a^{-1}(b^{-1}ab)\in A$; $(a^{-1}b^{-1}a)b\in B$.

6 Use $(a^{-1}b^{-1}ab)\theta=(a\theta)^{-1}(b\theta)^{-1}a\theta b\theta$ and $G'=\langle [a, b]: a\in A, b\in B\rangle$.

7 $H^{(n)}\leq G^{(n)}$ implies $H^{(n+1)}=[H^{(n)}, H^{(n)}]\leqslant[G^{(n)}, G^{(n)}]=G^{(n+1)}$. Show that $G^{(n)}\theta=(G\theta)^{(n)}$ implies $G^{(n+1)}\theta=[G^{(n)}\theta, G^{(n)}\theta]$, by induction.

8 (a) Assuming that $G^{(n)}=A^{(n)}\times B^{(n)}$ show that $G^{(n+1)}=\langle[(a, b), (u, v)]\rangle$ where $a, u\in A$ and $b, v\in B$. But $[(a, b), (u, v)]=([a, u], [b, v])$. (b) $G/(H\cap K)$ is isomorphic to a subgroup of $G/H\times G/K$ (by exercise 6.3.17).
Now use part (a) and exercise 7.

9 'If': Assume that each G_{i-1}/G_i is abelian. Then $G'_{i-1}\leq G_i$ (by 6.5.4). Now prove, by induction, that $G^{(i)}\leq G_i$ for each i and deduce that $G^{(k)}=\langle e\rangle$.

10 Taking $D_5=\langle b^ia^j: 0\leq i\leq 1, 0\leq j\leq 4\rangle$ we get $D_5>\langle a\rangle>\langle e\rangle$.

11 Take the (restricted) direct product of groups of increasing soluble length—see 6.5.6(iv).

12 Any such series $\mathbb{Z}=G_0>G_1>\cdots$ would have to have each G_i/G_{i+1} finite cyclic (since simple abelian—see exercise 5.8.7).

13 Take H, K of order 2 inside S_3. If $K\lhd G$ then $h_1k_1k_2^{-1}h_2^{-1}=h_1h_2^{-1}(h_2k_1k_2^{-1}h_2^{-1})\in HK$.

14 $C_{180}>C_{90}>C_{45}>C_{15}>C_5>\langle e\rangle$: $C_{180}>C_{60}>C_{12}>C_4>C_2>\langle e\rangle$ are two such.

15 Let $G/H>G_1/H>\cdots>G_k/H>H/H$ and $H>H_1>\cdots>H_l=\langle e\rangle$ be composition series for G/H and H respectively. Consider $G>G_1>\ldots>G_k>H>H_1>\ldots>H_l=\langle e\rangle$.

16 Such a sequence would imply the existence of a normal subgroup of order 2 or 3 in S_4. Contradiction.

17 Let $\theta: G\Rightarrow G/N$. Then $(G\theta)^{(i)}=G^{(i)}\theta=(G^{(i)}N)\theta=G^{(i)}N/N$. Since the length of G/N is m, we have $G^{(m)}N=N$, i.e. $G^{(m)}\leq N$. But then $N^{(n)}=\langle e\rangle$ implies that $G^{(m+n)}=\langle e\rangle$.

19 If $|G|=p_1^{\alpha_1}p_2^{\alpha_2}\ldots p_t^{\alpha_t}$, the p_i being primes then each composition series of G has $\alpha_1+\alpha_2+\cdots+\alpha_t$ terms—including G and $\langle e\rangle$. But $|H|<|G|$. (Or: let $G=G_0>G_1>\cdots>G_r=\langle e\rangle$ be a composition series for G. Show that the series $H=H_0\geqslant H\cap G_1\geqslant\cdots\geqslant H\cap G_r=\langle e\rangle$ becomes a composition series for H when the repeated terms are removed. Use the isomorphisms $H\cap G_i/H\cap G_{i+1}\cong H\cap G_i/(H\cap G_i)\cap G_{i+1}\cong(H\cap G_i)G_{i+1}/G_{i+1}$.)

20 C_3 and C_5 are Sylow $\{3, 5\}$ subgroups of A_5 since A_5 has no elements of orders $3^2, 5^2$ nor 15. Note that $5 \mid |A_5 : C_3|$ and $3 \mid |A_5 : C_5|$—so that neither index is a π'-number.

Exercises 6.6

1 Use exercise 5.8.7.

3 (Cf. exercise 6.3.19.) The number of Sylow q- (respectively p-) subgroups is of the form $1+kq$ (resp. $1+kp$) and divides p (resp. q). Hence there is one such (the latter since $p>q$ (!!)). Hence G is their direct product (why?). If $q \mid p-1$, then we still have $P \lhd G$ and G/P is abelian. Now use exercise 6.5.17.

4 [WLOG is forbidden since p^2q is not symmetrical in p and q.] Now G has $N=1+kq$ $(M=1+kp)$ Sylow q- (p-) subgroups where $N=1$ or p or p^2, $(M=1$ or $q)$. Thus if $M=1$ then there exists $P \lhd G$ such that $|P|=p^2$ so that G is soluble. If $M=q$ then $N \neq p$. Thus $N=1$ (in which case argue as in the case $P \lhd G$) or $N=p^2$. But then p^2 distinct q cycles leave just p^2-1 elements to form, with e, a single subgroup of order p^2.

5 All groups with orders less than 60 have orders of the form p^n, pq, p^2q, p^2q^2 except for 24, 30, 40, 42, 48, 54, 56, Orders 42, 48 and 56 are dealt with in 6.2.16. Now, if $|G|=24$, then G has a subgroup of index 3 and so there exists (exercise 5.10.19) a non-trivial homomorphism into S_3. If $|G|=40$ then G has $1+5k$ Sylow 5-subgroups.

6 If A_5 and N are distinct normal subgroups of index 2 in S_5, then $A_5 \cap N$ is a normal subgroup of index 2 in A_5. D_4 has two such subgroups.

7 Each (finite) even permutation on $\mathbb{Z}^+$ may be regarded as being in A_n for some n so that $P = \bigcup_{n=1}^{\infty} A_n$, with $A_r < A_s$ if $r<s$. If $N \lhd P$, then $N \cap A_n$ is normal in each A_n in particular for $n \geq 5$. If $N > \langle e \rangle$, then $N \cap A_t > \langle e \rangle$ for all $t >$ some t_0. Then, for $s > \max\{t_0, 5\}$ we have $\langle e \rangle < N \cap A_s \lhd A_s$ so that $N \geq A_s$. Hence $N \geq P$; i.e. P is simple.

8 Show that $G < S_n < A_{n+2}$ for some n. (See exercise 5.9.20.)

10 Let H be a group with composition series of length 35. Take G as a suitable A_n (see exercise 8).

11 Each group of order $n<60$ is soluble (exercise 5) and so has a composition series with prime cyclic factors. For $n=60$ compare the composition series of C_{60} and A_5.

In solutions to Chapter 7 we may sometimes assume a given polynomial is monic without expressly saying so.

Exercises 7.2

1 The method reduces $x^{11}-1$ to a quintic which may (or may not) be soluble by radicals.

2 $\alpha=\sqrt{(1+\sqrt{2})}\Rightarrow\alpha^2-1=\sqrt{2}$, so $f(x)=x^4-2x^2-1$ is the minimum polynomial of α. This has two non-real zeros. But $E_2\subset\mathbb{R}$.

4 (ii) x^4-10x^2+1 has zeros $\pm\sqrt{(5\pm\sqrt{24})}$. $\mathbb{Q}\subset\mathbb{Q}(\sqrt{24})\subset\mathbb{Q}(\sqrt{(5+\sqrt{24})})=S_f$ gives the required splitting field and radical tower. (Why is $\sqrt{(5-\sqrt{24})}$ in $\mathbb{Q}(\sqrt{(5+\sqrt{24})})$?). (iii) $\sin 72°=\sqrt{\dfrac{5+\sqrt{5}}{8}}=\alpha$, say. Then $\mathbb{Q}(\mathrm{i}\alpha)$ contains $-\dfrac{5+\sqrt{5}}{8}$ and hence $\dfrac{\sqrt{5}-1}{4}=\cos 72°$. Hence $S_f=\mathbb{Q}(\mathrm{i}\alpha)\supset\mathbb{Q}(\sqrt{5})\supset\mathbb{Q}$. (Why is S_f *equal* to $\mathbb{Q}(\mathrm{i}\alpha)$?). (iv) Use exercise 4.6.7 to reduce the problem to a cubic and the methods of Section 5.2 to solve the cubic. Then continue as in (vii)!!. (vi) the zeros are $\cos\theta+\mathrm{i}\sin\theta$ where $\theta=t\pi/6$, $t=1, 3, 5, 7, 9, 11$. Thus i is a zero of x^6+1 and so $S_f=\mathbb{Q}\left(\mathrm{i},\dfrac{\sqrt{3}}{2},\dfrac{1}{2}\right)=\mathbb{Q}(\mathrm{i},\sqrt{3})$. (vii) Determine $\alpha=\sqrt[3]{\left(\dfrac{1}{2}-\mathrm{W}\dfrac{31}{108}\right)}$ as in Section 5.2. Then $S_f=\mathbb{Q}(\alpha+\beta,\omega\alpha+\omega^2\beta,\omega^2\alpha+\omega\beta)\subseteq\mathbb{Q}\left(\sqrt{\dfrac{31}{108}},\alpha,\omega\right)$. (Note that $\beta=-1/3\alpha$.)

5 Calculate $(x+\sqrt{2}+\sqrt{3})(x+\sqrt{2}-\sqrt{3})(x-\sqrt{2}+\sqrt{3})(x-\sqrt{2}-\sqrt{3})$ (for a surprise!). Show that $\sqrt{2},\sqrt{3}\in\mathbb{Q}(\sqrt{2}+\sqrt{3})$ (why?) so that $\mathbb{Q}(\sqrt{2},\sqrt{3})=\mathbb{Q}(\sqrt{2}+\sqrt{3})$. Note also that $\pm\sqrt{2}\pm\sqrt{3}\in\mathbb{Q}(\sqrt{2}+\sqrt{3})$ (why?). [In particular $\mathbb{Q}(\sqrt{2}+\sqrt{3})=\mathbb{Q}(\sqrt{(5+\sqrt{24})})$ - cf. exercise 4]. For (ii) try replacing x by $x+1$.

7 Use $\mathbb{R}\subset\mathbb{C}$ and 4.8.1. For the next part see exercises 7.9.

9 (a) Use exercise 4.5.5; (b) $x^3-\hat{3}$ is irreducible over $\mathbb{Z}_7$. Let α be a zero in $GF(7^3)$ (see pp. 166, 168). Then, in fact, $x^3-\hat{3}=(x-\alpha)(x-2\alpha)(x-4\alpha)$ in $GF(7^3)[x]$.

11 Each element of S is a ratio of polynomials in the α_i with coefficients in F. If the α_i are the zeros of $f(x)=f_0+f_1x+\cdots+f_mx^m$ then, clearly, so too are the $\sigma(\alpha_i)$. Hence, if $S=S_f$, each $\sigma(\alpha_i)$ is one of the α_j. Thus σ permutes the α_i.

12 (a) In 7.2.1 $[E_i:E_{i-1}]\leq n_i$ (cf. 4.3.4). Hence $[R:F]\leq n_1n_2\ldots n_s$ is finite. $\mathrm{Gal}(R/F)$ is finite since, with $R=F(r_1,r_2,\ldots,r_s)$, each r_i must map to a zero of its minimum polynomial in $F[x]$. (b) By the proof of 4.2.1, $f(x)$ factorises as a product of a linear factor and a polynomial $f_1(x)$, say, of degree $n-1$ in $E[x]$, where $[E:F]\leq n$. Now use induction on n.

14 Let $\sigma\in\mathrm{Gal}(S_f/F)$ and α be a zero of $f_i(x)$. Then $\sigma(\alpha)$ is zero of $f_i(x)$. If $\sigma(\alpha)$ is also a zero of $f_j(x)$ then $x-\alpha\,|\,\gcd(f_i(x),f_j(x))$ (why not $\sigma(\alpha)$?) in $S_f[x]$. But $1=u(x)f_i(x)+v(x)f_j(x)$ for suitable $u(x), v(x)\in F[x]$. (Why?) Substituting α gives $1=u(\alpha)\cdot 0+v(\alpha)\cdot 0$.

16 The modification is needed since $F(\gamma_1)$, $F(\gamma_2)$ need not be *equal*—only isomorphic.

17 (a) Naïvely, just 'change each 1 to a 2'. A more highbrow way to establish Λ is to map $F_1[x] \xrightarrow{\bar{\lambda}} F_2[x] \xrightarrow{\rho} F_2[x]/[g_2(x)]$ and show that the kernel of $\bar{\lambda}\circ\rho$ is the ideal $[g_1(x)]$.
(b) (Loosely) Any factorisation of $g_2(x)$ in $F_2[x]$ can be 'dragged back' via $\bar{\lambda}^{-1}$ to an equivalent factorisation of $g_1(x)$ in $F_1[x]$.

18 $g_2(x)=x_2+\sqrt{2}$ so $S_2=\mathbb{Q}(i\sqrt[4]{2})$. μ sends $\sqrt[4]{2}$ to $\pm i\sqrt[4]{2}$.

19 (a) The other zeros (in $\mathbb{C}$) of the minimum polynomial of $\sqrt{(1+\sqrt{2})}$ are $\sqrt{(1-\sqrt{2})}$, $-\sqrt{(1+\sqrt{2})}$, $-\sqrt{(1-\sqrt{2})}$. Thus a non-identity automorphism of $\mathbb{Q}(\sqrt{(1+\sqrt{2})})$ must take $\sqrt{(1+\sqrt{2})}$ to $-\sqrt{(1+\sqrt{2})}$. This then implies that $\sqrt{2}$ is sent to $\sqrt{2}$. So answers are: No; None!

20 Follow the hint—or show that a basis for L over K is mapped, by σ, to a set of elements of L which are independent over K. Thus the dimension of $\sigma(L)$ over K is at least as great as that of L over K. Hence $\sigma(L)\subsetneqq L$ is impossible.

Exercises 7.3

1 Try to get the answer by examining $\mathbb{Q}(\sqrt[3]{2})$ and $\mathbb{Q}(\sqrt[4]{2})$.

2 $\mathbb{Q}(r)$ $(\subset\mathbb{C})$ contains all the zeros $r, r^2, \ldots, r^{p-1}$ of the polynomial $f(x) = x^{p-1}+\cdots+x+1$, which is irreducible over $\mathbb{Q}$. Use 7.2.3′ with $F_1=F_2=\mathbb{Q}$ and $f_1(x)=f_2(x)=g_1(x)=g_2(x)=f(x)$.

3 Since $-\omega\neq\omega, \omega^2$ the answer is ...?

4 The answer to one part is 'yes' and to the other 'no'. Which is which?

5 Informally!: (i) Identity and $i\sqrt{3}\to -i\sqrt{3}$; (iii) 8 of them. Briefly: $i\to\pm i$; $\sqrt[4]{2}\to\pm\sqrt[4]{2}$ or $\pm i\sqrt[4]{2}$. (v) The minimum polynomial of $i\sqrt[4]{3}$ over $\mathbb{Q}(\sqrt{3})$ is $x^2+\sqrt{3}$. Hence there are two automorphisms (one being complex conjugation). (vii) Note that $\mathbb{Q}(i\sqrt{3})=\mathbb{Q}(\omega)$ $\left(\omega=\dfrac{-1+i\sqrt{3}}{2}\right)$. Hence there are three automorphisms given by: $\sqrt[3]{5}\to\alpha\sqrt[3]{5}$ $(\alpha=1,\omega,\omega^2)$.

6 (a) There are two ways of extending $\lambda:\mathbb{Q}\to\mathbb{Q}$ to an automorphism μ of $\mathbb{Q}(\sqrt{2})$ to $\mathbb{Q}(\sqrt{2})$ (viz. $\sqrt{2}\to\pm\sqrt{2}$) and then two ways of extending each such μ to an automorphism ν of $\mathbb{Q}(\sqrt{2},\sqrt{3})$. E.g. if $\mu(\sqrt{2})=-\sqrt{2}$, then ν sends $(\sqrt{2},\sqrt{3})$ to $(-\sqrt{2},\sqrt{3})$ or $(-\sqrt{2},-\sqrt{3})$.
(b) There are 8 automorphisms of $\mathbb{Q}(\sqrt{2},\sqrt{3},\sqrt{5})$ [but note, not of $\mathbb{Q}(\sqrt{2},\sqrt{3},\sqrt{6})$ so take care!].

7 (i) Looks as though the answer *must* be $3\times3\times2$.

8 Calling the zeros r, ωr, $\omega^2 r$ one finds that, for example, the map given by $\omega \to \omega^2$, $r \to \omega r$ yields the permutation $\begin{pmatrix} r & \omega r & \omega^2 r \\ \omega r & r & \omega^2 r \end{pmatrix}$.

9 (ii) 6 (cf. x^3-2); (iii) 8 ($\alpha \to \alpha$, $i\alpha$, $-\alpha$, $-i\alpha$: $i \to i$, $-i$ where $\alpha = \sqrt[4]{3}$).

10 (i) 4. (The zeros of x^4+1 are ξ, ξ^3, ξ^5, ξ^7, where ξ is a primitive eighth root of 1.) Since x^4+1 is irreducible over $\mathbb{Q}$ (exercise 1.9.11), ξ can map to any ξ^j in S_f. The image of ξ determines the rest. (iii) x^3+2x-1 is irreducible over $\mathbb{Q}$ and has only one real zero a, say. (Why?). So we have $\mathbb{Q}(a, \beta, \bar{\beta}) = S_f$. By 7.2.3′ there exists an automorphism σ, say, mapping a to β. There is also an automorphism τ, say, of order 2 due to complex conjugation. See which permutations of the three zeros arise, recalling that $\text{Gal}(S_f/\mathbb{Q}) \leq S_3$ since $f(x)$ has degree three.

12 $\mathbb{Q} \subset \mathbb{Q}(\zeta) \subset \mathbb{Q}(\zeta, \sqrt[p]{2})$ where ζ ($\neq 1$) is a primitive pth root of unity and $\sqrt[p]{2}$ is the (unique) real pth root of 2. $|\text{Gal}(\mathbb{Q}(\zeta)/\mathbb{Q})| = p-1$ and x^p-2 is irreducible over $\mathbb{Q}(\zeta)$. Hence, to each automorphism μ of $\mathbb{Q}(\zeta)$ there are p automorphisms ν of $\mathbb{Q}(\zeta, \sqrt[p]{2})$ extending μ. Thus $|\text{Gal}(S_f/\mathbb{Q})| = p(p-1)$.

13 Take $R = \mathbb{Q}(\sqrt{2}, \sqrt[3]{5}) \supset \mathbb{Q}(\sqrt{2}) \supset \mathbb{Q}$. So $[R:\mathbb{Q}] = 6 \neq |\text{Gal}(R/\mathbb{Q})|$. For (i), note that $S_f = \mathbb{Q}(\sqrt{2}, \sqrt[3]{5}, \omega)$. Hence $|\text{Gal}(S_f/\mathbb{Q}(\sqrt{2}))| = 6$.

14 The minimum polynomial is x^4-2x^2-1 (see exercise 7.2.2.). It is irreducible over $\mathbb{Q}$ and splits over $\mathbb{Q}(\sqrt{(\sqrt{2}+1)}, i\sqrt{(\sqrt{2}-1)})$ $[= \mathbb{Q}(\sqrt{(\sqrt{2}+1)}, i)$—why?]. Hence $|\text{Gal}(S_f/\mathbb{Q})| = 8$.

15 $\text{Cos}\, 72° = \dfrac{\sqrt{5}-1}{4}$. So $\mathbb{Q}(\cos 72°) = \mathbb{Q}(?)$?

Exercises 7.4

1 $x^2 - 2ex + (e^2 + \pi^2)$. $|\text{Gal}(S_f/\mathbb{R})| = 2$—since $S_f = \ldots$?

2 (ii) Let $F = \text{Fix}\,\Sigma$. If $x \in G$ then $x = \sigma_1^{\pm 1}\sigma_2^{\pm 1} \ldots \sigma_k^{\pm 1}$ where the $\sigma_i \in \Sigma$. But each σ_i and σ_i^{-1} belongs to $\text{Aut}(L)$ and fixes F elementwise. So then, rather trivially, does x. Thus $\text{Fix}\, G \supseteq \text{Fix}\,\Sigma$.

4 (i) $\mathbb{Q}(\sqrt{2})$; (iii) The answer is *not* $\mathbb{Q}(\sqrt[3]{2})$ (why not?)

5 The given field is the splitting field over $\mathbb{Q}$ of $(x^3-2)(x^4-3)$. Use 7.4.8.

6 $\mathbb{Q}$, $\mathbb{Q}(\sqrt{2})$, $\mathbb{Q}(\sqrt{3})$, $\mathbb{Q}(\sqrt{2}, \sqrt{3})$. Any more? What about $\mathbb{Q}(\sqrt{2}+\sqrt{3})$?

7 (i) $\mathbb{Q}(x^2)$; (ii) $\mathbb{Q}$; (iii) $\mathbb{Q}$.

8 If $\alpha \in \text{Fix}(\langle \cup G_i \rangle)$ then, certainly $\alpha \in \text{Fix}\, G_i$ for each i. Hence $\alpha \in \cap(\text{Fix}\, G_i)$. The converse is just as easy. For the other part try $G_1 = \langle \sigma \rangle$, $G_2 = \langle \tau \rangle$ where σ and τ are as in exercise 7.

9 $\sigma^2(\xi)=\xi^{25}$. Hence σ^2 is the identity map on S_f. Clearly σ fixes ξ^2 and hence also ....

10 See exercises 7.9.

11 (iii) Let $\alpha=a_0+a_1r+\cdots+a_{p-2}r^{p-2}$ be a fixed element. If σ_i is the automorphism given by $\sigma_i(r)=r^i$, then $\sigma_i(\alpha)$ has coefficient of r^i equal to a_1 (since $1, r, \ldots, r^{p-2}$ form a ...? for $\mathbb{Q}(r)$ over $\mathbb{Q}$). Hence

$$\alpha=a_0+a_1(r+\cdots+r^{p-2})=a_0-a_1 \text{ (why?)} \in \mathbb{Q}.$$

(To use 7.48 you must show x^p-1 separable over $\mathbb{Q}$.)

12 (vi) $R=\mathbb{Q}(\sqrt[8]{2})$ is a real field and so only two of the zeros of x^8-2 are in R. Clearly, then, $\text{Fix}(\text{Gal}(R/F))=\mathbb{Q}(\sqrt[4]{2})$.

13 (v) For $|\text{Gal}(R/F)|=6$ see exercise 7.3.6: for $[R:F]=6$ use the basis 1, ω, ω^2, r, ωr, $\omega^2 r$.

14 (i) No; (ii) $|\text{Gal}(R/\mathbb{Q})|=4$.

15 By definition, each element of $\text{Gal}(L/K)$ is an automorphism of L which extends the identity map on K. Then 7.4.3, with $K_1=K_2=K$ and $L_1=L_2=L$, suffices.

16 This is straightforward (once each $f(x)$ is proved separable) since, for example, in (iv) $1, \sqrt{2}, \sqrt{3}, \sqrt{6}$ is a basis for $\mathbb{Q}(\sqrt{2}, \sqrt{3})=S_f$ over $\mathbb{Q}(=F)$.

18 Let $g(x)$ be an irreducible factor of $f(x)=x^k+\cdots+x+1$. Since $f(x)$ is of degree k and has k distinct zeros in its splitting field $S_f=\mathbb{Q}(e^{2\pi i/(k+1)})$, we deduce that $g(x)$ can have no repeated zeros in *its* splitting field $S_g\subseteq S_f$. (This is much improved in 7.5.1!)

19 Note that $F\subset E$, $S_f(F)\subseteq S_f(E)$. Let $g_E(x)$ be an irreducible factor of $f(x)$ *in* $E[x]$ and let $g_E(x)=(x-\alpha_1)\ldots(x-\alpha_m)$ with the α_i in $S_f(F)$. Let $g_F(x)$ be that irreducible factor of $f(x)$, *in* $F[x]$, which has α_1 as a zero in $S_f(F)$. Then $g_E(x)|g_F(x)$ in $E[x]$. (Otherwise $1=r(x)g_E(x)+s(x)g_F(x)$ for suitable $r(x)$, $s(x)$ $\in E[x]$—and then $1=0$ in $S_f(E)$. [Cf. exercise 7.2.14.]) Thus $g_E(x)|g_F(x)$ in $S_f(E)[x]$. But by hypothesis, $g_F(x)=(x-\beta_1)\ldots(x-\beta_n)$ with distinct β_j from $S_f(E)$. Thus the α_i are some of the β_j and so are pairwise distinct in $S_f(E)$.

22 If α is a zero of x^p-t in S_f, then $\alpha^p=t$. Thus, in $S_f[x]$, $x^p-t=x^p-\alpha^p$ $=(x-\alpha)^p$.

Exercises 7.5

1 If $f(x)=(g_1(x))^{\alpha_1}\ldots(g_m(x))^{\alpha_m}$ in $F[x]$ with the $g_i(x)$ being distinct and irreducible, then $f(x)/d(x)=g_1(x)\ldots g_m(x)$.

3 (The first part becomes trivial if one uses exercise 9.) Let $g(x)\in L[x]$ be irreducible and let α be a zero of $g(x)$ in M. Let $M_\alpha(x)$ be the minimum polynomial of α *over K*. Show $g(x)|M_\alpha(x)$ in $L[x]$ (cf. exercise 7.4.19). But $M_\alpha(x)$ factors into linear factors in $M[x]$. (Why?) Hence so does $g(x)$. For the last part let $h(x)$ be irreducible in $K[x]$ and have a zero, α say, in L. Since $h(x)$ splits into linear factors in $M[x]$ we only have to show each zero β, say, lies in L. But there exists $\sigma \in \mathrm{Gal}(M/K)$ such that $\sigma(\alpha)=\beta$ (why?). Hence $\beta \in L$ (why?)

5 $F \supseteq K$ so $[L:F]<\infty$. If $[L:F]>1$ let $g(x)$ be the minimum polynomial of $\alpha \in L \backslash F$. Then consider, as in 7.5.2, $h(x)=\prod_\sigma (x-\sigma(\alpha))$ as σ ranges over $\mathrm{Gal}(L/K)$. Continue—as in 7.5.2!—deducing that $g(x)|h(x)$ etc. (see 7.6.3).

6 Clearly L is separable over K. To show M is separable over L let $\alpha \in M$ and let $g(x)$ (resp. $h(x)$) be its minimum polynomial over L (resp. K). Show $g(x)|h(x)$ in $L[x]$ (cf. exercise 7.4.19). Let S_L be a splitting field for $h(x)$ regarded as an element of $L[x]$. Then S_L contains S_K—a splitting field for $h(x)$ regarded as an element of $K[x]$. In $S_K[x]$, $h(x)=(x-\alpha_1)\dots(x-\alpha_n)$ with distinct α_i. How does $g(x)$ factorise in $S_L[x]$?

7 That Artin's definition implies ours follows from exercise 5. Conversely take S to be the splitting field of x^2-t over $\mathbb{Z}_2(t)$. The field S is normal in our sense over $\mathbb{Z}_2(t)$ {look at $[S:\mathbb{Z}_2(t)]$}, but not separable over $\mathbb{Z}_2(t)$ (exercise 7.4.22). Indeed $\mathrm{Gal}(S/\mathbb{Z}_2(t))=\langle e\rangle$ and so Fix ...?

8 (ii)$\rightarrow$(iii) is just 7.4.7; (iii)$\rightarrow$(iv): If $\mathrm{Fix}(\mathrm{Gal}(L/K))=F \supsetneq K$, then $[L:K] =|\mathrm{Gal}(L/K)|=|\mathrm{Gal}(L/F)|$ (why?) $\leq [L:F]$ (by 7.4.4). But $[L:K]<[L:F]$ is impossible; (iv)$\rightarrow$(ii) is essentially 7.5.2. For: $|\mathrm{Gal}(L/K)|<\infty$ (as in exercise 7.2.12). Now let $g(x)$ be any polynomial in $K[x]$ with a zero β_1 in L. Forming $h(x)=\prod(x-\beta_i)$ as in 7.5.2, show that $g(x)$ is a product of distinct linear factors in $L[x]$.

10 Clearly $[L_1 \cap L_2 : K]<\infty$. Let $g(x) \in K[x]$ be irreducible and have a zero in $L_1 \cap L_2$. Then $g(x)=(x-\alpha_1)..(x-\alpha_n)=(x-\beta_1)\dots(x-\beta_n)$ for $\alpha_i \in L_1$, $\beta_j \in L_2$. Thus the α_i belong to $L_1 \cap L_2$. For $L_1 \cup L_2$: by exercise 9, L_1 and L_2 are splitting fields of $f_1(x), f_2(x)$, say, over K. Hence $\langle L_1 \cup L_2\rangle$ is a splitting field over K of $f_1(x)f_2(x)$.

11 The hypotheses imply that L is the splitting field over K of $M_{\alpha_1}(x)\dots M_{\alpha_n}(x)$. Now use exercise 9.

12 A non-separable splitting field? That of x^2-t over $\mathbb{Z}_2(t)$. For the other part, there are three examples in 7.3.1. (See 7.4.2 if stuck.)

13 (a) Even in $K[x]$, if $f(x)$ (of degree ≥ 1) has a repeated zero in its splitting field, then $(f(x), f'(x)) \neq 1$. This is impossible (why?) unless $f'(x)=0$, that is, unless $f(x)$ has the given form. Conversely: Suppose $f(x)$ is as given. If $f(x)=(x-\alpha)g(x)$ then $0=f'(x)=g(x)+(x-\alpha)g'(x)$. Hence $x-\alpha | g(x)$. (b) If $|K|=p^t$ then, in (a) each $a_i=(a_i^{p^{t-1}})^p$. But

$$b_0^p+b_1^p x^p+\cdots+b_n^p x^{np}=(b_0+b_1x+\cdots+b_nx^n)^p$$

has repeated factors.

14 $\phi(a+b)=(a+b)^p=a^p+b^p$ etc. Hence ϕ is a homomorphism. Trivially ϕ is 1–1 (why?) and hence onto if K is finite.

Exercises 7.6

1 $G(F(H))=\mathrm{Gal}(L/\mathrm{Fix}\,H)\geq H$ is immediate. Hence $F(G(F(H)))\subseteq F(H)$. But $F(G(M))=\mathrm{Fix}(\mathrm{Gal}(L/M))\supseteq M$ is also immediate. Choosing $M=F(H)$ we get . . . ?

2 Write H for $\mathrm{Gal}(L/K)$—which is finite. Then, by 7.6.3, L is a Galois extension of $\mathrm{Fix}\,H=F$. 7.6.4 then gives the required map.

4 There must be such a k by exercise 3. In the second part k can be chosen to be one of 1, $\sqrt{2}$, $\sqrt{3}$, $\sqrt{6}$ (why?)

5 The k_i may be chosen as non-trivial solutions to the system of m equations in n unknowns (with $m<n$).

6 Trivially $(H=)\mathrm{Gal}(K/\langle E_1\cup E_2\rangle)\leq \mathrm{Gal}(K/E_1)\cap\mathrm{Gal}(K/E_2)=H_1\cap H_2$. If $H\lneqq H_1\cap H_2\lneqq H_1, H_2$ then, by 7.6.4, there would be a field E such that $\langle E_1\cup E_2\rangle\supsetneqq E\supsetneqq E_1, E_2$, which is impossible.

7 Since $K=\mathrm{Fix}\,G=\mathrm{Fix}\langle G,\psi\rangle$ we have, by 7.6.3, $|G|=|L:\mathrm{Fix}\,G]$ $=[L:\mathrm{Fix}\langle G,\psi\rangle]=|\langle G,\psi\rangle|$. Hence $\psi\in G$.

9 The 'lower' (i.e. smaller) field is $\mathbb{Q}(ir^2)$; the 'upper' (i.e. smaller) group is $\{\sigma_1,\sigma_3\}$.

10 I'll leave the pictures to you! That for (iii) is pretty beastly! The subgroups joined to $\langle e\rangle$ are $\langle\sigma^2\rangle$, $\langle\sigma^3\rangle$, $\langle\sigma^i\tau\rangle$ $(i=0, 1, 2, 3, 4, 5)$. Those joined to $\mathrm{Gal}(S_f/\mathbb{Q})$ are $\langle\sigma\rangle$, $\langle\sigma^2,\tau\rangle$, $\langle\sigma^3,\tau\rangle$, $\langle\sigma^2,\sigma\tau\rangle$, $\langle\sigma^3,\sigma\tau\rangle$, $\langle\sigma^3,\sigma^2\tau\rangle$, where, briefly, $\sigma(\sqrt[6]{2}\to\varepsilon\sqrt[6]{2}:\varepsilon\to\varepsilon)$ and $\tau(\sqrt[6]{2}\to\sqrt[6]{2}:\varepsilon\to\varepsilon^5)$, ε being a primitive sixth root of 1—so that $\tau^{-1}\sigma\tau=\sigma^{-1}$.

11 (i) is very like 7.6.5; (ii) the zeros are $\pm r$, $\pm ir$ where $r=\sqrt{2}$. Working as in 7.6.5 shows that $\mathrm{Gal}(\mathbb{Q}(i,\sqrt{2})/\mathbb{Q})=C_2\times C_2$. The lattices look like

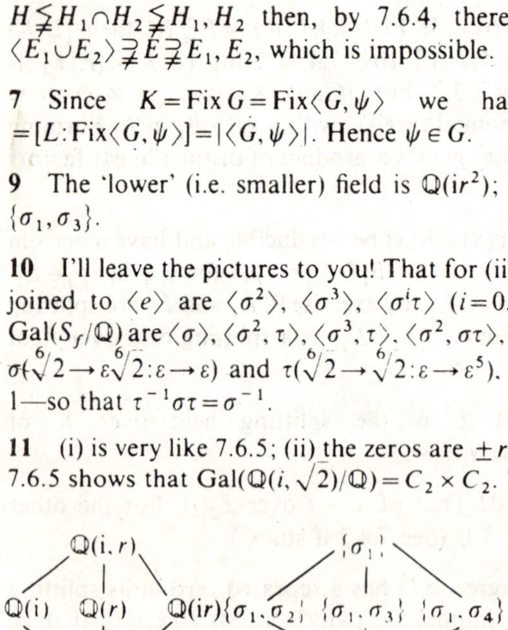

where

	σ_1	σ_2	σ_3	σ_4
r	r	$-r$	r	$-r$
i	i	i	$-$i	$-$i

12 S_f is also the splitting field of $(x^2+2x-1)(x^3-2)$. Hence $S_f=S_g(\sqrt[3]{2})$ where S_g is the splitting field of $g(x)=x^2+2x-1$. Thus $\mathrm{Gal}(S_f/\mathbb{Q})$ is $C_2\times S_3$.

13 $x^4+\hat{2}=(x+\hat{1})(x+\hat{2})(x^2+\hat{1})$ in $\mathbb{Z}_3$. So $GF(9)$ is a splitting field for $f(x)$. Since $GF(9)$ is a Galois extension of degree 2 over $GF(3)$ we find that $\mathrm{Gal}(S_f/\mathbb{Z}_3)$ has order 2.

15 As many as there are subgroups of $C_2 \times C_2 \times C_2$! (Why?)

16 $S_f=\mathbb{Q}(\zeta)$ where $\zeta=e^{2\pi i/8}$ has minimum polynomial x^4+1 with zeros ζ, ζ^3, ζ^5, ζ^7. Define σ, τ by $\sigma(\zeta)=\zeta^3$, $\tau(\zeta)=\zeta^5$. Show that $\mathrm{Gal}(S_f/\mathbb{Q})=$ Id, $\sigma, \tau, \sigma\tau\}$ Since $\sigma(a+b\zeta+c\zeta^2+d\zeta^3)=a+b\zeta^3-c\zeta^2+d\zeta$ we see that $\mathrm{Fix}\{\sigma\}=\{a+b(\zeta+\zeta^3):a,b\in\mathbb{Q}\}$ Find $\mathrm{Fix}\{\tau\}$, $\mathrm{Fix}\{\sigma\tau\}$ similarly.

17 The statement on fields follows from the fundamental 1–1 correspondence and the analogous results concerning cyclic groups (see 5.8.2(i)).

Exercises 7.7

1 Surely $[\theta(S_f):\theta(F)]=[S_f:F]\geq[\theta(S_f):F]$.

2 $n=9$.

3 $\mathrm{Gal}(L/K)$ is trivial, since x^5-5 has only one zero in L. Thus $\mathrm{Gal}(K/\mathbb{Q})$ $=S_3=\mathrm{Gal}(L/\mathbb{Q})$ and $\mathrm{Fix}(\mathrm{Gal}(L/\mathbb{Q}))=\mathbb{Q}(\sqrt[5]{5})\supset\mathbb{Q}=\mathrm{Fix}(\mathrm{Gal}(K/\mathbb{Q}))$.

4 The minimum polynomial of $\alpha=\sqrt[3]{(3+\sqrt{2})}$ over $\mathbb{Q}$ is x^6-6x^3+7 (see exercise 7.3.16). It has one other real zero, $\beta=\sqrt[3]{(3-\sqrt{2})}$, and four non-real ones. But $\beta\notin\mathbb{Q}(\alpha)$. Hence $\mathrm{Gal}(\mathbb{Q}(\alpha)/\mathbb{Q})=\langle e\rangle$. Note that $\mathrm{Gal}(\mathbb{Q}(\sqrt{2})/\mathbb{Q})=C_2$.

5 T is the least field containing V and $S_f=F(\alpha_1,\ldots,\alpha_n)$. Since $V\supseteq F$, $T=V(\alpha_1,\ldots,\alpha_n)$. If $\alpha\in T\backslash V$ then there exists $\theta\in\mathrm{Gal}(R/F)=\mathrm{Gal}(R/V)$ such that $\theta(\alpha)\neq\alpha$. But θ acts as an automorphism $\bar{\theta}$ (say) on T (since $\theta(V)\subseteq V$ and $\theta(S_f)\subseteq S_f$) and so $\bar{\theta}(\alpha)\neq\alpha$.

6 $\mathbb{Q}(\sqrt{2},\sqrt{3})$, $\mathbb{Q}(\sqrt{2},\sqrt{3},\sqrt{5})$, etc.

7 (This generalises 7.7.3 slightly.) M is a splitting extension of K (exercise 7.5.9) and hence of L and $\sigma(L)$. Now apply 7.2.4.

8 Let $\alpha=\sqrt{(1+\sqrt{2})}$. Then $M_\alpha(x)=x^4-2x^2-1$ which has zeros $\pm\sqrt{(1+\sqrt{2})}$, $\pm\sqrt{(1-\sqrt{2})}$. But $\sqrt{(1-\sqrt{2})}\in\mathbb{Q}(\sqrt{(1+\sqrt{2})},i)$ (look at $(\sqrt{(1+\sqrt{2})})^{-1}$). On the contrary $\sqrt{(1-\sqrt{3})}\notin\mathbb{Q}(\sqrt{(1+\sqrt{3})},i)$.

9 L is a splitting extension of K (exercise 7.5.9) and so, as in 7.7.2, each $\theta\in\mathrm{Gal}(M/K)$ restricts to an element $\bar{\theta}$ of $\mathrm{Gal}(L/K)$. $\theta\to\bar{\theta}$ provides a homomorphism of $\mathrm{Gal}(M/K)$ into $\mathrm{Gal}(L/K)$ with kernel the set of those automorphisms of M which fix L.

10 By definition, $m(x)$ splits in $S_m=R(u_1,\ldots,u_t)=F(r_1,\ldots,r_s,u_1,\ldots,u_t)$. Since $r_1,\ldots,r_s,u_1,\ldots,u_t$ are all zeros of $m(x)$ in S_m, we can see that S_m is a splitting field for $m(x)$ *over* F.

Exercises 7.8

1 The minimum polynomial $N_{r_i}(x)$, say, divides $x^{p_i}-a_i$ and $M_{r_i}(x)$ and has degree >1 (why?). Hence two of its zeros are $\zeta^j\alpha$, $\zeta^k\alpha$, where ζ is a primitive p_ith root of unity, α is a p_ith root of a_i and $j\neq k$. This shows that $\zeta\in S_M$.

2 $x^{12}-1=(x^6-1)(x^6+1)=(x^6-1)(x^2+1)(x^4-x^2+1)$. Hence $\Phi_{12}(x)=x^4-x^2+1=\prod(x-\xi^i)$, where the ξ^i $(i=1, 5, 7, 11)$ are the primitive 12th roots of 1.

3 $S_f=\mathbb{Q}(\xi)$, where ξ is a primitive 12th root of 1. The other zeros of $\Phi_{12}(x)$ are ξ^5, ξ^7, ξ^{11} as in exercise 2. The elements of $G=\text{Gal}(S_f/\mathbb{Q})$ are Id, ρ, σ, τ, where $\rho(\xi)=\xi^5$, $\sigma(\xi)=\xi^7$ and $\tau(\xi)=\xi^{11}$. One shows easily that $\rho^2=\sigma^2=\tau^2=\text{Id}$ and that $\tau=\rho\sigma$ to deduce that $G\cong C_2\times C_2$. The fields are $\mathbb{Q}$, $\mathbb{Q}(\xi)$, $\mathbb{Q}(\sqrt{-1})$, $\mathbb{Q}(\sqrt{-3})$ and $\mathbb{Q}(\sqrt{3})$. [Hint: For Fix σ use $1+\xi^4=\xi^2$, for example.]

4 $\Phi_n(x)=\prod(x-\xi^i)$ where ξ^i runs over all primitive nth roots of 1. The maps σ_j given by $\sigma_j(\xi)=\xi^j$ are the elements of the Galois group G, say, and the map $\Lambda: G\to\mathbb{Z}_n$ given by $\Lambda(\sigma_j)=j \pmod n$ establishes the asserted isomorphism. (Λ is clearly 1–1 and is onto since, if $(i, n)=1$, then ξ^i is a primitive nth root of 1 and so, for that value of i, σ_i exists.)

5 (ii) $S_f=\mathbb{Q}(\xi)$ where ζ is a 10th root of 1 whose minimum polynomial is $(x-\xi)(x-\xi^3)(x-\xi^7)(x-\xi^9)=(x^5+1)/(x+1)=x^4-x^3+x^2-x+1$. (Why is this irreducible over $\mathbb{Q}$?) Hence $\text{Gal}(S_f/\mathbb{Q})=\{\sigma_1, \sigma_3, \sigma_7, \sigma_9\}$, where $\sigma_i(\xi)=\xi^i$. Since $\sigma_3^2=\sigma_9\neq\sigma_1$ and $\sigma_3^3=\sigma_7$ we have $\text{Gal}(S_f/\mathbb{Q})=C_4$.

6 For 7.8.3 the map $\phi_j\to j \pmod n$ is a 1–1 homomorphism of $\text{Gal}(L/K)$ into $\langle\mathbb{Z}_n, +\rangle$. The other assertion follows as in exercise 4 with the use of exercise 4.5.5.

7 $S_f=\mathbb{Q}(\omega, \alpha)$ where ω is a primitive pth root of 1 and α is the (or a) real pth root of a. The elements of $\text{Gal}(S_f/\mathbb{Q})$ are the maps σ_{ij} defined by $\sigma_{ij}(\omega)=\omega^i$, $\sigma_{ij}(\alpha)=\omega^j\alpha$ $(1\le i\le p-1, 0\le j\le p-1)$. Now $\sigma_{kl}\sigma_{ij}(\omega)=\omega^{ik}$ whilst $\sigma_{kl}\sigma_{ij}(\alpha)=\sigma_{kl}(\omega^j\alpha)=\omega^{jk}\omega^l\alpha$. Therefore $\sigma_{kl}\sigma_{ij}=\sigma_{ik, jk+l}$. Now associate σ_{ij} with the transformation τ_{ij} on $\mathbb{Z}_p$ given by $\tau_{ij}(y)=iy+j$.

8 Cf. exercise 7.6.10.

9 Exercise 7.7.9 says that if $K\subseteq L\subseteq M$ and L is normal over K, then $\text{Gal}(M/L)\lhd\text{Gal}(M/K)$. Here $S_f=\mathbb{Q}(\xi, r)\supset\mathbb{Q}(\xi)\supset\mathbb{Q}$, where r is a real nth root of a and ξ is a primitive nth root of 1. Now use 7.8.3 and 7.8.2 on $\text{Gal}(S_f/\mathbb{Q}(\xi))$ and $\text{Gal}(\mathbb{Q}(\xi)/\mathbb{Q})$.

10 Let ξ be a primitive mnth root of 1. Then $u=\xi^n$ and $v=\xi^m$ are primitive mth and nth roots of 1. Define ρ_i by $\rho_i(\xi)=\xi^i$ where $(i, mn)=1$. Then ρ_i induces σ_i, τ_i given by $\sigma_i(u)=u^i$, $\tau_i(v)=v^i$—elements of $\text{Gal}(S_f/\mathbb{Q})$ and $\text{Gal}(S_g/\mathbb{Q})$ respectively. Show that the map $\rho_i\to(\sigma_i, \tau_i)$ of $\text{Gal}(S_h/\mathbb{Q})$ to $\text{Gal}(S_f/\mathbb{Q})\times\text{Gal}(S_g/\mathbb{Q})$ is 1–1 and a homomorphism. It is *onto* since $|\text{Gal}(S_h/\mathbb{Q})|=\phi(mn)=\phi(m)\phi(n)$. (Why?)

Exercises 7.9

2 Since the a_i are independent over F, exercise 4.7.3 shows that we may identify $F(a_0, a_1, \ldots, a_4)$ with $F(s_1, s_2, \ldots, s_5)$ of exercise 1. Thus $S_f \cong L$ as in exercise 1. Hence $\text{Gal}(S_f/F(a_0, a_1, \ldots, a_4)) \cong \text{Gal}(L/K) \cong S_5$.

3 It doesn't mention over which field the polynomial isn't soluble by radicals! Indeed *all* polynomials in $\mathbb{Q}[x]$ *can* be solved by radicals over $\mathbb{R}$, but, of course, it is not true that all polynomials in $\mathbb{Q}[x]$ can be solved by radicals *over* $\mathbb{Q}$.

4 Each group G is isomorphic to a subgroup of some S_n (Theorem 5.9.6). Now use 7.6.4.

5 We need five real numbers which are completely algebraically independent over $\mathbb{Q}$. Take $a_0 = \pi$, say (since π satisfies no polynomial equation over $\mathbb{Q}$ (see (C) on p.174). Now take a_1 to be any real number which is a zero of no polynomial in $\mathbb{Q}(\pi)[x] = \mathbb{Q}(a_0)[x]$. Such an a_1 exists since $\mathbb{Q}$, $\mathbb{Q}(\pi)$, $\mathbb{Q}(\pi)[x]$ and, also, the set of all (real) zeros of polynomials in $\mathbb{Q}(\pi)[x]$ are all countable sets, whereas $\mathbb{R}$ is not. Continue in this way to find a_2, a_3, a_4. Then apply exercise 1.

Exercises 7.10

1 'Only if' is the normality part of 7.7.3 (or see exercise 7.7.9). 'If' is really 7.10.3 with F, K, E replacing $F_{i-1}, F_i, S_f(\zeta)$. [Or see exercise 5 below.]

2 The only such N is $\mathbb{Q}(\omega)$ since S_3 has only one normal subgroup other than S_3 and $\langle e \rangle$.

3 $\text{Gal}(S_f/\mathbb{Q}) = C_6$ (exercise 7.8.6). C_6 has a normal subgroup $H = \{\text{Id}, \sigma\}$ of order 2, σ being complex conjugation. Then $[\text{Fix}\, H : \mathbb{Q}] = 3$. Suppose $\text{Fix}\, H = \mathbb{Q}(\beta)$, where $\beta \in \mathbb{C}$ and $\beta^n \in \mathbb{Q}$ $(n \geq 3)$. [Why just one term in this radical tower?] If $M(x)$ is the minimum polynomial of β over $\mathbb{Q}$ then $M(x)$ has degree 3 and divides $x^n - \beta^n$. Hence $M(x)$ has zeros β, $\xi^i\beta$, $\xi^j\beta$, where ξ is an nth root of 1. Then $\xi^i, \xi^j \in \text{Fix}\, H$. This is impossible since $\text{Fix}\, H \subseteq \mathbb{R}$ (why?)

4 Assume that $K = \mathbb{Q}(\beta)$, where $\beta^p \in \mathbb{Q}$, is normal over $\mathbb{Q}$. Then $G = \text{Gal}(K/\mathbb{Q}) = C_p$. Look at the $\sigma(\beta)$ $(\sigma \in G)$ which are distinct pth roots of β.

5 'If': Suppose $\alpha(K) = T$. Then $\alpha\sigma\alpha^{-1}(T) = T$. Hence $T \subseteq \text{Fix}(\text{Gal}(M/L))$, i.e. $T \subseteq L$. (Why equality?)

6 For $\text{Aut}(F) \cong S_3$ see exercise 5.3.1(l). That $k \in \text{Fix}\, S$ is trivial—and then the rest is easy, too.

7 Let a, b be any two zeros of $f(x)$ in S_f. According to 7.2.3′ and 7.2.4 the identity automorphism on F extends to one between $F(a)$ and $F(b)$, which itself extends to an automorphism of S_f.

8 Extend R to a *normal* radical tower N, say, over F. Since $f(x)$ has one zero in $E(\subseteq R \subseteq N)$ we deduce that $f(x)$ splits in $N[x]$.

9 The result stated implies that if the zeros of $f(x)$ are labelled $1, 2, \ldots, p$ then the elements of $\mathrm{Gal}(S_f/F)$ form a subgroup S, containing all $\sigma_{u,v}$, in the solution of exercise 7.8.7. Now show that if $\sigma_{u,v}(a) = a$ and $\sigma_{u,v}(b) = b$ then $u = 1$ and $v = 0$. This proves that the only element of $\mathrm{Gal}(S_f/F)$ fixing (F and) a and b is e_G. But $\mathrm{Fix}(\langle e_G \rangle) = S_f$.

10 On adding the δ_i use the equality $1 + (\varepsilon^j) + \cdots + (\varepsilon^j)^{p-1} = 0$.

11 There is no contradiction: $f(x)$ *is* soluble if *all* elements of $\mathbb{R}$ are given to you 'for free'—but not necessarily so if the only 'given' elements are those in $\mathbb{Q}(a_0, \ldots, a_4)$.

12 (a) Choose $L \supset K$ such that $\mathrm{Gal}(L/K)$ is: in (i) C_{10}; in (ii) D_5. (D_5 is a subgroup of the group of exercise 7.8.7 when $p = 5$.)
(b) (ii) $S_f = \mathbb{Q}(i)$; (iii) try exercise 7.8.10; (iv) $\mathbb{Q}(\sqrt{2} + \sqrt[3]{5}) = \mathbb{Q}(\sqrt{2}, \sqrt[3]{5})$—cf. exercise 7.3.13. For (i), the only normal subfield other than $\mathbb{Q}$ and $\mathbb{Q}(\sqrt[3]{3}, \omega)$ is $\mathbb{Q}(\omega)$.

13 (i) and (ii) factorise (how?) In (iii) replace x by $-2y$.

Exercises 7.11

1 Use: Each permutation can be written as a product of disjoint cycles.

2 The real (sorry!!) point is: complex conjugation maps M *to itself*.

3 If one has zeros $\pm\sqrt{2} \pm \sqrt{3}$ what do you think the other is?

4 No! This $f(x)$ factorises in $\mathbb{Q}[x]$.

5 What about using a cubic, irreducible over $\mathbb{Q}$ and with three real zeros, one being negative: now replace x by x^2?

6 Regard $\sigma \in \mathrm{Gal}(S/F)$ as a permutation of the α_i. Clearly $\sigma(\delta) = \delta$ or $-\delta$ according as σ is an even/odd permutation. For (i) use $\sigma(\delta^2) = \sigma(\delta)\sigma(\delta)$. For (ii) note that $\mathrm{Gal}(S/F) \leq A_n$ iff each σ fixes δ, i.e. iff $\delta \in \mathrm{Fix}(\mathrm{Gal}(S/F)) = ?$

7 Combine exercises 7.9.1 and 7.11.6.

9 The values suggested in the text for c, d does the first! The second is similar.

10 True. By exercise 7.10.9, if more than one zero is real then ....

11 (i), (ii): Look for quadratic factors—there being no linear ones. (iv) to (ix) are not soluble. (The derivative of (ix) has two rational zeros which can be found by the rational root test. Investigate (ix) at these turning points.)

12 (a) If ϕ has order d then $\phi^d(\alpha)=\alpha$ for each $\alpha \in L$. But $\phi^d(\alpha)=\alpha^{p^d}$. Consequently $\alpha^{p^d-1}=e$. Now K^* is cyclic of order p^n-1 (exercise 4.5.5). Hence $d=n$. It follows that Aut(L) is a cyclic group of order n generated by ϕ.

Exercises 7.12

2 Note that if $[L:K]<\infty$ then $L=K(\alpha_1,\ldots,\alpha_n)$ for (finitely many) suitable $\alpha_i \in L$ (with each α_i algebraic over K). Now apply exercise 1 $n-1$ times.

3 (a) (iii) Recall that the splitting field is $\mathbb{Q}(\sqrt[3]{2},\omega)$.

4 7.10.2.

5 Suppose that $[L:K]=n$ and that, by exercise 2, $L=K(\alpha)$. Then α has minimum polynomial of degree n over K (see 4.3.4). By 7.2.3′ there are exactly $n(=[L:K])$ ways of extending the identity map on K to an element of Gal$(L/K)\ldots$.

Bibliography

Despite its length this bibliography has no pretentions to completeness, as a visit to your university/polytechnic/college library will show. All the items listed are either mentioned in the text or are items you might enjoy reading just for fun. The point of producing such a list is neither to dazzle nor depress you with its length but to give you a wide choice of where to start reading.

Because of its length the list has been split into subsections. In some cases items listed under one heading could easily have gone under several others.

There exist a number of biographies/obituaries in book and paper form of several of the mathematicians referred to in the text. We have not included mention of these as most can be tracked down through [134].

Algebra

1 Adamson, Iain T. *Introduction to Field Theory*. Oliver and Boyd, Edinburgh, 1964.

2 Albert, A A (ed). *Studies in Modern Algebra*. Mathematical Association of America, 1963.

3 Artin, Emil. *Galois Theory*. University of Notre Dame Press, Indiana, 1964.

4 Bastida, Julio R. *Field Extensions and Galois Theory*. Addison-Wesley, California, 1984.

5 Birkhoff, Garrett and MacLane, Saunders. *A Survey of Modern Algebra* (3rd edition). Macmillan, New York, 1965.

6 Burton, David M. *A First course in Rings and Ideals*. Addison-Wesley, Reading, Massachusetts, 1970.

7 Childs, Lindsay. *A Concrete Introduction to Higher Algebra*. Springer-Verlag, New York, 1979.

8 Clifford, A H and Preston, G B. *The Algebraic Theory of Semigroups*, Vols 1, 2. American Mathematical Society, Rhode Island, 1961.

9 Cohn, P M. *Algebra* Vols 1, 2. Wiley, London, 1974 and 1977.

10 Coxeter, H S M and Moser, W O J. *Generators and Relations for Discrete Groups*. Springer-Verlag, Berlin, 1965.

11 Dehn, Edgar. *Algebraic Equations*. Columbia University Press, New York, 1930.

12 Dickson, L E. *Algebraic Theories*. Dover Publications, New York, 1959.

13 Fraleigh, John B. *A First Course in Abstract Algebra*. Addison-Wesley, Reading, Massachusetts, 1967.

14 Gaal, Lisl. *Classical Galois Theory: with Examples*. Markham, Chicago, 1971.
15 Garling, D J H. *A Course in Galois Theory*. Cambridge University Press, Cambridge, 1986.
16 Gorenstein, Daniel. *Finite Simple Groups: An Introduction to their Classification*. Plenum Press, New York, 1982.
17 Hadlock, Charles Robert. *Field Theory and its Classical Problems*. Carus Monograph 19. Mathematical Association of America, 1978.
18 Herstein, I N. *Topics in Algebra* (2nd edition). Wiley, New York, 1975.
19 Herstein, I N. *Noncommutative Rings*. Carus Monograph 15. Mathematical Association of America, 1968.
20 Jacobson, Nathan. *Lectures in Abstract Algebra* Vols 1, 2, 3. Van Nostrand, Princeton, NJ, 1951–64.
21 Jacobson, Nathan. *Basic Algebra I, II*. Freeman, San Francisco, 1974–80.
22 Kaplansky, I. *Commutative Rings*. University of Chicago Press, Chicago, 1974.
23 Ledermann, W. *Introduction to Group Theory*. Oliver and Boyd, Edinburgh, 1973.
24 McCoy, Neal H. *Fundamentals of Abstract Algebra*. Allyn and Bacon, Boston, 1973.
25 McCoy, Neal H. *Rings and Ideals*. Carus Monograph 8. Mathematical Association of America, 1948.
26 McCoy, Neal H and Berger, Thomas R. *Algebra: Groups, Rings and other Topics*. Allyn and Bacon, Boston, 1977.
27 Macdonald, Ian D. *The Theory of Groups*. Oxford University Press, Oxford, 1968.
28 MacLane, Saunders and Birkhoff, Garrett. *Algebra*. Macmillan, New York, 1967.
29 Magnus, Wilhelm, Karrass, Abraham and Solitar, Donald. *Combinatorial Group Theory: Presentations of Groups in terms of Generators and Relations*. Wiley–Interscience, New York, 1966.
30 Meldrum, J D P. *Near-rings and Their Links with Groups*. Pitman, London, 1985.
31 Miller, G A, Blichfeldt, H F and Dickson, L E. *Theory and Application of Finite Groups*. Dover, New York, 1961.
32 Mostow, George D, Sampson, Joseph H, and Meyer, Jean-Pierre. *Fundamental Structures of Algebra*. Holt, Rinehart and Winston, New York, 1966.
33 Pilz, Günter. *Near-Rings: the Theory and its Applications*. North-Holland, Amsterdam, 1977.
34 Rotman, Joseph J. *The Theory of Groups. An Introduction* (2nd edition). Allyn and Bacon, Boston, 1973.
35 Sawyer, W W. *A Concrete Approach to Abstract Algebra*. Freeman, San Francisco, 1959.
36 Scott, W R. *Group Theory*. Prentice-Hall, Englewood Cliffs, 1964.
37 Shapiro, Louis. *Introduction to Abstract Algebra*. McGraw-Hill, London, 1975.
38 Stewart I. *Galois Theory* (2nd edition). Chapman and Hall, London, 1989.

39 van der Waerden, B L. *Algebra* Vols 1, 2 (7th edition). Ungar, New York, 1970.

40 Weber, Heinrich. *Lehrbuch der Algebra*. Friedrich Vieweg und Sohn, Braunschweig, 1895–6.

41 Zariski, Oscar and Samuel, Pierre. *Commutative Algebra* Vols 1, 2. Van Nostrand, New York, 1958.

Number theory (including Number Systems)

42 Allenby, R B J T and Redfern, E J. *Introduction to Number Theory with Computing*. Edward Arnold, London, 1989.

43 Burton, David M. *Elementary Number Theory*, Revised Printing, Allyn and Bacon, Boston, 1980.

44 Carmichael, Robert D. *Theory of Numbers and Diophantine Analysis*. Dover Publications Inc., 1914.

45 Davenport, H. *Higher Arithmetic: An Introduction to the Theory of Numbers*. Hutchinson, London, 1962.

46 Edwards, Harold M. *Fermat's Last Theorem. A Genetic Introduction to Algebraic Number Theory*. Springer-Verlag, New York, 1977.

47 Hardy, G H and Wright, E M. *An Introduction to the Theory of Numbers* (4th edition). Oxford University Press, Oxford, 1960.

48 Isaacs, G L. *Real Numbers. A Development of Real Numbers in an Axiomatic Set Theory*. McGraw-Hill, New York, 1968.

49 Le Veque, William J. *Topics in Number Theory* Vols 1, 2. Addison-Wesley, Reading, Massachusetts, 1956.

50 Mendelson, Elliot. *Number Systems and the Foundations of Analysis*. Academic Press, New York, 1973.

51 Pollard, Harry. *The Theory of Algebraic Numbers*. Carus Monograph 9. Mathematical Association of America, 1961.

52 Ribenboim, Paulo. *Algebraic Numbers*. Wiley–Interscience, New York, 1972.

53 Ribenboim, Paulo. *13 Lectures on Fermat's Last Theorem*. Springer-Verlag, New York, 1980.

54 Shanks, Daniel. *Solved and Unsolved Problems in Number Theory*. Spartan Books, Washington DC, 1962.

55 Stewart, Ian and Tall, David. *Algebraic Number Theory*. Chapman and Hall, London, 1979.

56 Uspensky, J V and Heaslett, M A. *Elementary Number Theory*. McGraw-Hill, New York, 1939.

Logic, set theory

57 Exner, Robert M and Rosskopf, Myron F. *Logic in Elementary Mathematics*. McGraw-Hill, New York, 1959.

58 Halmos, Paul R. *Naive Set Theory*. Van Nostrand, New York, 1960.

59 Rosser, J Barkley. *Logic for Mathematicians*. McGraw-Hill, New York, 1953.
60 Stoll, Robert R. *Sets, Logic and Axiomatic Theories*. Freeman, San Francisco, 1961.

Applications (in a broad sense)

61 Birkhoff, Garrett and Bartee, Thomas C. *Modern Applied Algebra*. McGraw-Hill, New York, 1970.
62 Boardman, A D, O'Connor, D E and Young, P A. *Symmetry and its Applications in Science*. McGraw-Hill, London, 1973.
63 Buerger, M J. *Elementary Crystallography*. Wiley, New York, 1956.
64 Buerger, Martin J. *Crystal-structure Analysis*. Wiley, New York, 1960.
65 Ellis, Andrew and Treeby, Terence. *Algebraic Structure*. John Murray, London, 1974.
66 Gillman, L and Jerison, M. *Rings of Continuous Functions*. Van Nostrand, Princeton, NJ, 1960.
67 Hadley, George. *Linear Algebra*. Addison-Wesley, Reading, Massachusetts, 1961.
68 Liebeck, Hans. *Algebra for Scientists and Engineers*. Wiley, London, 1969.
69 Lockwood, E H and Macmillan, R H. *Geometric Symmetry*. Cambridge University Press, Cambridge, 1978.
70 Miller Jr, Willard. *Symmetry Groups and their Applications*. Academic Press, New York, 1972.
71 Murnaghan, Francis D. *The Theory of Group Representations*. The Johns Hopkins Press, Baltimore, 1938.
72 Schonland, David S. *Molecular Symmetry*. Van Nostrand, London, 1965.
73 Schwarzenberger, R L E. *N-Dimensional Crystallography*. Research Notes in Mathematics, 41. Pitman, London, 1980.
74 South, G F. *Boolean Algebra and Its Uses*. Van Nostrand Reinhold, London, 1974.
75 Whitesitt, J Eldon. *Boolean Algebra and its Applications*. Addison-Wesley, Reading, Massachusetts, 1961.

History

76 Bell, Eric Temple. *Development of Mathematics* (2nd edition). McGraw-Hill, New York, 1945.
77 Bell, Eric Temple. *Men of Mathematics*. Simon and Schuster, New York, 1962.
78 Bell, Eric Temple. *Mathematics, Queen and Servant of Science*. G Bell, London, 1952.
79 Bourbaki, Nicolas. *Eléments d'Histoire des Mathématiques*. Hermann, Paris, 1960.
80 Boyer, Carl B. *A History of Mathematics*. Wiley, New York, 1968.

81 Crowe, Michael J. *A History of Vector Analysis*. University of Notre Dame Press, Indiana, 1967.
82 Dickson, L E. *History of the Theory of Numbers* Vols 1, 2, 3. Carnegie Institute of Washington, Washington, 1919.
83 Edwards, Harold M. *Galois Theory*. Springer-Verlag, New York, 1984.
84 Eves, Howard. *An Introduction to the History of Mathematics* (3rd edition). Holt, Rinehart and Winston, New York, 1969.
85 Kline, Morris. *Mathematical Thought from Ancient to Modern Times*. Oxford University Press, New York, 1972.
86 Ore, Oystein. *Number Theory and its History*. McGraw-Hill, New York, 1948.
87 Novy, Lubos. *Origins of Modern Algebra*. Noordhoff International, Netherlands, 1973.
88 Struik, Dirk J. *A Concise History of Mathematics*. G Bell, London, 1954.
89 van der Waerden, B L. *A History of Algebra*. Springer-Verlag, Berlin, Heidelberg, 1985.

Papers

The following is just a sample. I could have quadrupled the list with no difficulty. I leave you the excitement of discovering some of these other papers (in the journals mentioned below and elsewhere) for yourself.

The abbreviations *AMM* and *AHES* used below stand for American Mathematical Monthly and Archive for History of Exact Sciences.

90 Aschbacher, Michael. The Classification of the Finite Simple Groups. *The Mathematical Intelligencer* **3**, 2, 1981, 59–65.
91 Ayoub, Raymond G. Paolo Ruffini's Contributions to the Quintic. *AHES* **23**, 1980, 253–77.
92 Birkhoff, Garrett. Current Trends in Algebra. *AMM* **80**, 1973, 760–82.
93 Cohn, P M. Unique Factorisation Domains. *AMM* **80**, 1973, 1–18.
94 Cohn, P M. Rings of Fractions. *AMM* **78**, 1971, 596–615.
95 Conway, J H. Monsters and Moonshine. *The Mathematical Intelligencer* **2**, 4, 1980, 165–71.
96 Cooke, Roger. Letter to the editor. *AMM* **91**, 1984, 382.
97 Dickson, L E. Fermat's Last Theorem and the Origin and Nature of the Theory of Algebraic Numbers. *Annals of Mathematics*, Series 2, **18**, 1916–17, 161–87.
98 Dieudonné, J. The Historical Development of Algebraic Geometry. *AMM* **79**, 1972, 827–66.
99 E, H. A Short History of the Fields Medal. *The Mathematical Intelligencer* **1**, 3, 1978, 127–9.
100 Edwards, Harold M. The Genesis of Ideal Theory. *AHES* **23**, 1980, 321–78.
101 Fefferman, C. An Easy Proof of the Fundamental Theory of Algebra. *AMM* **74**, 1967, 854–5.
102 Fisher, Charles S. The Death of a Mathematical Theory: a Study in the Sociology of Knowledge. *AHES* **3**, 1966–7, 137–59.

103 Hamburg, Robin Rider. The Theory of Equations in the 18th Century: The Work of Joseph Lagrange. *AHES* **16**, 1976, 17–36.
104 Hawkins, Thomas. The Origins of The Theory of Group Characters. *AHES* **7**, 1971, 142–70.
105 Hawkins, Thomas. The Theory of Matrices in the 19th Century. *Proc. Internat. Congress of Mathematicians*, Vancouver, 1974. Vol. 2 Canadian Math Congress, 1975, 561–70.
106 Henkin, Leon. On Mathematical Induction. *AMM* **67**, 1960, 323–38.
107 Hungerford, Thomas W. A counterexample in Galois Theory. *AMM* **97**, 1990, 54–7.
108 Kiernan, B Melvin. The Development of Galois Theory from Lagrange to Artin. *AHES* **8**, 1971–2, 40–154.
109 Koppelman, Elaine. The Calculus of Operations and the Rise of Abstract Algebra. *AHES* **8**, 1971–2, 155–242.
110 Mead D G. The Missing Fields. *AMM* **94**, 1987, 871–2.
111 Miller, G A. History of the Theory of Groups to 1900. *The Collected Works of George Abram Miller*. University of Illinois, Urbana, 1935–59.
112 Moran, W and Pym, J S. On the Construction of the Real Number System. *Mathematics Magazine*, **43**, 1970, 257–9.
113 Motzkin, Th. The Euclidean Algorithm. *Bulletin Amer. Math. Soc.* **55**, 1949, 1142–6.
114 Pierpont, James. Lagrange's Place in the Theory of Substitutions. *Bulletin Amer. Math. Soc.* **1**, 1895, 196–204.
115 Pierpont, James. Early History of Galois' Theory of Equations. *Bulletin Amer. Math. Soc.* **4**, 1898, 332–40.
116 Rothman, Tony. Genius and Biographers: The fictionalization of Evariste Galois. *AMM* **89**, 1982, 84–10.
117 Ruchte, M F and Ryden, R W. A Proof of the Uniqueness of Factorisation in the Gaussian Integers. *AMM* **80**, 1973, 58–9.
118 Samuel, Pierre. Unique Factorization. *AMM* **75**, 1968, 945–52.
119 Samuel, Pierre. About Euclidean Rings. *Journal of Algebra* **19**, 1971, 282–301.
120 Schwarzenberger R L E. The 17 plane symmetry groups. *Mathematical Gazette*, **58**, 1974, 123–31.
121 Seidenberg, A. Did Euclid's Elements, Book I, develop Geometry Axiomatically? *AHES* **14**, 1974–5, 263–95.
122 Simmons, G J. The Number of Irreducible Polynomials of Degree n over $GF(p)$. *AMM* **77**, 1970, 743–5.
123 Stark, H M. On the Problem of Unique Factorization in Complex Quadratic Fields. *Proc. of Symposia in Pure Mathematics*, **XII**. American Math. Soc., 1969, 41–56.
124 Stewart, Ian. The Truth about Venn Diagrams. *Mathematical Gazette* **60**, 1976, 47–54.
125 Szabó, Árpád. Greek Dialectic and Euclid's Axiomatics. *Problems in the Philosophy of Mathematics* (Imre Lakatos, ed.). North-Holland, Amsterdam, 1967.
126 van der Waerden,, B L. Die Galois-Theorie von Heinrich Weber bis Emil Artin. *AHES* **9**, 1972, 240–8.

127 Weyl, Hermann. A Half-Century of Mathematics. *AMM* **58**, 1951, 523–53.
128 Wigner, Eugene P. Symmetry Principles in Old and New Physics. *Bulletin Amer. Math. Soc.* **74**, 1968, 793–815.
129 Wilder, Raymond L. The Role of the Axiomatic Method. *AMM* **74**, 1967, 115–27.
130 Wilson, J C. A Principal Ideal Ring that is not a Euclidean Ring. *Math. Magazine*, **46**, 1973, 74–8.
131 Youschkevitch, A P. The Concept of Function up to the Middle of the 19th Century. *AHES* **16**, 1976, 37–85.

Other references

132 Mal'cev, A I. *Groups and Other Algebraic Systems in Mathematics: its Contents, Methods and Meaning* Vol 3. MIT Press, Cambridge, Massachusetts, 1963.
133 Newman, James R (ed). *The World of Mathematics* Vols 1, 2, 3, 4. Allen and Unwin, London, 1960.
134 *Dictionary of Scientific Biography*. Charles Scribner's Sons, New York, 1970–80.
135 *Encyclopedic Dictionary of Mathematics*. MIT Press, Cambridge, Massachusetts, 1977.

Notation

Because of the limited number of (sensible) symbols available to us, it is not uncommon for one symbol to be used to denote totally distinct concepts. For example, note that here the symbol $(y_1, y_2, \ldots, y_n)$ will denote (i) a gcd of the elements $y_1, y_2, \ldots, y_n$ or (ii) an ordered n-tuple or (iii) a cyclic permutation, depending on the context. This is at worst irritating since there will rarely be any cause for confusion.

Other notation used includes:

(i) $|S|$, to denote the number of elements in a (finite) set or group. (Cf. Definition 5.3.7.)
(ii) $M_n(X)$, where X is a ring, to denote the set (ring) of all $n \times n$ matrices with entries from X. [For $n = 2$, $X = \mathbb{C}$ see Definition 3.2.5.]
(iii) $\langle x, y; x^2 = 1, xy^2 = y^2x \rangle$ (etc.), to denote the group generated by elements x, y subject only to the relationships $x^2 = 1$ and $xy^2 = y^2x$ (etc.). See p. 219 and pp. 238–41.

Index